Simulation
Methods for
Polymers

edited by
Michael Kotelyanskii
Rudolph Technologies, Inc.
Flanders, New Jersey, U.S.A.

Doros N. Theodorou
National Technical University of Athens
Athens, Greece

CRC Press
Taylor & Francis Group
Boca Raton London New York

CRC Press is an imprint of the
Taylor & Francis Group, an **informa** business

First published 2004 by Marcel Dekker, Inc.

Published 2019 by CRC Press
Taylor & Francis Group
6000 Broken Sound Parkway NW, Suite 300
Boca Raton, FL 33487-2742

© 2004 by Taylor & Francis Group, LLC
CRC Press is an imprint of Taylor & Francis Group, an Informa business

First issued in paperback 2019

No claim to original U.S. Government works

ISBN 13: 978-0-367-44657-4 (pbk)
ISBN 13: 978-0-8247-0247-2 (hbk)

Visit the Taylor & Francis Web site at
http://www.taylorandfrancis.com

and the CRC Press Web site at
http://www.crcpress.com

Library of Congress Cataloging-in-Publication Data
A catalog record for this book is available from the Library of Congress.

Simulation
Methods for
Polymers

Preface

In the last 20 years, materials modeling and simulation has grown into an essential component of research in the chemical and pharmaceutical industries. Because the field of process design is already quite mature, competitive advances in chemical synthesis and separation operations occur primarily through the development and use of material systems with tailored physicochemical characteristics (e.g., metallocene catalysts for polyolefin production, polymeric membranes offering a more favorable combination of permeability and selectivity for separations, and environmentally acceptable solvents with prescribed thermophysical properties). Furthermore, there is a shift of emphasis from process to product design, which is intimately related to materials behavior. With the current trend toward nanotechnology, the scientist or engineer is often called on to develop new, often hierarchical material structures with key characteristics in the 0.1–10-nm length scale, so as to benefit from the unique mechanical, electronic, magnetic, optical, or other properties that emerge at this scale. Materials that develop such structures through self-assembly processes, or modify their structure in response to environmental conditions, are frequently sought.

Meeting these new design challenges calls for a quantitative understanding of structure–property–processing–performance relations in materials. Precise development of this understanding is the main objective of materials modeling and simulation. Along with novel experimental techniques, which probe matter at an increasingly fine scale, and new screening strategies, such as high-throughput experimentation, modeling has become an indispensable tool in the development of new materials and products. Synthetic polymers and biopolymers, either by themselves or in combination with ceramics and metals, are central to contemporary materials design, thanks to the astonishing range of properties that can be achieved through manipulation of their chemical constitution, molecular organization, and morphology.

Polymer modeling and simulation has benefited greatly from advances in computer hardware. Even more important, however, has been the development of new methods and algorithms, firmly based on the fundamental physical and chemical sciences, that permit addressing the wide spectrum of length and time scales that govern structure and motion in polymeric materials. It is by now generally accepted that the successful solution of materials design problems calls for hierarchical, or multiscale, modeling and simulation, involving a judicious combination of atomistic (<10 nm), mesoscopic ($10–1000$ nm), and macroscopic methods. How to best link these methods together, strengthen their fundamental underpinnings, and enhance their efficiency are very active problems of current research.

This book is intended to help students and research practitioners in academia and industry become active players in the fascinating and rapidly expanding field of modeling and simulation of polymeric materials. Roughly five years ago, we decided to embark on an effort to produce a "how-to" book that would be coherent and comprehensive, encompass important recent developments in the field, and be useful as a guide and reference to people working on polymer simulation, such as our own graduate students. Rather than attempt to write the whole book ourselves, we chose to draft a list of chapters and then recruit world-class experts to write them. Contributors were instructed to make their chapters as didactic as possible, incorporating samples of code, where appropriate; make reference to key works, especially review papers and books, in the current literature; and adhere to a more or less coherent notation. In our editing of the chapters, we tried to enhance uniformity and avoid unnecessary repetition. The result is gratifyingly close to what we envisioned. We hope that the reader will find the progression of chapters logical, and that the occasional switches in style and notation will serve as attention heighteners and perspective broadeners rather than as sources of confusion.

Chapter 1 introduces basic elements of polymer physics (interactions and force fields for describing polymer systems, conformational statistics of polymer chains, Flory mixing thermodynamics, Rouse, Zimm, and reptation dynamics, glass transition, and crystallization). It provides a brief overview of equilibrium and nonequilibrium statistical mechanics (quantum and classical descriptions of material systems, dynamics, ergodicity, Liouville equation, equilibrium statistical ensembles and connections between them, calculation of pressure and chemical potential, fluctuation equations, pair distribution function, time correlation functions, and transport coefficients). Finally, the basic principles of "traditional" molecular simulation techniques (Monte Carlo (MC), molecular dynamics (MD), Brownian dynamics (BD), transition state theory (TST)-based analysis, and simulation of infrequent events) are discussed.

Part I focuses on the calculation of single-chain properties in various environments. The chapter by E. D. Akten et al. introduces the Rotational Isomeric State (RIS) model for calculating unperturbed chain properties from atomistic conformational analysis and develops systematically an illustrative example based on head-to-head, tail-to-tail polypropylene. The chapter by Reinhard Hentschke addresses MD and BD simulations of single chains in solution. It reviews Monte Carlo sampling of unperturbed chain conformations and calculations of scattering from single chains and presents a coarse-graining strategy for going from atomistic oligomer or short helix simulations, incorporating solvent effects, to RIS-type representations and persistence lengths.

Part II addresses lattice-based Monte Carlo simulations. The chapter by Kurt Binder et al. provides excellent motivation for why such simulations are worth undertaking. It then discusses the Random Walk (RW), Non-Reversal Random Walk (NRRW), and Self-Avoiding Random Walk (SAW) models, presents algorithms for sampling these models, and discusses their advantages and limitations. In the chapter by Tadeusz Pakula, algorithms for the Monte Carlo simulation of fluids on a fully occupied lattice are discussed. Applications to macromolecular systems of complex architecture and to symmetrical block copolymers are presented to illustrate the power and generality of the algorithms, and implementation details are explained.

Part III, by Vagelis Harmandaris and Vlasis Mavrantzas, addresses molecular dynamics simulations. Following an exposition of basic integration algorithms, an example is given of how MD can be cast in an unconventional ensemble. State-of-the-art multiple time integration schemes, such as rRESPA, and constraint simulation algorithms are then discussed. Examples of using MD as part of a hierarchical strategy for predicting polymer melt viscoelastic properties are presented, followed by a discussion of parallel MD and parallel tempering simulations.

Part IV, by Tushar Jain and Juan de Pablo, introduces the Configurational Bias (CB) method for off-lattice MC simulation. Orientational CB is illustrated with studies of water in clay hydrates, and CB for articulated molecules is discussed as part of expanded grand canonical MC schemes for the investigation of chains in slits and of the critical behavior of polymer solutions. Topological CB is introduced and the combined use of CB and parallel tempering is outlined.

Part V, by Andrey Dobrynin, focuses on simulations of charged polymer systems (polyelectrolytes, polyampholytes). Chains at infinite dilution are examined first, and how electrostatic interactions at various salt concentrations affect conformation is discussed, according to scaling theory and to simulations. Simulation methods for solutions of charged polymers at finite concentration, including explicitly represented ions, are then presented. Summation methods for electrostatic interactions (Ewald, particle-particle particle mesh, fast multipole method) are derived and discussed in detail. Applications of simulations in understanding Manning ion condensation and bundle formation in polyelectrolyte solutions are presented. This chapter puts the recent simulations results, and methods used to obtain them, in the context of the state of the art of the polyelectrolyte theory.

Part VI is devoted to methods for the calculation of free energy and chemical potential and for the simulation of phase equilibria. The chapter by Thanos Panagiotopoulos provides a lucid overview of the Gibbs Ensemble and histogram reweighting grand canonical MC methods, as well as of the NPT + test particle, Gibbs–Duhem integration, and "pseudo-ensemble" methods. CB and expanded ensemble techniques are discussed, and numerous application examples to phase equilibria and critical point determination are presented. The chapter by Mike Kotelyanskii and Reinhard Hentschke presents and explains a method for performing Gibbs ensemble simulations using MD.

In Part VII, Greg Rutledge discusses the modeling and simulation of polymer crystals. He uses this as an excellent opportunity to introduce principles and techniques of solid-state physics useful in the study of polymers. The mathematical description of polymer helices and the calculation of X-ray diffraction patterns from crystals are explained. Both optimization (energy minimization, lattice dynamics) and sampling (MC, MD) methods for the simulation of polymer crystals are then discussed. Applications are presented from the calculation of thermal expansion, elastic coefficients, and even activation energies and rate constants for defect migration by TST methods.

Part VIII focuses on the simulation of bulk amorphous polymers and their properties. The chapter by Jae Shick Yang et al. applies a combined atomistic (energy minimization)-continuum (finite element) simulation approach to study plastic deformation in glassy bisphenol-A-polycarbonate. This is an excellent example of multiscale modeling for addressing mechanical behavior at long time- and length-scales, also discussed in the next section. It also serves as an introduction to the simulation of glassy polymers and of materials under strain. The chapter by Wolfgang Paul et al. addresses polymer dynamics in the melt state: how it can be tracked with atomistic and coarse-grained models using MD, and what insights it can provide about glass transition phenomena. An overview is presented of the calculation of static scattering patterns, as well as of the observables of nuclear magnetic resonance (NMR) relaxation and of neutron-scattering measurements from MD simulation trajectories. Segmental dynamics and terminal dynamics, as revealed by MD simulations of oligomeric polyethylenes, are compared against Mode Coupling Theory and Rouse model predictions. In the chapter by Mike Greenfield, TST-based methods for the prediction of sorption and diffusion of small molecules in amorphous polymers are thoroughly discussed. The general approach followed to obtain the diffusivity, based on atomistic TST-based determination of rate constants for individual jumps executed by the penetrant in the polymer matrix and subsequent use of these rate constants within a kinetic MC simulation to track displacement at long times, is another good example of hierarchical modeling. Three TST-based methods (frozen polymer, average fluctuating polymer, and explicit polymer) for the calculation of rate constants are presented, with examples. The intricacies of performing TST analyses in generalized coordinates using the flexible, rigid, or infinitely stiff polymer models are also explained.

Part IX is devoted to bridging length- and time-scales through multiscale modeling, whose importance has already been stressed above and brought forth in some of the earlier sections. The chapter by Ulrich Suter and his colleagues discusses the key issue of coarse-graining, whereby detailed atomistic representations can be mapped onto computationally more manageable models with fewer degrees of freedom, without loss of significant information. Examples of coarse-graining detailed polyethylene models into bond fluctuation (discussed in Part I) and bead-and-spring models are presented. Simultaneous atomistic/continuum calculations conducted on different scales are also explained, with emphasis on combined finite element/molecular simulation schemes for tracking inelastic deformation in polymer solids (introduced in Part VIII). The chapter by Manuel Laso and Hans Christian Öttinger is devoted to CONNFFESSIT, a nontraditional multiscale method for simulating polymer flows that combines finite elements with stochastic simulation of coarse-grained molecular models. After a dense review of the general field of computational rheology, algorithms and codes for tracking the particles in the finite element mesh, integrating the particle stochastic equations of motion, and reconstructing meshes are treated thoroughly. The chapter by Wouter den Otter and Julian Clarke introduces dissipative particle dynamics (DPD), a mesoscopic method for tracking the temporal evolution of complex fluid systems that fully accounts for hydrodynamic interactions. After a basic description of DPD and the problems to which it has been applied, the question is taken up of mapping real systems, or atomistic models thereof, onto the coarse-grained models employed by DPD. Applications of DPD for the simulation of polymer solution dynamics and microphase separation in block copolymers are presented. Last but not least, the chapter by Andre Zvelindovsky et al. gives an account of their dynamic density functional theory (DFT), a mesoscopic functional Langevin approach for tracking morphology development in complex soft-matter systems. The theoretical underpinnings of the approach are explained, and applications are presented from pattern-formation phenomena in complex copolymer systems and solutions of amphiphilic molecules at rest, under shear, in the presence of chemical reactions, or confined by solid surfaces.

We are grateful to all the contributors for their dedication and painstaking work and for sharing our vision of a book on simulation methods for polymers. Our families are thanked for their patience and understanding of all the time we dedicated to the book, rather than sharing it with them. We hope that the book may prove to be of considerable value to students and practitioners of polymer simulation in academia and industry.

Michael Kotelyanskii
Doros N. Theodorou

Contents

Contents

Contributors

E. Demet Akten Department of Polymer Science, The University of Akron, Akron, Ohio, U.S.A.

J. Baschnagel Institut Charles Sadron, Strasbourg, France

K. Binder Institut für Physik, Johannes-Gutenberg-Universität Mainz, Mainz, Germany

J. H. R. Clarke Department of Chemistry, University of Manchester Institute of Science and Technology, Manchester, United Kingdom

W. K. den Otter Department of Applied Physics, University of Twente, Enschede, The Netherlands

J. J. de Pablo Department of Chemical Engineering, University of Wisconsin–Madison, Madison, Wisconsin, U.S.A.

Andrey V. Dobrynin Institute of Materials Science and Department of Physics, University of Connecticut, Storrs, Connecticut, U.S.A.

Mark D. Ediger Department of Chemistry, University of Wisconsin–Madison, Madison, Wisconsin, U.S.A.

J. G. E. M. Fraaije Gorlaeus Laboratoria, Leiden University, Leiden, The Netherlands

Michael L. Greenfield Department of Chemical Engineering, University of Rhode Island, Kingston, Rhode Island, U.S.A.

Vagelis A. Harmandaris Institute of Chemical Engineering and High-Temperature Chemical Processes, and University of Patras, Patras, Greece

Reinhard Hentschke Physics Department and Institute for Materials Science, Bergische Universität, Wuppertal, Germany

T. S. Jain Department of Chemical Engineering, University of Wisconsin–Madison, Madison, Wisconsin, U.S.A.

Won Ho Jo Department of Fiber and Polymer Science, Seoul National University, Seoul, Korea

Michael Kotelyanskii Rudolph Technologies, Inc., Flanders, New Jersey, U.S.A.

Manuel Laso Department of Chemical Engineering, Polytechnic University of Madrid, Madrid, Spain

Wayne L. Mattice Department of Polymer Science, The University of Akron, Akron, Ohio, U.S.A.

Vlasis G. Mavrantzas Institute of Chemical Engineering and High-Temperature Chemical Processes, and University of Patras, Patras, Greece

M. Müller Institut für Physik, Johannes-Gutenberg-Universität Mainz, Mainz, Germany

Hans Christian Öttinger Eidgenössische Technische Hochschule (ETH), Zurich, Switzerland

Tadeusz Pakula Max-Planck-Institute for Polymer Research, Mainz, Germany, and Technical University of Lodz, Lodz, Poland

Athanassios Z. Panagiotopoulos Department of Chemical Engineering, Princeton University, Princeton, New Jersey, U.S.A.

Wolfgang Paul Institut für Physik, Johannes-Gutenberg-Universität Mainz, Mainz, Germany

Gregory C. Rutledge Department of Chemical Engineering, Massachusetts Institute of Technology, Cambridge, Massachusetts, U.S.A.

Serge Santos Department of Materials, Eidgenössische Technische Hochschule (ETH), Zurich, Switzerland

G. J. A. Sevink Gorlaeus Laboratoria, Leiden University, Leiden, The Netherlands

Grant D. Smith University of Utah, Salt Lake City, Utah, U.S.A.

Ulrich W. Suter Department of Materials, Eidgenössische Technische Hochschule (ETH), Zurich, Switzerland

Doros N. Theodorou Department of Materials Science and Engineering, School of Chemical Engineering, National Technical University of Athens, Athens, Greece

Jae Shick Yang Department of Fiber and Polymer Science, Seoul National University, Seoul, Korea

Do Y. Yoon School of Chemistry, College of Natural Science, Seoul National University, Seoul, Korea

A. V. Zvelindovsky Gorlaeus Laboratoria, Leiden University, Leiden, The Netherlands

1
Background

MICHAEL KOTELYANSKII   Rudolph Technologies, Inc., Flanders, New Jersey, U.S.A.

DOROS N. THEODOROU   National Technical University of Athens, Athens, Greece

I. BASIC CONCEPTS OF POLYMER PHYSICS

Polymers consist of very large molecules containing thousands or millions of atoms, with molecular weights of hundreds of thousands g/mol or more. They can be synthesized through polymerization or copolymerization reactions from a wide variety of monomers. These large molecules can have the simple topology of linear chains, most common for synthetic polymers, or they can be rings, helices, combs, stars, or large networks. A macroscopic piece of rubber, such as an automobile tire, is a network made of cross-linked polymer molecules; it can be regarded as a single molecule built from small monomer units, each unit containing tens to hundreds of atoms.

The large variety of chemical constitutions and molecular architectures of polymeric materials is responsible for the wide range of properties they exhibit. A great number of contemporary technological applications rely on the peculiar mechanical properties of polymers. Most remarkable among those is rubber elasticity, i.e., the ability of a material to deform to many times its original size without breaking and to return to its original shape when the stress is released. Also important are the toughness and ductility exhibited by semicrystalline polymers, glassy polymers and blends, and the strength of oriented semicrystalline and liquid crystalline polymers; these properties, combined with light weight and processability, have formed a basis for a great number of plastic, fiber, and structural material applications. The rheological properties of polymers are key to the design of processing operations such as extrusion, blow molding, and film blowing, whereby they are shaped into the multitude of products we use in everyday life and in advanced technologies. Permeability properties of polymers are

1

important in the design of membranes for blood dialysis, water desalination by reverse osmosis, and industrial gas separations, as well as in packaging materials with barrier properties towards atmospheric gases. The surface and interfacial properties of polymers and copolymers are critical to their widespread use as adhesives, stabilizers of emulsions and suspensions, compatibilizers of blends, coatings with controlled wettability, and biocompatible materials. The optical and electronic properties of polymers are important to common products, such as transparent packaging film and Plexiglas^{TM} windows, as well as to emerging applications such as polymeric light-emitting diodes and optical switches.

In most cases, polymer materials with unique and valuable properties are discovered by chance, or as a result of many years of painstaking trial-and-error experimentation. The ability to design materials tailored for particular applications is the major challenge of polymer materials science. In recent years, computer simulations have proved to be very helpful in meeting this challenge.

Analytical theory can only solve very simplified models of polymers, leaving out many details of the polymer molecular structure, and even these simplified models can be solved only using certain approximations. Computer simulations allow study of a simplified model directly and thus identification of whether discrepancies between theory and experiment are due to the simplifications of the model, to the approximations used in solving the theory, or to both. The simplified models help identify properties that are more general and can be observed for polymers with sometimes very different chemical structure, but having similar chain topology, or flexibility, or charge distribution.

More realistic models that account for the details of chemical structure of particular polymers help identify relationships between the "chemical personality" of each polymer and the values of particular properties. The use of realistic models has been popular in biological applications.

Polymer simulations started with lattice models, as exemplified by the pioneering work of Wall and collaborators, first on single chains in the 1950s [1] and then on multichain systems [2], and also by the work of Alexandrowicz and Accad [3]. Off-lattice Monte Carlo simulations of single chains were presented by Lal in the late 1960s [4]. Continuum simulations of multichain systems appeared in the 1970s, as exemplified by the work of De Vos and Bellemans [5], by the pioneering molecular dynamics study of liquid butane by Ryckaert and Bellemans [6], and by the work of Balabaev, Grivtsov, and Shnol' [7]. In 1980 Bishop et al. [8] presented continuum simulations of a bulk polymer with chains consisting of freely-jointed Lennard-Jones segments. An early molecular dynamics simulation of polyethylene using a realistic model for both bonded and nonbonded

interactions was presented in 1979 by Weber and Helfand [9], while
Vacatello et al. used a similar model to explore liquid triacontane in 1980
[10]. In parallel, important theoretical investigations pertaining to the
configuration-space distribution of polymer models were conducted by
Fixman [11] and by Go and Scheraga [12,13].

A. Interactions in Polymer Systems

In molecular systems, the potential energy function $\mathcal{V}(\mathbf{r})$ includes *bonded*
interactions between atoms connected by chemical bonds, and *nonbonded*
interactions between atoms of different molecules, or between atoms of the
same molecule which are not chemically bonded, for example between the
atoms of nonadjacent monomers of the same polymer chain.

Bonded interactions depend on the deviations of the chemical bond
lengths and bond angles from their equilibrium values, as well as on the
values of the dihedral (torsion) angles. The simplest and most often used
approach is to represent the bonded energy as a sum of three separate
contributions, described below:

$$\mathcal{V}_B = \sum_b \mathcal{V}_b + \sum_\theta \mathcal{V}_\theta + \sum_\phi \mathcal{V}_\phi \tag{1}$$

The chemical bonds are very stiff, deviations from the equilibrium bond
length usually being much less than a tenth of an Å at room temperature.
Thus, the energy due to bond stretching is described by a harmonic
potential, proportional to the square of the deviation of the bond length
from its equilibrium value:

$$\mathcal{V}_b = \frac{1}{2} k_b (l - l_0)^2 \tag{2}$$

The bond angles are less stiff than the bond lengths; nevertheless, at usual
temperatures they normally do not deviate by more than a few degrees from
their equilibrium values. The bond angle potential is also fairly well
approximated by a harmonic potential

$$\mathcal{V}_\theta = \frac{1}{2} k_\theta (\theta - \theta_0)^2 \tag{3}$$

Torsional potentials describe the energy change due to rotation around a
bond. This energy originates from interactions between the atoms connected

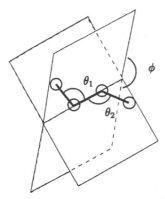

FIG. 1 Bond angles (θ_1, θ_2), and torsion angle (ϕ) in the butane molecule. For simplicity, only the carbon atoms are shown.

to the atoms linked by the bond, and possibly spin–spin interactions between the binding electrons from the adjacent bonds. The torsion angles define the conformation of the polymer molecule. The torsional potential is a periodic function* of the torsion angle ϕ. Its minima correspond to the conformations where the molecule likes to spend most of its time. For butane, shown in Fig. 1, the minima correspond to the trans-$(\phi = \pi)$ and the two gauche-$(\phi = \pm \pi/3)$ conformations.†

Typically, the torsional potential is represented as a Fourier series

$$V_\phi = \sum_n A_n \cos(n\phi) \tag{4}$$

Nonbonded interactions between skeletal or pendant atoms connected to the two ends of the bond, whose relative position changes when ϕ is changed, may be considered as separate nonbonded components of the energy or may be incorporated as part of V_ϕ.

The constants k_b, l_0, θ_0, A_n differ for different kinds of bonds. They are usually adjusted to fit experimental data or results of quantum chemical

---

*Full 360 degrees rotation around the bond does not change the energy. If some of the pendant atoms happen to be the same, the period can be less than 360 degrees.
†The reader should be careful about the conventions used in measuring torsion angles. The old convention places $\phi = 0$ at trans, the newer convention places $\phi = 0$ in the energetically most unfavorable conformation where all four carbon atoms are on the same plane, atoms 1 and 4 being on the same side of the middle bond between 2 and 3 [14]. In both conventions, ϕ is taken as positive when it corresponds to rotation in a sense that would lead to unscrewing if the middle bond were a right-handed screw.

calculations, or both. More sophisticated representations of the bonded interactions, designed to better fit experimental data and *ab initio* calculation results, may include cross-terms that simultaneously depend on bond lengths, bond angles and torsions [15].

Nonbonded interactions are typically modeled as electrostatic interactions between partial charges on the atoms, London dispersion forces due to correlated fluctuations of the electronic clouds of the atoms, and exclusion forces at short distances. They depend on the distance between the atoms $r_{ij} = |\mathbf{r}_i - \mathbf{r}_j|$, and are represented as a sum of Coulomb and Lennard-Jones potentials.

$$V_{NB} = \sum_{ij} V_{LJ}(r_{ij}) + \sum_{ij} V_q(r_{ij}) \tag{5}$$

$$V_{LJ}(r_{ij}) = 4\epsilon\left(\left(\frac{\sigma}{r_{ij}}\right)^{12} - \left(\frac{\sigma}{r_{ij}}\right)^{6}\right) \tag{6}$$

$$V_q(r_{ij}) = -\frac{1}{4\pi\epsilon_0}\frac{q_i q_j}{r_{ij}} \tag{7}$$

Here ϵ and σ are parameters dependent on the type of atoms. ϵ is the well depth of the Lennard-Jones potential achieved at $r_{ij} = r_0 = \sqrt[6]{2}\sigma$. The r^{-6} attractive term in the Lennard-Jones (LJ) potential finds theoretical justification in the 1930 quantum mechanical perturbation theory calculation of London. Recent modeling work has shown that an exponential term, $A\exp(-kr)$, for excluded volume interactions provides a more satisfactory representation of thermodynamic properties than the r^{-12} term of the LJ potential. Such an exponential term for the repulsions also has better theoretical justification.

ϵ_0 is the dielectric permittivity of free space, equal to 8.854×10^{-12} C/(m V) in SI units and to $1/(4\pi)$ in cgs units. Potential parameters are not just different for different elements, they also change depending on what molecule or group the atoms belong to. For example, the values typically used for the Lennard-Jones ϵ for a carbon in an aromatic ring are higher by almost a factor of two than the corresponding values for an aliphatic carbon in the polyvinyl chain backbone. Partial charges on atoms are determined by the electron density distribution within the molecule. They are usually obtained from quantum *ab initio* or semiempirical calculations and sometimes adjusted to reproduce experimentally measured multipole moments of low-molecular weight analogs.

The set of parameters describing the interatomic interactions is often referred to as a "force field." Most popular force fields [15–17] are adjusted to fit the experimental and quantum calculation results over a wide range of organic compounds. Parameters for the same atoms may be quite different between different force fields.

The way the potential energy function is split into various kinds of interactions is quite arbitrary and may vary from one force field to another. Two different force fields may give very similar energy for the same molecule in the same conformation, but the individual contributions: torsional, nonbonded, bond angles, etc. may be quite different in different force fields. This is why it is generally dangerous to "mix" force fields, i.e., take, say, the Lennard-Jones parameters from one force field and bonded parameters from another.

In polymer simulations energy is usually expressed in kcal/mol or kJ/mol, distances in Å and angles in radians or degrees, so the dimensions of the constants are the following—k_b: kcal/(mol Å^2), k_θ: kcal/(mol rad^2), A_n: kcal/mol, ϵ: kcal/mol, r_{ij}, σ, l, and l_0: Å, charges q_i: multiples of the elementary charge $e = 1.6022 \times 10^{-19}$ C.

B. Simplified Polymer Chain Models

From typical values of bonded and nonbonded interaction parameters one can conclude that the fastest motions in a polymer system are the chemical bond stretching vibrations, their frequencies typically being on the order of 10^{14} Hz; bond angle bending vibrations and torsional librations are about 10 times slower. Conformational (e.g., gauche $\leftrightarrow$ trans) isomerizations over the free energy barriers separating torsional states occur with rates on the order of 10^{11} Hz or slower at room temperature. The characteristic time for the end-to-end distance of a sequence of monomeric units to lose memory of its orientation through these elementary motions grows rapidly with the length of the sequence. The characteristic time for a chain in a melt of moderate molecular weight (say, $C_{10,000}$ linear polyethylene) to diffuse by a length commensurate to its size and thus forget its previous conformation can well exceed 1 ms.

Most interesting polymer properties are observed at frequency scales of 10^6 Hz and lower* This means that, in order to compare atomistic simulation results to experiment, one should be able to reproduce the model behavior at time scales spanning more than 10 orders of magnitude! This task would be daunting even for a computer 10,000 times faster than the most powerful supercomputers available today.

---

*These are the typical working frequency ranges for routine dielectric relaxation and rheological experiments. More sophisticated techniques, such as NMR relaxation and inelastic neutron scattering, are sensitive to higher frequencies up to 10^{12} Hz.

Very large model systems, which are often necessary to track the morphological characteristics responsible for the peculiar properties of polymers, also present a great challenge in polymer simulations. Detailed atomistic multichain polymer models used today seldom contain more than a few thousands of atoms, although domain decomposition strategies on parallel machines offer the possibility of going up to millions of atoms.

General computer simulation techniques useful for both low-molecular weight molecules and polymers have been covered in some excellent textbooks. We will briefly go over the fundamentals of these techniques in this chapter. The major focus of the book, however, is on specific approaches developed to handle the long time- and length scale challenges of polymer problems. By necessity, these techniques must be hierarchical ones, involving several levels of description, each designed to address a specific window of time- and length scales. Going from one level to another should entail a systematic coarse-graining procedure, wherein the detailed information from more fundamental (shorter time- and length scale) levels of modeling is built into some key parameters invoked by the more macroscopic levels.

Analytical polymer theories usually address simplified models of polymer chains, which capture universal features such as the chain topology, flexibility, etc. Despite lacking many fine details, such models still manage to predict, sometimes even quantitatively, many physical properties of polymer networks, solutions, and melts. When a simplified general model turns out to be capable of describing a particular polymer property or phenomenon, this means that it successfully captures the relevant physics. Such results provide valuable understanding of which particular features (e.g., chain length, architecture, stiffness) are mainly responsible for a particular property. Many polymer-specific effects and properties, such as rubber elasticity, the viscoelastic rheological response of melts in the terminal region, and overall molecular shapes in dilute solutions and melts in the bulk and next to nonadsorbing surfaces, are similar for polymers of different chemical structures. They can be rationalized with relatively simple arguments based on the decrease of entropy associated with chain extension and on environment-dependent excluded volume interactions between segments.

Figure 2 shows the most popular simplified models of polymer chains [18,19].

The freely-jointed chain model (Fig. 2a) is the simplest; it has been described as the "ideal gas" of polymer physics, as all interactions between the chain segments, except the chemical bonds connecting them, are neglected. The model represents a chain as a sequence of steps, or rectilinear statistical (Kuhn) segments, connected together at their ends. Each statistical segment represents a sequence of several chemical bonds,

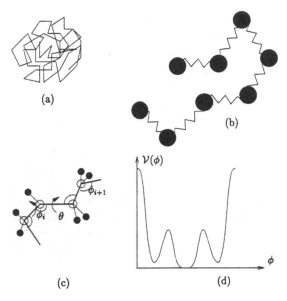

FIG. 2 The most common simplified models of polymer chains: freely-jointed chain (a), bead-spring chain (b), rotational isomer model (c), and the typical torsional potential $\mathcal{V}(\phi)$ for butane molecule (d).

depending on the conformational stiffness of the chain. The steps can point in any direction, the directions of different steps being completely independent, and the chain can intersect itself. Clearly, this model is identical to the random flight model. It can easily be shown that the probability to observe a certain value of the vector **R** connecting the two ends of a sufficiently long freely-jointed chain obeys a Gaussian distribution with zero average and mean squared value

$$\langle \mathbf{R}^2 \rangle \equiv \langle R^2 \rangle = N_K b_K^2 \tag{8}$$

for a chain of N_K statistical segments, each of length b_K. This is one of the most fundamental results of polymer science; its derivation and corollaries can be found in polymer textbooks [18–20]. The radius of gyration tensor, characterizing the average shape of the chain, and its properties can also be calculated. Once the distribution is known, all the statistical and thermodynamic properties of the ideal chain can be obtained.

A real chain whose conformation is governed only by local (effective) interactions along its contour can be mapped onto a freely-jointed chain by requiring that Eq. (8) is satisfied, and also that the contour length of the freely-jointed chain, $N_K b_K$, matches that of the real chain at full extension.

The distribution of the distance between two points in the freely-jointed chain is also well approximated by a Gaussian distribution, but with the chain length N_K in Eq. (8) replaced by the number of statistical segments between the points. This makes a freely-jointed chain self-similar on various length scales. If we look at it with different spatial resolutions it looks the same, as long as the resolution is much larger than b_K and less than $N_K^{1/2} b_K$.

The bead-spring model (Fig. 2b) represents a polymer chain as a collection of beads connected by elastic springs. It, too, is a coarse-grained model. The coarse-graining is based on the polymer chain self-similarity, with a single bead corresponding to a chain fragment containing several monomers. Springs reproduce the Gaussian distribution of separations between monomers connected through a large number of chemical bonds. The spring constant is given by $(3k_B T)/\langle R_{sp}^2 \rangle$, where $\langle R_{sp}^2 \rangle$ is the mean square end-to-end distance of the actual chain strand represented by the spring. The spring reproduces the entropic free energy rise associated with the reduction of conformations of a strand as its two ends are pulled apart.

The rotational isomer model (Fig. 2c) makes a step from the freely-jointed chain towards the geometry of the real polymer. It invokes fixed, realistic values for the chemical bond lengths ℓ and the chemical bond angles θ, and a potential energy function $\mathcal{V}(\phi_\in, \phi_\ni, \ldots)$. Figure 2d schematically shows the torsional potential for a butane molecule, containing only three backbone C–C bonds, where it is a function of only one angle. The rotational isomer model addresses *unperturbed* chains, i.e., chains whose conformation remains unaffected by nonlocal interactions between topologically distant atoms along the backbone. Under unperturbed conditions, $\mathcal{V}(\phi_2, \phi_3, \ldots)$ can be written as a sum of terms, each depending on a small number (usually two) of adjacent torsion angles. A rotational isomer chain of sufficient length also follows the scaling relation of Eq. (8), which can be written more specifically in terms of the number of chemical bonds N along the backbone and the length of a bond ℓ as [20,18]:

$$\langle \mathbf{R}^2 \rangle = C_\infty N \ell^2 \tag{9}$$

The *characteristic ratio* C_∞ characterizes chain flexibility. It depends on the θ and torsional potential and is determined by the chemical structure of the monomers [20]. The rotational isomeric state (RIS) model, introduced by P.J. Flory [20] is essentially an adaptation of the one-dimensional Ising model of statistical physics to chain conformations. This model restricts each torsion angle ϕ to a discrete set of states (e.g., trans, gauche$^+$, gauche$^-$), usually defined around the minima of the torsional potential of a single bond, $\mathcal{V}(\phi)$ (see Fig. 2d). This discretization, coupled with the locality of interactions, permits calculations of the conformational partition function, the C_∞, and conformational averages of many properties by matrix algebra, once a set of statistical weights is available for all pairs of torsional states assumed by successive bonds. The latter weights can be determined by atomistic conformational analysis of oligomers. Extensions to chains containing several types of chemical bonds and to nonlinear chemical architectures have been made. The RIS model is described in detail in chapter 2 by E.D. Akten, W.L. Mattice, and U.W. Suter, and in the book [14].

Unperturbed conditions are realized experimentally in bulk amorphous polymers and in dilute solutions in specific (rather bad) solvents at specific temperatures, defining the so-called Θ point. For a chain dissolved in a good solvent, nonlocal interactions cannot be neglected; polymer segments strive to maximize their favorable contacts with solvent molecules, leading to a repulsive intersegment potential of mean force. These repulsions prohibit chain self-intersections. The simplest mathematical model describing this effect is the "self-avoiding" random walk (SAW). The SAW problem can only be solved numerically using computer simulations. The end-to-end distance of a SAW follows the scaling $\langle R^2 \rangle \propto N^\nu$ with ν (the excluded volume exponent) close to 0.6, i.e., significantly different from the value 0.5 characteristic of unperturbed chains. This is easy to understand, as the repulsion between monomers leads to an increase in coil size.

All major simulation techniques described in the rest of this book are applicable to these simplified models, as well as to the detailed atomistic models. Simulations of the simplified models, however, require much less computer power. They are helpful in studying phenomena where universal chain characteristics, such as chain length, flexibility, and topology are of interest.

C. Unperturbed Polymer Chain

As already mentioned, a popular descriptor of the overall size of a polymer chain is the mean squared end-to-end distance $\langle R^2 \rangle$. For a chain of $N+1$ identical monomers, linked by N bonds of length ℓ, the end-to-end vector is a sum of bond vectors $\ell_i = \mathbf{r}_i - \mathbf{r}_{i-1}$, and its mean square is:

$$\langle \mathbf{R}^2 \rangle = \left\langle \left(\sum_{i=1}^{N} \ell_i \right)^2 \right\rangle = 2 \sum_{i=1}^{N-1} \sum_{j=i+1}^{N} \langle \ell_i \cdot \ell_j \rangle + \sum_{i=1}^{N} \langle \ell_i \cdot \ell_i \rangle$$

$$= 2 \sum_{i=1}^{N-1} \sum_{k=1}^{N-i} \langle \ell_i \cdot \ell_{i+k} \rangle + N\ell^2 \tag{10}$$

For very long chains, ignoring end effects, we can write

$$\langle \mathbf{R}^2 \rangle = N\ell^2 + 2\ell^2 \sum_{k=1}^{N-1}(N-k)\langle \boldsymbol{\ell}_1 \cdot \boldsymbol{\ell}_{k+1} \rangle / \ell^2$$

$$= N\ell^2 + 2N\ell^2 \sum_k (1 - k/N)C(k) \tag{11}$$

Here $C(k)$ is a bond direction correlation function. As the number of monomers separating two bonds increases, their directions become uncorrelated, and hence $C(k)$ is a decreasing function of k. For a random walk, $C(k)=0$ for all $k \geq 1$. For an unperturbed chain, $C(k)$ falls exponentially with increasing k at large k and $C = 1 + 2\sum_k C(k)$.

The mean squared end-to-end distance of a linear unperturbed chain of N monomers is proportional to N. As $\mathbf{R}$ is a sum of a large number of independent random vectors, the probability density for the end-to-end separation vector to have a certain value $\mathbf{R}$ is given by a Gaussian distribution:

$$\rho_N(\mathbf{R}) = \left(\frac{3}{2\pi\langle R^2 \rangle}\right)^{3/2} \exp\left(-\frac{3}{2}\frac{\mathbf{R}^2}{\langle R^2 \rangle}\right) \tag{12}$$

The quantity $-k_B T \ln \rho_N(\mathbf{R})$ is a conformational free energy associated with the fact that a given end-to-end vector $\mathbf{R}$ can be realized through a large number of conformations.

When the chain is deformed, the distribution of end-to-end distance vectors will change towards less probable values, thus reducing the polymer chain entropy and causing an increase in free energy.

A rubber is a network built of cross-linked chains; the above reasoning explains, at least qualitatively, the mechanism of rubber elasticity.

The conformation of a chain of $N+1$ segments, numbered from 0 to N, can also be characterized by the mean square radius of gyration:

$$\langle R_g^2 \rangle = \frac{1}{(N+1)^2}\left\langle \sum_{i=0}^{N-1}\sum_{j=i+1}^{N}(\mathbf{r}_i - \mathbf{r}_j)^2 \right\rangle \tag{13}$$

or by the hydrodynamic radius

$$R_h = \frac{1}{2}\left[\frac{1}{(N+1)^2}\left\langle \sum_{i=0}^{N-1}\sum_{j=i+1}^{N}\frac{1}{|\mathbf{r}_i - \mathbf{r}_j|} \right\rangle\right]^{-1} \tag{14}$$

measured by elastic and inelastic light scattering techniques, respectively. The former is related to the coil size and shape, while the latter describes a coil's hydrodynamic properties, as inelastic light scattering measures the self-diffusion coefficient of polymer coils in dilute solution.

It can be shown, that, for large chain lengths, both hydrodynamic and gyration radii of the polymer coil can be expressed in terms of the average end-to-end distance [19]:

$$\langle R_g^2 \rangle = \frac{1}{6} \langle R^2 \rangle$$

$$R_h = \left(\frac{3\pi}{128} \right)^{1/2} \langle R^2 \rangle^{1/2} \tag{15}$$

D. Mixing Thermodynamics in Polymer–Solvent and Polymer–Polymer Systems

The fact that a polymer molecule consists of a large number of monomers linked together in a chain is responsible for various properties, which are specific to polymers in solution and in the melt. Polymer solutions have very different properties, depending on concentration. As the concentration increases, a polymer solution changes from the dilute to the semidilute, to the concentrated regimes.

When the concentration is very small, polymer coils do not contact each other. The effect of the polymer on the solution properties is additive, consisting of single-coil contributions.

In the semidilute solution the monomer volume fraction is still very low, but the polymer coils begin to overlap and entangle with each other. This is possible because the monomer concentration in the Gaussian coil is relatively low. Indeed, N monomers in a good solvent occupy a volume of the order of magnitude of $R^3 \propto \ell^3 N^{9/5}$, so that the average monomer density in a coil is of the order of $\ell^{-3} N^{-4/5}$. The coils will begin to touch when the polymer volume fraction in the solution becomes $\phi^* \propto N^{-4/5}$. This concentration, which can be very low for large N, marks the boundary between dilute and semidilute regimes.

Concentrated solutions with polymer fraction on the order of 1 behave pretty much like polymer melts.

Polymers are in general much less soluble and miscible with each other, compared to the corresponding monomers under the same conditions. Miscibility or solubility is defined by the balance of the entropy gain and energy loss or gain upon mixing. Consider mixing substance A with substance B. If the interactions between molecules of different components (AB interactions) are stronger than between molecules of the same kind (AA and BB interactions), mixing is favorable at all temperatures. Such a system is always miscible. Usually this case is realized when there are some specific interactions (e.g., acid–base, hydrogen-bonding) between A and B. More common is the case where AB interactions are energetically less favorable than AA and BB interactions. This is typically the case for nonpolar substances held together by London dispersion forces. In this case, mixing occurs above a certain temperature, when the entropic contribution $-T\Delta S$ to the free energy of mixing overcomes the unfavorable positive energy of mixing. The entropy gain upon mixing is due mainly to translational entropy of the molecules. In the mixture, each A and B molecule can occupy the whole volume occupied by the mixture, while in the phase-separated case each molecule is localized in the volume occupied by the particular phase.

In polymer chains large numbers of monomers are linked together, therefore, the translational entropy gain upon mixing per unit mass or volume is much smaller than in the case of mixing unlinked monomers. The entropy gained by a monomer in a liquid of unlinked monomers is now gained by each polymer molecule, and thus the entropy gained per mole of monomers is many times less in the polymer case. One of the most fundamental results of polymer thermodynamics, first derived (independently) by Flory and Huggins [20,21] through a lattice-based mean field theory, states that the free energy change per mole of segments upon mixing polymer chains made up of A-type segments with chains made up of B-type segments of lengths N_A and N_B, respectively is:

$$\frac{\Delta G_{\mathrm{mix}}}{RT} = \frac{1}{N_A}\phi_A\ln(\phi_A) + \frac{1}{N_B}(\phi_B)\ln(\phi_B) + \chi\phi_A\phi_B \tag{16}$$

In the theory, A and B segments are envisioned as equal in volume, occupying sites on a lattice of coordination number Z. ϕ_A is the volume fraction of A-type segments in the system, $\phi_B = 1 - \phi_A$ is the volume fraction of B-type segments. $\chi = (Z/k_BT)(\varepsilon_{AB} - 1/2\varepsilon_{AA} - 1/2\varepsilon_{BB})$ is the Flory interaction parameter, describing the interaction energy difference between unlike- and like- segment interactions occupying adjacent sites on the lattice. The parameter χ is positive for most polymer mixtures. From Eq. (16) it is seen that the entropy gain per mole of segments upon mixing polymers is smaller than the corresponding gain for mixing monomers ($N_A = N_B = 1$) by a factor equal to the inverse chain length. This explains the difficulty in mixing different polymers. For $N_A = N_B = N$, Eq. (16) predicts an upper critical solution temperature which scales proportionally with the chain length, χN assuming a value of 2 at the critical point. Traditionally, a strategy for achieving miscible blends has been to use systems capable of developing specific interactions, which can make χ negative.

As we discussed above, polymer chains are not rigid bodies, and there is an entropy associated with the different polymer chain conformations. Polymer chains usually contain very large numbers of monomers (thousands or more), and, provided that environmental conditions ensure the same excluded volume exponent ν, all polymers with the same chain topology are similar at the length scales comparable to the chain radius of gyration. Flory's model is the simplest; it assumes that chain conformation does not change upon mixing, and therefore there is no conformational entropy contribution in Eq. (16). This contribution is, however, very important in many cases and has to be taken into account to explain the thermodynamics and phase behavior of mixtures and solutions where chain conformation is different in different phases, for example in systems containing block copolymers, chains next to surfaces or in restricted geometries, in the swelling of polymer gels etc. [19,22,23].

E. Polymer Chain Dynamics

The motion of a large polymer molecule is quite complex. Even though individual atoms move about with the same equilibrium distribution of speeds as if they were disconnected, their motion is constrained by the chemical bonds keeping the chain together. Longer and longer parts of the chain need longer and longer times to rearrange, and quantities depending on overall conformation, such as the end-to-end distance or the radius of gyration, take a very long time to forget their original values.

A general characteristic of polymer liquids is their viscoelastic response to flow. Because of the long relaxation times associated with large-scale conformational rearrangements, chains subjected to a fast flow field are oriented, or even unraveled by the flow. Part of the energy imparted by the flow is stored as elastic free energy and is released upon cessation of the flow, when chains spring back to their unperturbed conformations causing macroscopic recoil phenomena. Polymeric fluids have memory, and this is of paramount importance to their processing and applications.

The dynamics of a coil in a low-molecular weight melt can be described well by an ingeniously simple model, the Rouse model [18]. This model represents a polymer chain as a set of beads connected by harmonic springs (compare Fig. 2b). The beads move as (tethered) Brownian particles subject to random forces and to frictional forces proportional to their velocity exerted from their environment (see Section I.F). For a linear polymer chain the Rouse model predicts that the longest relaxation time (time needed for the chain center of mass to diffuse by a length commensurate to $\langle R^2 \rangle^{1/2}$, approximately equal to the time needed for the end-to-end vector **R** to lose memory of its original orientation) is proportional to the chain length squared N^2. The viscosity of a melt in the molecular weight range described by the Rouse model is also proportional to N and the self–diffusivity of a chain is proportional to N^{-1}.

The Rouse model has been extended to deal with the dynamics of chains in dilute solution [18]. In solution a moving bead perturbs the solvent flow around another bead, leading to effective, so-called hydrodynamic, interactions. The Zimm model generalizes the Rouse model by taking hydrodynamic interactions into account.

As the molecular weight of a polymer melt increases, it is envisioned that entanglement constraints arise between different polymer coils, making relaxation even slower. Experimentally, it is found that the viscosity of the melt rises with increasing chain length as $N^{3.4}$. The reptation theory of polymer dynamics [18] postulates that long chains move in a Rouse-like fashion along their contour (primitive path), while motion normal to the primitive path is constrained by entanglements. The reptation model in its original formulation predicts a relaxation time and a melt viscosity which scale as N^3, and a chain self-diffusivity which scales as N^{-2}. Several extensions and refinements of the reptation model have appeared.

The slow dynamics of polymers presents a formidable challenge for computer simulations, as on the one hand the most interesting polymer-specific phenomena occur at time scales comparable with the relaxation times of the whole chain, but on the other hand to reproduce the motion correctly one has to solve the equations of motion with a time step smaller than the period of the fastest monomer motions. Using coarse-grained models in the simulation is one way in which one may try to overcome this problem.

F. Glass Transition Versus Crystallization

Low-molecular weight liquids usually crystallize upon cooling. Very fast cooling rates are usually required to form a glass (amorphous solid). Contrary to that, because of their very slow relaxation, polymer melts are much more viscous, and the glassy state is the most common for solid polymers. Many polymers, such as atactic vinyl polymers, cannot crystallize due to their irregular stereostructure. In polymers crystallized from the melt, the crystalline phase is usually only around 10–50%. The degree of crystallinity and the morphology of semicrystalline polymers are highly dependent on the conditions of temperature and processing flow under which they are formed. A high degree of crystallinity is usually obtained by cooling the polymer from a highly oriented liquid crystal state, or by stretching.

Semicrystalline polymers also exhibit very different behavior under deformation. When stretched, crystals of low-molecular weight substances usually break at an elongation of several percent. Semicrystalline polymers can deform up to 100% without breaking; on the contrary, the elastic modulus increases with increasing elongation (strain-hardening effect). This is because, at high deformations, polymer chains in the amorphous regions, which may grow at the expense of the crystalline regions, begin to stretch and align along the direction of deformation, leading to a higher elastic modulus.

Our fundamental understanding of glass formation, even in low-molecular weight fluids, is still incomplete. The mode coupling theory [24,25] describes the falling out of ergodicity at a critical temperature which is substantially higher than the glass temperature T_g, where macroscopic manifestations of solidification are seen in volume, enthalpy, and elastic constant measurements.

From a practical viewpoint, one can say that glass formation occurs because the characteristic time of segmental motions (motions associated with the α-mode of relaxation measurements) is a very strongly increasing function of temperature. As temperature is decreased, segmental motions become slower and slower, until there comes a point where the rate of segmental motion cannot keep up with the rate of change in temperature. The system is no longer capable of exploring its entire configuration space over ordinary time scales and is confined within a small region of that space, containing at least one local minimum of the potential energy $\mathcal{V}$. Macroscopically, a change in slope is seen in the specific volume vs. temperature and specific enthalpy vs. temperature curves. The glass temperature, T_g, depends, to a certain extent, on cooling rate. It is typically reported for cooling rates on the order of 1 K/min.

A glassy material is not in thermodynamic equilibrium. Its physical properties change gradually with time ("physical ageing"). The time scales of these changes, however, are enormous a few decades of degrees below T_g, so that the material can be regarded as a dimensionally stable solid for all practical purposes.

The temperature dependence of the characteristic time τ of molecular motions responsible for the glass transition is strongly non-Arrhenius. Above T_g it is described by the Vogel–Fulcher law

$$\tau(T) = \tau_0 \exp \frac{T_A}{T - T_V} \tag{17}$$

with T_A (activation temperature) and T_V (Vogel temperature) being constant.

The ratio of characteristic times at two temperatures $T \geq T_o$, $a_T = \tau(T)/\tau(T_o)$, (shift factor) consequently follows the Williams–Landel–Ferry (WLF) equation

$$\log a_T = -C_1 \frac{T - T_o}{T - T_o + C_2} \tag{18}$$

When the reference temperature T_o is chosen as the calorimetric glass temperature, T_g, the constants fall in relatively narrow ranges for all polymers: $C_1 = 14$ to 18 and $C_2 = 30$ to 70K [26]. At the glass temperature, τ is on the order of 1 min.

II. STATISTICAL MECHANICS

A. Trajectories in Phase Space

In classical mechanics, the state of a molecular system is described by positions of atomic nuclei and their time derivatives—velocities (or momenta). Each state represents a point in the multidimensional space spanned by positions and momenta, which is termed the *phase space* of the system. The position vectors of the atoms, or the set of generalized coordinates providing the same information as atomic positions, are called *degrees of freedom*; they span the *configuration space* of the system. The space spanned by the momenta (or generalized momenta) of the degrees of freedom is called *momentum space*. The evolution of the system's microscopic state with time can be represented as a set of state points, corresponding to successive moments in time. This set defines a line in phase space, which constitutes the system's *dynamical trajectory*. The following Fig. 3 shows the simplest example of state points and trajectories in phase space. Notice that not just any line can represent a trajectory. The line crossed out in Fig. 3 cannot be a trajectory. It corresponds to the impossible situation where the coordinate is increasing with time, while the velocity is negative.

B. Classical and Quantum Mechanics

In both quantum and classical mechanics, a system is defined by its degrees of freedom and by its potential and kinetic energy functions. In a quantum mechanical description, electrons, in addition to nuclei, are included among the degrees of freedom. Given the potential and kinetic energies as functions

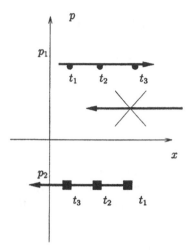

FIG. 3 State points at different moments in time $t_1 < t_2 < t_3$ and trajectory in the phase space $\{x, p\}$ for two free particles traveling along the x axis with different velocities $x_i = v_i$ and momenta $p_i = m_i v_i$. Particle 2 goes to the left ($p_2 < 0$). The crossed out trajectory is impossible.

of the degrees of freedom, we can write the equations of motion and, in principle, we should be able to predict the system's motion under any given initial conditions by solving these equations. Unfortunately, the situation is not so simple for at least two reasons. First, in the majority of cases, it turns out that the equations are too complex to be solved, even numerically. Imagine that to describe the motion of $1 \, \text{cm}^3$ of water classically we need to store positions and velocities for about 10^{22} molecules and solve the same number of equations. Just the storage is beyond the capacity of the most powerful computer available today.

A second difficulty is that equations of motion are highly nonlinear and therefore lead to instability with respect to initial conditions. *Small* uncertainties in the initial conditions may lead to *large* deviations in the trajectories. Two trajectories starting at almost the same state point will diverge exponentially with time, a small uncertainty in the initial coordinates or velocities making the motion completely unpredictable. If we are trying to describe a molecular system in a cubic container of side L, when the uncertainty in the calculated particle coordinates becomes of the same order of magnitude as the characteristic model size L, the prediction becomes totally worthless.

The latter, however, can be turned into an advantage. The inherent mechanical instability and complexity of the motion justifies a *statistical* description of the system. In contrast to the *deterministic* approach, i.e., trying to follow the motion of every molecule as closely as possible, the *statistical* approach concentrates on the *average* properties calculated over the system's trajectory or over phase space.* As was demonstrated by the founding fathers of statistical mechanics, L. Boltzmann and J.W. Gibbs in the late 19th century, these trajectory or phase space averages are directly related to the experimentally observed thermodynamic properties.

In general, molecular motion should be described using the laws of quantum mechanics. In quantum mechanics dynamical trajectories themselves are probabilistically defined entities. The state of the system is described by a probability amplitude function, Ψ, which depends on coordinates and, possibly, spin states of all nuclei and electrons present in the system. $\Psi^*\Psi$ is the probability density for observing the system in a particular point in phase space. Motion of the system, or in other words its change in state with time, is described by the time-dependence of the Ψ-function. It is determined by solving the Schrödinger equation:

$$\sum_i -\frac{\hbar^2}{2m_i} \nabla^2_{\mathbf{r}_i} \Psi + \mathcal{V}\Psi \equiv \hat{\mathcal{H}}\Psi = \frac{\partial\Psi}{\partial t} \tag{19}$$

Here, m_i is the mass of particle i, $\mathcal{V}$ is the potential energy, a function of the positions of electrons and nuclei, and $\hat{\mathcal{H}}$ is the Hamilton operator.[‡]

Equation (19) is a partial differential equation. The number of its variables equals the total number of electron and nuclear coordinates and spin states, or in other words the total number of degrees of freedom. Solving this equation is a very difficult, virtually impossible task, but fortunately it can be simplified by the following two approximations.

First, the nuclei of almost all atoms, except maybe hydrogen and helium, are heavy enough for the classical approximation to describe their motion sufficiently well at normal temperatures.

Electrons, on the other hand, being several thousand times lighter than nuclei, move much faster. The second approximation is to assume that the electrons adjust practically instantaneously to the current positions of the nuclei. Given a particular set of nuclear coordinates, the electron state is adjusted to make the total potential energy of the system minimal with respect to the electron state. This approximation is called *adiabatic*, or

---

[†]The equivalence of these two averages is the essence of a very important system property called *ergodicity*, which deserves a separate subsection further in this chapter.

[‡]A hat is used here to emphasize that, in quantum mechanics, the Hamiltonian is an operator acting on the Ψ function and not just a function of particle coordinates and momenta, as in classical mechanics.

the Born–Oppenheimer approximation; in engineering language, it is a "quasi-steady state" approximation for the electrons. The adiabatic approximation may not be applicable when electron excitation processes are important, for example in the case of chemical reactions or conformational changes induced by the absorption of light.

The adiabatic approximation is very helpful because, if the electron states are adjusted to the positions of the nuclei, the system's state can be described by the coordinates of the nuclei alone. The number of degrees of freedom is then significantly reduced. Accordingly, in the Schrödinger equation, $\mathcal{V}$ is replaced by the effective potential $\mathcal{V}(\mathbf{r}_n) = \mathcal{V}(\mathbf{r}_n, \psi_{e0}(\mathbf{r}_n))$, where $\psi_{e0}(\mathbf{r}_n)$ describes the electron state providing the lowest potential energy for the given set of nuclear coordinates $\mathbf{r}_n$, and is thus also a function of the nuclear coordinates. The accuracy of $U(\mathbf{r}_n)$ determines the accuracy of quantities calculated from computer simulations.

An approach for tracking electronic degrees of freedom in parallel with a numerical integration of the classical equations of motion for the nuclei, and therefore determining $\mathcal{V}(\mathbf{r}_n)$ "on the fly," has been devised by Car and Parrinello [27]. This extended ensemble molecular dynamics method, termed "*ab initio* molecular dynamics," solves the electronic problem approximately using the Kohn–Sham formulation of Density Functional Theory. This approach proved useful for covalent systems; it still has to be applied to the systems where the properties of interest are defined by Lennard-Jones interactions.

C. Classical Equations of Motion

When the interactions in the molecular system are known, the classical description can be cast in one of three different forms: Hamilton, Newton, or Lagrange equations of motion. Consider a molecular system with potential energy $\mathcal{V}(q_1, q_2, \ldots, q_{N_f})$ and kinetic energy $\mathcal{K}(q_1, q_1, \ldots, q_{N_f}, \dot{q}_1, \dot{q}_2, \ldots, \dot{q}_{N_f})$ or $\mathcal{K}(q_1, q_2, \ldots, q_{N_f}, p_1, p_2, \ldots, p_{N_f})$, where q_i are generalized coordinates with $\dot{q}_i$ being their time derivatives, and p_i are generalized momenta. The configuration space and momentum space are both N_f-dimensional, where N_f is the total number of degrees of freedom. The phase space is $2N_f$-dimensional. When Cartesian coordinates of the atoms are used as degrees of freedom, a triplet of q_i stands for the position vector of an atom, $\mathbf{r}_k$, and a triplet of p_i stands for an atomic momentum vector $\mathbf{p}_k = m_k \dot{\mathbf{r}}_k = m_k \mathbf{v}_k$. In this case, the kinetic energy can simply be written as

$$\mathcal{K} = \sum_i \frac{p_i^2}{2m_i} = \sum_k \frac{\mathbf{p}_k^2}{2m_k} = \sum_i \frac{1}{2} m_i \dot{q}_i^2 = \sum_k \frac{1}{2} m_k \mathbf{v}_k^2$$

The three different forms of the equations describing the motion of atoms are:

- Hamilton equations

$$\mathcal{H}(q_1, q_2, \ldots, q_{N_f}, p_1, p_2, \ldots, p_{N_f}) = \mathcal{K} + \mathcal{V}$$

$$\frac{\partial q_i}{\partial t} = \frac{\partial H(q_1, q_2, \ldots, q_{N_f}, p_1, p_2, \ldots, p_{N_f})}{\partial p_i}$$

$$\frac{\partial p_i}{\partial t} = -\frac{\partial \mathcal{H}(q_1, q_2, \ldots, q_{N_f}, p_1, p_2, \ldots, p_{N_f})}{\partial q_i}$$

(20)

- Newton equations

$$\frac{d\mathbf{r}_i}{dt} = \mathbf{v}_i$$

$$m_i \frac{d\mathbf{v}_i}{dt} = \mathbf{f}_i$$

$$\mathbf{f}_i = -\frac{\partial \mathcal{V}(\mathbf{r}_1, \mathbf{r}_2, \ldots, \mathbf{r}_{N_f/3})}{\partial \mathbf{r}_i}$$

(21)

- Lagrange equations

$$\mathcal{L}\left(q_1, q_2, \ldots, q_{N_f}, \dot{q}_1, \dot{q}_2, \ldots, \dot{q}_{N_f}\right) = \mathcal{K} - \mathcal{V}$$

$$\dot{q}_i = \frac{dq_i}{dt}$$

$$\frac{d}{dt}\frac{\partial \mathcal{L}}{\partial \dot{q}_i} = \frac{\partial \mathcal{L}(q_1, q_2, \ldots, q_N, \dot{q}_1, \dot{q}_2, \ldots, \dot{q}_N)}{\partial q_i}$$

(22)

In the above, $\mathcal{H}(q_1, q_2, \ldots, q_{N_f}, p_1, p_2, \ldots, p_{N_f})$ and $\mathcal{L}(q_1, q_2, \ldots, q_{N_f}, \dot{q}_1, \dot{q}_2, \ldots, \dot{q}_{N_f})$ stand for the Hamiltonian and Lagrangian functions of the system, respectively. As an illustration, let us consider these equations for two examples. The first is a monatomic fluid, and the second is a harmonic oscillator.

A monatomic fluid is governed by a potential energy function of the form $\mathcal{V}(\mathbf{r}_1, \mathbf{r}_2, \ldots, \mathbf{r}_N)$. The degrees of freedom are the $3N$ Cartesian coordinates $\{\mathbf{r}_1, \mathbf{r}_2, \ldots, \mathbf{r}_N\}$. The Newton equations are:

$$\frac{d\mathbf{r}_i}{dt} = \mathbf{v}_i$$

$$m_i \frac{d\mathbf{v}_i}{dt} = \mathbf{f}_i$$

$$\mathbf{f}_i = -\frac{\partial \mathcal{V}}{\partial \mathbf{r}_i} \equiv -\nabla_{\mathbf{r}_i} \mathcal{V} \tag{23}$$

If we use $\{\mathbf{r}_1, \mathbf{r}_2, \ldots, \mathbf{r}_N\}$ as the generalized coordinates, the Lagrangian is a function $\mathcal{L}(\mathbf{r}_1, \mathbf{r}_2, \ldots, \mathbf{r}_N, \dot{\mathbf{r}}_1, \dot{\mathbf{r}}_2, \ldots, \dot{\mathbf{r}}_N)$, and the Lagrange equations of motion are:

$$\dot{\mathbf{r}}_i = \frac{d\mathbf{r}_i}{dt}$$

$$\mathcal{L}(\mathbf{r}_1, \mathbf{r}_2, \ldots, \mathbf{r}_N, \dot{\mathbf{r}}_1, \dot{\mathbf{r}}_2, \ldots, \dot{\mathbf{r}}_N) = \mathcal{K} - \mathcal{V} = \sum_{i=1}^{N} \frac{1}{2} m_i \dot{\mathbf{r}}_i^2 - \mathcal{V}(\mathbf{r}_1, \mathbf{r}_2, \ldots, \mathbf{r}_N)$$

$$\frac{d}{dt} \frac{\partial \mathcal{L}}{\partial \dot{\mathbf{r}}_i} = m_i \frac{d^2 \mathbf{r}_i}{dt^2} = \frac{\partial \mathcal{L}}{\partial \mathbf{r}_i} = -\nabla_{\mathbf{r}_i} \mathcal{V} \tag{24}$$

The Hamiltonian function and Hamilton equations are:

$$\mathcal{H}(\mathbf{r}_1, \mathbf{r}_2, \ldots, \mathbf{r}_N, \mathbf{p}_1, \mathbf{p}_2, \ldots, \mathbf{p}_N) = \mathcal{K} + \mathcal{V}$$

$$\mathcal{H} = \sum_{i=1}^{N} \frac{\mathbf{p}_i^2}{2m_i} + \mathcal{V}(\mathbf{r}_1, \mathbf{r}_2, \ldots, \mathbf{r}_N)$$

$$\frac{\partial \mathbf{r}_i}{\partial t} = \frac{\partial \mathcal{H}}{\partial \mathbf{p}_i} = \frac{\mathbf{p}_i}{m_i}$$

$$\frac{\partial \mathbf{p}_i}{\partial t} \equiv m_i \frac{d\mathbf{v}_i}{dt} = \frac{\partial \mathcal{H}}{\partial \mathbf{r}_i} = -\nabla_{\mathbf{r}_i} \mathcal{V} \tag{25}$$

In the special case where $\mathcal{V}(\mathbf{r}_1, \mathbf{r}_2, \ldots, \mathbf{r}_N) = 0$ for all configurations, the above equations describe an ideal monatomic gas. The ideal gas particles travel with their velocities not changing with time. This is because they do not interact with each other or with anything else. The ideal gas is a limiting case of a system with no interactions; it is a simple system, whose statistical mechanics can be solved exactly. It is used as a reference to build more complex systems, where interactions are non-negligible.

Another very important analytically solvable case is the harmonic oscillator. This term is used for a mechanical system in which potential energy depends quadratically on displacement from the equilibrium position. The harmonic oscillator is very important, as it is an interacting system (i.e., a system with nonzero potential energy), which admits an analytical solution. A diatomic molecule, linked by a chemical bond with potential energy described by Eq. (2), is a typical example that is reasonably well described by the harmonic oscillator model. A chain with harmonic potentials along its bonds (bead-spring model), often invoked in polymer theories such as the Rouse theory of viscoelasticity, can be described as a set of coupled harmonic oscillators.

The harmonic oscillator is particularly important, because *any* mechanical system in the vicinity of stable equilibrium can be approximated by a harmonic oscillator. If the deviations from equilibrium are small, one can set the origin at the equilibrium point and expand the potential energy in powers of the displacement. To take a simple example, for a pair of identical atoms interacting via the Lennard-Jones potential [Eq. (6)], if the distance between the atoms r is close to the equilibrium value $r_0 = \sigma\sqrt[6]{2}$, we can expand the potential in powers of $u = r - r_0$

$$V_{LJ} = 4\epsilon\left(\left(\frac{\sigma}{r_0(1 + u/r_0)}\right)^{12} - \left(\frac{\sigma}{r_0(1 + u/r_0)}\right)^6\right)$$
$$= \epsilon\left(-1 + 36\left(\frac{u}{r_0}\right)^2 + O\left(\left(\frac{u}{r_0}\right)^3\right)\right)$$
$$\approx V_0 + \frac{ku^2}{2}$$
$$V_0 = -\epsilon, \qquad k = \frac{72\epsilon}{r_0^2} \tag{26}$$

If $u = x_1 - x_2 - r_0$, assuming that the y and z coordinates of both atoms are zero (Fig. 4), Newton's equations of motion become:

$$m\ddot{x}_1 = -\frac{\partial V}{\partial x_1} = -k(x_1 - x_2 - r_0)$$
$$m\ddot{x}_2 = -\frac{\partial V}{\partial x_2} = k(x_1 - x_2 - r_0)$$
$$m\ddot{y}_i = 0, \qquad i = 1, 2$$
$$m\ddot{z}_i = 0, \qquad i = 1, 2 \tag{27}$$

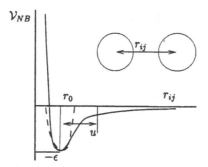

FIG. 4 Harmonic approximation (dashed line) for the Lennard-Jones potential (solid line).

Adding and subtracting the first and second equations gives the following two equations:

$$m\ddot{x}_1 + m\ddot{x}_2 = 0$$
$$m\ddot{u} = -2ku \qquad (28)$$

The first equation describes the motion of the "dimer" center of mass $x_c = (mx_1 + mx_2)/2m$. Its left hand side, $m\ddot{x}_1 + m\ddot{x}_2$, equals the total mass $2m$ times the center of mass acceleration. As long as there is no net external force applied, the acceleration is zero. However, if the initial velocities of the atoms $v_1(t=0)$ and $v_2(t=0)$ are such that the center of mass velocity $v_c = [mv_1(t=0) + mv_2(t=0)]/2m$ is not zero, the value of v_c does not change with time, and the "dimer" travels as a whole with constant speed. The second equation describes the motion of the atoms relative to each other, as u measures the deviation of their separation from its equilibrium value r_0. The solution of the second equation,

$$u(t) = A\cos(\omega t) + B\sin(\omega t)$$
$$\omega^2 = \frac{2k}{m} \qquad (29)$$

with A and B defined from the initial conditions:

$$A = u(t = 0); \qquad B\omega = \dot{u}(t = 0) = v_1(t = 0) - v_2(t = 0) \qquad (30)$$

describes oscillations around the equilibrium value $u = 0$ with period $2\pi/\omega$. A typical period of the vibrations due to Lennard-Jones interactions is on the order of several picoseconds.

To derive the Lagrange and Hamilton equations of motion we need expressions for the kinetic and potential energies:

$$K = \frac{m_1\dot{\mathbf{r}}_1^2}{2} + \frac{m_2\dot{\mathbf{r}}_2^2}{2}$$

$$V = \frac{k}{2}(|\mathbf{r}_1 - \mathbf{r}_2| - r_0)^2 \tag{31}$$

Here we used the harmonic approximation for potential energy (26) and set the zero energy level at $V_0 = -\epsilon$. Choosing the x axis along the line connecting the atoms, following the Lagrangian or Hamiltonian formalism, given by the Eqs. (20) or (22), and remembering that, for simplicity, we consider here the case $m_1 = m_2 = m$, one can arrive at the Newton equations of motion (27). When Cartesian coordinates are used, Newton, Lagrange, and Hamilton equations result in exactly the same differential equations for the coordinates.

Newton equations, however, do require use of Cartesian coordinates, while Lagrange and Hamilton equations do not. Let us see how the Lagrange and Hamilton formulations can help us derive the more convenient set of Eqs. (28) from the very beginning. All we have to do is choose the center of mass coordinate x_c and the oscillation amplitude u as generalized coordinates, describing the motion along the x direction in which we are interested. We leave the Cartesian coordinates y_1, z_1 and y_2, z_2 for the other four degrees of freedom. The potential energy is simply $V = ku^2/2$. The kinetic energy is a little more tricky:

$$K = \frac{m_1(\dot{y}_1^2 + \dot{z}_1^2)}{2} + \frac{m_2(\dot{y}_2^2 + \dot{z}_2^2)}{2}$$

$$+ \frac{(m_1 + m_2)v_c^2}{2} + \frac{1}{2}\left(\frac{1}{m_1} + \frac{1}{m_2}\right)^{-1}\dot{u}^2 \tag{32}$$

Again, using the above equations in the Lagrange (20) or Hamilton (22) formalism, working out through all the necessary derivatives, and using $m_1 = m_2 = m$, one arrives directly at Eqs. (28) for u and x_c. Figure 5 shows the harmonic oscillator's trajectories, corresponding to different initial conditions.

The harmonic oscillator model is particularly useful in the study of crystalline solids. Potential energy of the crystal can be expanded around the state of mechanical equilibrium under given macroscopic dimensions (quasiharmonic approximation). By diagonalizing the Hessian matrix of second derivatives of the potential energy with respect to atomic

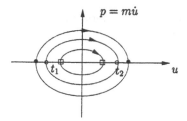

FIG. 5 Trajectories of the harmonic oscillator in its two-dimensional phase space, and positions at different moments of time $t_2 > t_1$, but $t_2 - t_1 < T$.

displacements, vibrational motion can be analyzed as a set of $3N$ independent harmonic oscillators characterized by a spectrum of frequencies. Lower frequencies correspond to longer-range, collective motions. The Debye model is a good model for this spectrum in the region of low frequencies [28].

Observing that Lagrange, Hamilton, and Newton equations of motion lead to exactly the same differential equations for particle coordinates, one might conclude that the three different forms were invented by arrogant mathematicians to create more confusion and show off their sophistication. This is not at all the case, however. Hamilton and Lagrange equations do reduce to the simple Newton's law, known from high-school physics, only when the system's motion is described using Cartesian coordinates, and when potential energy is a function of coordinates only, but not momenta or velocities, and the kinetic energy does not depend on coordinates, but is a function of only momenta or velocities. This choice is not always the most convenient, as we tried to demonstrate above for the harmonic oscillator approximation to the Lennard-Jones "dimer." The general theorem concerning the equivalence of Lagrange and Hamilton equations is beyond the scope of this book; the interested reader is referred to the excellent books by Arnold [29], Goldstein [30] or any other classical mechanics text. The power of Lagrange or Hamilton equations is realized in cases where the generalized, not necessarily Cartesian coordinates, are more convenient to use. As the chemical bond lengths are usually fixed in simulations of macromolecules, the bond angles and torsion angles (Fig. 1) are a natural choice as generalized coordinates. Fixing all or some bond angles in addition to the bond lengths is also practiced in some cases.

Torsion angles are often used as sole coordinates in biological applications [12,31–33] in which one is interested in refining conformations of a single protein or DNA molecule, or their complexes, and where large pieces of the molecules, containing a hundred or more atoms, can be considered moving as rigid bodies. Movements of these rigid fragments can be described by changing a few torsion angles instead of hundreds of Cartesian coordinates. This approach is sometimes used to simulate atomistic models of polymer chains [34,35]. If bond lengths and bond angles are considered constant, the configuration of the system can be described by the torsion angles, the position of the first atom and the orientation of the first bond of each chain. It is possible to describe polyvinyl or polycarbonate chains of a hundred or so monomers and more than a thousand atoms with just a few hundred variables instead of several thousand Cartesian coordinates. Such fixed bond length and bond angle models are particularly effective in energy minimization calculations (molecular mechanics) used to generate representative configurations for glassy polymers. In dynamic simulations it is advisable to fix the bond lengths but let the skeletal bond angles be flexible, to allow for cooperativity in the motion of bond angles and torsion angles.

As good things never come for free, the reduction in the number of degrees of freedom when generalized coordinates are used is traded for more complex equations of motion. The kinetic energy depends on both angles and angular velocities, and solving the Lagrange equations in generalized coordinates requires computationally expensive matrix inversions. Besides, the nonbonded potential energy is usually a function of Cartesian coordinates of the atoms, and it is still necessary to perform transformations between the generalized and Cartesian atom coordinates every time the energy and force are evaluated.

The Lagrange equations of motion allow another useful alternative. For instance, if the freely-jointed chain is described in Cartesian coordinates, the monomer motions are constrained by the constant bond lengths b_K. For a polymer chain of $N_K + 1$ segments and N_K bonds there is a set of constraint equations,

$$g_i = (\mathbf{r}_{i+1} - \mathbf{r}_i)^2 - b_K^2 = 0 \tag{33}$$

for every bond i. The Lagrangian is modified by including additional terms for the constraints:*

$$\mathcal{L} = \mathcal{K} - \mathcal{V} + \sum_{i=1}^{N_K} \lambda_i g_i \tag{34}$$

---

*This technique is usually called the Method of Lagrange Multipliers in physical and mathematical literature.

The set of Lagrange multipliers λ_i is added to the set of $3(N_K + 1)$ Cartesian coordinates tracked during the numerical solution of the equations of motion. Accordingly, N_K equations for the λ_i's should be derived from (22) in addition to the $3(N_K + 1)$ equations for the Cartesian coordinates. These additional equations are equivalent to the constraint equations (33). The physical meaning of the Lagrange multiplier associated with a bond length constraint is the magnitude of the force that is exerted on the particles connected by the bond in order to maintain the bond length fixed. Bond length constraint equations are solved iteratively in the SHAKE and RATTLE molecular dynamics algorithms, popular in polymer simulations. These algorithms will be described in detail in the following chapters.

Naturally, the Lagrange multiplier approach can be generalized to constrain bond angles, or any other geometrical characteristics, e.g., distances between particular atoms, as in protein or DNA structure refinement, based on the experimental information obtained from X-ray or NMR experiments.

Solving the Lagrange equations of motion in the presence of holonomic constraints for bond lengths and bond angles amounts to sampling the "rigid" polymer model; here, the constraints are considered as being imposed from the beginning. Alternatively, one can consider the bonds and bond angles as being subject to harmonic potentials and take the limit of the properties as the force constants of these potentials are taken to infinity ("flexible model in the limit of infinite stiffness"); this model is sampled by Monte Carlo simulations with constant bond lengths and bond angles. The two models differ in their kinetic energy function and have nonequivalent statistical mechanics. One can make the rigid model sample the configuration space of the flexible model in the limit of infinite stiffness by adding to the potential energy function the "Fixman" potential [13,11].

D. Mechanical Equilibrium, Stability

Equation (21) tells us that force equals negative potential energy gradient, and therefore points in configuration space tend to move in the direction of decreasing potential energy. The position of (mechanical) equilibrium is defined as a point where all forces equal zero, hence the potential energy gradient must be zero at equilibrium. This, in turn, implies that the equilibrium point must be an extremum of the potential energy as a function of the generalized (or Cartesian) coordinates.

Stability of the equilibrium point requires that the system returns back to the equilibrium state in response to a small perturbation. This excludes maxima and saddle points, leaving only potential energy minima as candidates for stable equilibrium points. Indeed, if we make a small displacement from a maximum or from a saddle point, we will find a point with potential energy lower than its value at the extremum.

As the force is directed in the direction of decreasing potential energy, it points *away from the extremum* near the maximum or saddle point, pulling the system *out of* a state of unstable equilibrium. Only in the vicinity of the potential energy minima is the potential energy always higher than at the extremum itself, and the force is directed back towards equilibrium.

Let us consider for example a system with one degree of freedom, q. If the system has more than one degree of freedom, its Hessian matrix of second derivatives can be diagonalized, and the problem is reduced to several independent one-dimensional problems. The kinetic energy being $\mathcal{K} = \frac{1}{2}m\dot{q}^2$, the potential energy can be expanded in Taylor series in terms of the deviation u from the equilibrium point q_0:

$$V = V(q_0) + \left.\frac{\partial V}{\partial q}\right|_{q_0} u + \frac{1}{2}\left.\frac{\partial^2 V}{\partial q^2}\right|_{q_0} u^2 + \mathcal{O}(u^3) \tag{35}$$

As q_0 is an equilibrium point, the first derivative is zero and the leading term is quadratic. Setting $k = \left.\partial^2 V/\partial q^2\right|_{q_0}$, the equation of motion becomes

$$m\ddot{u} = -ku$$
$$u(t) = A\exp(-\bar{\omega}t) + B\exp(\bar{\omega}t) \tag{36}$$

with coefficients A and B determined by the initial conditions $\dot{u}(t=0) = v_0$, $u(t=0) = u_0$, and $\bar{\omega} = \sqrt{-a/m}$. If $a < 0$, ω is real, and the solution for u contains a term that is exponentially increasing with time, so that, given a small initial deviation from equilibrium, u_0, the deviation will increase exponentially with time. The equilibrium is unstable in this case.

If, on the contrary, a is positive, $\bar{\omega}$ is imaginary ($\bar{\omega} = i\omega$). As in the case of the harmonic oscillator (29), the solution becomes

$$u(t) = A'\sin(\omega t) + B'\cos(\omega t) \tag{37}$$

which describes bounded oscillations around the equilibrium configuration. The equilibrium is stable. These results are in agreement with what we see in the two-dimensional example of Fig. 6.

When there is more than one degree of freedom, in the vicinity of the equilibrium, the Hessian matrix of second derivatives of the potential energy can be diagonalized and the problem can essentially be reduced to a set of

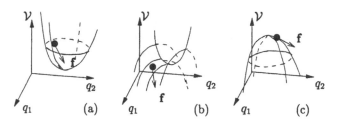

FIG. 6 Different potential energy extrema. Minimum (a), saddle point (b), and maximum (c). Force $\mathbf{f} = -\nabla \mathcal{V}$ points along the direction of the fastest potential energy decrease.

independent one-dimensional problems by a linear coordinate transformation $\mathbf{u} \to \bar{\mathbf{u}}$. In terms of the new coordinates, in the vicinity of the stationary point, the potential energy $\mathcal{V} = \frac{1}{2} \sum_i k_i \bar{u}_i^2$. When all k_i are positive, the potential energy has a minimum and the equilibrium is stable. When at least one of k_i is negative the equilibrium is unstable, as any small initial perturbation along the direction $\bar{u}_i$ will grow exponentially with time, taking the system away from equilibrium. A maximum of the potential energy occurs when all k_i are negative.

Notice that, if we consider two trajectories starting from two points that are close to each other in phase space in the vicinity of the stable equilibrium point, the difference between the two trajectories will always remain of the same order of magnitude as the initial difference; the two trajectories will remain about as close to each other as their initial points are.

In contrast to this, if the two points are in the vicinity of an unstable equilibrium point, the difference between the trajectories will grow exponentially with time, and the trajectories will diverge. Thus, in the vicinity of an unstable equilibrium point, the small uncertainty in the initial condition will grow exponentially, with characteristic time of $\bar{\omega}^{-1}$. Any attempt to predict the system motion for a time much longer than that will fail. Notice, also, that the same argument applies to any two trajectories, as soon as they are not confined to the immediate vicinity of the stable equilibrium. If the system is unharmonic, as almost all systems are, and its trajectories are not confined to the vicinity of a stable equilibrium, the trajectories are exponentially divergent.

E. Statistical Description, Ergodicity

In principle, in computer simulations we could specify the initial coordinates and velocities for all atoms in the system, follow the trajectory by solving the equations of motion, and calculate some properties of interest from this particular realization of the system's trajectory. A real experiment is very different; we have no control over the positions, velocities or spin states of each atom or electron in the molecules. So, it is natural to question if any relationship exists between the results of such calculations and the values observed in real experiments. Material properties, such as the equation of state, the elastic moduli, thermal conductivity, viscosity, etc., originate from molecular motions and interactions. The relationship between molecular geometry and interactions and macroscopic properties is established by statistical mechanics. Statistical mechanics tells us that it is really not necessary, and moreover it does not even make any sense, to know the exact system state at any moment of time to predict its observable properties. The observable properties, such as pressure, temperature, strain, heat conductivity, diffusion coefficients, polarization, etc., are related to the average values of different functions, which in turn depend on the system microscopic state, calculated along many different trajectories with different initial conditions. Therefore, the relevant quantity to look for is the *probability* to observe a particular value of the energy, polarization, or any other observable of interest.

The probabilistic approach of statistical mechanics is justified by the fact that, except for some special cases, such as the harmonic oscillator or the ideal gas, system trajectories in phase space are very complex and virtually unpredictable, as discussed in the previous section.

The very important property of ergodicity states that the time averages along dynamical trajectories of an equilibrium system are equivalent to phase space averages. In other words, if we are interested in macroscopic thermodynamic properties, such as pressure, temperature, stress, average polarization, etc., it is not necessary to follow the system dynamics exactly. It is sufficient to sample enough points in phase space and to calculate the proper average. Ergodicity is based on the assumption (provable for some Hamiltonians) that any dynamical trajectory, given sufficient time, will visit all "representative" regions in phase space, the density distribution of points in phase space traversed by the trajectory converging to a stationary distribution.

Observed equilibrium properties are time averages over long dynamical trajectories:

$$\lim_{t \to \infty} \frac{1}{t} \int_0^t \mathcal{A}(\mathbf{q}(\tau), \mathbf{p}(\tau)) \, d\tau = \langle \mathcal{A} \rangle_{obs} \tag{38}$$

By virtue of ergodicity, they can also be calculated as averages over phase space, with respect to an appropriate equilibrium probability density distribution.

Ergodicity is very important for the statistical mechanical and thermodynamic descriptions to be valid for particular systems; it has to be checked for each particular case. It holds for systems where the characteristic molecular relaxation times are small in comparison to the observation time scale. Polymeric and other glasses represent a classical example of nonergodic systems. They can be trapped, or "configurationally arrested," within small regions of their phase space over very long times.

In the context of equilibrium simulations, it is always important to make sure that the algorithm used in the simulation is ergodic. In other words, that no particular region in phase space is excluded from sampling by the algorithm. Such an exclusion would render the simulation wrong, even if the simulated object itself is ergodic. As an extreme example consider an algorithm to simulate various conformations of the freely-jointed chain in three dimensions, which, for some reason, such as a programming error, never selects bonds parallel to the z axis. Evidently, many representative conformations will be erroneously excluded. Of course, such a programming error is easy to find and fix. Often the situation can be more complex, however. It is quite common for lattice simulation algorithms to sample only the odd or even numbered lattice sites, or to be unable to find their way out of particular configurations, especially in two dimensions [36].

In simulation practice, ergodicity of the system can and should be checked through reproducibility of the calculated thermodynamic properties (pressure, temperature, etc.) in runs with different initial conditions.

F. Microscopic and Macroscopic States

The *microscopic* state of the system defines coordinates, momenta, spins for every particle in the system. Each point in phase space corresponds to a microscopic state. There are, however, many microscopic states, in which the states of particular molecules or bonds are different, but values of the macroscopic observables are the same. For example, a very large number of molecular configurations and associated momenta in a fluid can correspond to the same number of molecules, volume, and energy. All points of the harmonic oscillator phase space that are on the same ellipse in Fig. 5 have the same total energy.

The set of values of the macroscopic observables, such as temperature, pressure, average polarization or magnetization, average chain end-to-end distance, etc., describes the system's *macroscopic* state. One macroscopic state combines all the microscopic states that provide the same values of the macroscopic observables, defined by the macroscopic state.

The probability to observe a certain macroscopic state of the system (or, in other words, certain values for the observables) equals the sum of the probabilities of all the corresponding microscopic states.

G. Probability Distribution of the Microscopic States. Statistical Ensembles

We use the abbreviation Γ to denote the set of coordinates and momenta corresponding to one microscopic state.* If a Cartesian description is used, $\Gamma = (\mathbf{r}_1, \mathbf{r}_2, \ldots, \mathbf{r}_N, \mathbf{p}_1, \mathbf{p}_2, \ldots, \mathbf{p}_N)$. The probability density for finding a system in the vicinity of Γ will be denoted as $\rho(\Gamma)$. $\rho(\Gamma)$ depends on the macroscopic state of the system, i.e., on the macroscopic constraints defining the system's size, spatial extent, and interactions with its environment. A set of microscopic states distributed in phase space according to a certain probability density is called and *ensemble*. According to the ergodicity hypothesis we can calculate the observables of a system in equilibrium as averages over phase space with respect to the probability density of an equilibrium ensemble. Equation (38) for the average can be rewritten as

$$
\langle \mathcal{A}_{obs} \rangle = \lim_{t \to \infty} \frac{1}{t} \int_0^t A(\mathbf{q}(\tau), \mathbf{p}(\tau)) \, d\tau
$$

$$
= \langle \mathbf{A}(\Gamma) | \rho(\Gamma) \rangle = \int \mathcal{A}(\Gamma) \rho(\Gamma) \, d\Gamma \tag{39}
$$

H. Liouville Equation

Imagine that we start with a set of identical systems, whose states are distributed in phase space according to a density distribution $\rho(\Gamma)$ at time $t = 0$, and let the systems move according to their equations of motion. The ensemble constituted by the systems (points in phase space) evolves in time. As the systems evolve, the density distribution $\rho(\Gamma)$ should, in general, change with time. However, systems just move, no new systems are created, and none of the systems is destroyed. Therefore, there should be a conservation law for the probability density, similar to the continuity equation (mass conservation) of hydrodynamics. The conservation law in phase space is called the *Liouville theorem*. It states that the total time

---

*Or to one point in phase space.

derivative of the probability density is zero. For a classical system of N_f degrees of freedom q_i and momenta p_i, the Liouville equation is:

$$\frac{\partial \rho(\Gamma, t)}{\partial t} + \sum_i \left[\dot{q}_i \frac{\partial}{\partial q_i} + \dot{p}_i \frac{\partial}{\partial p_i} \right] \rho(\Gamma, t) = 0 \tag{40}$$

Combining with the Hamiltonian equations of motion, (20), one obtains:

$$\frac{\partial \rho(\Gamma, t)}{\partial t} + \sum_i \left[\frac{\partial \mathcal{H}}{\partial p_i} \frac{\partial}{\partial q_i} - \frac{\partial \mathcal{H}}{\partial q_i} \frac{\partial}{\partial p_i} \right] \rho(\Gamma, t) = 0 \tag{41}$$

which is the Liouville equation.

For an equilibrium system there should be no explicit time dependence for ρ, i.e., a stationary distribution should be reached by the ensemble.

The main postulate of statistical mechanics states that, for an equilibrium system of given mass, composition, and spatial extent, all microstates with the same energy are equally probable [37,38]. This postulate, along with the ergodic hypothesis, can be justified on the basis of the mixing flow in phase space exhibited by the dynamical trajectories of real systems [28]. It means that all microstates of an isolated system, which does not exchange energy and material with its environment, should occur equally often.

The equilibrium probability density in phase space for a system with total energy E_0 is therefore given by:

$$\rho(\Gamma) \sim \delta(\mathcal{H}(\Gamma) - E_0) \tag{42}$$

The $\delta(x)$ is a Dirac δ-function, which is nonzero only when its argument equals zero. Here it selects those microscopic states Γ that have total energy* $\mathcal{H}$ equal to E_0.

The above ensemble of systems with constant number of particles N, occupying constant volume V, with the total energy E conserved, is called the *microcanonical*, or (NVE) ensemble.

I. Partition Function, Entropy, Temperature

A very important measure of the probability distribution of an equilibrium ensemble is the partition function Q. This appears as a normalizing factor in the probability distribution defined by the ensemble.

---

*Remember the definition of the Hamiltonian, Eq. (20).

For the microcanonical ensemble:

$$Q_{NVE} = \sum_{\Gamma} \delta(\mathcal{H}(\Gamma) - E_0)$$

$$\rho(\Gamma) = Q_{NVE}^{-1} \delta(\mathcal{H}(\Gamma) - E_0)$$

$$\langle A \rangle = Q_{NVE}^{-1} \sum_{\Gamma} \delta(\mathcal{H}(\Gamma) - E_0) \mathcal{A}(\Gamma) \tag{43}$$

The summation over states $\sum_{\Gamma}$ here and in the previous equations is used for the quantum case, where microscopic states are discrete and $\rho(\Gamma)$ has the meaning of a probability. For one-component classical systems, on which we have mainly focused so far, the sum should be replaced by the integral:

$$\sum_{\Gamma} \rightarrow \frac{1}{N!} \frac{1}{h^{3N}} \int d\Gamma$$

$$d\Gamma = \prod_{i=1}^{N} d^3 r_i d^3 p_i \tag{44}$$

$N!$ here takes care of the indistinguishability of particles of the same species. For a multicomponent system of N_1 particles of type 1, N_2 particles of type 2, ..., $N!$ should be replaced by $N_1! N_2! \ldots$. h is Planck's constant. It describes the phase space volume occupied by one state and renders the product $dr\ dp$ dimensionless.

The partition function defines the thermodynamic potential. The partition function is a function of the thermodynamic state variables that are kept constant in the definition of the equilibrium ensemble. The expression of the thermodynamic potential in terms of these state variables constitutes a fundamental equation of thermodynamics.

The proper thermodynamic potential for the microcanonical ensemble is the entropy:

$$-S/k_B = -\ln Q_{NVE} \tag{45}$$

where $k_B = R/N_A$ is the Boltzmann constant. The argument of the logarithm on the right hand side of Eq. (45) is just the number of states with the energy E_0. We therefore have a statistical thermodynamic definition of entropy as a quantity proportional to the logarithm of the number of microscopic states under given N, V, E [37,38]. The proper way to simulate the microcanonical ensemble is to numerically solve the equations of motion (20) or (21) or (22) for the closed system. This is done in the simplest versions of molecular dynamics simulation.

If the system of N particles is contained in the constant volume V but is allowed to exchange energy with its environment through diathermal walls, its energy is not constant any more. The energy is fluctuating, but the *temperature* is constant [37,38]. Temperature describes the probability distribution of energy fluctuations. Such a system is represented by the *canonical* or NVT ensemble. The probability density is given by the Gibbs distribution:

$$\rho_{NVT}(\Gamma) = Q_{NVT}^{-1} \exp\left(-\frac{\mathcal{H}(\Gamma)}{k_B T}\right)$$

$$Q_{NVT} = \frac{1}{N!} \frac{1}{h^{3N}} \int d\Gamma \exp\left(-\frac{\mathcal{H}(\Gamma)}{k_B T}\right)$$

$$\frac{A}{k_B T} = -\ln(Q_{NVT}) \tag{46}$$

A, the Helmholtz energy, is a thermodynamic potential for the canonical ensemble. Q_{NVT}, often symbolized simply as Q, is the canonical partition function. The last of Eqs. (46) defines a fundamental equation in the Helmholtz energy representation by expressing A as a function of N, V, T. Often in classical systems it is possible to separate the energy contributions that depend on the momenta only (kinetic energy $\mathcal{K}$) from the potential energy $\mathcal{V}$, which depends only on the coordinates. When Cartesian coordinates are used as degrees of freedom, for example, the partition function can be factorized as:

$$Q_{NVT} = \frac{1}{N!} \frac{1}{h^{3N}} \int \prod_{i=1}^{N} d^3 p_i \exp\left(-\frac{\mathcal{K}}{k_B T}\right) \int \prod_{i=1}^{N} d^3 r_i \exp\left(-\frac{\mathcal{V}}{k_B T}\right)$$

$$= Q_{NVT}^{ig} Q_{NVT}^{ex} \tag{47}$$

into two contributions—the partition function of the ideal gas:

$$Q_{NVT}^{ig} = \frac{V^N}{N! \Lambda^{3N}}$$

$$\Lambda = \left(\frac{h^2}{2\pi m k_B T}\right)^{1/2} \tag{48}$$

(Λ being the thermal or de Broglie wavelength), and the excess part:

$$Q_{NVT}^{ex} = V^{-N} \int \prod_{i=1}^{N} d^3 r_i \exp\left(-\frac{\mathcal{V}}{k_B T}\right) \tag{49}$$

It is also customary to use the configurational integral, defined as

$$Z_{NVT} = \int \prod_{i=1}^{N} d^3 r_i \exp\left(-\frac{\mathcal{V}}{k_B T}\right) \tag{50}$$

instead of the Q_{NVT}^{ex}.

As a consequence, all the thermodynamic properties can be expressed as a sum of an ideal gas part and an excess part. All specifics of systems are included in the latter, and more attention is usually focused on it. In fact, in Monte Carlo simulations the momentum part of the phase space is usually omitted, and all calculations are performed in configuration space. The ideal gas contribution is added after the simulations to compare to experiments or to other simulations.

Another important consequence of Eq. (47) is that the total average kinetic energy is a universal quantity, independent of the interactions in the system. Indeed, computing the average of $\mathcal{K} = \sum_{i=1}^{N} \mathbf{p}_i^2 / 2m$ with respect to the probability distribution of Eq. (46) and using the factorization of Eq. (47) we obtain that $\langle \mathcal{K} \rangle = 3/2\, N k_B T$ or, more generally, $\langle \mathcal{K} \rangle = 1/2\, N_f k_B T$ for a system of N_f degrees of freedom. If the kinetic energy can be separated into a sum of terms, each of which is quadratic in only one momentum component, the average kinetic energy per degree of freedom is $1/2\, k_B T$. The last result is a special case of the *equipartition theorem* which characterizes classical systems [38,37]. It is often used in simulation practice to test for system equilibration.

When, in addition to the energy, a system is allowed to exchange volume with the environment by being contained in (partly) movable boundaries, as for instance is the gas in a vessel covered by a piston, both the volume and energy are fluctuating, but the temperature and pressure are constant. Such a system is represented by the isothermal–isobaric or *NPT* ensemble. The probability density is given by the distribution:

$$\rho_{NPT}(\Gamma, V) = Q_{NPT}^{-1} \exp\left(-\frac{\mathcal{H}(\Gamma) + PV}{k_B T}\right)$$

$$Q_{NPT} = \int dV/V_0 \; Q_{NVT} \exp\left(-\frac{PV}{k_B T}\right)$$

$$Q_{NPT} = \frac{1}{N!} \frac{1}{h^{3N}} \frac{1}{V_0} \int dV \int d\Gamma \exp\left(-\frac{\mathcal{H}(\Gamma) + PV}{k_B T}\right)$$

$$\frac{G}{k_B T} = -\ln(Q_{NPT}) \tag{51}$$

The last of Eqs. (51) defines a thermodynamic fundamental equation for $G = G(N, P, T)$ in the Gibbs energy representation. Note that passing from one ensemble to the other amounts to a Legendre transformation in macroscopic thermodynamics [39]. V_0 is just an arbitrary volume used to keep the partition function dimensionless. Its choice is not important, as it just adds an arbitrary constant to the free energy. The NPT partition function can also be factorized into the ideal gas and excess contributions. The configurational integral in this case is:

$$Z_{NPT} = \int dV \exp\left(-\frac{PV}{k_B T}\right) \int \prod_{i=1}^{N} d^3 r_i \exp\left(-\frac{\mathcal{V}}{k_B T}\right) \tag{52}$$

Some authors also include the $N!$ and V_0 factors in the configurational integral. A simulation in the isothermal–isobaric ensemble should be conducted with the volume allowed to change, but the way these changes are implemented must provide for the proper probability density distribution given by Eq. (52).

The *grand canonical* ensemble describes a system of constant volume, but capable of exchanging both energy and particles with its environment. Simulations of open systems under these conditions are particularly useful in the study of adsorption equilibria, surface segregation effects, and nanoscopically confined fluids and polymers. Under these conditions, the temperature and the chemical potentials μ_i of the freely exchanged species are specified, while the system energy and composition are variable. This ensemble is also called the μVT ensemble. In the case of a one-component system it is described by the equilibrium probability density

$$\rho_{\mu VT}(\Gamma, N) = Q_{\mu VT}^{-1} \exp\left(-\frac{\mathcal{H}(\Gamma) - \mu N}{k_B T}\right)$$

$$Q_{\mu VT} = \sum_N Q_{NVT} \exp\left(\frac{\mu N}{k_B T}\right)$$

$$Q_{\mu VT} = \sum_N \frac{1}{N!} \frac{1}{h^{3N}} \exp\left(\frac{\mu N}{k_B T}\right) \int d\Gamma \exp\left(-\frac{\mathcal{H}(\Gamma)}{k_B T}\right)$$

$$\Omega/k_B T = -\ln(Q_{\mu VT}) \tag{53}$$

$Q_{\mu VT}$, often symbolized as $\Xi_{\mu VT}$, is the grand partition function. Ω is the grand potential. For bulk systems, in the absence of interfaces, it can be shown that $\Omega = -PV$ [37]. Simulations in the grand canonical ensemble should allow the total number of particles in the system and its energy to fluctuate. These fluctuations should occur according to the $\rho_{\mu VT}(\Gamma, N)$ probability distribution.

Notice that, to formulate the *NPT* and μVT ensembles starting from the *NVT* ensemble, we had to add one more product term to the numerator of the exponent ($-PV$ and μN, respectively). This product contains the variable of the original ensemble that is allowed to fluctuate in the new ensemble (V and N for the *NPT* and μVT, respectively) times its conjugate thermodynamic variable or "field," which is kept constant in the new ensemble ($-P$ and μ, respectively). The partition function of the new ensemble is derived by integration or summation of the partition function of the original (*NVT*) ensemble over all values of the variable allowed to fluctuate. Of course, these observations are by no means just a coincidence.* Remember that an ensemble is equivalent to a particular phase space probability density distribution. Each of these ensembles can be considered as a superset of another ensemble: the canonical ensemble is a set of microcanonical ensembles with different energies weighted by the factor $\exp(-\mathcal{H}/(k_B T))$; the *NPT* ensemble is a set of *NVT* ensembles with different volumes weighted by $\exp(-PV/(k_B T))$; the grand canonical ensemble is a set of canonical ensembles with different numbers of particles weighted with $\exp(\mu N/(k_B T))$. In the thermodynamic limit (system increasing in size to become macroscopic, while all intensive variables are kept constant), the integration or summation operation by which one passes from one ensemble to the other is equivalent to a Legendre transformation.

To appreciate the generality of this approach, consider, for instance, a system of N molecular dipoles at constant volume V. We can introduce an ensemble to describe this system under constant external electric field $\mathbf{E}$, at constant temperature T. The energy contribution due to the electric field equals the negative product of the field and the total dipole moment of the system, $-V\mathcal{P} \cdot \mathbf{E}$, where $\mathcal{P}$ is the electric polarization of a configuration. The probability density and the partition function for this *ENVT-ensemble* is:

$$\rho_{ENVT}(\Gamma) = Q_{ENVT}^{-1} \exp\left(-\frac{\mathcal{H}(\Gamma) - V\mathcal{P} \cdot \mathbf{E}}{k_B T}\right)$$

$$Q_{ENVT} = \int d\Gamma \exp\left(\frac{V\mathcal{P} \cdot \mathbf{E}}{k_B T}\right) \exp\left(-\frac{\mathcal{H}(\Gamma)}{k_B T}\right) \tag{54}$$

The macroscopic polarization $\mathbf{P} = \langle \mathcal{P} \rangle$ resulting from this field will no longer be zero for nonzero $\mathbf{E}$.

---

*More detailed and rigorous discussion can be found in [38,37,40], or any other general thermodynamics or statistical mechanics textbook.

Many more different ensembles can be introduced; we shall see some of them in the following chapters. When designing a new simulation technique, it is always important to understand which ensemble it actually simulates, as this determines how to interpret the simulation results.

With all the above said, it is natural to expect, that equations of motion, phase space, statistical ensemble, and other abstract mathematical concepts we spend so much effort describing in this and previous sections, do indeed have something to do with computer simulations. In fact, various simulation techniques are nothing else but methods for the numerical solution of the statistical mechanics given a Hamiltonian $\mathcal{H}(\Gamma)$. They allow for the realization of these abstract concepts. It is through the principles of statistical mechanics, which we have briefly described above, that the numbers produced by the computer simulation program are linked to the results of real-life experiments and to the properties of real materials.

III. PROPERTIES AS OBTAINED FROM SIMULATIONS. AVERAGES AND FLUCTUATIONS

According to statistical mechanics, physically meaningful results, comparable to experiment, are obtained by calculating phase or configuration space averages, with different microscopic states weighted according to the proper ensemble probability density distribution. Different simulation techniques are essentially different ways of sampling the system's phase space with the proper probability density in order to calculate the thermodynamic averages.

Given a way to generate microscopic states according to the equilibrium ensemble distribution, we can calculate any thermodynamic average simply as an average over these states. This is how thermodynamic averages are calculated in Monte Carlo simulations. The simplest example is of course the configurational part of the internal energy in thermodynamics, which equals the average potential energy over the set of configurations generated in the Monte Carlo simulation run. Kinetic energy is trivial as soon as the equipartition theorem holds; its average is just $1/2\ k_B T$ for each degree of freedom, and can be added after the run to compare the total average energy to the experiments.

Any equilibrium property that has a microscopic quantity associated with it, such as polarization, stress, average orientation, average specific volume, average polymer chain conformation, or density in the isobaric or grand canonical ensembles, can be calculated as a simple average of values for the corresponding microscopic parameter over the random configurations generated by the Monte Carlo algorithm, according to the desired ensemble probability distribution.

A. Pressure

Pressure is one of the most important thermodynamic properties. According to the previous subsection, to calculate pressure from a simulation we need a function of the coordinates and momenta of the particles, whose average equals the thermodynamic pressure.

In equilibrium thermodynamics, pressure is defined as a negative derivative of the Helmholtz energy with respect to volume:

$$P = -\frac{\partial A}{\partial V}\bigg|_{N,T} = k_B T \frac{1}{Q}\frac{\partial Q}{\partial V} = k_B T \frac{1}{Z}\frac{\partial Z}{\partial V} \tag{55}$$

$$\frac{\partial Z}{\partial V} = \frac{\partial}{\partial V}\int_V \cdots \int_V \exp[-\mathcal{V}(\mathbf{r}_1, \ldots, \mathbf{r}_N)/(k_B T)]d^3 r_1 \cdots d^3 r_N$$

Here we consider, for simplicity, a classical system described in terms of Cartesian coordinates. In this case the kinetic energy contribution to the partition function is independent of the coordinates and volume, and Eq. (55) allows us to consider only the configurational integral Z [Eq. (50)] instead of the full partition function Q. To calculate the derivative of the integral with respect to volume, we also assume, for simplicity, that particles are in a cubic container* with edge length $V^{1/3}$. Now let us make a transformation to the normalized coordinates s_i: $\{\mathbf{r}_1, \ldots, \mathbf{r}_N\} = \{V^{1/3}\mathbf{s}_1, \ldots, V^{1/3}\mathbf{s}_N\}$. This transformation changes the integration over the volume to an integration over a cube with sides of unit length.

$$\frac{\partial Z}{\partial V} = \frac{\partial}{\partial V}\left(V^N \int_0^1 \cdots \int_0^1 \exp[-\mathcal{V}(\mathbf{r}_1, \ldots, \mathbf{r}_N)/(k_B T)]d^3 s_1 \ldots d^3 s_N\right) \tag{56}$$

$$= NV^{N-1}\int_0^1 \cdots \int_0^1 \exp[-\mathcal{V}(\mathbf{r}_1, \ldots, \mathbf{r}_N)/(k_B T)]d^3 s_1 \ldots d^3 s_N$$

$$+ \frac{V^N}{k_B T}\int_0^1 \cdots \int_0^1 \exp[-\mathcal{V}(\mathbf{r}_1, \ldots, \mathbf{r}_N)/(k_B T)]$$

$$\times \left[\sum_{k=1}^N -\frac{\partial \mathcal{V}}{\partial \mathbf{r}_k} \cdot \mathbf{r}_k \frac{1}{3}V^{-1}\right]d^3 s_1 \cdots d^3 s_N$$

$$= \frac{N}{V}Z - \frac{1}{3}\frac{Z\left\langle \sum\limits_{k=1}^N -(\partial \mathcal{V}/\partial \mathbf{r}_k)\cdot\mathbf{r}_k\right\rangle}{V k_B T}$$

---

*The following argument is obviously valid for any complex container shape, but the calculations are much longer. In the case of two dimensions the factor 1/3 is to be replaced by 1/2.

So, from Eq. (55),

$$PV = Nk_BT - \frac{1}{3}\left\langle \sum_{k=1}^{N} -\frac{\partial \mathcal{V}}{\partial \mathbf{r}_k} \cdot \mathbf{r_k} \right\rangle \tag{57}$$

Usually, $\mathcal{V}$ depends on distance vectors $\mathbf{r}_i - \mathbf{r}_j$ between sites. Then [41] one can rewrite Eq. (57) as

$$PV = Nk_BT - \frac{1}{3}\left\langle \sum_{i=1}^{N-1} \sum_{j=i+1}^{N} -\frac{\partial \mathcal{V}}{\partial(\mathbf{r}_i - \mathbf{r}_j)} \cdot (\mathbf{r}_i - \mathbf{r}_j) \right\rangle \tag{58}$$

The quantity $\partial \mathcal{V}/\partial(\mathbf{r}_) - \mathbf{r}_)$ can be taken as a definition of the force $\mathbf{f}_{kl}$ on site k due to site l, even in cases where $\mathcal{V}$ is not merely a sum of pairwise-additive site–site potentials [41]. Then,

$$PV = Nk_BT - \frac{1}{3}\left\langle \sum_{i=1}^{N-1} \sum_{j=i+1}^{N} \mathbf{f}_{ij} \cdot (\mathbf{r}_i - \mathbf{r}_j) \right\rangle \tag{59}$$

Equation (59) is useful in simulations where periodic boundary equations are employed. In the presence of periodic boundary conditions Eq. (57) should not be used [41]. The product of the force acting on a particle times its position vector is called *virial*, so Eqs. (57)–(59) are forms of the (atomic) *virial theorem* for the pressure.

Equation (59) remains valid in the presence of constraints, such as fixed bond lengths and/or bond angles. In this case, if i and j are involved in a constraint, $\mathbf{f}_{ij}$ is the Lagrange multiplier force acting on i from j due to the constraint. If the "rigid model" is sampled in the simulation and properties of the "flexible model in the limit of infinite stiffness" are desired, $\mathbf{f}_{ij}$ must additionally incorporate contributions from the Fixman potential.

In molecular systems one can write an equation similar to (59), with $\mathbf{f}_{ij}$ being the *total* force exerted on molecule i from molecule j (i.e., the sum of all site–site interactions between the molecules) and $\mathbf{r}_i - \mathbf{r}_j$ being the difference between the position vectors of the *centers of mass* of the two molecules (molecular virial equation [41]). The molecular virial equation is particularly convenient to use in systems of chain molecules, because it does not require knowledge of intramolecular bonded or constraint forces.

B. Chemical Potential

A very important theorem due to Widom [42] allows expression of the excess chemical potential of a component in a fluid (i.e., the chemical potential of the component minus the chemical potential it would have if it were an ideal gas under the temperature and molar density with which it is present in the fluid) as an ensemble average, computable through simulation. The Widom theorem considers the virtual addition of a "test particle" (test molecule) of the component of interest in the fluid at a random position, orientation, and conformation. If $\mathcal{V}^{test}$ is the potential energy change that would result from this addition,

$$\mu_i^{ex} = -k_B T \left[\ln \left\langle \exp \left(-\frac{\mathcal{V}^{test}}{k_B T} \right) \right\rangle - \ln \left\langle \exp \left(-\frac{\mathcal{V}^{ig}}{k_B T} \right) \right\rangle \right] \tag{60}$$

where the first average is taken over all configurations of the fluid and all positions, orientations, and conformations of the added test molecule, while the second average is a weighted average of the intramolecular energy of one molecule over all conformations adopted in the ideal gas state. Equation (60) holds in the canonical ensemble. Analogous expressions for the NVE and NPT ensembles have been derived [43]. Also, bias schemes that can deal with the difficulties of randomly inserting long flexible molecules in the simulated phase of interest have been developed [44].

The estimation of the chemical potential is also possible through virtual "deletion" of a real particle (inverse Widom scheme), provided the bias associated with creating a "hole" in the fluid as a result of the removal is correctly accounted for [45].

C. Fluctuation Equations

The heat capacity, isothermal compressibility, and other thermodynamic properties that correspond to second derivatives of the free energy, can be calculated from the fluctuations in different ensembles [43,38,37]. Let us consider the heat capacity in the NVT ensemble, as the simplest example. With $E = \langle \mathcal{H} \rangle$ the internal energy, computed as an ensemble average of the Hamiltonian, we have

$$c_V = \frac{\partial E}{\partial T} \bigg|_{N,V} = \frac{\partial \langle \mathcal{H} \rangle}{\partial T} \bigg|_{N,V} \tag{61}$$

$$\langle \mathcal{H} \rangle = \frac{1}{Q} \frac{1}{h^{3N} N!} \int \mathcal{H} \exp(-\mathcal{H}/k_B T) \, d\Gamma$$

$$\frac{\partial \langle \mathcal{H} \rangle}{\partial T} = -\frac{1}{Q^2} \frac{\partial Q}{\partial T} \frac{1}{h^{3N} N!} \int \mathcal{H} \exp(-\mathcal{H}/(k_B T)) \, d\Gamma$$

$$+ \frac{1}{Q} \frac{1}{h^{3N} N!} \int \exp(-\mathcal{H}/(k_B T)) \left(\frac{\mathcal{H}^2}{k_B T^2} \right) d\Gamma$$

$$\frac{\partial Q}{\partial T} = \frac{1}{h^{3N} N!} \int \exp(-\mathcal{H}/(k_B T)) \left(\frac{\mathcal{H}}{k_B T^2} \right) d\Gamma = \frac{1}{k_B T^2} \langle H \rangle Q$$

$$c_V = \frac{\langle \mathcal{H}^2 - \langle \mathcal{H} \rangle^2 \rangle}{k_B T^2} = \frac{(\delta \mathcal{H})^2}{k_B T^2}$$

Similar manipulations in the *NPT* ensemble lead to equations for the isobaric heat capacity c_p and isothermal compressibility κ_T

$$\langle [\delta(\mathcal{H} + PV)]^2 \rangle_{NPT} = k_B T^2 c_P \tag{62}$$

$$\langle (\delta V)^2 \rangle_{NPT} = V k_B T \kappa_T$$

More fluctuation expressions can be derived using similar manipulations [38,37,43].

As the expressions for averages are different in different ensembles, so are the relationships between the free energy derivatives and fluctuations. The fluctuation equation for c_V, (61), is not valid in the *NPT* or any other ensemble. Likewise, Eqs. (62) are valid in the *NPT* ensemble only.

D. Structural Properties

One of the main goals of molecular simulations since their very early days was to obtain information about structure together with the thermodynamic properties, given the interaction potentials between the atoms.

The most important structural characteristics are the density distribution and correlation functions. The *n*-particle distribution function $g^{(n)}(\mathbf{r}_1, \dots, \mathbf{r}_n)$ is defined so that $\rho^n g^{(n)}(\mathbf{r}_1, \dots, \mathbf{r}_n)$, with ρ being the mean density of particles in a system, is the probability density for finding a particle at position $\mathbf{r}_1$, a particle at position $\mathbf{r}_2, \dots$, a particle at position $\mathbf{r}_n$.

Particularly important is the pair distribution function, $g^{(2)}(\mathbf{r}_1, \mathbf{r}_2)$, or simply $g(\mathbf{r}_1, \mathbf{r}_2)$. In a homogeneous system this is only a function of the interparticle distance vector $\mathbf{r} = \mathbf{r}_2 - \mathbf{r}_1$. The quantity $\rho g(\mathbf{r})$ can be interpreted as the mean local density of particles at position $\mathbf{r}$ relative to a particle that is taken as reference. The quantity $h(\mathbf{r}) = g(\mathbf{r}) - 1$ is called the (total) pair correlation function. If the system is also spherically symmetric, as in the case of an unperturbed liquid or glass, g and h depend only on the *magnitude* of the separation, $r = |\mathbf{r}|$. This of course does not apply to oriented polymers, or to a polymer melt under flow. In this case g is also called the *radial distribution function, g(r)*.

The structure of low density gas phases and crystalline solid states is easy to describe, because the correlation functions for them are easy to obtain. In low density gases, intermolecular interactions are negligible and there are no correlations between the particles. $g(\mathbf{r})$ is 1 for all $\mathbf{r}$. At the other extreme, in crystalline solids correlations are very strong but, due to the lattice symmetry, the structure and correlation functions can be easily determined. Correlation functions are particularly important for dense disordered systems such as liquids and glasses, and hence for polymers, for which the crystalline state is rather the exception than the norm. Different models used in classical liquid state theory are essentially different approximations to calculate the correlation functions. All thermodynamic properties can be expressed in terms of the structure correlation functions and interaction potentials [37,46].

The Fourier transform of the pair correlation function, the structure factor, can be measured experimentally by X-ray, neutron, or light scattering techniques [37,38,43,47]. Moreover, in the simple and often used approximation, whereby all the potential energy is expressed as a sum of pair interaction $u(r)$ over all pairs of atoms, knowing $g(r) \equiv g^{(2)}(r)$ allows one to calculate all thermodynamic properties.

Energy and pressure, in an isotropic system for instance, are:

$$E = \frac{3}{2}Nk_BT + \frac{1}{2}N\rho \int_0^\infty u(r)g(r)4\pi r^2 \, dr \tag{63}$$

$$PV = Nk_BT - \frac{1}{6}N\rho \int_0^\infty r\frac{du(r)}{dr}g(r)4\pi r^2 \, dr$$

Essentially, integration with the $g(r)$ replaces sums over all pairs, and similar expressions can be derived for averages of any pairwise function.

Calculating the pair distribution function in a simulation is straightforward; all we need to do is count the number of atom pairs separated by a distance in the range from r to $r + \delta r$, and then normalize it. Usually, $g(r)$ is normalized by the number of pairs $N_{ig}(r, r+\delta r)$ that would be observed in an ideal gas of the same density, so that, in the limit of large distances, $r \rightarrow \infty$, where correlations disappear, $g(r) \rightarrow 1$. A typical FORTRAN code to calculate $g(r)$ from 0 to `rmax` with the resolution `deltar` is given below. The separation distance histogram is calculated for `nbins = rmax/deltar` bins, for a model of N particles in a box with sides `boxx`, boxy, and boxz along the x, y, and z directions, respectively, with the atomic coordinates stored in the arrays x(i),y(i),z(i).*

```
integer nbins, ibin, N
double precision hist(0:nbins)
double precision g(0:nbins)
double precision x(N),y(N),z(N)
double precision boxx,boxy,boxz,invbx,invby,invbz
double precision anorm, pi
double precision volume, rmax, deltar, dist

pi=dacos(-1.0d0)
.................................
volume=boxx*boxy*boxz
anorm = N*(N-1)/volume * 4.d0/3.d0 * pi
invbx=1.d0/boxx
invby=1.d0/boxy
invbz=1.d0/boxz
nbins=rmax/deltar

do i=1,N-1
 do j=i+1,N
  dx = x(i)-x(j)
  dx = dx - boxx*anint(dx*invbx)
  dy = y(i)-y(j)
  dy = dy - boxy*anint(dy*invby)
  dz = z(i)-z(j)
  dz = dz - boxz*anint(dz*invbz)
  dist= sqrt(dx*dx+dy*dy+dz*dz)
  ibin= int(dist/deltar)
  hist(ibin) = hist(ibin)+2.d0
 enddo
enddo
do ibin=1,nbins
  r= ibin*deltar
  g(ibin)=hist(ibin)/anorm/(r**3-(r-deltar)**3)
enddo

.................
```

The above code fragment would be run for each configuration of the simulation run, and the corresponding arrays would have to be averaged over the configurations.

---

*This code is not optimized for performance. The periodic boundary conditions are taken into account using the intrinsic function anint().

E. Time Correlation Functions. Kinetic Properties

Transport coefficients, such as the diffusion coefficient, the viscosity, and the thermal conductivity describe a system's response in the time domain to time-dependent perturbations. If the perturbation is not very strong, the response depends linearly on the perturbation and is described by a corresponding transport coefficient. For instance, if a concentration gradient is created in the system, the system will respond by developing a mass flux proportional to the magnitude of the concentration gradient, which tends to equalize the concentrations everywhere in the system—this is the well known Fick's law, and the proportionality coefficient is the diffusion coefficient. If a velocity gradient is imposed on a liquid, the liquid responds with a stress (momentum flux) which is proportional to the velocity gradient, the proportionality coefficient being the viscosity. Similarly, when an electric field is applied to a conductor, the conductor responds with an electric current density (charge flux) proportional to the electric field, and the proportionality coefficient is the conductivity. This approach is applicable to a wide variety of different phenomena; it is called "linear response" theory, and is described in detail in [40,37,38]. The coefficients of proportionality between the relaxation rate and the applied driving force are called *kinetic coefficients*.

Of course, one way to determine kinetic coefficients from simulation is to simulate the perturbation and measure the system's response directly. This is a feasible approach in some systems, which will be discussed in one of the following chapters.

One of the most important principles of linear response theory relates the system's response to an externally imposed perturbation, which causes it to depart from equilibrium, to its equilibrium fluctuations. Indeed, the system response to a small perturbation should not depend on whether this perturbation is a result of some external force, or whether it is just a random thermal fluctuation. Spontaneous concentration fluctuations, for instance, occur all the time in equilibrium systems at finite temperatures. If the concentration c at some point of a liquid at time zero is $\langle c \rangle + \delta c(\mathbf{r}, t)$, where $\langle c \rangle$ is an average concentration, concentration values at time $t + \delta t$ at $\mathbf{r}$ and other points in its vicinity will be affected by this. The relaxation of the spontaneous concentration fluctuation is governed by the same diffusion equation that describes the evolution of concentration in response to the external imposition of a compositional heterogeneity. The relationship between kinetic coefficients and correlations of the fluctuations is derived in the framework of linear response theory. In general, a kinetic coefficient is related to the integral of the time correlation function of some relevant microscopic quantity.

The time correlation function of two quantities $\mathcal{A}$ and $\mathcal{B}$, which are functions of the phase space point Γ, is defined as:

$$C_{AB}(t) = \langle \delta\mathcal{A}(\Gamma(0))\delta\mathcal{B}(\Gamma(t))\rangle \tag{64}$$

Here $\delta\mathcal{A} = \mathcal{A} - \langle\mathcal{A}\rangle$ and $\delta\mathcal{B} = \mathcal{B} - \langle\mathcal{B}\rangle$ are the deviations of $\mathcal{A}$ and $\mathcal{B}$ from their average values. The brackets indicate averaging over an ensemble of systems prepared under the same macroscopic initial conditions and subject to the same macroscopic constraints; here, equilibrium averages will be considered. When $\mathcal{A}$ and $\mathcal{B}$ are the same function, c_{AA} is called the auto-correlation function of $\mathcal{A}$. A normalized correlation function c_{AB} can be defined as

$$c_{AB} = \frac{C_{AB}}{\sqrt{\langle(\delta\mathcal{A})^2\rangle\langle(\delta\mathcal{B})^2\rangle}} \tag{65}$$

The absolute value of this function is bounded between zero and one.

The self-diffusion coefficient D, for instance, is related to the velocity autocorrelation function. In three dimensions

$$D = \frac{1}{3}\int_0^\infty dt\langle\mathbf{v}_i(0)\cdot\mathbf{v}_i(t)\rangle \tag{66}$$

where $\mathbf{v}_i$ is a center of mass velocity of a single molecule. The shear viscosity is related to the correlations of the off-diagonal ($\alpha\neq\beta$) components of the instantaneous stress tensor [compare Eq. (57)] $\mathcal{P}_{\alpha\beta}$:

$$\mathcal{P}_{\alpha\beta} = \frac{1}{V}\left(\sum_i p_{i\alpha}p_{i\beta}/m_i + \sum_i r_{i\alpha}f_{i\beta}\right)$$

$$\eta = \frac{V}{k_BT}\int_0^\infty dt\langle\mathcal{P}_{\alpha\beta}(0)\mathcal{P}_{\alpha\beta}(t)\rangle \tag{67}$$

Here, α,β stand for the Cartesian components x, y, z of the corresponding quantity—force acting on the particle $\mathbf{f}_i$, particle coordinate $\mathbf{r}_i$, or momentum $\mathbf{p}_i$.

Equations (66) and (67) are known as Green–Kubo relations for the transport coefficients.

Alternative, mathematically equivalent expressions, known as Einstein relations, may be obtained by carrying out the integration by parts:

$$D = \lim_{t \to \infty} \frac{1}{6t} \left\langle \left| \mathbf{r}_i(t) - \mathbf{r}_i(0) \right|^2 \right\rangle \tag{68}$$

$$\eta = \lim_{t \to \infty} \frac{1}{2t} \frac{V}{k_B T} \langle (\mathcal{L}_{\alpha\beta}(t) - \mathcal{L}_{\alpha\beta}(0))^2 \rangle \tag{69}$$

$$\mathcal{L}_{\alpha\beta} = \frac{1}{V} \sum_i r_{i\alpha} p_{i\beta} \tag{70}$$

It is easy to see that $\mathcal{P}_{\alpha\beta}$ is a time derivative of $\mathcal{L}_{\alpha\beta}$.

The above equations provide two alternative routes for calculating kinetic coefficients from simulations of a system at equilibrium. Averages in the above equations are ensemble averages, hence the results are ensemble-sensitive. The time correlation functions contain more information than just the kinetic coefficients. The Fourier transforms of time correlation functions can be related to experimental spectra. Nuclear magnetic resonance (NMR) measures the time correlation functions of magnetization, which is related to the reorientation of particular bonds in the polymer molecule; inelastic neutron scattering experiments measure the time correlation functions of the atom positions; infrared and Raman scattering spectroscopies measure the time correlation function of dipole moments and polarizabilities of the molecules.

Using the Einstein relations is generally a more robust way to calculate the kinetic coefficients. One just has to calculate the average on the right hand side of the Eqs. (68), (69) as a function of t for long enough times, when it becomes approximately linear in time, and to determine the slope; in contrast, application of the Green–Kubo relations requires long dynamic simulations to accumulate the "tails" of time correlation functions with sufficient precision.

Usually, in computer simulations the time correlation functions are calculated after the run is finished, using the information saved during the run (i.e., in a "postprocessing" stage). One typically saves the instantaneous values of the quantities of interest during the run in one or several files on the hard disk, which are processed after the simulation run.

A direct way to calculate the time correlation function from the values saved during the simulation run* is just to literally implement its definition. Suppose we have $M + 1$ values of $\mathcal{A}(t)$ and $\mathcal{B}(t)$, obtained at the regular time intervals $m\delta t$, where m is an integer running from 1 to M, stored in the

---

*Or calculated from the saved configurations.

arrays a and b. To calculate the time correlation function $C_{AB}(t)$ one would have to scan the array, picking up pairs of values a(1) and b(n), calculating their product, and saving it in the appropriate location in the array c(j), j=|n−1|, where the values of $C_{AB}(j\delta t)$ are to be stored. Afterwards, the accumulated quantities have to be normalized by the number of a and b pairs used for each value of 1. The following code fragment implements the subroutine that takes the values of $A(\sqcup)$ and $B(\sqcup)$ stored in the arrays a, b, and returns array c, containing the normalized correlation function values for times from 0 to mcor δt. It also calculates the averages of $A(t)$ and $B(t)$ a_av and b_av and their root mean squared deviations delta2a, delta2b, to obtain normalized correlation functions. A and B are available for time slices 0 to m.

```
subroutine correl(a,b,c,a_av,b_av,delta2a,delta2b,mcor,m)

double precision a(0:m),b(0:m),c(0:mcor)
double precision a_av, b_av,a2av, b2av, delta2a,delta2b
double precision aux1,aux2, ai
integer mcor, count (0:mcor), maxt, m, i

a_av=0.0d0
b_av=0.0d0
a2av=0.0d0
b2av=0.0d0

c zero c and count
  do i=0,m
    c(i)=0.0d0
    count(i)=0.0d0
  end do

c double loop through the arrays
  do i=0,m
    ai=a(i)
    a_av=a_av+ai
    b_av=b_av+b(i)

    maxt=min(i+mcor,m)
    do j=i,maxt
      1=j-i
      count(1)=count(1)+1
      c(1)=c(1)+ai*b(j)
    end do
  end do
c normalize averages
  aux = 1.0d0/dble(m+1)

  a_av=a_av*aux
  b_av=a_av*aux
```

```
   a2av=a2av*aux
   b2av=b2av*aux

   delta2a=sqrt(a2av-a_av*a_av)
   delta2b=sqrt(a2av-b_av*b_av)
c normalize corr. function
   aux1=a_av*b_av
   aux2=1.0d0/(delta2a*delta2b)

   do i=0,m
     c(i)=(c(i)/count(i) - a_av*b_av)*aux2
   end do

   return
   end
```

For most of the modern high-speed CPUs, multiplication takes significantly fewer cycles than division. This is why division is replaced by multiplication whenever possible. Also, the values unchanged inside the loops are calculated in advance outside of the loop.

An alternative method for calculating the time correlation function, especially useful when its spectrum is also required, involves the fast Fourier transformation (FFT) algorithm and is based on the convolution theorem, which is a general property of the Fourier transformation. According to the convolution theorem, the Fourier transform of the correlation function C_{AB} equals the product of the Fourier transforms of the correlated functions:

$$\hat{C}_{AB}(\nu) = \hat{A}(\nu) * \hat{B}(\nu) \tag{71}$$

The direct Fourier transforms $\hat{A}(\nu)$ and $\hat{B}(\nu)$ should be calculated by a Fourier transform routine that can be found in many scientific program libraries. Then the Fourier transform of the correlation function is calculated using (71). In the case where the autocorrelation function is calculated $(A = B)$, this Fourier transform represents a frequency spectrum of the system associated with the property A and can be related to experimental data, as discussed above. The correlation function in the time domain is then obtained by an inverse Fourier transformation. Fast Fourier transformation routines optimized for particular computer architectures are usually provided by computer manufacturers, especially for the parallel or vector multiprocessor systems.

IV. MONTE CARLO SIMULATIONS

In a Monte Carlo simulation many microscopic states are sampled using the generation of random numbers, and averages are calculated over these states. It is this random nature of the technique which is responsible for its name. The Monte Carlo method was developed in the Los Alamos National Laboratory in New Mexico in the late 1940s–early 1950s and its inventors named it after Monte Carlo, capital of Monaco, a small country in Southern Europe famous for its casinos.*

Let us first consider a very simple example—calculation of the area of a complex shape, such as the polygon Ω in Fig. 7. The simplest way would be to set up a mesh that is fine enough to provide the required accuracy, and to calculate the number of mesh points inside the polygon. If the complex shape happens to be a hole in the fence, this can be done by covering the fence with chicken wire and counting the number of knots inside the hole and the number of knots in a rectangular section *ABCD* of the fence which completely surrounds the hole. The ratio of these two numbers times the area of the rectangular section gives the area of the polygon.

Alternatively, one might draw a rectangle *ABCD* of known size around the polygon and start throwing points inside *ABCD* at random

FIG. 7   Area calculations.

---

*Maybe, if the technique had been developed several years later, it could have been called Las Vegas.

(i.e., according to a uniform distribution in *ABCD*). Clearly, the probability for a randomly chosen point to fall inside the polygon equals the ratio of the hatched area inside the polygon to the known area of the rectangle. Again, in the spirit of Fig. 7, one might equally well do this by randomly shooting, or throwing darts, at the brick fence *ABCD* and counting the number of shots that went through the hole Ω. It is important to note that the shooter must be blind, or at least very inexperienced, to guarantee the randomness of the shots. Naturally, the accuracy of the area estimate increases with the number of trial points. According to statistics, the error is inversely proportional to the square root of the total number of trials. This technique is the simplest variant of the Monte Carlo method. It can easily be generalized to calculate an integral of some arbitrary function $f(x, y)$ over a complex-shaped region. Indeed, if we were to calculate not just the area of the hatched polygon but an integral of function $f(x, y)$ over the polygon Ω, we could do this by first extending the integration region to the whole rectangular region *ABCD* and, secondly, by redefining the function $F(x, y)$ to be equal to $f(x, y)$ inside and on the border of Ω and $F(x, y) = 0$ everywhere else. Now by throwing M random points in the rectangle $(x, y)_i$ we can estimate:

$$\int_\Omega dx \, dy f(x, y) \cong \frac{1}{M} \sum_{i=1}^{M} F((x, y)_i) \tag{72}$$

Of course, if Ω accounts for a very small fraction of *ABCD*, the random technique becomes increasingly wasteful. Imagine that the function $F(x, y)$ is nonzero over less than 1% of the total area *ABCD*. Then, on the average, 99 out of 100 shots will contribute zeroes to the average, and only one out of a hundred will deliver a nontrivial contribution.

To take a polymer example, consider simulations of the bead-spring model of the linear polymer chain from Fig. 2, at temperature T. The probability density to find it in a particular conformation Γ, specified by the set of coordinates of its N monomers, $\Gamma = \{\mathbf{r}_1, \mathbf{r}_2, \ldots, \mathbf{r}_N\}$, is a product of probability densities for the bonds [compare Eq. (12)]; it can be considered as being a canonical distribution of the form of Eq. (46) (here we use Γ to denote a point in configuration space, not in phase space).

$$\rho(\Gamma) = \left(\frac{3}{2\pi} \right)^{3/2} \frac{1}{b^3} \exp\left[-\frac{3}{2b^2} \sum_{i=2}^{N} (\mathbf{r}_i - \mathbf{r}_{i-1})^2 \right] \tag{73}$$

with b^2 the mean square distance between successive beads. The quantity $\mathcal{V}(\Gamma) = (3k_B T / 2b^2) \sum_{i=2}^{N} (\mathbf{r}_i - \mathbf{r}_{i-1})^2$ can be thought of as an elastic "energy" of the springs, which, as we have pointed out, is entirely of entropic origin. The mean squared end-to-end distance $R^2 = |\mathbf{r}_N - \mathbf{r}_1|^2$ should be calculated as:

$$\langle R^2 \rangle = \int d\Gamma \rho(\Gamma) |\mathbf{r}_N - \mathbf{r}_1|^2 \tag{74}$$

The simplest Monte Carlo scheme for simulating this chain, as described above, would generate random conformations Γ, scattered with uniform probability throughout configuration space; each time it would select the N bead positions randomly within a three-dimensional domain whose dimensions are much larger than b. Thus, it would calculate an average of the product $\rho(\Gamma) |\mathbf{r}_N - \mathbf{r}_1|^2$ over all realizations of the chain.

Most of the randomly generated configurations, corresponding to random positions of the monomers, however, will have a very small weight, as the distances between connected beads will in most cases be much longer than b. The probability density of each of these states $\rho(\Gamma)$ will then be negligibly small, and so will be its contribution to the right hand side of Eq. (74). This simplest Monte Carlo scheme will then be spending most of the time calculating zeroes, almost never finding a configuration in which all the distances between connected monomers are on the order of b, which would contribute substantially to the average. This simple approach is very inefficient.

There is a solution, though. Imagine that we have an algorithm that will generate microscopic states using pseudorandom numbers, but with a nonuniform probability density, visiting the relevant lower-energy states more often and almost avoiding the high-energy states. The perfect choice would be to sample the phase space with the probability density $\rho(\Gamma)$ directly. Then, ensemble averages can be calculated simply as averages over the generated conformations:

$$\langle R^2 \rangle = \frac{1}{M} \sum_{i=1}^{M} |\mathbf{r}_N - \mathbf{r}_1|_i^2 \tag{75}$$

The core part of the Metropolis Monte Carlo simulation technique is the algorithm to sample the microscopic states according to the required ensemble probability distribution. The Monte Carlo algorithm simulates a stochastic process (Markov chain) producing a sequence of points or states in configuration space, in which the choice of the next state depends only on the current state. The original Metropolis (or MR^2T^2) Monte Carlo algorithm, proposed by Metropolis et al. [48], was for sampling the canonical (*NVT*) ensemble. The algorithm generates a sequence of random phase space points according to the following rules: (a) if the state generated at step k is Γ_k with energy value $\mathcal{V}(\Gamma_k)$, the candidate for the new state Γ_{new} is generated randomly with some distribution of "attempt" probabilities $\alpha(\Gamma_k \to \Gamma_{new})$. In the case of an atomic or molecular system this is usually realized by adding a random displacement to one or several randomly chosen atoms. The average displacement is of course zero, and its components are most often just uniformly distributed inside the interval between $-\delta_{max}$ and δ_{max} in each coordinate direction. The new state is accepted with the probability $P_{acc} = max(1, \exp[-(\mathcal{V}(\Gamma_{new}) - \mathcal{V}(\Gamma_k))/(k_B T)])$. If the Γ_{new} has lower energy than the current $\mathcal{V}(\Gamma_k)$, it is always accepted; if it has higher energy, it can still be accepted with probability exponentially decreasing with the energy difference. If Γ_{new} is accepted, $\Gamma_{k+1} = \Gamma_{new}$. Otherwise the old state is counted once again as a new state, $\Gamma_{k+1} = \Gamma_k$.

Given this description of the Metropolis algorithm, how do we make sure that it does reproduce a canonical ensemble distribution?

A. Microreversibility

Let us assume that the probability density for being at the phase space point Γ after completion of step k in the Metropolis algorithm is $\rho_M^{(k)}(\Gamma)$. Let the probability to go from one point Γ to another Γ' in one Monte Carlo step be $W(\Gamma \to \Gamma')$. This function W, describing how to proceed from one state to the next, completely defines the algorithm. The following equation describes the evolution of the probability density during our random process:*

$$\rho_M^k(\Gamma) - \rho_M^{k-1}(\Gamma) = \sum_{\Gamma'} \rho_M^{k-1}(\Gamma')W(\Gamma' \to \Gamma)$$

$$- \sum_{\Gamma'} \rho_M^{k-1}(\Gamma)W(\Gamma \to \Gamma') \tag{76}$$

The equation is actually quite simple; it is similar to the Liouville equation (41). As we are looking for a stationary algorithm, producing results independent of the initial conditions, we want ρ_M to depend only on Γ, not time, and hence we will set the left hand side to zero. The first sum on the right hand side describes the "influx" to Γ, the probability to come there from other states in step k. It is a sum$^+$ over all other states Γ' of the product of

---

*This equation is very important in the theory of random processes; it has many different names, of which *master equation* is most appropriate. We will not cover this subject in depth but rather refer the interested reader to the excellent book by van Kampen [49].
$^+$Or an integral, if the phase space is continuous.

the probability to be at Γ' at step $k-1$, times the probability of transition $\Gamma' \to \Gamma$ at the kth step. The second sum is very similar to the first; it equals the total probability to leave Γ. Again it is expressed as a sum of the probabilities to go from Γ to any other point Γ' at step k over all other points Γ'. How can we satisfy Eq. (76) with zero left hand side? Notice that the right hand side is essentially a sum over different pairs (Γ, Γ'). One possibility is to require the contribution from each pair to be zero. This leads to the following equation, which describes the *detailed balance*, also called *microreversibility* condition:

$$\rho_M(\Gamma')W(\Gamma' \to \Gamma) = \rho_M(\Gamma)W(\Gamma \to \Gamma')$$
$$\frac{W(\Gamma' \to \Gamma)}{W(\Gamma \to \Gamma')} = \frac{\rho_M(\Gamma)}{\rho_M(\Gamma')} \tag{77}$$

Equation (77) requires that the flux between any two points in phase space be the same in both forward and reverse directions. We have to emphasize that Eq. (77) is *not equivalent* to (76). It provides a *sufficient* but *not necessary* condition for the latter to hold. In simple terms, this means that, if detailed balance is satisfied, Eq. (76) is always satisfied, too. However, it may also be possible to make Eq. (76) true with ρ_M and W not satisfying (77).

It is easy to check that detailed balance holds for the Metropolis algorithm described above with $\rho_M(\Gamma)$ equal to the canonical ensemble distribution in configuration space. In the Metropolis algorithm, $W(\Gamma \to \Gamma') = \alpha(\Gamma \to \Gamma')min(1, \rho_M(\Gamma')/\rho_M(\Gamma))$, where $\alpha(\Gamma \to \Gamma')$ is the probability of attempting a move from Γ to Γ' and $P_{acc}(\Gamma \to \Gamma') = min(1, \rho_M(\Gamma')/\rho_M(\Gamma))$ is the probability of accepting that move. In the classical Metropolis algorithm, the attempt step is designed such that $\alpha(\Gamma \to \Gamma') = \alpha(\Gamma' \to \Gamma)$. In the NVT ensemble, the acceptance probability is $min(1, \exp[-(\mathcal{V}' - \mathcal{V})/(k_B T)])$, as described above. One can easily verify that the detailed balance equation (77) is satisfied under these conditions.

In recent years it has been realized that the sampling of configuration space may be more efficient if a nonsymmetric attempt probability matrix $\alpha(\Gamma \to \Gamma')$ is utilized. In that case, the correct acceptance criterion for ensuring microscopic reversibility is

$$P_{acc}(\Gamma \to \Gamma') = min\left(1, \frac{\alpha(\Gamma' \to \Gamma)\rho_M(\Gamma')}{\alpha(\Gamma \to \Gamma')\rho_M(\Gamma)}\right) \tag{78}$$

Furthermore, it may be convenient to conduct the moves by introducing random changes not in the original set of coordinates Γ (e.g., Cartesian coordinates) with respect to which the probability density $\rho_M(\Gamma)$ is known, but rather in a set of generalized coordinates $\tilde{\Gamma}$. The correct acceptance criterion is then

$$P_{acc}(\Gamma \to \Gamma') = min\left(1, \frac{\alpha(\tilde{\Gamma}' \to \tilde{\Gamma})\rho_M(\Gamma')J(\tilde{\Gamma}' \to \Gamma')}{\alpha(\tilde{\Gamma} \to \tilde{\Gamma}')\rho_M(\Gamma)J(\tilde{\Gamma} \to \Gamma)}\right) \tag{79}$$

where $J(\tilde{\Gamma} \to \Gamma)$ is the Jacobian of transformation from the set of coordinates $\tilde{\Gamma}$ to the set of coordinates Γ; $J(\tilde{\Gamma} \to \Gamma)$ times a differential volume in $\tilde{\Gamma}$-space gives the corresponding volume in Γ-space. The presence of the Jacobian takes care of the fact that equal volumes in $\tilde{\Gamma}$-space may not correspond to equal volumes in Γ-space.

Since both MC and molecular dynamics methods are available for predicting the equilibrium properties of material systems, and since molecular dynamics can provide kinetic in addition to thermodynamic properties, one may ask why MC is used at all. The answer is that, by appropriate selection of the moves employed, MC can achieve a sampling of configurations many orders of magnitude more efficient than that provided by MD. This means that the MC simulation can converge to well-equilibrated averages in a small fraction of the time that would be required by MD. Another advantage of MC is that it can readily be adapted to simulations with variable numbers of particles (e.g., grand canonical and Gibbs ensemble simulations) and is therefore very convenient for phase equilibrium calculations. Such simulation schemes will be described in the remainder of this book.

V. MOLECULAR DYNAMICS (MD)

At first glance Molecular Dynamics looks like a simplistic brute force attempt to literally reproduce what we believe is happening in the real world. Given a set of the initial velocities and coordinates, it tries to integrate the equations of motion numerically. Following the trajectory, MD provides information necessary to calculate various time correlation functions, the frequency spectra, diffusion coefficients, viscosity, and other dynamic properties. It also calculates thermodynamic properties ($P, T, \ldots$) as time averages at equilibrium.

In the simplest version of MD, the Newton (21) or Lagrange equations (22) are integrated for a closed system, in which the volume, total energy, and number of particles are conserved. This simulates the *microcanonical*, *NVE* ensemble. Both kinetic and potential energies fluctuate in the microcanonical ensemble but their sum remains constant.

For many problems, however, it is more convenient to keep the temperature, pressure, or chemical potential constant, instead of the total energy, volume, and number of particles. Generalizations of the molecular dynamics technique to virtually any ensemble have been developed, and they will be discussed in the following chapters. Of course, constant temperature MD does not conserve the total system energy, allowing it to fluctuate as is required at constant temperature. Similarly, volume is allowed to fluctuate in constant pressure molecular dynamics. The trick is to make these quantities fluctuate in a manner consistent with the probability distribution of the desired ensemble.

There are two different ways of accomplishing this. The first is to introduce some random perturbations to the system, which emulate its interaction with the environment. For constant temperature, for instance, this can be done by randomly [50] changing momenta by some random increments drawn from the Gaussian distribution. This mimics collisions between the molecules in the system and the virtual thermostat particles. The choice of parameters for the Gaussian distribution of momentum increments is determined by the required temperature.

Alternative techniques were proposed by Nosé [51] and Hoover [52]. Interaction with an external thermostat is described by an extra degree of freedom s, with associated momentum variable p_s. Extra potential energy $\mathcal{V}_s = (f+1)k_B T \ln(s)$ and kinetic energy $\mathcal{K}_s = 1/2 \, Q_s \dot{s}^2$ are added. Here f is the number of the degrees of freedom in the original system and Q_s is a thermal inertia parameter, with dimensions of energy times time squared. The degree of freedom s is a scaling factor between real time and simulation time, $s = dt_{\text{simulation}}/dt_{\text{real}}$. A "simulation time clock" is introduced, in addition to the real time clock, s being an instantaneous ratio of simulation time to real time intervals; for a well-designed simulation, s should fluctuate around 1. In the Nosé version the extended system is conservative, and its equations of motion can be derived following either a Lagrangian or a Hamiltonian formalism:

$$Q\ddot{s} = \sum_i m_i v_i^2 s - (f+1)k_B T/s$$

$$\ddot{\mathbf{r}} = \mathbf{f}/(ms^2) - 2\dot{s}\dot{\mathbf{r}}/s \qquad\qquad (80)$$

The dots denote derivatives with respect to simulation time. The total energy, including the kinetic and potential energy associated with the additional degree of freedom, is conserved. It can be shown that the "original system" samples the canonical ensemble. Hoover [52] proposed a mathematically equivalent scheme in which the coupling parameter is a friction factor, whose dynamics is governed by the difference between the current kinetic energy and its average at the desired temperature T.

$$\dot{\mathbf{r}} = \mathbf{p}/m$$
$$\dot{\mathbf{p}} = \mathbf{f} - s\mathbf{p}$$
$$\dot{s} = \frac{1}{Q_s}\left(\sum_i m_i v_i^2/2 - f k_B T/2\right) \tag{81}$$

In the Hoover scheme, the distinction between real time and simulation time disappears. Again here, f is the number of degrees of freedom and Q_s is a parameter that determines how strongly the system interacts with the "thermostat." It is chosen by trial and error for a particular system. If it is too large, the system dynamics is strongly perturbed and is dominated by the coupling to the "thermostat." If Q_s is too small, the coupling is weak, and equilibration takes too long.

For the above techniques it is possible to prove that they reproduce the canonical ensemble distribution.

There are, however, some not so rigorous but still very practical methods, based on velocity rescaling. One "obvious" and the most crude method to simulate constant temperature in MD would be to just rescale all the velocities, so that the kinetic energy corresponds to the desired temperature according to $\mathcal{K} = 1/2\ N_f k_B T$, where N_f is the number of degrees of freedom. This, of course, is a caricature of the real NVT ensemble, where kinetic energy is allowed to fluctuate and is not always constant. Even though this approach is totally unsuitable for production runs, it is quite practical and is often used to equilibrate the model at a required temperature, starting from an initial configuration. Some gentler variations on this theme have been suggested in [53]. There, the kinetic energy is not kept constant at every step. Velocities are rescaled by a factor *chi*, which is dependent on the difference between the current and desired kinetic energy values. Deviations of the kinetic energy from the desired value relax to that value with a predefined characteristic time:

$$\chi = \left(1 + \frac{\delta t}{t_T}\left(\frac{T}{\mathcal{T}} - 1\right)\right)^{1/2}$$
$$\mathcal{T} = \frac{1}{f/k_B}\sum_i m_i v_i^2 \tag{82}$$

Here δt is the time step and t_T is a parameter that determines how strong the perturbation is. Similar to the parameter Q_s in the Nosé and Hoover schemes, it has to be chosen by trial and error, so that the perturbation of the original system dynamics is not too strong, but at the same time is strong enough to equilibrate the system. This scheme is reminiscent of the method of Hoover (81). The important difference between them is that here it is the rescaling factor itself which depends on the difference between the instant and desired temperatures, while in Eq. (81) it is its time derivative. Unlike the stochastic and Nosé–Hoover methods, velocity rescaling with the factor described by Eq. (82) does not rigorously sample the canonical ensemble. Nevertheless, this method is widely used in practice for equilibration purposes. It was also shown to provide results similar to Hoover's thermostat (81) [54].

Every technique that involves velocity rescaling contains a parameter, adjustments in which determine how strong is the perturbation to the system dynamics. An inappropriate choice can lead to very interesting phenomena, which unfortunately can be passed unnoticed if special care is not taken. Rescaling particle velocities makes the average kinetic energy correspond to the required temperature, but still relies on the system's internal mechanisms to redistribute the kinetic energy between different degrees of freedom to provide equipartitioning. Usually the energy is redistributed due to the interactions between the particles and due to the anharmonicity of the system. Such mechanisms are absent in the extreme cases of the ideal gas and of the system of coupled harmonic oscillators, discussed earlier in this chapter. As there are no interactions between the particles in the former case and no interactions between the vibrational modes in the latter, the energy of each degree of freedom remains at its initial value. Imagine that some initial coordinates and velocities are assigned to the particles in the ideal gas. If the kinetic energy happens to be less than $\mathcal{K}_0 = 3/2 \, Nk_BT_{set}$, the scaling factor for the velocities will be greater than one, and "hot" faster particles will get more energy than slower particles. Similarly, in the harmonic oscillator, "hotter" vibrational modes will gain more energy than the "colder" ones. If errors in the time integration scheme are such that the total energy decreases with time,* together with velocity rescaling, they may lead to an unequal kinetic energy distribution between different degrees of freedom. Of course, the modeled systems are never ideal, and there is always interaction between different degrees of freedom, which redistributes energy between them. However, care must be taken as, if these channels are not fast enough,

---

*This is common for many stable integration algorithms. It is a well known fact that when the most popular Verlet algorithm is used to numerically integrate the equations of motion, in very long MD runs the system gets colder, due to the integration errors.

undesired effects may occur [55]. One of the most common is the so-called "flying ice cube" effect. The translational motion of the system as a whole is often decoupled from other degrees of freedom, as potential energy does not depend on it. There is no way for energy exchange between this degree of freedom and the rest. Velocity rescaling can lead to the energy being pumped into the overall translational motion, while being taken away from other degrees of freedom. Nominally the kinetic energy of the system is $1/2Nk_BT$, but most of it is in the translational motion of the system's center of mass. With no kinetic energy left in the other modes, the system looks like a frozen cube flying in space at a rather high velocity. This can be taken care of by correcting for the center of mass motion during the run. If the initial velocities are chosen at random, there is always some nonzero center of mass velocity component, which has to be subtracted in the beginning of the run.* Overall center of mass motion can also accumulate in very long MD runs due to the rounding errors.

With this said we see that, when using any one of the thermostats described by Eqs. (80), (81), and (82), the time constant chosen should not be too small to allow the energy to redistribute equally between the different degrees of freedom after it is artificially pumped into the system or taken from it by velocity rescaling.

VI. BROWNIAN DYNAMICS

The Brownian Dynamics technique is based on the Langevin equation of motion, originally proposed to describe the Brownian motion of a heavy, for instance colloidal, particle in a solvent. The Brownian particle is subjected to a random force from many collisions with solvent molecules. In addition, the solvent exerts a "friction" force which is assumed to be proportional to the particle's velocity.

$$\mathbf{p} = -\xi\mathbf{p} + \mathbf{f}_r(t) \tag{83}$$

Here ξ is a friction coefficient and $\mathbf{f}_r$ is the random force, which is also assumed to exhibit no directional preference: $\langle \mathbf{f}_r(t) \rangle = \mathbf{0}$, to be uncorrelated with the particle's momentum: $\langle \mathbf{f}_r(t) \cdot \mathbf{p}(0) \rangle = 0$ for all t, and to be uncorrelated with previous and future values of itself, $\langle \mathbf{f}_r(t) \cdot \mathbf{f}_r(t') \rangle = R_o\delta(t - t')$.

The Langevin equation (83) is a stochastic differential equation. It can be derived formally from the full set of dynamical equations describing the

---

*Unless it affects the goal of the simulation, of course.

particle and solvent system by projecting out the motion of the fast solvent particles [56,57].

Multiplying both sides of Eq. (83) by $\mathbf{p}(0)$ and averaging over all stochastic trajectories consistent with the same initial conditions, taking into account the delta-correlation of the random force, we obtain:

$$\frac{d}{dt}\langle \mathbf{p}(0) \cdot \mathbf{p}(t) \rangle = -\xi \langle \mathbf{p}(0) \cdot \mathbf{p}(t) \rangle$$

$$\langle \mathbf{p}(0) \cdot \mathbf{p}(t) \rangle = \langle \mathbf{p}^2 \rangle \exp(-\xi t) \tag{84}$$

We see that the velocity autocorrelation function of a free Brownian particle falls exponentially with time. The time required for an arbitrary initial velocity distribution of the particle to settle down to the Maxwell–Boltzmann form corresponding to the temperature of the bath (thermalization of velocities) is on the order of ξ^{-1}.

For the diffusion coefficient we calculate the mean square displacement of the particle:

$$\langle (\mathbf{r}(t) - \mathbf{r}(0))^2 \rangle = \frac{1}{m^2} \int_0^t dt' \int_0^t dt'' \langle \mathbf{p}(t'') \cdot \mathbf{p}(t') \rangle$$

$$= 2 \frac{\langle \mathbf{p}^2 \rangle}{m^2} \int_0^t dt' \int_0^{t'} dt'' \exp(-\xi(t' - t''))$$

$$= 2 \frac{\langle \mathbf{p}^2 \rangle}{m^2} \int_0^t dt' \exp(-\xi t') \int_0^{t'} \exp(\xi t'') \, dt''$$

$$= 2 \frac{\langle \mathbf{p}^2 \rangle}{m^2} \frac{t}{\xi} - 2 \frac{\langle \mathbf{p}^2 \rangle}{m^2} \frac{1}{\xi^2} (1 - \exp(-\xi t)) \tag{85}$$

In the limit of long times ($\xi t \gg 1$), invoking the equipartition theorem for kinetic energy,

$$\lim_{t \to \infty} \langle (\mathbf{r}(t) - \mathbf{r}(0))^2 \rangle = 2 \frac{\langle \mathbf{p}^2 \rangle}{m^2} \frac{t}{\xi} = 6 \frac{k_B T}{m} \frac{t}{\xi} = 6Dt \tag{86}$$

[compare Eq. (68)]. So, the diffusion coefficient of the particle D is related to the friction constant ξ as

$$D = \frac{k_B T}{m \xi} \tag{87}$$

Equation (87) is an Einstein relation for Brownian motion. There is a connection between the friction factor ξ and the quantity R_o, measuring the mean squared magnitude of the random force. Solving Eq. (83) just like a deterministic ordinary differential equation gives

$$\mathbf{p}(t) = \mathbf{p}(0)\exp(-\xi t) + \exp(-\xi t)\int_0^t \exp(\xi t')\mathbf{f}_r(t')\,dt' \tag{88}$$

Squaring and taking the ensemble average,

$$\langle \mathbf{p}^2(t)\rangle = \langle \mathbf{p}^2(0)\rangle \exp(-2\xi t) + 2\exp(-2\xi t)\int_0^t \exp(\xi t')\langle \mathbf{f}_r(t')\cdot\mathbf{p}(0)\rangle dt'$$

$$+ \exp(-2\xi t)\int_0^t dt' \int_0^t dt'' \exp[\xi(t'+t'')]\langle \mathbf{f}_r(t')\cdot\mathbf{f}_r(t'')\rangle \tag{89}$$

Using the properties $\langle \mathbf{f}_r(t')\cdot\mathbf{p}(0)\rangle = 0$ and $\langle \mathbf{f}_r(t')\cdot\mathbf{f}_r(t'')\rangle = R_o\delta(t'-t'')$, we obtain

$$\langle \mathbf{p}^2(t)\rangle = \langle \mathbf{p}^2(0)\rangle \exp(-2\xi t) + \frac{R_o}{2\xi}[1 - \exp(-2\xi t)] \tag{90}$$

As $t \to \infty$, the particle will become thermally equilibrated with the bath and so, by equipartition, $\lim_{t\to\infty}\langle \mathbf{p}^2\rangle = 3mk_BT$. We have already used this in deriving (86). On the other hand, Eq. (90) gives $\lim_{t\to\infty}\langle \mathbf{p}^2\rangle = R_o/2\xi$. Equating,

$$R_o = 6\xi mk_BT\langle \mathbf{f}_r(t)\cdot\mathbf{f}_r(t')\rangle = 6\xi mk_BT\delta(t-t') \tag{91}$$

This is the fluctuation-dissipation theorem. It requires that the mean energy given to the particle by the random force be equal to the mean energy taken out of the particle by the frictional forces with its environment. For a more rigorous formulation, the reader is referred to [58].

If the particle is in an external potential, or if there are many interacting particles, another deterministic force term describing this interaction is added to the right hand side of the Langevin equation:

$$\dot{\mathbf{p}} = -\xi\mathbf{p} - \nabla\mathcal{V} + \mathbf{f}_r(t)$$
$$\mathbf{p} = \dot{\mathbf{r}}/m \tag{92}$$

Let us try to derive an equation for the evolution in time of the phase space probability density $\rho(\mathbf{p},\mathbf{r},t)$ for a random process generated by numerical integration of the Langevin Eq. (92) with some time step δt. As in the Monte Carlo master Eq. (76), we should write down the balance of the incoming and outgoing fluxes for the phase space point $\Gamma = \{\mathbf{p}, \mathbf{r}\}$ after the time increment δt, taking into account that the coordinate increment equals $\Delta \mathbf{r} = (\mathbf{p}/m)\delta t$ [59] and the momentum increment equals $\Delta \mathbf{p} = -(\xi\mathbf{p} - \nabla\mathcal{V})\delta t + \bar{\Delta}\mathbf{p}$, with $\bar{\Delta}\mathbf{p}$ the momentum increment due to the random force. $P(\Delta \mathbf{p})$ is the probability density of the momentum increment $\Delta\mathbf{p}$

$$\rho(\mathbf{p}, \mathbf{r}, t + \delta t) = \rho(\mathbf{p}, \mathbf{r}, t) - \int d^3\Delta p \, \rho(\mathbf{p}, \mathbf{r}, t) \, P(\Delta \mathbf{p})$$

$$+ \int d^3\Delta p \, \rho(\mathbf{p} - \Delta\mathbf{p}, \mathbf{r} - \Delta\mathbf{r}, t) \, P(\Delta\mathbf{p}) \tag{93}$$

The first two terms on the right hand side cancel each other, as $\rho(\mathbf{p}, \mathbf{r}, t)$ can be pulled out from the integral and $\int d^3\Delta p P(\Delta \mathbf{p}) = 1$. Expanding ρ around $\{\mathbf{p}, \mathbf{r}\}$ under the integral of the third term on the right hand side,

$$\rho(\mathbf{p}, \mathbf{r}, t + \delta t) = \int d^3\Delta p P(\Delta\mathbf{p})\rho(\mathbf{p}, \mathbf{r}, t) - \sum_i \delta t \frac{p_i}{m}\frac{\partial\rho}{\partial r_i} \int d^3\Delta p P(\Delta\mathbf{p})$$

$$- \sum_i \frac{\partial\rho}{\partial p_i} \int d^3\Delta p P(\Delta\mathbf{p})\Delta p_i$$

$$+ \frac{1}{2}\sum_{i,j}\left[\frac{\partial^2\rho}{\partial p_i \partial p_j} \int d^3\Delta p P(\Delta\mathbf{p})\Delta p_i \Delta p_j \right.$$

$$+ (\delta t)^2 \frac{\partial^2\rho}{\partial r_i \partial r_j}\frac{p_i p_j}{m^2} \int d^3\Delta p P(\Delta\mathbf{p})$$

$$\left. + \delta t \frac{\partial^2\rho}{\partial r_i \partial p_j}\frac{p_i}{m} \int d^3\Delta p P(\Delta\mathbf{p})\Delta p_j \right] \tag{94}$$

The indices i, j run over all coordinate directions. Taking into account the following relationships, which are a consequence of $\langle\bar{\Delta}\mathbf{p}\rangle = 0$ and the lack of correlation between the random force and momentum or position:

$$\int d^3\Delta p P(\Delta\mathbf{p})\Delta p_i = \delta t(-\xi p_i - \nabla_i\mathcal{V})$$

$$\int d^3\Delta p P(\Delta\mathbf{p})\Delta p_i \Delta p_j = (\delta t)^2(-\xi p_i - \nabla_i\mathcal{V})(-\xi p_j - \nabla_j\mathcal{V}) + 1/3\langle(\bar{\Delta}\mathbf{p})^2\rangle\delta_{ij}$$

$$\tag{95}$$

we obtain

$$\rho(\mathbf{p}, \mathbf{r}, t + \delta t) - \rho(\mathbf{p}, \mathbf{r}, t) = \delta t \left[-\sum_i \frac{p_i}{m} \frac{\partial \rho}{\partial r_i} + \sum_i \nabla_i V \frac{\partial \rho}{\partial p_i} + \sum_i \frac{\partial}{\partial p_i} \xi p_i \rho \right]$$
$$+ \frac{1}{6} \sum_i \frac{\partial^2 \rho}{\partial p_i^2} \langle \bar{\Delta} \mathbf{p}^2 \rangle$$
$$+ \frac{1}{2} (\delta t)^2 \sum_{ij} \frac{p_i}{m} \frac{\partial}{\partial r_i} \left[\frac{\partial \rho}{\partial p_j} (\xi p_j - \nabla_j V) + \frac{\partial \rho}{\partial r_j} \frac{p_j}{m} \right]$$

$$(96)$$

In the limit of $\delta t \to 0$, using the fact that $\langle \bar{\Delta} \mathbf{p}^2 \rangle = 6 \delta t m k_B T \xi$ fluctuation-dissipation theorem, compare Eq. (91), Eq. (96) reduces to:

$$\frac{\partial \rho(\mathbf{p}, \mathbf{r}, t)}{\partial t} = -\sum_i \frac{p_i}{m} \frac{\partial \rho}{\partial r_i} + \sum_i \nabla_i V \frac{\partial \rho}{\partial p_i} + \sum_i \frac{\partial}{\partial p_i} \xi p_i \rho + m k_B T \xi \sum_i \frac{\partial^2 \rho}{\partial p_i^2}$$

$$(97)$$

which is a Fokker–Planck equation [59]. By direct substitution one can check that the Boltzmann distribution

$$\rho(\mathbf{p}, \mathbf{r}, t) = \text{Const.} \exp \left[-\left(\frac{\mathbf{p}^2}{2m} + V \right) / (k_B T) \right]$$

is its stationary solution. This means that a Brownian dynamics method solving the Langevin equation reproduces the canonical ensemble. Notice, however, that such numerical schemes are *approximate*, with accuracy on the order of (δt^2). This error is called *discretization* error. To obtain exact canonical ensemble averages from a Brownian dynamics simulation it is necessary to run simulations with different time steps and to extrapolate the results to $\delta t \to 0$ [58].

Another interesting and important fact is that it is really not necessary for the random momentum increments to be drawn from a Gaussian distribution. Any distribution with zero mean and appropriate mean square deviation satisfying the condition $\langle \bar{\Delta} p_i^2 \rangle = 2 m k_B T \xi \delta t$ along each coordinate direction will be appropriate.

Brownian dynamics was first introduced by Ermak et al. [60,61] and is commonly used to simulate polymers, ions, or colloidal particles in solution. The monomer dynamics is modeled by the Langevin equation, with the solvent modeled as the medium providing the frictional and random forces.

If the evolution of the system over times $\Delta t \gg 1/\xi$ is of interest (thermalization of velocities in each time step—"high friction limit"), the friction term is dominant and the inertia term on the left hand side of the Langevin equation can be neglected. Then the equation becomes first order:

$$\dot{\mathbf{r}} = \frac{D}{k_B T}(-\nabla \mathcal{V} + \mathbf{f}_r(t)) \tag{98}$$

The configuration-space probability distribution in this case is defined by the Smoluchowski equation, which can be derived in the same way as the Fokker–Planck equation above.

$$\frac{\partial}{\partial t}\rho(\mathbf{r}, t) - \frac{D}{k_B T}\nabla(\nabla\mathcal{V}\rho(\mathbf{r}, t)) = D\nabla^2\rho(\mathbf{r}, t) \tag{99}$$

This equation describes diffusion in the external potential $\mathcal{V}(\mathbf{r})$. If the potential is constant (or zero), it reduces to the usual diffusion equation.

VII. TECHNIQUES FOR THE ANALYSIS AND SIMULATION OF INFREQUENT EVENTS

Molecular dynamics is a useful technique for probing transport and relaxation processes in materials, provided the relevant time correlation functions decay appreciably over times shorter than 100 ns. Unfortunately, many dynamical processes in real-life materials, especially polymers, are governed by time scales appreciably longer than this. Techniques alternative to "brute force" MD must be developed to predict the kinetics of such processes from molecular constitution.

Many important dynamical processes in materials occur as successions of infrequent events. The material system spends most of its time confined within relatively small regions or "states" in its configuration space, which are surrounded by high (relative to $k_B T$) energy barriers. Only infrequently does the system jump from one state to another through a fluctuation that allows it to overcome a barrier. Once initiated, the jump process occurs quite quickly (on a time scale that can be followed by MD). The mean waiting time between jumps, however, is very long. A MD, or even BD, simulation would exhaust itself tracking the relatively uninteresting motion of the system as long as it is confined in a few states, but would be unable to sample a sufficient number of the jumps. Examples of infrequent event processes include conformational transitions of chains in solution and in the melt, diffusion of small molecules in glassy polymers, structural relaxation in the glassy state, the formation of stable nuclei at the onset of crystallization, and the conversion from reactants to products in a chemical reaction.

By shifting attention towards the energy barriers that must be overcome for a transition between states to occur, special techniques for the analysis and simulation of infrequent events manage to calculate a rate constant for the transition utilizing the machinery of common MD and MC simulation. These techniques are based on the principles of Transition-State Theory.

Envision a system whose free energy as a function of a generalized coordinate q looks as shown in Fig. 8. Defining an appropriate reaction coordinate in an N_f-dimensional configuration space is in itself an interesting problem, which we will address briefly below. The free energy plotted in Fig. 8 incorporates the effects of thermal fluctuations in directions orthogonal to the reaction coordinate. The barrier at $q^{\ddagger}$ defines two states, centered around local free energy minima: A state A with $q < q^{\ddagger}$ and a state B with $q > q^{\ddagger}$. Macroscopically, if the barrier is high relative to $k_B T$, there will be a wide range of times that are long relative to the correlation times

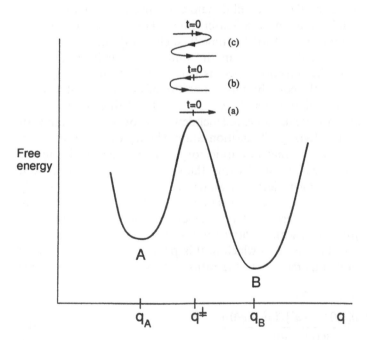

FIG. 8 Free energy profile as a function of the reaction coordinate q showing two states separated by a barrier. The lines in the upper part of the figure show different ways in which dynamical trajectories can cross the barrier region.

governing the motion of the system in each of the states, but still very short compared to the residence time of the system in either state (time scale separation). Over this range of times, the probabilities of occupancy p_A, p_B of the states evolve according to the master equations:

$$\frac{dp_A}{dt} = -k_{A \to B} p_A + k_{B \to A} p_B$$

$$\frac{dp_B}{dt} = -k_{B \to A} p_B + k_{A \to B} p_A \tag{100}$$

A microscopic expression for the rate constant $k_{A \to B}$ is [62]

$$k_{A \to B}(t) = \frac{\langle \dot{q}(0)\delta(q^{\ddagger} - q(0))\, \theta(q(t) - q^{\ddagger})\rangle}{\langle \theta(q^{\ddagger} - q)\rangle} \tag{101}$$

where δ and θ stand for a Dirac delta function and a Heaviside function, respectively. The numerator in Eq. (101) is an average rate of motion along the reaction coordinate taken over all dynamical trajectories that cross the barrier at time 0 and end up in state B after time t. Trajectories crossing the barrier in the direction from A to B at time 0, such as (a) and (c) in Fig. 8, contribute positively to the average in the numerator, while trajectories crossing in the opposite direction, such as (b), contribute negatively. The denominator is simply the equilibrium probability of occupancy of state A. When time scale separation holds, the $k_{A \to B}$ calculated through Eq. (101) turns out to exhibit a plateau value, independent of t, over a wide range of (not too short) times. Eyring's Transition-State Theory (TST) rests on an approximation: It assumes that any trajectory crossing the barrier in the direction from A to B will eventually lead the system to state B. That is, it ignores barrier recrossing trajectories along which the system ultimately thermalizes in the state in which it originated [such as trajectory (b) in Fig. 8]. In more formal terms, TST replaces $q(t)$ in the numerator of Eq. (101) with $q(0^+)$, i.e., with its limit for $t \to 0^+$, at which the system is bound to be in the state towards which $\dot{q}(0)$ is pointing. The TST estimate of the rate constant thus reduces to the ratio of two equilibrium ensemble averages:

$$\begin{aligned} k_{A \to B}^{TST} &= \frac{\langle \dot{q}(0)\, \theta[q(0^+) - q^{\ddagger}]\, \delta[q^{\ddagger} - q(0)]\rangle}{\langle \theta(q^{\ddagger} - q)\rangle} \\ &= \frac{\langle \dot{q}\theta(\dot{q})\delta[q^{\ddagger} - q(0)]\rangle}{\langle \theta(q^{\ddagger} - q)\rangle} \end{aligned} \tag{102}$$

$k_{A \to B}^{TST}$ emerges essentially as an average velocity in the direction from A to B along the reaction coordinate, which can readily be computed in most cases from the momentum-space distribution of the system, times the ratio of probabilities of being at the barrier and of residing in the origin state A. The latter ratio can be calculated by MC or constraint MD methods designed to sample the barrier region.

It is customary to express the rate constant $k_{A \to B}$ of Eq. (101) in terms of its TST estimate, $k_{A \to B}^{TST}$, as

$$k_{A \to B}(t) = k_{A \to B}^{TST} \kappa(t) \tag{103}$$

where the "dynamical correction factor" or "transmission coefficient" $\kappa(t)$ is given by:

$$\kappa(t) = \frac{\langle \dot{q}(0) \, \delta[q^{\ddagger} - q(0)] \, \theta(q(t) - q^{\ddagger}) \rangle}{\langle \dot{q}(0) \, \delta[q^{\ddagger} - q(0)] \, \theta(q(0^+) - q^{\ddagger}) \rangle} \tag{104}$$

In a two-state system with time scale separation, $\kappa(t)$ quickly settles to a plateau value within a time comparable to the correlation time of molecular velocities and much shorter than $1/k_{A \to B}$. This value is less than unity, for TST overestimates the rate constant by neglecting dynamical recrossing events. $k(t)$ can be calculated straightforwardly by generating a number of MD trajectories initiated on the barrier with an equilibrium distribution of velocities resulting in a $\dot{q}$ value pointing from A to B. Each trajectory quickly thermalizes in state A or state B. The fraction of trajectories which thermalize in state B provides the plateau value of κ. A variant of the method is to initiate the trajectories with random velocities, integrate them both forward and backward in time, and count the fraction of them that are effective in bringing about an $A \to B$ or $B \to A$ transition [63,64].

The possibility of calculating $k_{A \to B}^{TST}$ and κ with modest computational means offers a way out of the problem of long times in the case of dynamical processes occurring as sequences of infrequent events. Note, however, that, to apply TST-based approaches, one must have a good idea of the states and reaction coordinates, i.e., one must already know something about the (free) energy landscape of the system.

To illustrate these ideas with a simple example, consider a one-dimensional particle of mass m moving in an external field $\mathcal{V}(q)$ of the form depicted by the curve in Fig. 8. In this case, the ordinate is merely potential energy, rather than free energy. The field $\mathcal{V}(q)$ could be generated by the atoms of a medium, in which the particle is diffusing. We will implicitly assume that there is a mechanism for weak energy exchange between the particle and the medium (heat reservoir), as a result of which the particle velocity $\dot{q}$ is distributed according to the requirements of equilibrium at temperature T. By virtue of the separability of momentum and configuration-space distributions, and realizing that positive and negative values of $\dot{q}$ are equally probable, we can recast Eq. (102) for this problem as

$$
\begin{aligned}
k_{A \to B}^{TST} &= \frac{1}{2} \frac{\langle |\dot{q}| \rangle \langle \delta[q^{\ddagger} - q)] \rangle}{\langle \theta(q^{\ddagger} - q) \rangle} \\
&= \left(\frac{k_B T}{2\pi m} \right)^{1/2} \frac{\exp[-\mathcal{V}(q^{\ddagger})/k_B T]}{\int_{q \in A} \exp[-\mathcal{V}(q)/k_B T] \, dq}
\end{aligned}
\tag{105}
$$

The integral in the denominator of Eq. (105) is taken over the entire state A, to the left of $q^{\ddagger}$. At low temperatures, the major contribution to this integral comes from the immediate vicinity of the energy minimum q_A, where the function $\mathcal{V}(q)$ can be approximated by its Taylor expansion around q_A truncated at the second order term [compare Eq. (35)]. Setting $k_A = \partial^2 \mathcal{V}/\partial q^2|_{q_A}$, we obtain

$$
\begin{aligned}
\int_{q \in A} \exp\left[-\frac{\mathcal{V}(q)}{k_B T} \right] dq &\simeq \exp\left[-\frac{\mathcal{V}(q_A)}{k_B T} \right] \int_{-\infty}^{+\infty} \exp\left[-\frac{k_A (q - q_A)^2}{2 k_B T} \right] dq \\
&= \left(\frac{2\pi k_B T}{k_A} \right)^{1/2} \exp\left[-\frac{\mathcal{V}(q_A)}{k_B T} \right]
\end{aligned}
\tag{106}
$$

Using Eq. (106) in Eq. (105), we are led to the *harmonic approximation* for the transition rate constant, $k_{A \to B}^{TST, HA}$.

$$
\begin{aligned}
k_{A \to B}^{TST, HA} &= \frac{1}{2\pi} \left(\frac{k_A}{m} \right)^{1/2} \exp\left[-\frac{\mathcal{V}(q^{\ddagger}) - \mathcal{V}(q_A)}{k_B T} \right] \\
&= v^A \exp\left[-\frac{\mathcal{V}(q^{\ddagger}) - \mathcal{V}(q_A)}{k_B T} \right]
\end{aligned}
\tag{107}
$$

The harmonic approximation to the rate constant emerges as a product of the natural frequency v^A of oscillation of the system within the well of the origin state A times the Boltzmann factor of the barrier height measured from the minimum of state A.

An ingenious analysis of the transition rate in systems such as the one of Fig. 8, which takes explicitly into account energy exchange with the degrees of freedom not participating in the reaction coordinate, was carried out by Kramers [59]. In Kramers' analysis, these degrees of freedom are envisioned as comprising a Brownian bath which exerts Langevin and frictional forces on the degree of freedom q. The motion of q is described by a one-dimensional Langevin equation of the form (92). This, of course, presupposes that the motion of the bath degrees of freedom is fast relative to the evolution of q. The potential (or, more generally, free energy) function $\mathcal{V}(q)$ is approximated by its Taylor expansion truncated at the second order term both around the bottom of the well of the origin state (q_A) and around the top of the barrier $q^{\ddagger}$. Symbolizing by ξ the friction coefficient, as in Eq. (92), by κ_A the curvature of the potential at q_A, as above, and by $k^{\ddagger} > 0$ the opposite of the curvature of the potential at the barrier, $k^{\ddagger} = -\partial^2 \mathcal{V}/\partial q^2|_{q^{\ddagger}}$, Kramers' result for the rate constant is:

$$k_{A \to B} = \nu^A \left(\sqrt{\zeta^2 + 1} - \zeta \right) \exp\left(-\frac{\mathcal{V}(q^{\ddagger}) - \mathcal{V}(q_A)}{k_B T} \right) \tag{108}$$

where $\nu^A = (1/2\pi)\sqrt{\kappa_A/m}$ [natural frequency of oscillation in the harmonic region of the well, as in Eq. (107), and $\zeta = (\xi/2)\sqrt{m/\kappa^{\ddagger}}$ is a dimensionless parameter which increases with the friction coefficient and decreases with the curvature of the potential at the barrier. When the coupling between the reaction coordinate and the other degrees of freedom is weak, $\zeta \ll 1$ and the Kramers result, Eq. (108), reduces to the harmonic approximation TST estimate, $k_{A \to B}^{TST,HA}$. Note that the transmission coefficient $\kappa = k_{A \to B}/k_{A \to B}^{TST} = \sqrt{\zeta^2 + 1} - \zeta$ predicted by Kramers' theory is always smaller than 1, as it should be. In the case $\zeta \gg 1$ ("strong friction limit"), the quantity $\sqrt{\zeta^2 + 1} - \zeta$ is well approximated by $1/2\zeta$ and the Kramers result becomes

$$k_{A \to B} = \frac{\nu^A}{\xi} \sqrt{\frac{k^{\ddagger}}{m}} \exp\left(-\frac{\mathcal{V}(q^{\ddagger}) - \mathcal{V}(q_A)}{k_B T} \right) \tag{109}$$

In this limit (strong coupling between q and the bath and/or flat barrier) $k_{A \to B}$ can be orders of magnitude lower than the TST estimate.

The exact location where $q^{\ddagger}$ is chosen affects the values of $k_{A \to B}^{TST}$ and κ, but not their product $k_{A \to B}$. Nevertheless, in practice every effort should be made to define the reaction path so that the transition-state theory estimate of the rate constant is minimized and the dynamical correction factor is maximized.

How does one define the reaction path and the transition state given the potential energy function of a system with many degrees of freedom? We will assume here that the configuration is described in terms of the N_f mass-weighted Cartesian coordinates $x_i = m_i^{1/2} r_i$ for all atoms. We will use the symbol x to refer to all these degrees of freedom collectively. "States," such as A and B in the examples above, are constructed around local minima of the potential energy, at which the gradient vector $g(x) = \nabla V(x)|_x$ is 0 and the Hessian matrix $H(x) = \partial^2 V / \partial x \partial x^T|_x$ is positive definite. Between two neighboring local minima x_A and x_B there will be at least one saddle point $x^\ddagger$, at which $g(x^\ddagger) = 0$ and the Hessian $H(x^\ddagger)$ has one negative eigenvalue with associated unit eigenvector $n^\ddagger$. This saddle point is the highest energy point on the lowest energy passage between x_A and x_B; it is usually a good choice for the *transition state* $\ddagger$ between A and B. The *reaction path* between A and B, along which the reaction coordinate is measured, is a line in N_f-dimensional space connecting x_A and x_B. To construct it, one initiates two steepest descent trajectories at $x^\ddagger$, one in the direction $+n^\ddagger$ and the other in the direction $-n^\ddagger$. Each such trajectory consists of small steps $dx = -(g(x)/|g(x)|) \, dq$ parallel to the direction of the local gradient vector and terminates in one of the two minima connected through $\ddagger$ (see Fig. 9). The *dividing surface* between states A and B is defined as an $(N_f - 1)$-dimensional hypersurface with equation $C(x) = 0$, with the following properties: (a) it passes through the saddle point, i.e., $C(x^\ddagger) = 0$; (b) at the saddle point it is normal to the eigenvector $n^\ddagger$ corresponding to the negative eigenvalue of the Hessian (and, therefore, to the reaction path); (c) at all points other than $x^\ddagger$, it is tangent to the gradient vector. Conditions (b) and (c) can be expressed mathematically as:

$$\left. \frac{\nabla C(x)}{|\nabla C(x)|} \right|_{x^\ddagger} = n^\ddagger \tag{110}$$

$$\nabla C(x) \cdot g(x) = 0, \qquad x \neq x^\ddagger \tag{111}$$

With this definition of the dividing surface, the TST estimate of the transition rate constant becomes [65]

$$k_{A \to B}^{TST} = \int_{x \in A} d^{N_f} x \int_{n(x) \cdot p > 0} d^{N_f} p \, [n(x) \cdot p] \, \delta[C(x)] |\nabla C(x)| \rho_{NVT}(x, p) \tag{112}$$

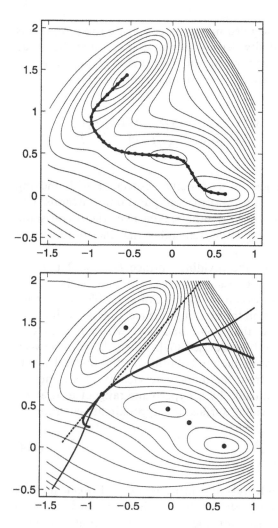

FIG. 9 Reaction paths and dividing surfaces in a two-dimensional configuration space. Hypersurfaces (lines in this case) of constant V are shown as thin closed curves. There are three minima with corresponding states A, B, and C, labeled from left to right. (Top) Reaction paths from A to B and from B to C are drawn as a thick line. Note that the direction of the reaction path may curve away from the straight line connecting two minima. (Bottom) The three minima and the two saddle points between them are shown as bold dots. The dividing surface (line) between states A and B is drawn as a solid line of medium thickness running from bottom left to top right. The hyperplane (line) approximation to the dividing surface is shown as a broken straight line. The thick curve traces an approximation to the dividing surface constructed through a local criterion. (After Ref. [65].)

where $\mathbf{p}$ is the vector of mass-weighted momenta conjugate to $\mathbf{x}$, $\rho_{NVT}(\mathbf{x})$ is the canonical ensemble probability density in phase space, and the Dirac delta function selects configurations on the dividing surface. Upon performing all momentum-space integrations, one obtains from Eq. (112)

$$k_{A \to B}^{TST} = \left(\frac{k_B T}{2\pi}\right)^{1/2} \left(\frac{\int_{\mathbf{x} \in A} d^{N_f} x \, \delta[\mathcal{C}(\mathbf{x})] |\nabla \mathcal{C}(\mathbf{x})| \exp[-\mathcal{V}(\mathbf{x})/k_B T]}{\int_{\mathbf{x} \in A} d^{N_f} x \exp[-\mathcal{V}(\mathbf{x})/k_B T]}\right) \qquad (113)$$

As in Eq. (105), $k_{A \to B}^{TST}$ emerges as a velocity associated with moving across the dividing surface times a ratio of configurational integrals, one taken over the dividing surface and the other taken over the origin state. [To reconcile the dimensions in Eq. (105), remember that the coordinates $\mathbf{x}$ are *mass-weighted*.]

In practice, the definition of the dividing surface $\mathcal{C}(\mathbf{x}) = 0$ that we introduced above may be difficult to use in systems with large N_f, because it is nonlocal. [If an analytical expression for $\mathcal{C}(\mathbf{x})$ is not available, one cannot judge whether a point $\mathbf{x}$ belongs on the dividing surface, unless one initiates a steepest descent trajectory at that point and sees where it ends up]. Approximate local criteria have been devised [65]. When the pass between states A and B is narrow, most of the contribution to the integral in the numerator of Eq. (113) comes from the immediate vicinity of $\mathbf{x}^\ddagger$, and the dividing surface can be approximated by a hyperplane tangent to it at $\mathbf{x}^\ddagger$. That is, one may use for the dividing surface the approximate equation

$$\mathcal{C}(\mathbf{x}) \simeq \mathbf{n}^\ddagger \cdot (\mathbf{x} - \mathbf{x}^\ddagger) = 0 \qquad (114)$$

The ratio of configurational integrals appearing in Eq. (113) can then be computed through a free energy perturbation technique which involves calculating free energy differences for the system being confined on a succession of hyperplanes normal to the reaction path [66].

If, in addition, a harmonic approximation is invoked for $\mathcal{V}(\mathbf{x})$, Eq. (113) leads to:

$$k_{A \to B}^{TST} = \frac{\prod_{i=1}^{N_f} \nu_i^A}{\prod_{i=2}^{N_f} \nu_i^\ddagger} \exp\left[-\frac{\mathcal{V}(\mathbf{x}^\ddagger) - \mathcal{V}(\mathbf{x}_A)}{k_B T}\right] \qquad (115)$$

where the eigenfrequencies ν_i^A are the square roots of the eigenvalues of the Hessian matrix of second derivatives of $\mathcal{V}$ with respect to the mass-weighted coordinates x in the origin state A and, similarly, $v_i^\ddagger$ are the square roots of the $N_f - 1$ positive eigenvalues of the Hessian matrix at the saddle point. Equation (115), first derived by Vineyard [67], is a useful generalization of Eq. (107). In case the eigenfrequencies v_i^A or $v_i^\ddagger$ are too high for the corresponding vibrational motion to be describable satisfactorily by means of classical statistical mechanics, a more appropriate form of Eq. (115) is

$$k_{A \to B}^{TST} = \frac{k_B T}{h} \frac{\prod_{i=1}^{N_f}[1 - \exp(-hv_i^A/k_B T)]}{\prod_{i=2}^{N_f}[1 - \exp(-hv_i^\ddagger/k_B T)]} \exp\left[-\frac{\mathcal{V}(x^\ddagger) - \mathcal{V}(x_A)}{k_B T}\right] \qquad (116)$$

(In the latter equation, classical partition functions for the harmonic oscillators representing the modes have been replaced with quantum mechanical ones. Zero point energy contributions are considered as being part of the $\mathcal{V}$'s.) The latter equation is a special case of the more general TST expression

$$k_{A \to B}^{TST} = \frac{k_B T}{h} \exp\left(-\frac{G^\ddagger - G_A}{k_B T}\right) \qquad (117)$$

where $G^\ddagger$ is the Gibbs energy of the system confined to the dividing surface and G_A is the Gibbs energy of the system allowed to sample the entire origin state.

The above discussion concerned ways of calculating the rate constant for a transition between two states, A and B. The evolution of real material systems (e.g., by diffusion or relaxation) often involves long sequences of transitions between different states. Once the rate constants are known between all pairs in a network of connected states, ensembles of dynamical trajectories for the system can readily be generated by Kinetic Monte Carlo simulation [64]. In such a simulation, the times for the next transition to occur are chosen from the exponential distribution of waiting times governing a Poisson process. If the rate constants are small, these times are long; the evolution of the system can be tracked over time periods many orders of magnitude longer than can be accessed by MD. Thus, the long-time problem of MD is eliminated. The challenge in such an approach is to identify the states and saddle points between them and to include all relevant degrees of freedom in the transition rate constant calculations.

VIII. SIMULATING INFINITE SYSTEMS, PERIODIC BOUNDARY CONDITIONS

Even the most powerful supercomputers available today can only handle up to about a million atoms. If we are interested in the properties of single molecules or small drops or clusters of diameter 100 Å or less, this is not a problem. Most of the time, however, one is interested in the properties of bulk materials.

When the simulated system is bounded by walls or by free surfaces, a substantial fraction of the atoms is located next to the surface, in an environment different from the bulk. Interactions with about half of the neighbors are replaced by interactions with the bounding wall, or are just absent. The effect of the surface is roughly proportional to the fraction of atoms in its vicinity, relative to the total number of atoms in the model. An obvious way to reduce the effect of the surface is to increase the system size, which is however limited by computational resources. This limitation was a lot more severe in the early days, when simulations were run on mainframes capable of handling only about a hundred particles. A very good solution to the problem was found back then and is still in wide use today.

System size effects are substantially reduced by using periodic boundary conditions [68,43,69]. A "primary" box, containing the particles, is replicated in space in all directions to form an infinite lattice of boxes with lattice parameters equal to the box lengths in the corresponding directions, L_x, L_y, and L_z. Each particle with coordinates r_i in the "primary" simulation box has an infinite number of periodic images with coordinates $r_i + n_x L_x e_x + n_y L_y e_y + n_z L_z e_z$, where (n_x, n_y, n_z) is a set of three integers ranging from minus to plus infinity and e_i are the unit vectors along the three coordinate directions.

Thus, a bulk material is simulated by an infinite periodic array of the simulation cells. Interactions are calculated taking into account the periodic images of particles. When a particle moves, all its periodic images move by exactly the same displacement. A two-dimensional example of periodic boundary conditions is shown in Fig. 10.

Unfortunately, periodic boundary conditions do not eliminate size effects completely. Imposing periodicity on the system cuts off its fluctuation spectrum. In the solid state the longest wavelength with which density fluctuations (sound) can propagate equals the box size. Density or composition fluctuations with characteristic lengths larger than the box size are impossible. Such long-range fluctuations play an important role in phase transitions. Thus, accurate simulations of phase transitions or phase coexistence, especially near critical points, require large system sizes and the use of finite size scaling techniques [70].

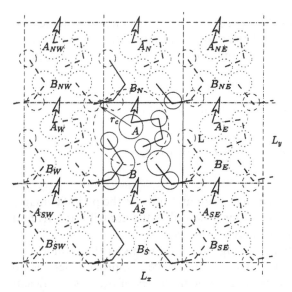

FIG. 10 Two-dimensional system with periodic boundary conditions.

If interparticle interactions are long range, the interactions between a particle and its own periodic images are substantial, and the symmetry of the lattice, artificially imposed by the periodic boundary conditions, affects properties of the otherwise isotropic system.

It is absolutely necessary to check for size effects on the simulation results by performing simulations with different model system sizes.

Periodic boundary conditions can be implemented in two different ways. One can keep track of the particles in the central (or primary) box, switching attention to the particle's periodic image entering the primary box from the opposite side whenever the particle crosses the box boundary. Alternatively, one can follow the coordinates of particles initially located inside the primary box and calculate coordinates of their images inside the primary box when necessary. The computational cost is about the same in both cases. The latter choice is more convenient when calculating diffusion coefficients, as the particle trajectory remains unperturbed, while one box length is added or subtracted to a particle coordinate when switching between the periodic images.

A. Calculating Energy and Forces with Periodic Boundary Conditions

The core part of any molecular dynamics, Brownian dynamics, or Monte Carlo simulation is the calculation of the interaction energy and forces.

In general, the energy in the periodic system encompasses interactions between all periodic images, and thus constitutes an infinite sum over all periodic images. The potential energy contribution from pairwise interactions $u(\mathbf{r}_i - \mathbf{r}_j)$ is given by:

$$\mathcal{V} = \sum_i \sum_{j>i} \sum_{n_x, n_y, n_z} u(\mathbf{r}_i - \mathbf{r}_j + n_x L_x \mathbf{e}_x + n_y L_y \mathbf{e}_y + n_z L_z \mathbf{e}_x) \tag{118}$$

$$+ \sum_i \sum_{n_x \neq 0, n_y \neq 0, n_z \neq 0} u(n_x L_x \mathbf{e}_x + n_y L_y \mathbf{e}_y + n_z L_z \mathbf{e}_z)$$

The sum over pairs of particles includes summation over the periodic images of the second particle. The second term describes the contribution from a particle interacting with its own periodic images.

If the interaction potential u is long range, all terms in the infinite sum should be taken into account. The "brute force" approach here is to truncate the summation at some large enough values of n_x, n_y, n_z. Efficient ways to do this for Coulombic interactions—the Ewald summation, fast multipole, and particle–mesh methods—will be described in the following chapters.

A much simpler technique is used for Lennard-Jones potentials or other short-range interactions. The Lennard-Jones potential, Eq. (6), drops down to only -0.016ϵ at $r_c = 2.5\sigma$. If we just cut the potential off at this distance, assuming that $u = 0$ at $r > r_c$, the error we make is less than 2%. We will see later, how it can be accounted for. If such a cutoff is used, the sums in Eq. (118) are limited only to the pairs of images that are closer to each other than the cutoff distance r_c. If r_c is smaller than one half of the smallest of L_x, L_y, or L_z, only the closest of images are taken into account in the first sum, and there is no second summation over the self-images. This approximation is called "the minimum image convention." The intermolecular potential energy is then given by:

$$\mathcal{V} = \sum_i \sum_{j>i} u(\mathbf{r}_i - \mathbf{r}_j + n'_x L_x \mathbf{e}_x + n'_y L_y \mathbf{e}_y + n'_z L_z \mathbf{e}_z) \tag{119}$$

In this equation, $n'_x n'_y n'_z$ stands for the set of integers that correspond to the pair of closest images of particles i and j. Notice that the circle centered at particle A (Fig. 10) with radius r_c may include a periodic image B_N of particle B, but not B itself.

The following fragment of FORTRAN code illustrates the implementation of the minimum image convention. It calculates the x, y, and z

components of the separation vector $\mathbf{r}_i - \mathbf{r}_j$, stored in the variables dx, dy, dz, given that particle coordinates are stored in the arrays x, y, z, and box sizes L_x, L_y, L_z are stored in the variables boxx, boxy, boxz.

```
. . . . . . . . . . . . . . . . . . . . . . . . . . . .
      invbx=1.d0/boxx
      invby=1.d0/boxy
      invbz=1.d0/boxz
. . . . . . . . . . . . . . . . . . . . . . . . . . . .
      dx = x(i)-x(j)
      dx = dx-boxx*anint(dx*invbx)
      dy = y(i)-y(j)
      dy = dy - boxy*anint(dy*invby)
      dz = z(i)-z(j)
      dz = dz - boxz*anint(dz*invbz)
. . . . . . . . . . . . . . . . . . . . . . . . . . . .
```

This piece of code is used throughout the modeling program with the periodic boundary conditions whenever the distance between two particles is needed: in the energy and force calculation, in the calculation of the pair correlation function, etc. The code uses the standard Fortran function* anint(a), which returns the nearest integer to the real or double precision number a. Note that anint(0.51)=1, anint(0.49)=0, anint(−0.51)=−1, anint(−0.49)=0.

To fold all the particles inside the "primary" simulation cube, $-L_x/2 < x_i < L_x/2$, $-L_y/2 < y_i < L_y/2$, $-L_z/2 < z_i < L_z/2$, we can use similar code:

```
. . . . . . . . . . . . . . . . . . . . . . . . . . . .
      invbx=1.d0/boxx
      invby=1.d0/boxy
      invbz=1.d0/boxz
. . . . . . . . . . . . . . . . . . . . . . . . . . . .
      x(i) = x(i) - boxx*anint(x(i)*invbx)
      y(i) = y(i) - boxy*anint(y(i)*invby)
      z(i) = z(i) - boxz*anint(z(i)*invbz)
. . . . . . . . . . . . . . . . . . . . . . . . . . . .
```

When a chemical bond crosses the box boundary, its periodic image reenters the box from the opposite side, as shown in Fig. 10. Unlike intermolecular interactions, the intramolecular interactions should be calculated between atoms of the same image of a molecule and not between

---

*anint(a) is also available in some C compiler libraries.

their periodic images. Nevertheless, it is also safe to use the minimum image convention for the intramolecular interactions. As chemical bonds are always shorter than half the box length, the convention will automatically provide the right separation vectors, no matter whether the particle coordinates are always folded back in the "primary" box, or whether the particle trajectory in space is preserved.

When the potential energy is cut off at distance r_c, interactions between particles separated by a distance larger than r_c are neglected. If these interactions are described by spherically symmetric pairwise potentials, the energy can be expressed in terms of the pair distribution function, as in Eq. (63). If the r_c is large enough, such that we can assume that there are no structural correlations beyond this distance $(g(r) \simeq 1)$, then the long-distance part of the integration can be replaced by the integration of the potential for $r > r_c$:

$$V = V_{sum} + N\rho \int_{r_c}^{\infty} g(r)u(r)4\pi r^2 \, dr \qquad (120)$$

$$\simeq V_{sum} + N\rho \int_{r_c}^{\infty} u(r)4\pi r^2 \, dr$$

$$PV = P_{sum}V - \frac{1}{6}N\rho \int_{r_c}^{\infty} g(r) \, r \frac{du(r)}{dr} 4\pi r^2 \, dr$$

$$\simeq P_{sum}V - \frac{1}{6}N\rho \int_{r_c}^{\infty} r \frac{du(r)}{dr} r 4\pi r^2 \, dr$$

V_{sum} and P_{sum} are the values obtained from explicit summation over the pairs within the cutoff radius r_c. Similar equations can be derived for any other property obtained via the summation over pairs of particles. The integrals on the right hand sides of the above equations can easily be calculated for any pair potential $u(r)$.*

Setting $g(r) = 1$ corresponds to replacing the material at long distances from the particle by a uniform background, with the average interaction properties. Thus, on the one hand one would like the cutoff radius to be not very large, to reduce the amount of calculations necessary for obtaining the energy, on the other hand r_c should be chosen large enough, so that structural correlations can be negligible at distances $r > r_c$. The latter assumption has to be checked by calculating $g(r)$.

Even though the interaction potential value $u(r_c)$ is quite small, the effect of the cutoff can be quite substantial. For instance, using the Lennard-Jones

---

*They might be nonconverging for the long-range potentials or in fewer than three dimensions.

(6-12) potential with cutoff $r_c = 2.5\sigma$ in the simple monatomic model of the noble gas lowers the critical temperature by about 30% [54] compared to the value obtained with a much longer cutoff or with the long-range corrections given by Eq. (120).

If the interaction is not pairwise, similar expressions can be obtained using the corresponding multiparticle correlation functions. If the interaction potential u depends on the direction of the separation vector $\mathbf{r}$ as well as on its length r, a multidimensional integration over the components of $\mathbf{r}$ is to be performed instead of the spherical averaging leading to integration over $r = |\mathbf{r}|$.

IX. ERRORS IN SIMULATION RESULTS

Although simulations are often associated with "theory," they have a lot in common with experiments. In particular, simulations are subject to errors.

There are systematic and statistical errors. The sources of the systematic errors are system size effects, inaccuracy of the interaction potentials, poor equilibration, deficiency in the random number generator, etc. Systematic errors are estimated by performing simulations with different system sizes, using different random number generators, and varying the interaction potentials, equilibration conditions, and starting configurations.

Properties calculated in computer simulations are obtained as statistical averages, and therefore are inherently subject to statistical errors. The average is calculated as a sum over many steps. According to the central limit theorem of probability theory, the average value obtained in the simulation is sampled from a Gaussian distribution, with average at the true mean. The variance of this distribution provides an estimate of the difference between the average estimate obtained in the simulation and the true mean value. According to statistics, given N uncorrelated values of $\mathcal{A}_i$ obtained during the simulation, with average $\langle \mathcal{A} \rangle_{run} = 1/N \sum_{i=1}^{N} \mathcal{A}_i$ and variance $\langle (\delta \mathcal{A})^2 \rangle_{run} = 1/N \sum_{i=1}^{N} (\mathcal{A}_i - \langle \mathcal{A} \rangle_{run})^2$, the variance of the estimated average $\langle \mathcal{A} \rangle$, $\sigma^2(\langle \mathcal{A} \rangle_{run})$, equals $\langle (\delta \mathcal{A})^2 \rangle_{run}/N$. The probability to observe a certain value of $\langle \mathcal{A} \rangle_{run}$ in a sample of N uncorrelated configurations is described by a Gaussian distribution with average at the true mean and variance $\langle (\delta \mathcal{A})^2 \rangle_{run}/N$. The expected error in the simulation estimate of $\mathcal{A}$ is thus $\langle (\delta \mathcal{A})^2 \rangle_{run}^{1/2}/N^{1/2}$.

The configurations obtained in successive steps of a molecular dynamics, Monte Carlo, or Brownian dynamics simulation are usually not very different from each other. Hence, one should expect the energy, virial, or any other function of the microscopic state obtained from the successive configurations to be correlated. These correlations disappear after some time, or number of steps. The correlation time depends on the modeled system and on the quantity evaluated. For instance, in solids or liquids, where collisions between particles are very frequent, velocities usually have much shorter correlation times than any structural characteristics. The correlation time is usually defined as an integral of the normalized time correlation function [43]: $\tau = \int_0^\infty c_{AA}(t)$. Values separated by at least one correlation time must be used to estimate the error in the average. If A-values are saved every time interval δt, each kth value, with $k = \tau/\delta t$, should be used. Thus, instead of N we have only N/k uncorrelated values, and $b = N/k$ must replace N in the expression for the expected error in the average. A common practice is to take averages over blocks of data of length k, $\langle A \rangle_i^{(k)}$, and use them as b independent values to estimate the variance (squared expected error in the simulation estimate) [43]:

$$\langle (\delta \langle A \rangle_{run})^2 \rangle = \frac{k}{N} \sum_{i=1}^b (\langle A \rangle_i^{(k)} - (\langle A \rangle_{run})^2) \tag{121}$$

The correct block size k that ensures statistical independence of the $\langle A \rangle_i^{(k)}$ can be determined through systematic analysis for different k [43].

To estimate errors in structural or other properties, expressed not just as numbers but as functions of either distance (e.g., $g(r)$), or time (e.g., time correlation functions), a similar analysis should be carried out for different values of the argument. For instance, when the time correlation function is calculated, much better statistics is obtained for short times, as there are much more data available. The time correlation function values for times comparable to the length of the simulation run are obtained with larger errors, as the average is taken over only a few pairs of values.

When the property calculated is a single particle property, such as the velocity autocorrelation function, or the particle's mean squared displacement, averaging over the particles does help a lot. Averaging over even a moderate number of particles of about 1000 decreases the error by more than an order of magnitude. Unfortunately, this is not possible for all properties. For instance, the calculation of viscosity calls for the stress correlation function [Eq. (67)], which is not a single particle property. This is why self-diffusion coefficients are usually estimated with much better accuracy than viscosity.

A major challenge in polymer simulations is to produce a large number of uncorrelated configurations, to sample various chain conformations. In molecular dynamics, each time step involves small changes in the monomer positions. Significantly changing the chain conformation involves large monomer displacements and requires many steps. The same thing happens in Monte Carlo simulations involving only small monomer displacements at each step. Global updates, involving large fragments of the polymer molecules, which are used, e.g., in Configurational Bias Monte Carlo techniques, help overcome this problem. Monte Carlo strategies for the rapid equilibration of polymer systems will be discussed in subsequent chapters of this book.

X. GENERAL STRUCTURE OF A SIMULATION PROGRAM

A typical computer simulation "experiment" consists of the following stages:

- Initial Configuration Generation
- Main Simulation Run
  - Read the initial configuration and the run parameters: total number of simulation steps, frequency of saving configurations, cutoff parameters, temperature, pressure, etc.
  - For the required number of simulation steps:
    * Advance coordinates (according to the chosen MD, MC, or BD algorithm)
    * Increment volume or number of particles if *NPT* or Grand Canonical ensemble simulation
    * Calculate averages monitored during the run (energy, atomic and molecular pressure, total momentum, atomic and molecular temperature)
    * Save configuration on disk for further analysis.
- Analysis of the obtained configurations. Calculate $g(r)$, time correlation functions, diffusion coefficients, etc.

Different simulation packages may have these stages implemented as one large program, or as separate modules, controlled by a common user interface.

Sophisticated commercial simulation packages usually also include a database of force field parameters, and prebuilt monomer units.

Generating an initial structure is fairly easy for the crystalline state, as one just has to arrange molecules according to the structure of the crystal. Liquid or glassy configurations of small molecules are usually generated by "melting" a crystalline structure. This happens quite fast when the molecules are small. For dense polymers it is impossible, due to the very long relaxation times. Polymer chains have a very specific conformation in the crystal, which is far from the random coil conformation prevailing in the polymer glass or melt. Therefore, special techniques are used to generate realistic models of polymer melts or glasses [34,71].

REFERENCES

1. Hiller, A.; Wall, F.T.; Wheeler, D.J. Statistical computation of mean dimensions of macromolecules. J. Chem. Phys. **1954**, *22*, 1036.
2. Wall, W.A.; Seitz, F.T. Simulation of polymers by self-avoiding, non-intersecting random chains at various concentrations. J. Chem. Phys. **1977**, *67*, 3722–3726.
3. Alexandrowicz, Z.; Accad, Y. Monte Carlo of chains with excluded volume: distribution of intersegmental distances. J. Chem. Phys. **1971**, *54*, 5338–5345.
4. Lal, M. Monte Carlo simulation of chain molecules I. Molec. Phys. **1969**, *17*, 57.
5. De Vos, E.; Bellemans, A. Concentration dependence of the mean dimension of a polymer chain. Macromolecules **1974**, *7*, 812–814.
6. Ryckaert, J.P.; Bellemans, A. Molecular dynamics of liquid butane near its boiling point. Chem. Phys. Lett. **1975**, *30*, 123–125.
7. Balabaev, N.K.; Grivtsov, A.G.; Shnol', E.E. Doklady Akad. Nauk SSSR **1975**, *220*, 1096–1098, in Russian.
8. Ceperley, H.L.; Frisch, H.L.; Bishop, M.; Kalos, M.H. Investigation of static properties of model bulk polymer fluids. J. Chem. Phys. **1980**, *72*, 3228–3235.
9. Weber, T.A.; Helfand, E. Molecular dynamics simulation of polymers. I. Structure. J. Chem. Phys. **1979**, *71*, 4760–4762.
10. Vacatello, M.; Avitabile, G.; Corradini, P.; Tuzi, A. A computer model of molecular arrangement in an n-paraffinic liquid. J. Chem. Phys. **1980**, *73*, 543–552.
11. Fixman, M. Classical statistical mechanics of constraints. A theorem and application to polymers. Proc. Nat. Acad. Sci. USA **1974**, *71*, 3050–3053.
12. Go, N.; Scheraga, H. Analysis of the contribution of internal vibrations to the statistical weights of equilibrium conformations of macromolecules. J. Chem. Phys. **1969**, *51*, 4751–4767.
13. Go, N.; Scheraga, H. On the use of classical statistical mechanics in the treatment of polymer chain conformation. Macromolecules **1976**, *9*, 535–542.
14. Mattice, W.; Suter, U.W. *Conformational Theory of Large Molecules, the Rotational Isomeric State Model in Macromolecular Systems*; J. Wiley: New York, 1994.
15. Insight II User Guide, version 2.3.0. Biosym Technologies: San Diego, 1993; http://www.accelrys.com.
16. van Gunsteren, W.; Berendsen, H.J.C. Computer simulation of molecular dynamics: methodology, applications and perspectives in chemistry. Angew. Chem. Int. Ed. Engl. **1990**, *29*, 992–1023.

17. Weiner, S.J.; Kollman, P.A.; Nguyen, D.T.; Case, D.A. An all atom force field for simulations of proteins and nucleic acids. J. Comput. Chem. **1986**, 7, 230, http://www.amber.ucsf.edu/amber/amber.html.

18. Doi, M.; Edwards, S.F. *The Theory of Polymer Dynamics*; Clarendon Press: Oxford, 1994.

19. Khokhlov, A.R.; Grosberg, A.Yu. *Statistical Physics of Macromolecules*; AIP Press: New York, 1994.

20. Flory, P.J. *Statistical Mechanics of Chain Molecules*; Hanser Publishers: Munich, 1989.

21. Ted Davis, H. *Statistical Mechanics of Phases, Interfaces, and Thin Films*; VCH: New York, 1996.

22. deGennes, P.-G. *Scaling Concepts in Polymer Physics*; Cornell University Press: Ithaca, 1979.

23. Fleer, G.J.; Cohen Stuart, M.A.; Scheutjens, J.M.H.M.; Cosgrove, T.; Vincent, B. *Polymers at Interfaces*; Chapman and Hall: London, 1993.

24. Götze, W.G. Recent tests of the mode-coupling theory for glassy dynamics. J. Phys.: Condens. Matter **1999**, 11, A1–A45.

25. Cummins, H.Z. The liquid-glass transition: a mode coupling perspective. J. Phys.: Condens. Matter **1999**, 11, A95–A117.

26. Strobl, G.R.; *The Physics of Polymers: Concepts for Understanding Their Structures and Behavior*; Springer-Verlag: Berlin, 1997.

27. Car, R.; Parrinello, M. Unified approach for molecular dynamics and density functional theory. Phys. Rev. Lett. **1985**, 55, 2471–2474.

28. Reichl, L.E. *A Modern Course in Statistical Physics*; University of Texas Press: Austin, 1980.

29. Arnold, V.I. *Mathematical Methods of Classical Mechanics*; Springer-Verlag: 2nd Ed., 1989.

30. Goldstein, H. *Classical Mechanics*; Addison–Wesley: 2nd Ed., 1980.

31. Kneller, G.R.; Hinsen, K. Generalized Euler equations for linked rigid bodies. Phys. Rev. **1994**.

32. Hinsen, K.; Kneller, G.R. Influence of constraints on the dynamics of poly-peptide chains. Phys. Rev. **1995**, E52, 6868.

33. Rice, L.M.; Brunger, A.T. Torsion angle dynamics—reduced variable conformational sampling enhances crystallographic structure refinement. Proteins **Aug 1994**, 19, 277–290.

34. Theodorou, D.N.; Suter, U.W. Detailed molecular structure of a vinyl polymer glass. Macromolecules **1985**, 18, 1467–1478.

35. Ludovice, P.J.; Suter, U.W. Detailed molecular structure of a polar vinyl polymer glass. In *Computational Modeling of Polymers*; Bicerano, J., Ed.; Marcel Dekker: New York, 1992; page 401.

36. Madras, N.; Sokal, A. Nonergodicity of local, length conserving Monte-Carlo algorithms for the self-avoiding walk. J. Statist. Phys. **1987**, 47(2), 573–595.

37. McQuarrie, D.A. *Statistical Mechanics*; Harper and Row: New York, 1976.

38. Landau, L.D.; Lifshitz, E.M. Statistical Physics, *volume 5 of Course of Theoretical Physics*. Pergamon Press: Oxford, 3rd edition, 1980.
39. Modell, M.; Reid, R.C. *Thermodynamics and Its Applications*; Prentice–Hall: Englewood Cliffs, 1983.
40. Chandler, D. *Introduction to Modern Statistical Mechanics*; Oxford University Press: New York, 1987.
41. Theodorou, D.N.; Dodd, L.R.; Boone, T.D.; Mansfield, K.F. Stress tensor in model polymer systems with periodic boundaries. Makromol. Chem. Theory Simul. **1993**, *2*, 191–238.
42. Widom, B.; Some topics in the theory of fluids. J. Chem. Phys. **1963**, *39*, 2808–2812.
43. Allen, M.P.; Tildesley, D.J. *Computer Simulation of Liquids*; Clarendon Press: Oxford, 1993.
44. Spyriouni, T.; Economou, I.G.; Theodorou, D.N. Thermodynamics of chain fluids from atomistic simulation: a test of the chain increment method for chemical potential. Macromolecules **1997**, *30*, 4744–4755.
45. Boulougouris, G.C.; Economou, I.G.; Theodorou, D.N. On the calculation of the chemical potential using the particle deletion scheme. Molec. Phys. **1999**, *96*, 905–913.
46. Hansen, J.-P.; McDonald, I.R. *Theory of Simple Liquids*; Academic Press: New York, 2nd edition, 1986.
47. Higgins, J.S.; Benoit, H.C. *Polymers and Neutron Scattering*; Clarendon Press: Oxford, 1994.
48. Metropolis, N.; Rosenbluth, A.W.; Rosenbluth, M.N.; Teller, A.H.; Teller, E. Equation of state calculations by fast computing machines. J. Chem. Phys. **1953**, *21*, 1087–1092.
49. van Kampen, N.G.; *Stochastic Processes in Physics and Chemistry*; Amsterdam, North Holland, 1992.
50. Andersen, H.-C. Molecular dynamics simulations at constant pressure and/or temperature. J. Chem. Phys. **1980**, *72*, 2384–2393.
51. Nosé, S. A molecular dynamics method for simulations in the canonical ensemble. Molec. Phys. **1984**, *52*, 255–268.
52. Hoover, W.G. Canonical dynamics: equilibrium phase space distribution. Phys. Rev. **1985**, *A31*, 1695–1697.
53. Berendsen, H.J.C.; Postma, J.P.M.; van Gunsteren, W.F.; Di Nola, A.; Haak, J.R. Molecular dynamics with coupling to an external bath. J. Chem. Phys. **1984**, *81*, 3684–3690.
54. Kotelyanskii, M.J.; Hentschke, R. Gibbs-ensemble molecular dynamics: liquid-gas equilibria for Lennard-Jones spheres and n-hexane. Molec. Simul. **1996**, *17*, 95–112.
55. Lemak, A.S.; Balabaev, N.K. A comparison between collisional dynamics and brownian dynamics. Molec. Simul. **1995**, *15*, 223–231.
56. Mazur, P.; Oppenheim, I. Molecular theory of Brownian motion. Physica **1970**, *50*, 241–258.

57. Deutsch, J.M.; Oppenheim, I. The concept of Brownian motion in modern statistical mechanics. Faraday discuss. Chem. Soc. **1987**, *83*, 1–20.

58. H.-C. Öttinger. *Stochastic Processes in Polymer Fluids: Tools and Examples for Developing Simulation Algorithms*; Springer: Berlin, 1996.

59. Kramers, H.A. Brownian motion in a field of force and the diffusion model of chemical reactions. Physica **1940**, *7*, 284–305.

60. Ermak, D.L.; Yeh, Y. Equilibrium electrostatic effects on behavior of polyions in solution: polyion–mobile ion interaction. Chem. Phys. Lett. **1974**, *24*, 243–248.

61. Ermak, D.L.; Buckholtz, H. Numerical integration of the Langevin equation: Monte Carlo simulation. J. Comput. Phys. **1980**, *35*, 169–182.

62. Chandler, D. Statistical mechanics of isomerization dynamics in liquids and the transition state approximation. J. Chem. Phys. **1978**, *68*, 2959–2970.

63. Voter, A.; Doll, J. Dynamical corrections to transition state theory for multi-state systems: surface self-diffusion in the rare-event regime. J. Chem. Phys. **1985**, *82*, 80–87.

64. June, R.L.; Bell, A.T.; Theodorou, D.N. Transition-state studies of xenon and SF_6 diffusion in silicalite. J. Phys. Chem. **1991**, *95*, 8866–8878.

65. Bell, T.A.; Sevick, E.M.; Theodorou, D.N. A chain of states method for investigating infrequent event processes in multistate, multidimensional systems. J. Chem. Phys. **1993**, *98*, 3196–3212.

66. Elber, R. Enhanced sampling in molecular dynamics—use of time dependent hartree approximation for a simulation of carbon-monoxide diffusion through myoglobin. J. Chem. Phys. **1990**, *93*, 4312–4321.

67. Vineyard, G.H. Frequency factors and isotope effects in solid state rate processes. J. Phys. Chem. Solids, **1957**, *3*, 121–127.

68. Born, M.; Von Karman, Th. Über schwingungen in raumgittern. Z. Physik, **1912**, *13*, 297–309.

69. Binder, K. editor. *The Monte Carlo Method in Condensed Matter Physics*; Springer: Berlin, 2nd edition, 1995.

70. Wilding, N.R.; Binder, K. Finite-size scaling for near-critical continuum fluids at constant pressure. Physica A **1996**, *231*, 439–447.

71. Kotelyanskii, M. Simulation methods for modeling amorphous polymers. Trends Polymer Sci. **1997**, *5*, 192–198.

2

Rotational Isomeric State (RIS) Calculations, with an Illustrative Application to Head-to-Head, Tail-to-Tail Polypropylene

E. DEMET AKTEN and WAYNE L. MATTICE   The University of Akron, Akron, Ohio, U.S.A.

ULRICH W. SUTER   Eidgenössische Technische Hochschule (ETH), Zurich, Switzerland

I. INTRODUCTION

The Rotational Isomeric State (RIS) model is excellent for the analysis of conformation-dependent physical properties of chain molecules in their unperturbed state. Its special strength is the incorporation of detailed information about the covalent structure (bond lengths, bond angles, torsion angles, and torsion potential energy functions) in a formalism that can be evaluated quickly by the smallest computer. The answer is the exact result for the specific model, as defined by the geometry and energies of the short-range intramolecular interactions.

The most frequently calculated property is the mean square unperturbed end-to-end distance, $\langle r^2 \rangle_0$. Other properties susceptible to rapid computation include the average of the end-to-end vector, $\langle \mathbf{r} \rangle_0$, and the mean square unperturbed radius of gyration, $\langle s^2 \rangle_0$. The viscosity of a dilute solution in a Θ solvent can be estimated from $\langle s^2 \rangle_0$ or $\langle r^2 \rangle_0$ via the equivalent sphere model for hydrodynamic properties. Several higher even moments, $\langle r^{2p} \rangle_0$ and $\langle s^{2p} \rangle_0$, $p = 2, 3, \ldots$, provide information about the shape (width, skewness) of the distribution functions for r^2 and s^2. When combined with information about the electronic charge distribution within individual rigid units, the RIS model can calculate the mean square dipole moment, $\langle \mu^2 \rangle_0$. Also accessible are optical properties that depend on the anisotropy of the polarizability tensor (optical anisotropy) in conjunction with **r** (stress-optical coefficient), **μ** (molar Kerr constant), or the magnetic susceptibility (Cotton–Mouton constant).

The speed of the calculations derives from the use of efficient generator matrix techniques that were introduced six decades ago [1]. These techniques were adapted to the study of macromolecules a decade later [2]. Therefore the description of conformation-dependent physical properties of macromolecules with the RIS model predates the widespread availability of computers and simulation methods that require enormous computational power for their implementation. A strong increase in the use of the RIS model over the decade from 1959 to 1969 culminated in the publication of Flory's classic book on the subject [3]. Some of the subsequent developments are contained in two books published in the 1990s. One book has an emphasis on self-instruction in the use of the RIS model [4]. The other book is a compilation, in standard format, of RIS models for over 400 polymers that have appeared in the literature through the mid-1990s [5].

II. THREE FUNDAMENTAL EQUATIONS IN THE RIS MODEL

A. The First Equation: Conformational Energy

This equation describes the conformation partition function, Z, as a serial product of statistical weight matrices, U_i, that incorporate the energies of all of the important short-range intramolecular interactions.

$$Z = U_1 U_2 \ldots U_n \tag{1}$$

There is one statistical weight matrix for each of the n bonds in the chain. The dimensions of the statistical weight matrices depend on the number of rotational isomers assigned to each bond. The continuous range for the torsion angle, ϕ, is replaced by a finite set of torsion angles, i.e., a continuous conformational integral is replaced by a discrete summation. The sum should be designed so that the terms include all regions of the local conformational space that have a significant population. Many unperturbed chains have pairwise interdependence of their bonds, causing each U_i to have a number of columns given by the number of rotational isomers at bond i (denoted ν_i), and a number of rows given by the number of rotational isomers at the preceding bond. Each element in U_i is the product of a statistical weight for a "first-order" interaction (which depends only on the state at bond i) and the statistical weight for a "second-order" interaction (which depends simultaneously on the states at bonds $i-1$ and i).

Each statistical weight is formulated as a Boltzmann factor with the appropriate energy and temperature. Sometimes there is inclusion of a preexponential factor that accounts for the differences in conformational entropy of the various rotational isomeric states.

The first equation incorporates all of the energetic information that will be employed in the calculation. It incorporates structural information only indirectly, via the assignment of the numbers of rotational isomers at the various bonds, and the assignment of numerical values for the statistical weights.

B. The Second Equation: Structure

The second equation describes a conformation-dependent physical property of a chain in a specific conformation as a serial product of n matrices, one matrix for each bond in the chain.

$$f = \mathbf{F}_1 \mathbf{F}_2 \ldots \mathbf{F}_n \tag{2}$$

There is no concern here about the probability of observation of this conformation; we are ignoring, for the moment, the energetic information contained in Z. Each $\mathbf{F}_i$ includes the information about the manner in which bond i affects f. The details, of course, depend on how we have chosen f. If f is the squared end-to-end distance, we need to know the length of bond i, the angle between bonds i and $i+1$, and the torsion angle at bond i. These three properties are denoted l_i, θ_i, and ϕ_i, respectively. The square of the end-to-end distance is written in terms of these local variables as

$$r^2 = 2 \sum_{k=1}^{n-1} \sum_{j=k+1}^{n} \mathbf{l}_k^T \mathbf{T}_k \mathbf{T}_{k+1} \ldots \mathbf{T}_{j-1} \mathbf{l}_j + \sum_{k=1}^{n} l_k^2 \tag{3}$$

where $\mathbf{l}_i$ denotes the bond vector, expressed in a local coordinate system for bond i,

$$\mathbf{l}_i = \begin{bmatrix} l_i \\ 0 \\ 0 \end{bmatrix} \tag{4}$$

and $\mathbf{T}_i$ is a transformation matrix. A vector expressed in the local coordinate system of bond $i+1$ is transformed into its representation in the local coordinate system of bond i through premultiplication by $\mathbf{T}_i$,

$$
\mathbf{T}_i = \begin{bmatrix} -\cos\theta & \sin\theta & 0 \\ -\sin\theta\cos\phi & -\cos\theta\cos\phi & -\sin\phi \\ -\sin\theta\sin\phi & -\cos\theta\sin\phi & \cos\phi \end{bmatrix}_i \tag{5}
$$

The single and double sums in Eq. (3) are evaluated exactly via Eq. (2) using

$$
\mathbf{F}_1 = \begin{bmatrix} 1 & 2\mathbf{l}_i^{\mathsf{T}}\mathbf{T}_1 & l_1^2 \end{bmatrix} \tag{6}
$$

$$
\mathbf{F}_i = \begin{bmatrix} 1 & 2\mathbf{l}_i^{\mathsf{T}}\mathbf{T}_i & l_i^2 \\ 0 & \mathbf{T}_i & \mathbf{l}_i \\ 0 & 0 & 1 \end{bmatrix}, \quad 1 < i < n \tag{7}
$$

$$
\mathbf{F}_n = \begin{bmatrix} l_n^2 \\ \mathbf{l}_n \\ 1 \end{bmatrix} \tag{8}
$$

C. The Third Equation: Conformational Energy Combined with Structure

The third equation combines the energetic information from Eq. (1) with the structural information in Eq. (2), in order to average f over all of the conformations in the RIS model. The combination is achieved in a generator matrix, $\mathbf{G}$.

$$
\langle f \rangle_0 = \frac{1}{Z} G_1 G_2 \ldots G_n \tag{9}
$$

$$
G_i = (\mathbf{U}_i \otimes \mathbf{I}_\mathbf{F})\text{diag}(\mathbf{F}_\alpha, \mathbf{F}_\beta, \ldots, \mathbf{F}_{\nu_i}) \tag{10}
$$

The generator matrix contains the energetic information in an expansion of the statistical weight matrix (via its direct product with an identity matrix of order equal to the number of columns in $\mathbf{F}$). This energetic information is multiplied onto the structural information (in the form of a block-diagonal matrix of the ν_i expressions for $\mathbf{F}$ for the rotational isomers at bond i). The efficiency of the calculation arises from the fact that the conformational average of the property of interest is obtained as a serial product of matrices via Eq. (9).

The remainder of this chapter is devoted to a case study. Readers who seek more detailed or general information on the origin, structure, and usage of the equations for Z, f, and $\langle f \rangle_0$ are referred to books on the RIS model [3,4].

III. CASE STUDY: MEAN SQUARE UNPERTURBED DIMENSIONS OF HEAD-TO-HEAD, TAIL-TO-TAIL POLYPROPYLENE

Polypropylene is an important polymer that is usually assembled with the monomer units arranged in head-to-tail sequence, producing -CH(CH$_3$)-CH$_2$- as the repeat unit. Several RIS models, based on three [6–12], four [13], five [14,15], or seven [16] states per rotatable bond, have been described for this common form of polypropylene. Head-to-head, tail-to-tail units have been considered briefly [11] in order to assess how the incorporation of a few such units, as defects, alters the unperturbed dimensions of the conventional head-to-tail polypropylene.

Two experimental investigations of head-to-head, tail-to-tail polypropylene, obtained by reduction of poly(2,3-dimethylbutadiene), give discordant results [17,18]. Values of 4.5 and 6.1–6.4, respectively, are obtained for the characteristic ratio, C_n, of samples of high molecular weight.

$$C_n \equiv \frac{\langle r^2 \rangle_0}{nl^2} \tag{11}$$

Can an analysis with the RIS model assist in the resolution of this discrepancy in the experimental results? A three-state RIS model for the head-to-head, tail-to-tail form of polypropylene, with a repeating sequence given by -CH$_2$-CH(CH$_3$)-CH(CH$_3$)-CH$_2$-, is constructed and analyzed here. The mean square unperturbed dimensions of exclusively head-to-head, tail-to-tail polypropylene are determined for chains with various stereochemical compositions and stereochemical sequences. The critical parameters in the model are identified, and the predicted mean square unperturbed dimensions are compared with experiment.

A. Construction of the RIS Model

Rotational Isomeric States. The model assumes three states (trans and gauche$^\pm$, abbreviated t and $g^\pm$) for each rotatable bond. The torsion angle, as defined on page 112 of Mattice and Suter [4], is 180° for a t placement. It increases with a clockwise rotation.

Short-Range Interactions.   The model is limited to first- and second-order interactions. The repulsive first-order interactions are variants of the one that occurs in the $g^{\pm}$ conformations of n-butane. The second-order interaction included is the one commonly described as the "pentane effect," which is generated by the repulsive interaction of the terminal methyl groups in the conformations of n-pentane with g placements of opposite sign. The statistical weight matrices are constructed under the simplifying assumption that methyl, methylene, and methine groups are equivalent insofar as first- and second-order interactions are concerned. This assumption reduces the number of distinguishable parameters in the model.

Description of the Stereochemical Sequence.   The stereochemical sequence is described using dl pseudoasymmetric centers, as defined on page 175 of Mattice and Suter [4]. The C–C bonds in the chain are indexed sequentially from 1 to n. A local Cartesian coordinate system is associated with each bond. Axis x_i lies along bond i, with a positive projection on that bond. Axis y_i is in the plane of bonds $i-1$ and i, with a positive projection on bond $i-1$. Axis z_i completes a right-handed Cartesian coordinate system. The chain atom at the junction of bonds $i-1$ and i is a d pseudoasymmetric center if it bears a methyl group with a positive coordinate along z_i; it is an l pseudoasymmetric center if this coordinate is negative.

General Form of the Statistical Weight Matrices.   First the statistical weight matrices are formulated in symbolic form. Later numerical values will be assigned to the various symbols.

The 3×3 statistical weight matrix for bond i has rows indexed by the state of bond $i-1$, columns indexed by the state of bond i, and the order of indexing is t, g^+, g^- in both cases. Each statistical weight matrix is assembled as $\mathbf{U}_i = \mathbf{V}_i \mathbf{D}_i$, where $\mathbf{V}_i$ is a 3×3 matrix that incorporates the statistical weights for second-order interactions, and $\mathbf{D}_i$ is a diagonal matrix that incorporates the statistical weights for first-order interactions. Efficient description of the interrelationships between selected pairs of matrices employs a matrix $\mathbf{Q}$ with the property $\mathbf{Q}^2 = \mathbf{I}_3$.

$$\mathbf{Q} = \begin{bmatrix} 1 & 0 & 0 \\ 0 & 0 & 1 \\ 0 & 1 & 0 \end{bmatrix} \tag{12}$$

Statistical Weight Matrices for the CH_2–CH_2 Bond.   This statistical weight matrix incorporates short-range interactions that occur in the fragment -CH-CH(CH$_3$)-CH$_2$—CH$_2$-CH-, where the longest dash denotes bond i. The matrix is denoted by $\mathbf{U}_{CC;x}$, where x denotes the stereochemistry (d or l) of the attachment of the methyl group. Figure 1 depicts this fragment when the methyl group produces an l pseudoasymmetric center. Internal rotation

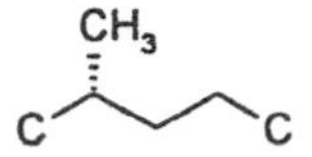

FIG. 1 Fragment considered in the construction of $U_{CC;l}$.

about bond i establishes the position of the terminal CH group with respect to the remainder of the fragment. A diagonal matrix incorporates the first-order interactions between the pair of bold CH groups in $-CH-CH(CH_3)-CH_2-CH_2-CH-$.

$$\mathbf{D}_{CC} = \text{diag}(1, \sigma, \sigma) \tag{13}$$

This matrix is independent of the stereochemistry of the attachment of the methyl group, because it only involves carbon atoms that are in the main chain. In Eq. (13), σ is the statistical weight for a $g^{\pm}$ placement at the CH_2-CH_2 bond, the t placement being assigned a statistical weight of 1.

The second-order interactions in $U_{CC;x}$ take place between the terminal bold CH group in $-CH-CH(CH_3)-CH_2-CH_2-CH-$ and the two groups in this fragment that are bold in the previous paragraph. Involvement of the methyl group in the second-order interactions demands two forms of $\mathbf{V}$, depending on whether the fragment contains a d or l pseudoasymmetric center.

$$\mathbf{V}_{CC;l} = \begin{bmatrix} 1 & 1 & \omega \\ 1 & \omega & \omega \\ 1 & \omega & 1 \end{bmatrix} \tag{14}$$

$$\mathbf{V}_{CC;d} = \mathbf{Q}\mathbf{V}_{CC;l}\mathbf{Q} \tag{15}$$

Here ω is the statistical weight for the second-order interaction of the terminal chain atoms when the conformation has two g placements of opposite sign (the ω in the 2,3 and 3,2 elements), or the equivalent interaction of the methyl group with the terminal CH (the remaining two ω, which occur in the 1,3 and 2,2 elements of $\mathbf{V}_{CC;l}$, but in the 1,2 and 3,3 elements of $\mathbf{V}_{CC;d}$). The full statistical weight matrices, incorporating both first- and second-order interactions, are

$$\mathbf{U}_{CC;l} = \mathbf{V}_{CC;l}\mathbf{D}_{CC} = \begin{bmatrix} 1 & \sigma & \sigma\omega \\ 1 & \sigma\omega & \sigma\omega \\ 1 & \sigma\omega & \sigma \end{bmatrix} \tag{16}$$

$$\mathbf{U}_{CC;d} = \mathbf{Q}\mathbf{U}_{CC;l}\mathbf{Q} \tag{17}$$

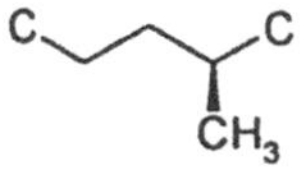

FIG. 2   Fragment considered in the construction of $U_{CM;l}$.

Statistical Weight Matrices for the CH_2–CH Bond.   This statistical weight matrix, denoted by $U_{CM;x}$, incorporates statistical weights for short-range interactions that occur in the fragment -CH-CH$_2$-CH$_2$—CH(CH$_3$)-CH-, where the longest dash denotes the bond that is now considered as bond i. An example of this fragment is depicted in Fig. 2. Rotation about bond i establishes the position of the bold CH and bold CH$_3$ in -CH-CH$_2$-CH$_2$—CH(CH$_3$)-CH- with respect to the remainder of the fragment. The first-order interactions now occur between the bold CH$_2$ and the other two bold groups. Involvement of the methyl group demands two forms of the matrix for the first-order interactions.

$$\mathbf{D}_{CM;l} = \text{diag}(1, 1, \tau) \tag{18}$$

$$\mathbf{D}_{CM;d} = \mathbf{Q}\mathbf{D}_{CM;l}\mathbf{Q} \tag{19}$$

Here τ is the statistical weight for two simultaneous first-order interactions, relative to the case where there is a single first-order interaction.

The second-order interactions occur between the initial CH and the other two bold groups in the fragment **-CH-CH$_2$-CH$_2$—CH(CH$_3$)-CH-**. Rotation about bond i alters the position of the methyl group, which was not true in the prior consideration of the second-order interactions that are incorporated in $U_{CC;x}$. The matrices of second-order interactions differ for the two bonds.

$$\mathbf{V}_{CM;l} = \begin{bmatrix} 1 & 1 & 1 \\ \omega & 1 & \omega \\ 1 & \omega & \omega \end{bmatrix} \tag{20}$$

$$\mathbf{V}_{CM;d} = \mathbf{Q}\mathbf{V}_{CM;l}\mathbf{Q} \tag{21}$$

The statistical weight matrices are

$$\mathbf{U}_{CM;l} = \begin{bmatrix} 1 & 1 & \tau \\ \omega & 1 & \tau\omega \\ 1 & \omega & \tau\omega \end{bmatrix} \tag{22}$$

$$\mathbf{U}_{CM;d} = \mathbf{Q}\mathbf{U}_{CM;l}\mathbf{Q} \tag{23}$$

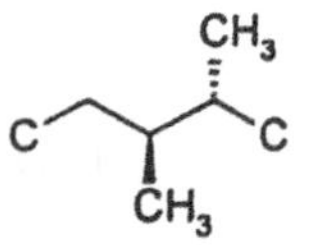

FIG. 3 Fragment considered in the construction of $U_{MM;ll}$.

Statistical Weight Matrices for the CH–CH Bond. This statistical weight matrix incorporates short-range interactions in the fragment -CH$_2$-CH$_2$-CH(CH$_3$)—CH(CH$_3$)-CH$_2$-, depicted in Fig. 3. The matrix is denoted by $U_{MM;xy}$, where x and y define the stereochemistry of the attachments of the two methyl groups, in the order that they occur in the fragment. The first-order interactions occur between pairs of bold groups in -CH$_2$-CH$_2$-CH(CH$_3$)—CH(CH$_3$)-CH$_2$-, with a member of the pair being selected from among the bold groups on either side of bond i. Involvement of two methyl groups demands four forms of the diagonal matrix for the first-order interactions. Two of these forms are identical.

$$\mathbf{D}_{MM;dd} = \mathbf{D}_{MM;ll} = \mathrm{diag}(1, \zeta, \zeta) \tag{24}$$

$$\mathbf{D}_{MM;dl} = \mathrm{diag}(\zeta, 1, \zeta) \tag{25}$$

$$\mathbf{D}_{MM;ld} = \mathbf{Q}\mathbf{D}_{MM;dl}\mathbf{Q} \tag{26}$$

Here ζ is the statistical weight for three simultaneous first-order interactions, relative to the case where there are two first-order interactions.

The second-order interactions occur between the initial methylene and the other two bold groups in -CH$_2$-CH$_2$-CH(CH$_3$)—CH(CH$_3$)-CH$_2$-. Involvement of only one of the methyl groups means that there are only two forms of the matrix of second-order interactions. These two matrices were encountered in the consideration of the previous bond.

$$\mathbf{V}_{MM;ll} = \mathbf{V}_{MM;dl} = \mathbf{V}_{CM;l} \tag{27}$$

$$\mathbf{V}_{MM;ld} = \mathbf{V}_{MM;dd} = \mathbf{V}_{CM;d} \tag{28}$$

The statistical weight matrices are

$$\mathbf{U}_{MM;ll} = \begin{bmatrix} 1 & \zeta & \zeta \\ \omega & \zeta & \zeta\omega \\ 1 & \zeta\omega & \zeta\omega \end{bmatrix} \tag{29}$$

$$\mathbf{U}_{MM;dd} = \mathbf{Q}\mathbf{U}_{MM;ll}\mathbf{Q} \tag{30}$$

$$\mathbf{U}_{\mathrm{MM};dl} = \begin{bmatrix} \zeta & 1 & \zeta \\ \zeta\omega & 1 & \zeta\omega \\ \zeta & \omega & \zeta\omega \end{bmatrix} \tag{31}$$

$$\mathbf{U}_{\mathrm{MM};ld} = \mathbf{Q}\mathbf{U}_{\mathrm{MM};dl}\mathbf{Q} \tag{32}$$

Statistical Weight Matrices for the CH–CH$_2$ Bond. This matrix incorporates statistical weights for short-range interactions in the fragment -CH$_2$-CH(CH$_3$)-CH(CH$_3$)—CH$_2$-CH$_2$-, depicted in Fig. 4. It is denoted by $\mathbf{U}_{\mathrm{MC};xy}$, where x and y define the stereochemistry of the attachments of the two methyl groups, in the order that they occur in the fragment. The first-order interactions involve the bold groups in -CH$_2$-**CH(CH$_3$)**-**CH(CH$_3$)**—CH$_2$-CH$_2$-, and the second-order interactions involve the bold groups in -**CH$_2$**-CH(CH$_3$)-CH(CH$_3$)—**CH$_2$**-CH$_2$-. All of the short-range interactions have been encountered previously, as have all of the matrices of first- and second-order interactions.

$$\mathbf{D}_{\mathrm{MC};ll} = \mathbf{D}_{\mathrm{MC};dl} = \mathbf{D}_{\mathrm{CM};d} \tag{33}$$

$$\mathbf{D}_{\mathrm{MC};dd} = \mathbf{D}_{\mathrm{MC};ld} = \mathbf{D}_{\mathrm{CM};l} \tag{34}$$

$$\mathbf{V}_{\mathrm{MC};ll} = \mathbf{V}_{\mathrm{MC};ld} = \mathbf{V}_{\mathrm{CC};l} \tag{35}$$

$$\mathbf{V}_{\mathrm{MC};dd} = \mathbf{V}_{\mathrm{MC};dl} = \mathbf{V}_{\mathrm{CC};d} \tag{36}$$

These matrices combine to give statistical weight matrices distinct from those obtained with the other bonds.

$$\mathbf{U}_{\mathrm{MC};ll} = \begin{bmatrix} 1 & \tau & \omega \\ 1 & \tau\omega & \omega \\ 1 & \tau\omega & 1 \end{bmatrix} \tag{37}$$

$$\mathbf{U}_{\mathrm{MC};dd} = \mathbf{Q}\mathbf{U}_{\mathrm{MC};ll}\mathbf{Q} \tag{38}$$

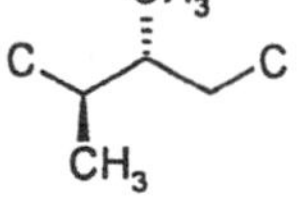

FIG. 4 Fragment considered in the construction of $\mathbf{U}_{\mathrm{MC};ll}$.

$$\mathbf{U}_{MC;ld} = \begin{bmatrix} 1 & 1 & \tau\omega \\ 1 & \omega & \tau\omega \\ 1 & \omega & \tau \end{bmatrix} \tag{39}$$

$$\mathbf{U}_{MC;dl} = \mathbf{Q}\mathbf{U}_{MC;ld}\mathbf{Q} \tag{40}$$

None of the statistical weight matrices for head-to-head, tail-to-tail polypropylene is found also in a three-state RIS model for the conventional head-to-tail polypropylene. This result follows immediately from the fact that none of the fragments in Figs. 1–4 is also a fragment found in head-to-tail polypropylene.

Conformational Partition Function. The conformation partition function for a chain of n bonds (see, e.g., pp. 77–83 of Ref. [4]) is given by Eq. (1) where $\mathbf{U}_1$, $\mathbf{U}_2$, and $\mathbf{U}_n$ adopt special forms.

$$\mathbf{U}_1 = \begin{bmatrix} 1 & 0 & 0 \end{bmatrix} \tag{41}$$

$$\mathbf{U}_2 = \begin{bmatrix} 1 & 1 & 1 \\ 1 & 1 & 1 \\ 1 & 1 & 1 \end{bmatrix} \mathbf{D}_2 \tag{42}$$

$$\mathbf{U}_n = \begin{bmatrix} 1 \\ 1 \\ 1 \end{bmatrix} \tag{43}$$

Any ω that appears in $\mathbf{U}_2$ is replaced by 1, because the chain has been initiated in a manner that eliminates any possibility of second-order interactions at bonds 1 and 2.

For long chains in which all of the pseudoasymmetric centers have the same chirality (isotactic chains), Z is dominated by the terms

$$\left(\mathbf{U}_{CC;l}\mathbf{U}_{CM;l}\mathbf{U}_{MM;ll}\mathbf{U}_{MC;ll}\right)^a \tag{44}$$

or

$$\left(\mathbf{U}_{CC;d}\mathbf{U}_{CM;d}\mathbf{U}_{MM;dd}\mathbf{U}_{MC;dd}\right)^a \tag{45}$$

where a is the number of successive configurational repeat units in the chain. Using Eqs. (17), (23), (30), and (38), along with $\mathbf{Q}^2 = \mathbf{I}_3$,

$$\left(\mathbf{U}_{CC;d}\mathbf{U}_{CM;d}\mathbf{U}_{MM;dd}\mathbf{U}_{MC;dd}\right)^a = \mathbf{Q}\left(\mathbf{U}_{CC;l}\mathbf{U}_{CM;l}\mathbf{U}_{MM;ll}\mathbf{U}_{MC;ll}\right)^a \mathbf{Q} \qquad (46)$$

which demonstrates the equivalence of the Z's for these two chains (they are enantiomers). The two chains with perfectly alternating (*dl* or *ld*) pseudo-asymmetric centers also have the same Z [which, however, is different from the Z in Eq. (46)] because the two structures are identical; they constitute different representations of a mesoform, which is syndiotactic.

B. Behavior of the RIS Model

1. Preliminary Estimates of the Values of the Statistical Weights

The foregoing matrices contain four distinct statistical weights. Three (σ, τ, and ζ) are statistical weights for first-order interactions that appear in the diagonal $\mathbf{D}$ matrices. The $\mathbf{V}$ matrices contain another statistical weight, ω, for a second-order interaction. Two of the statistical weights, σ and ω, are for interactions that are similar to the short-range interactions considered many years ago in RIS models for polyethylene. Model B of Abe et al. [19] estimates the corresponding energies as $E_\sigma = 1.8$–$2.5\,\text{kJ/mol}$ and $E_\omega = 7.1$–$8.0\,\text{kJ/mol}$. The first-order interaction with statistical weight τ occurs also in the head-to-tail variety of polypropylene. The corresponding energy has been estimated as $E_\tau = 3.8 \pm 1.7\,\text{kJ/mol}$ [9]. The statistical weight denoted here by ζ was previously estimated as 0.6, from a CNDO investigation of the conformations of 2,3-dimethylbutane [11].

A provisional set of values of statistical weights is defined at this point in the development of the model. Then the sensitivity of the model to each of the initial estimates can be determined. Our provisional initial set is $\sigma = 0.43$, $\tau = 0.22$, $\zeta = 0.6$, $\omega = 0.034$. These values are Boltzmann factors computed from the corresponding energies at 300K. For the geometry, we adopt a bond length (l) of 1.53 Å, a bond angle (θ) of 109.5° for all C-C-C angles in the backbone, and torsion angles (ϕ) of 180° and ±60° for the trans and gauche$^\pm$ states at each rotatable bond. The geometry enters the calculations via the matrices in Eqs. (6)–(8) [20].

2. Unperturbed Dimensions of Chains with Simple Stereochemical Sequences

The two simplest stereochemical sequences are those in which all pseudoasymmetric centers have the same chirality (either *ll...ll* or

$dd\ldots dd$), or where there is a strict alternation in chirality (either $ld\ldots ld$ or $dl\ldots dl$). The values of C_∞ can be calculated using the program in Appendix C of Mattice and Suter [4]. This short program computes the mean square unperturbed end-to-end distance, $\langle r^2 \rangle_0$, for a repeating unit of m bonds embedded in a chain of n bonds, using a special case of Eq. (VI-66) of Ref. [4], which for present purposes can be written as

$$\langle r^2 \rangle_0 = \frac{1}{Z} G_1 G_2 (G_3 \ldots G_m G_1 G_2)^{n/m-1} G_3 \ldots G_{m-1} G_m \qquad (47)$$

$$'Z = U_1 U_2 (U_3 \ldots U_m U_1 U_2)^{n/m-1} U_3 \ldots U_{m-1} U_m \qquad (48)$$

The internal bold italic G matrices in Eq. (47) are of dimensions 15×15, formulated as described in Eq. (10) (see, e.g., pp. 122–126 and 128–130 of Ref. [4]). The first and last matrices in Eqs. (47) and (48) take special forms of a row and column, respectively [20]. The $\langle r^2 \rangle_0$ are discussed here using the asymptotic limit for the dimensionless characteristic ratio, which is determined by linear extrapolation of C_n vs. $1/n$ (see pp. 19–20 of Ref. [4]).

$$C_\infty \equiv \lim_{n \to \infty} C_n \qquad (49)$$

Table 1 shows that identical results are obtained for a chain and its mirror image, as expected, but different C_∞ can be obtained for pairs of chains in which the stereochemical sequences are not mirror images. The relationship between the two distinct results in Table 1 can be rationalized qualitatively by identification of the preferred conformation of the sequence of four rotatable bonds. Identification can be achieved by inspection of the pattern of the appearance of σ, τ, ζ, and ω in the U matrices, realizing that each statistical weight has a value smaller than 1.

Equations (16), (22), (29), and (37), or, equivalently, Eqs. (17), (23), (30), and (38), show that the preferred conformation is $tttt$ when a single chirality is propagated along the chain. The all-trans conformation has a statistical weight of 1, and all other conformations have a statistical weight of less

TABLE 1 Characteristic Ratios with the Provisional Set of Statistical Weights

Sequence	U's (equation numbers)	C_∞
$ll\ldots ll = dd\ldots dd$	16, 22, 29, 37 (17, 23, 30, 38)	7.94
$ld\ldots ld$	16, 23, 31, 39	7.30
$dl\ldots dl$	17, 22, 32, 40	7.30

than 1. The preferred conformation becomes either ttg^+t or ttg^-t when there is a strict alternation in chirality of the pseudoasymmetric centers. The g conformation is preferred at the $CH_2CH(CH_3)$–$CH(CH_3)CH_2$ bond, due to the pattern of the statistical weights denoted by 1 and ζ in the **D** matrices in Eqs. (24)–(26), and hence in the **U** matrices in Eqs. (29)–(32). Therefore, on the basis of the preferred conformations, one would expect the first entry in Table 1 to have a higher C_∞ than the last two entries.

3. Sensitivity Tests for the Statistical Weights

The sensitivity of the unperturbed dimensions to each of the statistical weights was assessed using chains with the two distinct types of stereochemical sequences presented in Table 1. The values of C_∞ were recalculated by variation of one of the statistical weights, the other three statistical weights being retained at their provisional values. This information can be summarized in a tabular form as

$$\left(\frac{\partial \ln C_\infty}{\partial \ln x}\right)_{x_0} \simeq \frac{C_{\infty, x_0+\Delta x} - C_{\infty, x_0-\Delta x}}{2\Delta x} \frac{x_0}{C_{\infty, x_0}} \tag{50}$$

Here x is the statistical weight being varied, x_0 is its value in the provisional set, and all other parameters are held constant at their values in the provisional set. The values of this partial derivative are summarized in Table 2.

An increase in the value of any of the statistical weights produces a decrease in C_∞ if all pseudoasymmetric centers are either d or l. Changes in ζ affect C_∞ relatively more than comparable changes in any of the other three statistical weights. In contrast, a mixed pattern is seen with the chain in which there is strict alternation in the pseudoasymmetric centers. C_∞ increases with an increase in σ or ζ, or a decrease in τ or ω, with the greatest sensitivity seen upon a change in τ. Table 2 also shows how C_∞ is affected by an increase in all bond angles, or by increasing only that half of the bond

TABLE 2 Sensitivity to the Statistical Weights and Bond Angle

Parameter	Single chirality	Strictly alternating chirality
σ	−0.14	0.028
τ	−0.069	−0.106
ζ	−0.22	0.055
ω	−0.11	−0.067
All θ	3.8	4.6
θ at methylene	0.65	1.3

angles that is centered on a methylene group. Opening up the bond angles produces an increase in C_∞, as expected from the behavior of the simple freely rotating chain.

It is worthwhile to also consider the consequences of minor adjustments in the torsion angles assigned to the rotational isomers. The preferred conformations of n-butane have torsion angles of 180° and $\pm(60° + \triangle\phi)$, where $\triangle\phi$ is positive [19]. Displacement of the gauche states arises because of the weak onset at $\phi = \pm60°$ of the repulsive interaction characteristic of the cis state of n-butane, thereby displacing the dihedral angle for the gauche states slightly toward the trans state. Therefore the gauche states at the CH_2-CH_2 bond in head-to-head, tail-to-tail polypropylene can be adjusted to torsion angles of $\pm(60° + \triangle\phi)$. A displacement of similar origin occurs for the trans and one of the gauche states at the bonds between CH_2 and CH. For example, when all of the pseudoasymmetric centers are l, torsion angles associated with the columns of $U_{CM;l}$ are $180° - \triangle\phi$, $60° + \triangle\phi$, and $-60°$, and those associated with the columns of $U_{MC;ll}$ are $180° + \triangle\phi$, $60°$, and $-60° - \triangle\phi$. An increase in all of the $\triangle\phi$ by 1° increases C_∞ by about 1.5% when all pseudoasymmetric centers are of the same chirality, but the increase is only about 0.3% when the pseudoasymmetric centers alternate between d and l.

C. Comparison with Experiment

Arichi et al. [17] reported an estimate of 4.5 for the characteristic ratio of predominantly head-to-head, tail-to-tail polypropylene in isoamylacetate at 316K, which is the Θ temperature for this polymer–solvent system. This number is significantly smaller than the result of 5.9 (at 311K) reported for atactic head-to-tail polypropylene [21]. The sample studied by Arichi et al. was prepared by reduction of poly(2,3-dimethyl-1,3-butadiene). The parent polymer was composed predominantly (94%) of units arising from 1,4 addition, with the remaining 6% of the units arising from 1,2 addition. Upon reduction, the latter units yield chain atoms that bear both an isopropyl group and a methyl group.

Recently neutron scattering has been used to determine the mean square unperturbed dimensions for the polymer in its melt [18]. The samples were hydrogenated poly(2,3-dimethyl-1,3-butadiene), with deuterium introduced during the reduction. The parent polymers had 3% 1,2-units. The polymer had a characteristic ratio of 6.1–6.4 when studied over the range 27–167°C [18].

A characteristic ratio of 4.5 is lower than either number in Table 1. The values in Table 1 might be too high for an atactic polymer. The strictly alternating chirality imposes preferred conformations that are either $ttg^+t\ldots ttg^+t$ or $ttg^-t\ldots ttg^-t$, depending on whether the sequence is written as $ld\ldots ld$ or $dl\ldots dl$. In both of these preferred sequences, the gauche placement at every fourth bond is always of the same sign. In contrast, a random sequence of l and d, with equal amounts of each, would not perpetuate a gauche placement of a single sign at every fourth bond.

The consequences for the unperturbed dimensions are brought out in Table 3, which contains results for six distinguishable chains with equal numbers of l and d pseudoasymmetric centers, arranged in different repeating sequences. The first entry, taken from Table 1, is for the shortest such repeating sequence, where there is a strict alternation of l and d along the chain. The next two entries have a strict alternation of a pair of l's and a pair of d's. This sequence can be embedded in the chain in two distinct ways, which differ in whether two bonded carbon atoms with methyl substituents have the same or opposite chirality. The fourth entry has a repeating pattern of three l's followed by three d's. The fifth and sixth entries have a homopair (e.g., ll) followed by the opposite homopair (e.g., dd), and then a heteropair (e.g., ld), which can be embedded in the chain in two ways.

The results in Table 3 show that the strictly alternating dl polymer does indeed overestimate the C_∞ expected for a truly atactic polymer, but the size of the overestimate is small. If one averages over the first three entries in Table 3, the result is 6.77. The average is only slightly smaller (6.73) if one considers all repeating sequences of three $-CH_2-CH(CH_3)-CH(CH_3)-CH_2-$ units, assuming equal numbers of d and l pseudoasymmetric centers in each repeat unit, and equal probability that any pseudoasymmetric center is d or l. We infer that the C_∞ considering all possible sequences with equal numbers of d and l pseudoasymmetric centers cannot be smaller than 6, and

TABLE 3 Characteristic Ratios for Six Repeating Sequences, Each of Which Contains the Same Number of l and d Pseudoasymmetric Centers

Repeating sequence of pseudoasymmetric centers[a]	C_∞
(ld)	7.30
$(ld)(dl)$	6.03
$(ll)(dd)$	6.97
$(ll)(ld)(dd)$	6.94
$(ll)(dd)(ld)$	6.74
$(ld)(dl)(dl)$	6.32

[a]Parentheses enclose symbols for two bonded carbon atoms that each bear a methyl group.

is probably in the range 6.0–6.7. The predictions from the model are definitely larger than the experimental estimate of 4.5 [17], but are consistent with the experimental estimate of 6.1–6.4 [18].

1. Plausible Adjustments in Parameters

We now inquire whether reasonable adjustment in any of the parameters in the model reduces C_∞ for the atactic chain from 6.0–6.7 down to 4.5. First consider reasonable adjustments in the statistical weights. Table 2 shows that C_∞ for the chain with alternating chirality is more sensitive to τ than to any of the other statistical weights. An increase in τ will produce a decrease in C_∞. However, while it is conceivable that τ might be nearly as large as σ [22], there are strong physical arguments for assuming that $\tau \leq \sigma$, because two simultaneous repulsive interactions are unlikely to be weaker than two times a single interaction. An increase in τ to $\tau = \sigma$ reduces C_∞ by only about 10%, which is insufficient to explain the difference between the model and experimental result of 4.5. We conclude that reasonable adjustments in the statistical weights cannot account for the difference.

Reasonable adjustment in the bond angles would increase them above their provisional values of 109.5°, which would lead to an increase in C_∞. Similarly, any reasonable change in the torsion angles would assign $\triangle\phi \geq 0°$, which would also increase C_∞. It appears that no reasonable adjustment in the model, either in its energetics (σ, τ, ζ, ω) or in its geometry (θ, $\triangle\phi$), will bring the calculated C_∞ into the range reported in the experiments that find a characteristic ratio of 4.5. No adjustment is necessary in order to rationalize the experiment that reports a characteristic ratio in the range of 6.1–6.4.

The experiments were performed with a predominantly head-to-head, tail-to-tail chain that contained defects, arising from the 1,2 addition of 3 or 6% of the monomer units in the parent poly(2,3-dimethyl-1,3-butadiene). When short branches are present as defects in a polyethylene chain, they produce a decrease in C_∞, by disrupting the preferred conformation consisting of short sequences of trans placements [23]. The defects present in the chains examined in the experiment, illustrated in Fig. 5, may play

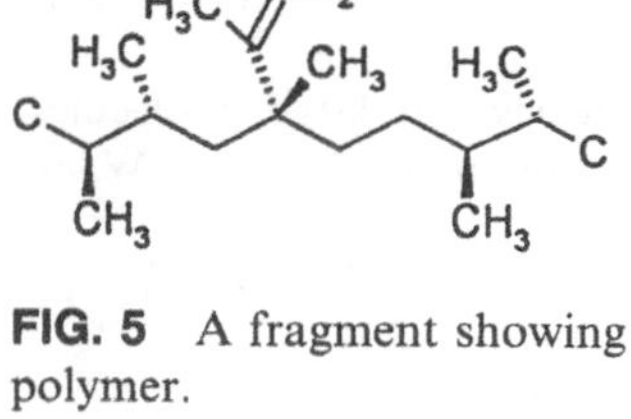

FIG. 5 A fragment showing the defect arising from the 1,2 addition in the parent polymer.

a similar role. However, it does not seem likely that an increase in defects from 3 to 6% could depress the characteristic ratio from 6.1–6.4 to 4.5.

D. Conclusion

The construction and analysis of a RIS model for a homopolymer have been illustrated by consideration of head-to-head, tail-to-tail polypropylene. Statistical weight matrices, incorporating first- and second-order inter-actions, are formulated for all bonds, and for all configurations of the attachments of the methyl groups to the chain. Preliminary estimates of the values of the statistical weights in these matrices could be obtained by reference to prior work, which in turn is based on conformational energy calculations for small hydrocarbons. Along with preliminary estimates of the geometry in the three rotational isomeric states for each rotatable bond, this information permits calculation of C_∞ for chains with simple repeating sequences.

The influence of the individual energetic and structural parameters on C_∞ is determined in the sensitivity tests.

The values of C_∞ specified by the model for an atactic homopolymer of head-to-head, tail-to-tail polypropylene are in excellent agreement with those reported by Graessley et al. [18]. Reasonable adjustments of the parameters in the RIS model cannot reduce C_n into the range reported by Arichi et al. [17].

ACKNOWLEDGMENT

The construction and analysis of the RIS model was supported by the Edison Polymer Innovation Corporation.

REFERENCES

1. Kramers, H.A.; Wannier, G.H. Phys. Rev. **1941**, *60*, 252.
2. Volkenstein, M.V. Dokl. Akad. Nauk SSSR **1951**, *78*, 879.
3. Flory, P.J. *Statistical Mechanics of Chain Molecules*; Wiley-Interscience: New York, 1969; reprinted with the same title by Hanser, München (Munich), 1989.
4. Mattice, W.L.; Suter, U.W. Conformational theory of large molecules. *The Rotational Isomeric State Model in Macromolecular Systems*; Wiley: New York, 1994.
5. Rehahn, M.; Mattice, W.L.; Suter, U.W. Adv. Polym. Sci. **1997**, *131/132*.
6. Abe, Y.; Tonelli, A.E.; Flory, P.J. Macromolecules **1970**, *3*, 294.
7. Biskup, U.; Cantow, H.J. Macromolecules **1972**, *5*, 546.

8. Tonelli, A.E. Macromolecules **1972**, *5*, 563.
9. Suter, U.W.; Pucci, S.; Pino, P.J. Am. Chem. Soc. **1975**, *97*, 1018.
10. Asakura, T.; Ando, I.; Nishioka, A. Makromol. Chem. **1976**, *177*, 523.
11. Asakura, T.; Ando, I.; Nishioka, A. Makromol. Chem. **1976**, *177*, 1493.
12. Alfonso, G.C.; Yan, D.; Zhou, A. Polymer **1993**, *34*, 2830.
13. Flory, P.J. J. Polym. Sci. Polym. Phys. **1973**, *11*, 621.
14. Heatley, F. Polymer **1972**, *13*, 218.
15. Suter, U.W.; Flory, P.J. Macromolecules **1975**, *8*, 765.
16. Boyd, R.H.; Breitling, S.M. Macromolecules **1972**, *5*, 279.
17. Arichi, S.; Pedram, M.Y.; Cowie, J.M.G. Eur. Polym. J. **1979**, *15*, 113.
18. Graessley, W.W.; Krishnamoorti, R.; Reichart, G.C.; Balsara, N.P.; Fetters, L.J.; Lohse, D.J. Macromolecules **1995**, *28*, 1260.
19. Abe, A.; Jernigan, R.L.; Flory, P.J. J. Am. Chem. Soc. **1966**, *88*, 631.
20. Flory, P.J. Macromolecules **1974**, *7*, 381.
21. Zhongde, X.; Mays, J.; Xuexin, C.; Hadjichristidis, N.; Schilling, F.C.; Bair, H.E.; Pearson, D.S.; Fetters, L.J. Macromolecules **1985**, *18*, 2560.
22. Mathur, S.C.; Mattice, W.L. Makromol. Chem. **1988**, *189*, 2893.
23. Mattice, W.L. Macromolecules **1986**, *19*, 2303.

3
Single Chain in Solution

REINHARD HENTSCHKE   Bergische Universität, Wuppertal, Germany

I. PHENOMENOLOGICAL FORCE FIELDS AND POLYMER MODELING

Perhaps the simplest theoretical approach to polymers in solution is molecular dynamics, i.e., the numerical integration of Newton's equations of motion, $m_i \ddot{\vec{r}}_i = -\vec{\nabla}_{\vec{r}_i} U(\vec{r}_1, \ldots, \vec{r}_n)$, where i extends over all n atoms in the simulation box. Common integrators like the Verlet or predictor corrector algorithms [1], yield trajectories consisting of positions and velocities of the interaction centers (usually the nuclei) stored at regular time intervals on the picosecond time scale. Positions and velocities can be tied to the thermodynamic quantities like pressure and temperature via the equipartition theorem. The forces in the above equations of motion may be derived from a potential of the form

$$
U = \sum_{\text{bonds}} f_b(b - b_o)^2 + \sum_{\substack{\text{valence} \\ \text{angles}}} f_\phi(\phi - \phi_o)^2
$$

$$
+ \sum_{\substack{\text{torsion} \\ \text{angles}}} \sum_{k=1}^{p} f_{\vartheta,k}[1 + \cos(m_k \vartheta - \gamma_k)]
$$

$$
+ \sum_{i<j} \left(\frac{A_{ij}}{r_{ij}^{12}} - \frac{B_{ij}}{r_{ij}^6} + \frac{q_i q_j}{r_{ij}} \right)
\tag{1}
$$

The first three terms describe potential energy variations due to bond (b), valence angle (ϕ), and bond torsion angle (ϑ) deformations. The remaining (nonbonding) terms are Lennard-Jones and Coulomb interactions between interaction sites separated by a distance $r_{ij} = |\vec{r}_i - \vec{r}_j|$. Nonbonded interactions usually exclude pairs of sites belonging to the same bond or valence angle. A model of this type is appropriate to study the dynamics of a system with maybe 10,000 interaction centers in a time window of about 10 ns, on current workstation computers. Figure 1 gives an impression of the typical size of such a system. Figure 2 on the other hand shows how slow (global) conformation changes may be, even for a short oligomer, compared to the accessible time window.

Clearly, Eq. (1) is a crude approximation to molecular interactions. It basically constitutes the simplest level on which the chemical structure of a specific polymer can still be recognized. The development of more complex molecular and macromolecular force fields is still ongoing, and the reader is referred to the extensive literature (a good starting point is Ref. [2]).

Fully atomistic molecular dynamics simulations quickly require excessive computer time as systems become large. This is despite the fact that the computational effort for large n, governed by the calculation of the nonbonded forces or interactions, scales as $O(n)$ for short-range interactions and as $O(n \ln n)$ for long-range interactions. This improvement over the naively expected $O(n^2)$ behavior (assuming simple pairwise additive interactions) is achieved by the use of various cell techniques, which are discussed in this book. Actually, a more severe restriction than the limited

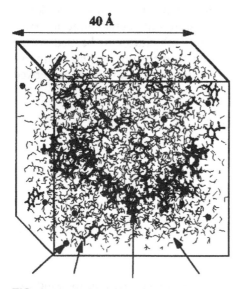

FIG. 1 Snapshot taken during the initial stage of a molecular dynamics simulation of oligo(vinylpyrrolidone) (20mer) in an ionic solution. The box indicates the cubic simulation volume. Periodic boundary conditions are used to extend the system to (quasi-)infinity. The arrows indicate (from left to right) Na^+, $C_6H_5SO_3^-$, Oligomer, and water.

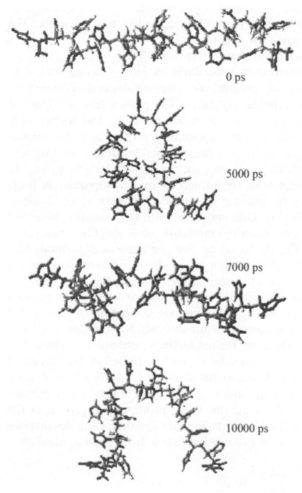

0 ps

5000 ps

7000 ps

10000 ps

FIG. 2 A sequence of conformations obtained for the oligomer in Fig. 1 in pure water at room temperature. The water molecules are omitted.

number of interaction centers is the time step in the integration algorithm, which is governed by the highest frequency in the system. Here multiple-time-step concepts [3–5] as well as the mapping of the real polymer onto a much simpler model with less complicated interactions (coarse graining) are used to speed up the calculations. Coarse graining reduces high frequencies effectively but the reverse procedure, i.e., going from the simplified model back to the specific polymer, is anything but straightforward [6,7]. Again, these techniques as well as sampling of conformation space with sophisticated Monte Carlo techniques, are discussed in this book, and the reader is referred to the pertinent chapters.

Force fields contain numerous parameters (here: f_b, b_o, f_ϕ, ϕ_o, $f_{\vartheta,k}$, m_k, γ_k, A_{ij}, B_{ij}, and q_i). Parameterization procedures employ training sets, e.g., amino acids in the case of proteins or relevant monomer-analogous molecules in the case of technical polymers. The parameters are adjusted using thermodynamic, spectroscopic, or structural data, and increasingly quantum chemical calculations (most importantly in the case of the torsion potential in the valence part of a force field) available for the training set.

Probably the most difficult parameters are the partial charges, q_i, usually located on the nuclei, since they are influenced by a comparatively large number of atoms in their vicinity. Let us consider the case of an oligomer in explicit solution (cf. Fig. 1). One may start with the vacuum values for the q_i determined for the training molecules via an ESP procedure (ElectroStatic Potential fitting) based on fast semiempirical methods like AM1 (Austin Model 1) [8]. Polarization effects due to solvent are included roughly in a mean field sense via a scaling factor multiplying the charges. The scaling factor may be determined by comparison with suitable experimental data (e.g., the density of a solution consisting of the solvent of interest and a monomer-analogous solute as function of temperature and solute concentration [9]). The same scaling factor is subsequently applied to the charges on the oligomer, which also can be determined in vacuum using AM1 (possibly including conformational averaging, which, if it is done correctly, of course requires information not available at this point). A more systematic and maybe in the long run more promising approach is the fluctuating charge method [10,11]. Here the q_i are dynamic variables just like the positions, $\vec{r}_i$. Their equations of motion follow from the Lagrangian

$$L = \frac{1}{2}\sum_m \sum_{i_m} m_{i_m} \dot{\vec{r}}_{i_m}^2 + \frac{1}{2}\sum_m \sum_{i_m} m_q \dot{q}_{i_m}^2 - U(\{r\},\{q\}) \tag{2}$$

The index m labels all molecules in the system, and i_m indicates the ith atom or charge center in the mth molecule. The quantity m_q is a mass parameter, which, for simplicity, is the same for all charges. The potential energy is

$$U(\{r\},\{q\}) = U^{\text{non-Coulomb}}(\{r\}) + E(\{r\},\{q\}) \tag{3}$$

Here $\{r\}$ means all positions, and $\{q\}$ means all partial charges. The first term includes all non-Coulomb contributions [cf. Eq. (1)]. The second term, the Coulomb part of the potential, is approximated in terms of a truncated expansion in q_i, $E(\{r\},\{q\}) = \sum_i (E_{io} + \chi_i^o q_i) + \frac{1}{2}\sum_{i,j} q_i q_j J_{ij}$. The quantities

E_{io} and χ_i^o are single atom parameters, and $J_{ij} = J_{ij}(r_{ij}; \zeta_i, \zeta_j)$ is a Coulomb integral (note: $J_{ij} \underset{r_{ij} \to \infty}{\to} r_{ij}^{-1}$) computed from two charge distributions modeled in terms of single Slater orbitals with the exponents ζ_i and ζ_j. The equations of motion, i.e.,

$$m_{i_m} \ddot{\vec{r}}_{i_m} = -\vec{\nabla}_{\vec{r}_{i_m}} U(\{r\}, \{q\}) \tag{4}$$

and

$$m_q \ddot{q}_{i_m} = -\mu_{i_m} - \lambda_m \tag{5}$$

follow from $\delta \int_{t_1}^{t_2} (L - \sum_m \lambda_m \sum_{i_m} q_{i_m}) \, dt = 0$, where the $\lambda_m = -N_m^{-1} \sum_{i_m} \mu_{i_m}$ are Lagrange parameters, i.e., the conditions $\sum_{i_m} q_{i_m} = 0$ ensure that the total charge on a molecule is zero. The quantity $\mu_{i_m} = \partial E / \partial q_{i_m}$ assumes the role of the chemical potential of the charges, and N_m is the total number of charge sites in the mth molecule.

The additional computational effort, compared to fixed partial charges, is rather minor, i.e., about 15% for simulations of neat water [10,11] (the authors consider the TIP4P and SPC/E water models). Figure 3 shows an application of this technique to oligo(ethyleneoxide) in water [12]. This simulation contained a 7mer immersed in 1000 water molecules at ambient

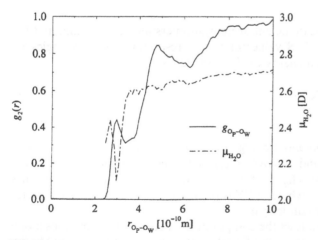

FIG. 3 Simulation of oligo(ethyleneoxide) in water using the fluctuating charge model. Solid line: pair distribution function, $g_2(r_{O_p-O_w})$, where $r_{O_p-O_w}$ denotes the distance between oxygen sites in the oligomer and in water. Broken line: corresponding magnitude of the water dipole moment, μ_{H_2O}.

conditions. The figure exemplifies the variation of the magnitude of the water dipole moment corresponding to the structure in the $O_p - O_w$ pair correlations. Here $r_{O_p - O_w}$ is the separation between oligomer oxygen atoms and water oxygen atoms. Note that the bulk dipole moment of this polarizable water model is 2.7 D. (The calculation included long-range interactions in terms of the particle mesh Ewald method.)

II. SOLVENT SPECIFIC POLYMER CONFORMATIONS IN SOLUTION BASED ON OLIGOMER SIMULATIONS

Probably the most powerful technique for sampling the conformation space of real polymers is the transfer matrix approach in combination with the RIS approximation discussed in the previous chapter. The approach was pioneered by Flory (e.g., [13]), but its potential for generating configurational averages was recognized very early (cf. the reference to E. Montrol in the seminal paper by Kramers and Wannier [14]). In the transfer matrix approach the partition function of a polymer is expressed in terms of a product of matrices, whose elements in the simplest case, e.g., polyethylene, have the form

$$t_{ij}^{(\nu\mu)} \equiv \exp\left[-\frac{1}{k_B T}\left(\frac{1}{2}U_M\left(\vartheta_i^{(\nu)}\right) + U_{MM}\left(\vartheta_i^{(\nu)}, \vartheta_j^{(\mu)}\right) + \frac{1}{2}U_M\left(\vartheta_j^{(\mu)}\right)\right)\right] \quad (6)$$

The lower indices indicate neighboring monomers along the chain, and the upper indices indicate RIS values of the corresponding torsion angles ϑ_i. Here the total potential energy of the polymer is

$$U = \sum_i U_M(\vartheta_i) + \sum_i U_{MM}(\vartheta_i, \vartheta_{i+1}) \quad (7)$$

where $U_M(\vartheta_i)$ is the potential energy of the isolated monomer (assumed to depend on ϑ_i only) and $U_{MM}(\vartheta_i, \vartheta_{i+1})$ is the coupling between adjacent monomers. For long chains the conformation free energy becomes $F_{conf} \approx -k_B TN\ln \lambda^{(max)}$, where $\lambda^{(max)}$ is the largest eigenvalue of the transfer matrix, and N is the chain length.

This approach also allows the computation of correlation functions along a polymer chain, e.g., the conditional probability, $p(\vartheta|\vartheta') = p(\vartheta\vartheta')/p(\vartheta')$, that a torsion angle ϑ' is followed by a torsion angle ϑ. Here $p(\vartheta')$ is the probability of occurrence for the main chain torsion angle ϑ'. Similarly, $p(\vartheta\vartheta')$ is the probability of occurrence for the neighbor pair $\vartheta\vartheta'$. Assuming the general case of m distinct rotational isomeric states $\left\{\vartheta^{(\nu)}\right\}_{\nu=1}^{m}$ this allows the following simple construction procedure for polymer conformations:

1. Compute a random number $z \in [0,1]$.
2. Following the torsion angle $\vartheta^{(k)}$ extend the chain with

   $\vartheta^{(1)}$ if $\qquad\qquad z \leq P(1|k)$

   $\vartheta^{(2)}$ if $\qquad\quad P(1|k) < z \leq P(1|k) + P(2|k)$

   $\vartheta^{(3)}$ if $\quad P(1|k) + P(2|k) < z \leq P(1|k) + P(2|k) + P(3|k)$

   $\vdots \quad \vdots \qquad\qquad\qquad\qquad \vdots$

   $\vartheta^{(m)}$ if $\qquad\qquad P(1|k) + \cdots + P(m-1|k) < z$

3. If $\vartheta^{(l)}$ is selected in (2) then generate the coordinates of the new bond vector, $\vec{b} = b\hat{b}$, via $\hat{b}_{i+1} = -\hat{b}_i \cos\phi + \hat{b}_i \times \hat{b}_{i-1} \sin\vartheta^{(l)} - (\hat{b}_i \times \hat{b}_{i-1}) \times \hat{b}_i \cos\vartheta^{(l)}$. Note that all valence angles ϕ are assumed to be identical, and that the cis conformation defines $\vartheta = 0$. Starting vectors may be arbitrary but not parallel.
4. Continue from (1).

It should be noted that in general the conformation of a polymer backbone is characterized by a periodic sequence of nonequivalent torsion angles, $\vartheta_A, \vartheta_B, \vartheta_C, \ldots, \vartheta_A, \vartheta_B, \ldots$, rather than by merely one (type of) torsion angle, ϑ, as in the present example. Nevertheless, the generalization is straightforward.

A *Mathematica* implementation of the above algorithm for $m=3$ (polyethylene) is:

```
CP[i_,j_]:=PP[[i,j]]/P[[j]];
k=1;
b1={1,0,0};
b2={0,1,0};
R={0,0,0};
f=112/180 Pi;
tab[n_]:=Line[Table[z=Random[ ];
If[z<=CP[1,k],l=1,If[z<=CP[1,k]+CP[2,k],l=2,l=3]];
k=1;
t=(2 l-1) Pi/3;
b3=N[-b2 Cos[f] + CrossProduct[b2,b1] Sin[t] -
  CrossProduct[CrossProduct[b2,b1],b2] Cos[t]];
b3=b3/Sqrt[b3.b3];
R=R+b3;
b1=b2;
b2=b3;
R,{n}]];
Show[Graphics3D[tab[10000]]]
```

CP is the above conditional probability computed from the normal probabilities P and PP. The indices i and j may assume the values 1, 2, and 3, here corresponding to the rotational isomeric states gauche($-$), trans, and gauche($+$).

An example for polyethylene, using the Jorgensen potential [15] to model the potential in Eq. (7), is illustrated in the upper panel of Fig. 4. Here $n(r, \Delta r)$ is the average number of *united atom* carbon pairs, i.e., pairs of effective methylene groups, divided by N, whose separation is between $r - \Delta r/2$ and $r + \Delta r/2$. Here N is the total number of methylene groups in the chain. For small r the distribution is discrete. The first pronounced peak

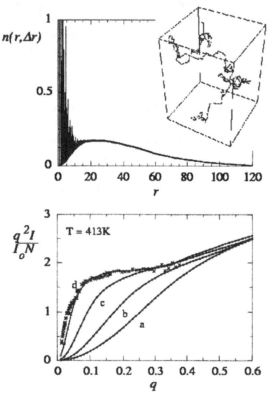

FIG. 4 (Top) Monomer–monomer distribution, $n(r, \Delta r)$, here truncated at 1, averaged over 100 conformations generated for $N = 1000$ at $T = 140°C$. The inset shows a single conformation constructed using the *Mathematica* program in the text. (Bottom) Corresponding Kratky plot of the scattering intensity, $q^2 I/(I_o N)$, vs. q for different chain lengths, i.e., $N = 20$ (a), 50 (b), 200 (c), 1000 (d). The crosses are experimental data taken from [34].

occurs at $r = 1$ corresponding to the bond length in the units used here [note: $n(1, \Delta r) = 2$]. As r increases the distribution becomes continuous. An important quantity based on $n(r, \Delta r)$ is the scattering intensity

$$\frac{I}{I_o} = N\left(1 + \sum_{i=1}^{i_{max}} \frac{\sin[q\Delta ri]}{q\Delta ri} n(\Delta ri, \Delta r)\right) \tag{8}$$

allowing the direct comparison with X-ray and neutron scattering experiments. Here I_o is a constant independent of the polymer structure, $r = \Delta r\,i$, and q is the magnitude of the scattering vector. For simplicity we have set the form factors for each scattering site equal to one. The scattering intensity corresponding to the above $n(r, \Delta r)$ is shown in the lower panel of Fig. 4. At large q all curves come together, because the relevant lengths ($\approx 2\pi q^{-1}$) are small compared to any of the chain lengths considered. The effect of the chain length is most pronounced for q values around 0.15, where a plateau develops as N is increased. It is worth emphasizing that in general scattering curves are more complex due to the more complex chemical architecture of most polymers in comparison to polyethylene [16]. With some experience, however, it is possible to relate many features of the scattering curve to underlying conformational features, e.g., helical conformations. In the present very simple example there are two easily distinguishable q-regimes (for large N), i.e., the linear increase beyond $q \approx 0.3$, which reflects the simple chemical structure of our model chain, and the initial increase with a subsequent development of a plateau below $q \approx 0.3$, which reflects the statistical chain behavior. In this q-regime the scattering intensity of a simple Gaussian chain is a good approximation to the shown scattering intensity (for large N). The Gaussian chain in particular yields the plateau $q^2 I/(I_o N) \rightarrow 12/C_N$ for large q (referring to the continuum limit, i.e., there is no atomic structure no matter how large q is). The quantity

$$C_N = \frac{\langle R_N^2 \rangle}{Nb^2} \tag{9}$$

is the characteristic ratio, and R_N is the end-to-end distance of the polymer consisting of N monomers. Note that R_N is a useful quantity to monitor during an oligomer simulation if one wants to decide whether configurational equilibration is possible or not. C_N, or rather its asymptotic value for $N \rightarrow \infty$, C_∞, is used commonly to characterize the shape or unperturbed dimension of polymers in different solvents in terms of a single number. Here the 1000mers come close to yielding a plateau at $C_\infty \approx C_{1000} \approx 7$. One may also obtain this value by direct evaluation of Eq. (9) using the above construction procedure to generate a large number of single chain conformations. The above value actually corresponds to polyethylene in a melt, an example which we have chosen for its simplicity. For most dilute polymer solutions one obtains larger values usually between 10 and 20. Notice that C_∞ obtained in this fashion describes the shape of a polymer on an intermediate length scale, on which self-avoidance has no significant effect yet. Because the above model includes short-range interactions only, a chain may intersect itself. Polymers modeled with this method eventually obey $\langle R_N^2 \rangle \propto N^{2\nu}$ with $\nu = 1/2$.

Powerful as the transfer matrix approach is, it suffers from certain deficiencies. Including coupling beyond nearest neighbors is difficult. Accounting for solvent effects also is difficult. In a crude approximation solvent effects may be included via an effective dielectric constant in a Coulomb term integrated into the coupling in Eq. (7) [17], which itself could be improved to include local solvent–polymer interactions in terms of effective interactions (e.g., [18]). On the other hand, it is of course possible to extract the probabilities $p(\vartheta)$ and $p(\vartheta\vartheta')$ directly from oligomer simulations in explicit solvent, and feed them into a construction procedure like the one outlined above [19]. Applications of this general idea to poly(isobutene) in vacuum (!) or to poly(vinylpyrrolidone) in water are discussed in [20] and [9]. Thus, other than in the usual transfer matrix calculations of these probabilities, molecular solvent effects are included on the length scale of the oligomer. Some care must be taken to diminish end effects, however. Moreover it is possible to include couplings beyond nearest neighbors, i.e., next-nearest neighbor effects may be introduced analogously via conditional probabilities of the type $p(\vartheta|\vartheta'\vartheta'') = p(\vartheta\vartheta'\vartheta'')/p(\vartheta'\vartheta'')$. An example can be found again in [9].

Before leaving the subject we briefly mention one common approximate calculation of C_N. The central assumption is the uncoupling of adjacent monomers along the chain, i.e., $U_{MM} = 0$, which allows us to derive the expression

$$C_N = \left((\mathbf{I} + \langle \mathbf{T} \rangle) \cdot (\mathbf{I} - \langle \mathbf{T} \rangle)^{-1} - \frac{2}{N} \langle \mathbf{T} \rangle \cdot (\mathbf{I} - \langle \mathbf{T} \rangle^N) \cdot \left[(\mathbf{I} - \langle \mathbf{T} \rangle)^{-1} \right]^2 \right)_{1,1} \quad (10)$$

which is discussed in many polymer textbooks. Here $\mathbf{I}$ is a unit matrix, and $\mathbf{T}$, in cases where the polymer backbone contains one type of torsion angle only (e.g., polyethylene), is given by

$$\mathbf{T} = \begin{pmatrix} -\cos\phi & \sin\phi & 0 \\ -\cos\vartheta\,\sin\phi & -\cos\vartheta\,\cos\phi & -\sin\vartheta \\ -\sin\vartheta\,\sin\phi & -\sin\vartheta\,\cos\phi & \cos\vartheta \end{pmatrix} \quad (11)$$

using the above convention. The index 1,1 indicates that C_N is the 1,1-element of the matrix. More complex polymer architectures may be accommodated by dividing the polymer into repeating sequences of effective bonds (chemical units), again assumed to be independent, where **T** is the product of the respective **T**-matrices for each effective bond. The conformation averages (e.g., $\langle\cos\phi\rangle$ or $\langle\cos\vartheta \sin\phi\rangle$) again may be computed directly from the corresponding oligomer simulations. Despite its crude approximations this method appears to yield reasonable first guesses for C_N [21].

Frequently the molecular structure and dynamics of the polymer–solvent interface also are of interest. Structural information may be derived via pair correlation functions of the general type considered in Fig. 3. For instance, $g_2(r)$ may be calculated via $g_2(r) = (4\pi r^2 \Delta r \rho)^{-1} n(r, \Delta r)$, where $n(r, \Delta r)$ is the average number of solvent atoms of a certain type within a spherical shell of radius r and thickness Δr centered on a polymer atom, and ρ is the density of the same solvent atoms in the bulk solvent. A dynamic quantity of interest is the average residence time of a solvent atom or molecule in a certain distance and/or orientation relative to a polymer atom or atom group, e.g., hydrogen bonds between certain polar groups in a polymer and water. Here $n(t, t_e; r, \Delta r)$ is the average number of solvent atoms or molecules again within a spherical shell (omitting the orientation dependence) at time t, under the condition that the same solvent molecules were present in the shell at $t = 0$. The quantity t_e is an excursion time allowing a solvent molecule to leave the shell and return within the excursion time without being discarded. The result may be fitted via $n(t, t_e; r, \Delta r) = \sum_{\nu=1}^{\nu_{max}} n_\nu e^{-t/\tau_\nu}$, which is sensible for $\nu_{max} \leq 2$ (cf. Ref. [9]). For large ν_{max}, i.e., a broad distribution of residence times, τ_ν, the dynamics assumes glasslike behavior.

There are important systems, where one is interested in the intrinsic stiffness of polymers, which in solution possess a regular superstructure—most commonly helical conformations. Here the above approaches are less useful, and a method called the segment method, is more applicable. A periodic section of the helix of interest is simulated in explicit solvent using molecular dynamics. The top and bottom of the segment are connected via suitable periodic boundary conditions. These are different from the usual periodic boundary conditions, because they include a covalent bond of the backbone. The simulation box should be large enough (a) to ensure that the solvent attains its bulk structure at large perpendicular distances from the helix, and (b) that the important undulatory wavelengths along the helix backbone should be included. Provided the simulation run is long enough one may extract sufficient independent segment (backbone) conformations to "Build up" realistic and solvent specific helix conformations of a desired molecular weight. "Build up" means that an existing piece of the helix may be extended by an extracted segment provided that a certain smoothness condition is fulfilled. The smoothness condition may be that the last torsion angle in the helix matched the first corresponding torsion angle in the segment fulfilling a certain accuracy condition. More complex conditions involving additional constraints can be employed also of course. Ideally there should be no correlation between k, where k is the kth segment conformation extracted from the simulation, and l, where l is the position of the segment along the constructed helix. Figure 5 shows high molecular weight fragments of poly(γ-benzyl-L-glutamate) (PBLG), a helical polypeptide, which were constructed in this fashion from the trajectory of a 27 Å-helix segment immersed in about 1000 dimethylformamide (DMF) molecules [22].

The contour flexibility of molecular helices, such as in the case of PBLG, is best described as persistent flexibility or homogeneous bend-elasticity. The measure of this type of flexibility is the persistence length, P, defined via

$$\langle \vec{u}(0) \cdot \vec{u}(s) \rangle = \exp[-s/P] \tag{12}$$

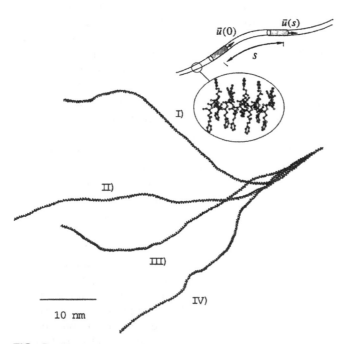

FIG. 5 Four PBLG backbone conformations (I–IV) constructed via the above segment method. The inset illustrates the transformation of the explicit local chemical structure into a persistent flexible rod.

where the $\vec{u}$ are unit vectors tangential to the backbone contour, and s is their separation along the contour [23] (cf. the inset in Fig. 5). Note that P in certain cases may be related to C_N [24]. Via direct application of Eq. (12) to contours like those shown in Fig. 5 one obtains $P \approx 1000\,\text{Å}$ for PBLG in DMF at $T = 313\text{K}$, in good accord with the experiments [22].

III. POLYMER CONFORMATIONS IN SOLUTION VIA DIRECT SIMULATION

Langevin dynamics is a method to simulate molecules in contact with a heat bath or solvent without considering the explicit structure of the solvent. An important application of this approach is conformation sampling for macromolecules in solution, because the expensive integration of the solvent trajectories is omitted—as well as solvent specific effects of course. Here the equations of motion are given by

$$\ddot{\vec{r}}_i = m_i^{-1}(-\vec{\nabla}_i U_{eff} + \vec{Z}_i) - \zeta_i \dot{\vec{r}}_i \tag{13}$$

U_{eff} is a potential of mean force of the solvent molecules including the explicit intramolecular interactions of the polymer. $\vec{Z}_i$ is a stochastic force simulating collisions of polymer site i with the solvent molecules. It is assumed that $\vec{Z}_i$ is uncorrelated with the positions and the velocity of the sites i. Moreover $\vec{Z}_i$ has a Gaussian distribution centered on zero with variance $\langle \vec{Z}_i(t) \cdot \vec{Z}_j(t') \rangle = 6 m_i \zeta_i k_B T_B \delta(t - t') \delta_{ij}$, where T_B is the temperature of the solvent, and ζ_i is a friction parameter. In practice the components α of $\vec{Z}_i$ are extracted from a Gaussian distribution with $\langle Z_{i,\alpha}(t) Z_{i,\alpha}(t') \rangle = 2 m_i \zeta_i k_B T_B / \Delta t$, where Δt is the integration time step. A simple leapfrog Verlet implementation of the equations of motion is the following

$$\vec{\upsilon}_i\left(t + \frac{1}{2}\Delta t\right) = \vec{\upsilon}_i\left(t - \frac{1}{2}\Delta t\right)(1 - \zeta_i \Delta t) + \frac{-\vec{\nabla}_i U_{eff} + \vec{Z}_i(t)}{m_i}\Delta t$$
$$+ O(\Delta t^2) \tag{14}$$
$$\vec{r}_i(t + \Delta t) = \vec{r}_i(t) + \vec{\upsilon}_i\left(t + \frac{1}{2}\Delta t\right)\Delta t + O(\Delta t^3)$$

A detailed analysis of different algorithms in the context of Langevin dynamics may be found in [25] (see also [26]).

A nice application of Eq. (14), which compares simulations for poly(ethyleneoxide) in the explicit solvents water and benzene with Langevin dynamics simulations, can be found in [27,28]. The authors use a modified Stokes relation, $\zeta_i = \lambda 4\pi R_{i,eff}\eta/m_i$, where λ is a parameter

$(0.01 < \lambda < 0.1)$ related to the transition rates between the rotational isomeric states, $R_{i,eff}$ is an effective radius corresponding to the solvent accessible surface of the atom or atom type, and η is the experimental viscosity coefficient of the solvent. It is probably fair to say that Langevin dynamics works for unspecific solvent interactions, whereas specific solvent interactions, e.g., hydrogen bonding, require explicit inclusion of molecular solvent.

Langevin dynamics in its above form is useful for sampling the conformations of a polymer, but it does not yield the correct dynamics. To see this we consider a simple polymer chain consisting of N monomers with masses m and total mass $M = Nm$. Using $\zeta_i = \zeta$ for simplicity the equation of motion of the polymer center of mass is $\vec{r}_{cm} = M^{-1} \sum_{i=1}^{N} \vec{Z}_i - \zeta \vec{r}_{cm}$ or $\vec{v}_{cm} = M^{-1} \sum_{i=1}^{N} \vec{Z}_i - \zeta \vec{v}_{cm}$. Note that the net force due to the gradient term in Eq. (13) vanishes. The analytic solution is $\vec{v}_{cm}(t) = M^{-1} \exp[-\zeta t] \int_o^t dt' \sum_{i=1}^{N} \vec{Z}_i(t') \exp[\zeta t']$ with $\vec{v}_{cm}(0) = 0$. Subsequent partial integration yields $\vec{r}_{cm}(t)$, and using the above expression for $\langle \vec{Z}_i(t) \cdot \vec{Z}_j(t') \rangle$ we obtain $\langle \vec{r}_{cm}(t)^2 \rangle = 6D_{cm}t$ for large t. The diffusion coefficient, $D_{cm} = k_B T (M\zeta)^{-1}$, is proportional to M^{-1}. Experimentally, however, one observes exponents in the range from 0.5 to 0.6 instead of 1 [29]. The reason for this discrepancy is the neglect of hydrodynamic interactions. The latter may be included via the Oseen tensor [30,31], leading to a modified dynamics [32]. For a discussion of hydrodynamic effects in the context of single chain dynamics see for instance [33].

REFERENCES

1. Allen, M.P.; Tildesley, D.J. *Computer Simulation of Liquids*; Oxford: Clarendon Press, 1990.
2. Lipkowitz, K.B.; Boyd, D.B. *Reviews in Computational Chemistry*; Weinheim: Wiley/VCH, 1990.
3. Street, W.B.; Tildesley, D.J.; Saville, G. Mol. Phys. **1978**, *35*, 639–648.
4. Tuckerman, M.E.; Berne, B.J. J. Chem. Phys. **1991**, *95*, 8362–8364.
5. Scully, J.L.; Hermans, J. Mol. Simul. **1993**, *11*, 67–77.
6. Tschöp, W.; Kremer, K.; Hahn, O.; Batoulis, J.; Bürger, T. Acta Polymerica **1998**, *49*, 75–79.
7. Tschöp, W.; Kremer, K.; Batoulis, J.; Bürger, T.; Hahn, O. Acta Polymerica **1998**, *49*, 61–74.
8. Leach, A.R. *Molecular Modeling*; Harlow: Addison Wesley Longman Limited, 1996.
9. Flebbe, T.; Hentschke, R.; Hädicke, E.; Schade, C. Macromol Theory Simul **1998**, *7*, 567–577.
10. Rick, S.W.; Stuart, S.J.; Berne, B.J. J. Chem. Phys. **1994**, *101*, 6141–6156.
11. Rick, S.W.; Berne, B.J. J. Am. Chem. Soc. **1996**, *118*, 672–679.

12. Stöckelmann, E. Molekulardynamische Simulationen an ionischen Grenzflächen. Ph.D. dissertation, Universität Mainz, Mainz, 1999.
13. Flory, P.J. Macromolecules **1974**, *7*, 381–392.
14. Kramers, H.A.; Wannier, G.H. Phys. Rev. **1941**, *60*, 252–262.
15. Jorgensen, W.L.; Madura, J.D.; Swenson, C.J. J. Am. Chem. Soc. **1984**, *106*, 6638–6646.
16. Kirste, R.G.; Oberthür, R.C. In *X-Ray Diffraction*; Glatter, O., Kratky, O., Eds.; Academic Press: New York, 1982.
17. Tarazona, M.P.; Saiz, E.; Gargallo, L.; Radic, D. Makromol. Chem. Theory Simul. **1993**, *2*, 697–710.
18. Chandler, D.; Pratt, L.R. J. Chem. Phys. **1976**, *65*, 2925–2940.
19. Mattice, W.L.; Suter, U.W. *Conformational Theory of Large Molecules*; John Wiley & Sons: New York, 1994.
20. Vacatello, M.; Yoon, D.Y. Macromolecules **1992**, *25*, 2502–2508.
21. Jung, B. Simulation der Kettenkonformation von Polymeren mit Hilfe der Konzepte der Molekulardynamik-Rechnungen. Ph.D. dissertation, Universität Mainz, Mainz, 1989.
22. Helfrich, J.; Hentschke, R.; Apel, U.M. Macromolecules **1994**, *27*, 472–482.
23. Landau, L.D.; Lifschitz, E.M. *Statistische Physik—Teil 1*; Akademie-Verlag: Berlin, 1979.
24. Cantor, C.R.; Schimmel, P.R. *Biophysical Chemistry. Part III: The Behavior of Biological Macromolecules*; W.H. Freeman and Company: New York, 1980.
25. Pastor, R.W.; Brooks, B.R.; Szabo, A. Mol. Phys. **1988**, *65*, 1409–1419.
26. Hünenberger, P.H.; vanGunsteren, W.F. In *Computer Simulation of Biomolecular Systems—Theoretical and Experimental Applications*; vanGunsteren, W.F., Weiner, P.K., Wilkinson, A.J., Eds.; Kluwer Academic Publishers: Dordrecht, 1997.
27. Depner, M. Computersimulation von Modellen einzelner, realistischer Polymerketten. Ph.D. dissertation, Universität Mainz, Mainz, 1991.
28. Depner, M.; Schürmann, B.; Auriemma, F. Mol. Phys. **1991**, *74*, 715–733.
29. Strobl, G. *The Physics of Polymers*; Heidelberg: Springer, 1997.
30. deGennes, P-G. *Scaling Concepts in Polymer Physics*. Ithaca: Cornell University Press, 1979.
31. Doi, M.; Edwards, S.F. *The Theory of Polymer Dynamics*; Oxford: Clarendon Press, 1986.
32. Ermak, D.L.; McCammon, J.A. J. Chem. Phys. **1978**, *69*, 1352–1360.
33. Dünweg, B.; Kremer, K. J. Chem. Phys. **1993**, *99*, 6983–6997.
34. Lieser, G.; Fischer, E.W.; Ibel, K. J. Polym. Sci., Polym. Lett. **1975**, *13*, 39–43.

4

Polymer Models on the Lattice

K. BINDER and M. MÜLLER Johannes-Gutenberg-Universität Mainz, Mainz, Germany

J. BASCHNAGEL Institut Charles Sadron, Strasbourg, France

I. INTRODUCTION

Simple random walks (RWs) and self-avoiding random walks (SAWs) on lattices have been used as models for the study of statistical properties of long flexible polymer chains, since shortly after the important sampling Monte Carlo method was devised [1]. Figure 1 explains the meaning of RWs, of SAWs, and of a further variant, the nonreversal random walk (NRRW), by giving examples of such walks on the square lattice [2]. The idea is that each site of the lattice which is occupied by the walk is interpreted as an (effective) monomer, and the lattice constant (of length a) connecting two subsequent steps of the random walk may be taken as an effective segment connecting two effective monomers. Thus, the lattice parameter a would correspond physically to the Kuhnian step length b_K. For a RW, each further step is completely independent of the previous step and, hence, the mean square end-to-end distance $\langle R^2 \rangle$ after N steps simply is

$$\langle R^2 \rangle = a^2 N \tag{1}$$

and the number of configurations (i.e., the partition function of the chain Z_N) is (z is the coordination number of the lattice)

$$Z_N = z^N \tag{2}$$

Although the RW clearly is an extremely unrealistic model of a polymer chain, it is a useful starting point for analytic calculations since any desired property can be calculated exactly. For example, for $N \gg 1$ the distribution

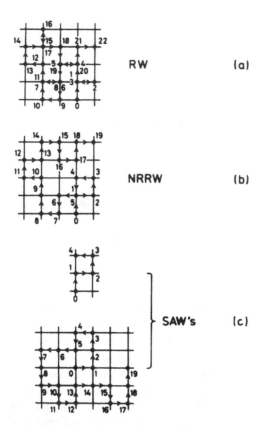

FIG. 1 (a) Construction of a 22-step random walk (RW) on the square lattice. Lattice sites are labeled in the order in which they are visited, starting out from the origin (0). Each step consists in adding at random an elementary lattice vector [$(\pm 1, 0)a$, $(0, \pm 1)a$, where a is the lattice spacing] to the walk, as denoted by the arrows. (b) Same as (a) but for a non-reversal random walk (NRRW), where immediate reversals are forbidden. (c) Two examples of self-avoiding walks (SAWs), where "visiting" any lattice site more than once is not allowed: trials where this happens in a simple random sampling construction of the walk are discarded. From Kremer and Binder [2].

of the end-to-end vector is Gaussian, as it is for the freely-jointed chain (see Chapter 1), i.e., (in $d = 3$ dimensions),

$$P(\mathbf{R}) = \left(\frac{3}{2\pi\langle\mathbf{R}^2\rangle}\right)^{3/2} \exp\left(-\frac{3}{2}\frac{\mathbf{R}^2}{\langle\mathbf{R}^2\rangle}\right) \tag{3}$$

which implies that the relative mean square fluctuation of R^2 is a constant,

$$\Delta_R^2 \equiv \frac{\langle (R^2)^2 \rangle - \langle R^2 \rangle^2}{\langle R^2 \rangle^2} = \frac{2}{3} \tag{4}$$

Equation (4) shows that one encounters a serious problem in computer simulations of polymers, which is not present for many quantities in simple systems (fluids of small molecules, Ising spin models, etc.; see Binder and Heermann [3]), namely the "lack of selfaveraging" [4]. By "selfaveraging" we mean that the relative mean square fluctuation of a property decreases to zero when the number of degrees of freedom (i.e., the number of steps N here) increases towards infinity. This property holds, e.g., for the internal energy E in a simulation of a fluid consisting of N molecules or an Ising magnet with N spins; the average energy is calculated as the average of the Hamiltonian, $E = \langle \mathcal{H} \rangle$, $\Delta_E^2 \equiv (\langle \mathcal{H}^2 \rangle - \langle \mathcal{H} \rangle^2)/\langle \mathcal{H} \rangle^2 \propto 1/N \to 0$ as $N \to \infty$, at least for thermodynamic states away from phase transitions. This selfaveraging does not hold for the mean square end-to-end distance of a single polymer chain, as Eq. (4) shows for the RW (actually this is true for more complicated polymer models as well, only the constant Δ_R may differ from $\sqrt{2/3}$, e.g., $\Delta_R \approx 0.70$ for the SAW in $d = 3$ [3]).

This lack of selfaveraging is one of the reasons why one is forced to use such extremely simplified models, as shown in Fig. 1, under certain circumstances. If it is necessary for the physical property under study to work with large chain lengths N or if a good accuracy is to be reached in the estimation of $\langle R^2 \rangle$ (and/or the gyration radius R_g, single-chain scattering function $S(\mathbf{k})$ at wavevectors $\mathbf{k}$ of the order $kR_g \approx 1$, etc.), a huge number of (statistically independent!) configurations of single chains need to be generated in the simulation. This is because the relative error of R^2 simply is $\Delta_R/$(number of generated chains)$^{1/2}$. Correspondingly, high precision estimates of R^2 for very long chains exist only for simple lattice models (see Sokal [5], also for a tutorial overview of programming techniques and a comprehensive comparison of different algorithms).

Another motivation why lattice models are useful for polymer simulation is the fact that analytical theories have often been based on lattice models and the resulting concepts are well defined in the lattice context. For instance, the Flory–Huggins theory of polymer mixtures [6] constructs the thermodynamic excess energy and the entropy of mixing precisely on the basis of a (many chain-) lattice model of the type of Fig. 1. Experimentally widely used quantities like the Flory–Huggins parameter χ have in the lattice context a well-defined meaning, and thus a critical test of this theory via simulations could be performed rather straightforwardly by Monte Carlo simulation of a corresponding lattice model [7]. Similarly, the

Gibbs–Di Marzio entropy theory of the glass transition [8,9], also based on a lattice description, has been critically examined by a corresponding simulation recently [10]. The key point is that not only has the "configurational entropy" a well-defined meaning for a lattice model, but also algorithms can be devised to accurately sample the entropy S of many-chain systems [10], while in general the entropy is not a straightforward output of computer simulations [3]. In addition, the simulation can then extract precisely the quantities used in the theoretical formulation from the simulated model as well, and hence a stringent test of the theory with no adjustable parameters whatsoever becomes possible.

A third motivation is that with present day computers even for medium chain length N of the polymers it is very difficult to study collective long-wavelength phenomena (associated with phase transitions, phase coexistence, etc.), since huge linear dimensions L of the simulation box and a huge number N_m of (effective) monomers in the simulation are required. For example, for a finite size scaling [11] study of the nematic–isotropic phase transition in a model for a melt of semiflexible polymers a lattice with $L = 130$ was used, and with the bond fluctuation model [12,13] at volume fraction of about $\phi \approx 1/2$ occupied sites this choice allowed for 6865 chains of length $N = 20$, i.e., 137,300 effective monomers, to be simulated [14]. Similarly, for a study of interfaces between unmixed phases of polymer blends a lattice model with $512 \times 512 \times 64$ sites was used, i.e., more than 16 million sites, and more than a million effective monomers (32,768 chains of length $N = 32$) [15]. Dealing with such large systems for more realistic off-lattice models would be still very difficult.

Finally, we emphasize that lattice models are a very useful testing ground for new algorithms that allow more efficient simulations of polymer configurations under suitable circumstances. Many of the algorithms that now are in standard use also for off-lattice models of polymers—such as the "slithering snake algorithm" [16,17], the "pivot algorithm" [18,19], the "configurational bias algorithm" [20] (see also Chapter 7 by T.S. Jain and J.J. de Pablo), and the "chain breaking algorithm" [21,22]—were first invented and validated for lattice models.

Why is there a need for these many different algorithms at all? Of course, there would be no need for these algorithms if one wanted to simulate only the simple random walk of Fig. 1a. But in fact, this is not a useful model for polymer chains in most cases, since it allows arbitrarily many effective monomers to sit on top of each other, ignoring excluded volume interactions completely. A slightly better variant of the RW is the NRRW, Fig. 1b, where immediate reversals are forbidden, but it is possible that loops are formed where the NRRW crosses itself. The NRRW is still straightforward to program—at each step there are $z - 1$ choices to proceed, from which one selects a choice at random. The partition function is hence simply $Z_N^{NRRW} = (z-1)^N$, and also $\langle R^2 \rangle$ can be worked out exactly. The problem is reminiscent of the RIS model of a single chain [23], and Chapter 2 by E.D. Akten, W.L. Mattice, and U.W. Suter, where three choices occur at each step, the long-range excluded volume along the backbone of a chain also being ignored. One believes that the Gaussian statistics, Eq. (3), which holds for both the RW and the NRRW, is true both in dense polymer melts, where the excluded volume interaction between monomers of the same chain is effectively screened by monomers of other chains [24] and in dilute solutions at the so-called Theta temperature θ, where the repulsive excluded volume interaction is effectively compensated by an attractive interaction between monomers. However, this does not mean that effective monomers can sit on top of each other at any point of the lattice to faithfully model these situations. In the cases of dense melts or theta solutions, Eq. (1) is not directly valid, but rather involves a nontrivial constant C_∞,

$$\langle R^2 \rangle = C_\infty a^2 N, \qquad N \to \infty \tag{5}$$

Therefore, the construction of appropriate equilibrium configurations of polymer chains in a lattice model is always a problem. For a dilute solution under good solvent conditions the excluded volume interaction between all monomers of the chain can no longer be ignored. It leads to a swelling of the chain with respect to its size in the melt or in theta solution. Thus, Eq. (5) does not hold, but has to be replaced by

$$\langle R^2 \rangle = C'_\infty a^2 N^{2\nu}, \qquad N \to \infty \tag{6}$$

where C_∞ is another constant, and the exponent ν differs from $1/2$, $\nu \simeq 0.59$ in $d = 3$ dimensions and $\nu = 3/4$ in $d = 2$ dimensions [24,5]. The behavior of Eq. (6) is reproduced by the SAW model (Fig. 1c), and one can also predict the corresponding chain partition function,

$$Z_N \propto N^{\gamma-1} z_{eff}^N, \qquad N \to \infty \tag{7}$$

where $z_{eff} < z - 1$ is a kind of "effective coordination number" that depends on the type of lattice, and the exponent γ takes the values $\gamma = 43/32$ in $d = 2$ dimensions [5] and $\gamma \simeq 1.16$ in $d = 3$ dimensions [25].

Equation (7) is the reason why the sampling of SAWs is so difficult. Note that all configurations of chains with N steps on the lattice that obey the excluded volume condition should have equal *a priori* probability to occur in the generated sample. A straightforward way to realize this is "simple random sampling": One carries out a construction of a RW, as in Fig. 1a, or better of a NRRW (see Fig. 2 for pseudo-codes), and whenever the walk intersects, thus violating the SAW constraint, the resulting configuration has to be discarded, and a new trial configuration is generated (see Fig. 3 for pseudo-code). Obviously, for large N the fraction of trials that is successful becomes very small, since it is just given by the ratio of the respective partition functions,

$$\text{fraction of successful constructions} = \frac{Z_N^{SAW}}{Z_N^{NRRW}} \propto N^{\gamma-1} \left[\frac{z_{eff}}{(z-1)}\right]^N \quad (8)$$

Since $z_{eff} < z - 1$ (e.g., $z_{eff} \approx 2.6385$ for the square lattice, $z_{eff} \approx 4.6835$ for the simple cubic lattice, see Kremer and Binder [2] for various other lattices), the fraction of successful constructions decreases exponentially with increasing N, namely as $\exp(-\lambda N)$ with $\lambda = -\ln[z_{eff}/(z-1)] \approx 0.1284$ (square lattice) and ≈ 0.08539 (simple cubic lattice). This problem that the success rate becomes exponentially small for large N is called the "attrition problem," λ is the so-called "attrition constant."

Of course, the problem gets even worse when considering many chains on a lattice: for any nonzero fraction ϕ of occupied lattice sites, the success rate of this simple sampling construction of self- and mutually avoiding walks will decrease exponentially with increase in the volume of the lattice. As a consequence, other methods for generating configurations of lattice models of polymers are needed. We discuss some of the algorithms that have been proposed in the following sections.

II. STATIC METHODS

A very interesting approach to overcoming the attrition problem was already proposed by Rosenbluth and Rosenbluth in 1955 [26] ("inversely restricted sampling"). The idea is to avoid failure of the simple sampling construction of the SAW by not choosing at random blindly out of $z - 1$ choices at each step, but by choosing the step in a biased way only out of those choices that avoid failure of the construction at this step. Consider a SAW of i steps on a lattice with coordination number z. To add the $(i + 1)$th step one first checks which of the $z_0 = z - 1$ sites are actually empty. If k_i (with $0 < k_i \le z_0$) sites are empty, one takes one of those with equal probability $1/k_i$ to continue the simple sampling construction. Only for $k_i = 0$ is the walk terminated and one has to start from the beginning. The probability of each N-step walk then is $P_N(\{\mathbf{r}_i\}) = \prod_{i=1}^{N}(1/k_i)$, $\mathbf{r}_i$ being the site of the ith monomer. One immediately sees that dense configurations of SAWs are more probable than less dense ones. To generate a sample of

```
Program RW                          Simple sampling of a random walk on a square lattice
begin
  for m = 1 to M do                 Repeat construction of RW of N steps M times for statistics
    x = 0; y = 0
    for n = 1 to N do                           Loop to generate a RW of N steps
      add-next-step(x, y)
    end
    do-analysis
  end
where                       Subroutine add-next-step: Add next step and return new end-point (x, y)
proc add-next-step(x, y) ≡
  ir = 4 * ranf + 1           Choose a random direction on the square lattice: ir = 1, 2, 3, 4
  if (ir = 1)     x = x + 1                                        step to the right
  elsif (ir = 2)  y = y + 1                                          step upward
  elsif (ir = 3)  x = x - 1                                        step to the left
  elsif (ir = 4)  y = y - 1                                        step downward
  fi.
end
```

```
Program NRRW              Simple sampling of a non-reversal random walk on a square lattice
begin
  no_of_nrrw = 0           Set counter number of successfully generated NRRW's of N steps
  for m = 1 to M do                                              Loop for statistics
    start:  x = 0; y = 0; x_pre = 0; y_pre = 0
                         Set mark start: Initialize NRRW and position of previous step (x_pre, y_pre)
    add-next-step(x, y)                                         Perform first step
    for n = 2 to N do
      x_tmp = x; y_tmp = y                           Store actual position (x, y) temporarily
      add-next-step(x, y)                                         Add new step
      if (x ≠ x_pre ∧ y ≠ y_pre) x_pre = x_tmp; y_pre = y_tmp
      elsif goto start
      fi          If no overlap with (x_pre, y_pre), accept step and update previous position
    end
    no_of_nrrw = no_of_nrrw + 1                             Update counter of NRRW's
    do-analysis
  end
end
```

FIG. 2 Pseudo-code for the simulation of a random walk (RW; code above the line) and a non-reversal random walk (NRRW; code below the line) via simple sampling. In both cases, the walks consist of N steps (innermost **for** loop), and the construction of the walks is repeated M times (outer **for** loop) to improve statistics. Whereas these repeated constructions always lead to a walk of N steps for the RW, they are not always completed successfully for the NRRW because immediate back-folding can occur. Then, the hitherto generated walk has to be discarded and the construction must be resumed from the beginning (jump to start). In both cases, a new step is appended to the walk by the subroutine *add-next-step*, in which *ranf* denotes a random number uniformly distributed between 0 and 1 i.e., $0 \leq ranf < 1$.

```
Program SAW                    Simple sampling of a self-avoiding random walk on a square lattice
begin
    no_of_saw = 0              Set counter for number of successfully generated SAW's of N steps
    integer lattice(-N:N, -N:N)
                               Define square lattice lattice(x,y): -N ≤ x,y ≤ N; x,y = integer
    for m = 1 to M do                                                Loop for statistics
        start:  x = 0; y = 0; lattice(-N:N, -N:N) = 0
                               Set mark start: Initialize SAW and all lattice sites by 0
        lattice(0,0) = 1            Occupy origin of the lattice with first monomer
        add-next-step(x,y)                                          Perform first step
        lattice(x,y) = 1            Occupy lattice with second monomer
        for n = 2 to N do
            add-next-step(x,y)                                        Add new step
            if (lattice(x,y) ≠ 0) goto start
            elsif lattice(x,y) = 1
            fi             If lattice(x,y) is occupied, restart construction from the beginning
        end
        no_of_saw = no_of_saw + 1                              Update counter of SAW's
        do-analysis
    end
end
```

FIG. 3 Pseudo-code for the simulation of a self-avoiding random walk (SAW) by simple sampling. For a SAW, it is necessary to keep track of all lattice sites that have already been occupied by a monomer during the construction of the walk, since one has to discard the hitherto obtained walk and restart the construction if the present step attempts to place a monomer on an already occupied lattice site (jump to start). A possibility to take into account the history of the walk is to define a (large) $(2N+1) \times (2N+1)$ lattice whose sites are initialized by 0 and updated to 1 if a monomer is successfully added. For more details and further algorithms see Binder and Heermann [3], Sokal [5], or the excellent textbook of Kinzel and Reents [62].

equally probable walks, one has to correct for this bias by not counting each chain with the weight $W_N = 1$ in the sampling, as one would do for simple sampling, but rather with a weight,

$$W_N(\{\mathbf{r}_i\}) = \prod_{i=1}^{N} k_i/z_0 \tag{9}$$

For large N, this weight varies over many orders of magnitude, however, and hence the analysis of the accuracy reached is rather difficult in practice. In fact, Batoulis and Kremer [27] demonstrated that the expected errors increase exponentially with increasing N.

In principle, this method can also be applied to multichain systems, but the problem of correcting for the bias becomes even more severe. In practice, one therefore has to resort to the "configurational bias" method which is an extension of the Rosenbluth sampling (see Chapter 7). But, inversely restricted sampling is still one of the possible options for not too large chain length N and volume fraction ϕ in order to generate a configuration of a multichain system that can be used as a starting point for a dynamic simulation method.

There are many variants of biased sampling methods for the SAW, for instance the idea of "scanning future steps" [28] but these techniques will remain outside of consideration here.

A different approach to overcoming attrition is the "dimerization method" [29], i.e., the idea of assembling long chains out of successfully generated short chains. Assume two (uncorrelated) walks of length $N/2$. Both walks are taken out of the $Z_{N/2} \propto (z_{eff})^{N/2} (N/2)^{\gamma-1}$ different configurations. The probability that they form one of the Z_N SAWs of N steps is simply [2,5]

$$P = \frac{Z_N}{\left(Z_{N/2}\right)^2} \propto \frac{N^{\gamma-1}}{(N/2)^{2(\gamma-1)}} = 2^{2(\gamma-1)} N^{-(\gamma-1)} \tag{10}$$

Thus the acceptance rate decreases only with a (small) power of N rather than exponentially.

Still another useful technique is the "enrichment method" [30]. This early work presented the first and compelling evidence that the exponent ν [Eq. (6)] has the value $\nu \approx 0.59$ ($d=3$) and $\nu \approx 0.75$ ($d=2$), and hence is a nice example of important discoveries made by Monte Carlo simulation.

The simple idea of the enrichment method is to overcome attrition by using successfully generated short walks of length s not only once, but p times: one tries to continue a chain of s steps to $2s$ steps by p different simple sampling constructions. Then, the number N_{2s} of $2s$ step chains is

$$N_{2s} = pN_s \exp(-\lambda s) \tag{11}$$

where λ is the attrition constant. This process is continued for blocks of s steps up to the desired chain length. One must fix p ahead of time to avoid a bias. The best choice is $p = \exp(\lambda s)$, since taking $p \ll \exp(\lambda s)$ does not reduce the attrition problem enough, while $p > \exp(\lambda s)$ would lead to an exploding size of the sample and all walks are highly correlated. The fact that the generated chains are not fully statistically independent and that therefore the judgement of the accuracy of the procedure is a subtle problem is the main disadvantage of the method.

There exist many extensions and variants for all these algorithms. For example, for a star polymer with s arms one can use an enrichment technique where one tries to add one step at each arm, and at each size of the star this attempt is repeated p times [31]. It is also possible to combine various methods suitably together. For example, Rapaport [32] combined the dimerization method with the enrichment technique, while Grassberger [33] combined the Rosenbluth–Rosenbluth method with enrichment techniques. With this so-called "Pruned-Enriched Rosenbluth Method" (PERM) simulations of polymers at the Theta temperature up to a chain length of a million were performed, in order to clarify the nature of logarithmic correction factors to Eq. (1) predicted by renormalization group theory [34]. Note that the SAW, modeling a swollen chain in good solvent, can be extended from an athermal situation to describe temperature-dependent properties by adding, for instance, a nearest neighbor energy ε that occurs if two (nonbonded) effective monomers occupy adjacent sites at the lattice. If a chain has n such nearest neighbor contacts in a sample generated by simple sampling, one takes the statistical weight of this chain configuration as proportional to $\exp(n\varepsilon/k_B T)$.

In PERM, enrichment is implemented in the Rosenbluth–Rosenbluth framework by monitoring the weight W_n of partially grown chains of n steps. If W_n exceeds some preselected upper threshold $W_n^>$, we make two or more copies of the chain, divide W_n by the number of copies made, place all except one onto a stack and continue with the last copy. In this way the total weight is exactly preserved, but it is more evenly spread on several configurations. This is done at every chain length n.

The last entry to the stack is fetched if the current chain has reached its maximal length N, or if we "prune" it . Pruning (the opposite of enrichment) is done when the current weight has dropped below some lower threshold $W_n^<$. If this happens, a random number r_n with prob $\{r_n = 0\}$=prob $\{r_n = 1\}$ = $1/2$ is drawn. If $r_n = 1$, we keep the chain but double its weight. If not, it is discarded ("pruned"), and one continues with the last entry on the stack. If the stack is empty, a new chain is started. When the latter happens, one says a new "tour" is started. Chains within one tour are correlated, but chains from different tours are uncorrelated.

III. DYNAMIC METHODS

Dynamic Monte Carlo methods are based on stochastic Markov processes where subsequent configurations $\mathbf{X}_\nu$ are generated from the previous one $(\mathbf{X}_1 \rightarrow \mathbf{X}_2 \rightarrow \mathbf{X}_3 \cdots)$ with some transition probability $W(\mathbf{X}_1 \rightarrow \mathbf{X}_2)$. To some extent, the choice of the basic move $\mathbf{X}_1 \rightarrow \mathbf{X}_2$ is arbitrary. Various methods, as shown in Fig. 4, just differ in the choice of the basic "unit of motion." Furthermore, W is not uniquely defined: We only require the principle of detailed balance with the equilibrium distribution $P_{eq}(\mathbf{X})$,

$$P_{eq}(\mathbf{X})W(\mathbf{X}_1 \rightarrow \mathbf{X}_2) = P_{eq}(\mathbf{X}_2)W(\mathbf{X}_2 \rightarrow \mathbf{X}_1) \tag{12}$$

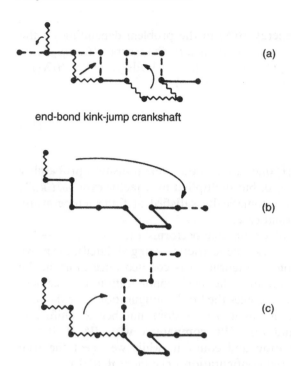

(a)

end-bond kink-jump crankshaft

(b)

(c)

FIG. 4 Various examples of dynamic Monte Carlo algorithms for SAWs: bonds indicated as wavy lines are moved to new positions (broken lines), other bonds are not moved. (a) Generalized Verdier–Stockmayer algorithms on the simple cubic lattice showing three types of motion: end-bond motion, kink-jump motion, crankshaft motion; (b) "slithering snake" ("reptation") algorithm; (c) "pivot" ("wiggle") algorithm.

In the athermal case (the standard SAW problem) each configuration has exactly the same weight (for single chains the normalized probability simply is $P_{eq}(\mathbf{X}) = 1/Z_N$). Then, Eq. (12) says that the probability to select a motion $\mathbf{X}_1 \to \mathbf{X}_2$ must be the same as the probabilty for the inverse motion, $\mathbf{X}'_1 \to \mathbf{X}_1$. One has to be very careful to preserve this symmetry in the actual realization of the algorithm, in particular if different types of move are carried out in the simulation (e.g., Fig. 4). This can be achieved by randomly choosing one of the possible moves and the new lattice sites to which it should shift the monomer (or monomers), and by rejecting this choice if it violates the excluded volume constraint. It would be completely wrong if one first checked for empty sites and made a random choice only between such moves that lead the monomers to these sites.

If there is an additional energy $\mathcal{H}(\mathbf{X})$ in the problem depending on the configuration $\mathbf{X}$, the equilibrium distribution is $P_{eq}(\mathbf{X}) = \exp(-\mathcal{H}(\mathbf{X})/k_B T)/Z$, where $Z = \sum_{\mathbf{X}} \exp(-\mathcal{H}(\mathbf{X})/k_B T)$. Hence, Eq. (12) leads to ($\delta\mathcal{H} \equiv \mathcal{H}(\mathbf{X}_2) - \mathcal{H}(\mathbf{X}_1)$ is the energy change caused by the move)

$$\frac{W(\mathbf{X}_1 \rightarrow \mathbf{X}_2)}{W(\mathbf{X}_2 \rightarrow \mathbf{X}_1)} = \exp\left(-\frac{\delta\mathcal{H}}{k_B T}\right) \tag{13}$$

Following Metropolis et al. [1] one can take the same transition probability as in the athermal case if $\delta\mathcal{H} \leq 0$, but multiplied by a factor $\exp(-\delta\mathcal{H}/k_B T)$ if $\delta\mathcal{H} > 0$. Then, Eq. (13) is automatically satisfied at finite temperature, if it was fulfilled in the athermal case.

Thus, at every step of the algorithm one performs a trial move $\mathbf{X}_1 \rightarrow \mathbf{X}_2$. If $W(\mathbf{X}_1 \rightarrow \mathbf{X}_2)$ is zero (excluded volume restriction being violated), the move is not carried out, and the old configuration is counted once more in the averaging. If $W(\mathbf{X}_1 \rightarrow \mathbf{X}_2)$ is unity, the new configuration is accepted, counted in the averaging, and becomes the "old configuration" for the next step. If $0 < W < 1$ we need a (pseudo-) random number x uniformly distributed between zero and one. We compare x with W: If $W \geq x$, we accept the new configuration and count it, while we reject the trial configuration and count the old configuration once more if $W < x$.

In the limit where the number of configurations M generated tends to infinity, the distribution of states $\mathbf{X}$ obtained by this procedure is proportional to the equilibrium distribution $P_{eq}(\mathbf{X})$, provided there is no problem with the ergodicity of the algorithm (this point will be discussed later). Then, the canonical average of any observable $A(\mathbf{X})$ is approximated by a simple arithmetic average,

$$\langle A \rangle \approx \overline{A} = \frac{1}{M - M_0} \sum_{\nu=M_0}^{M} A(\mathbf{X}_\nu) \tag{14}$$

where we have anticipated that the first M_0 configurations, which are not yet characteristic of the thermal equilibrium state that one wishes to simulate, are eliminated from the average. Both the judgements of how large M_0 should be taken and how large M needs to be chosen to reach some desired statistical accuracy of the result, are hard to make in many cases. Some guidance for this judgement comes from the dynamic interpretation of the Metropolis algorithm [5,35], to which we turn next. This interpretation also is very useful to evaluate the efficiency of algorithms.

We associate a (pseudo-) time variable $t' \equiv \nu/\tilde{N}$ with the label ν of successively generated configurations, where $\tilde{N}$ is the total number of monomers in the system, $\tilde{N} = $ (chain length $N + 1$) × (number of chains n_c). Then, $t_0 = M_0/\tilde{N}, t = M/\tilde{N}$, and Eq. (14) becomes a time average,

$$\overline{A} = \frac{1}{t - t_0} \int_{t_0}^{t} A(t') \, dt' \tag{15}$$

Since a move $\mathbf{X}_1 \rightarrow \mathbf{X}_2$ typically involves a motion of a single monomer or of a few monomers only (see Fig. 4), we have chosen one Monte Carlo step (MCS) per monomer as a time unit, i.e., every monomer has on average one chance to move.

The precise interpretation of the "dynamics" associated with the Monte Carlo procedure is that it is a numerical realization of a Markov process described by a master equation for the probabilty $P(\mathbf{X}, t)$ that a configuration $\mathbf{X}$ occurs at time t,

$$\frac{d}{dt} P(\mathbf{X}, t) = - \sum_{\mathbf{X}_2} W(\mathbf{X}_1 \rightarrow \mathbf{X}_2) P(\mathbf{X}, t) + \sum_{\mathbf{X}_2} W(\mathbf{X}_2 \rightarrow \mathbf{X}_1) P(\mathbf{X}_2, t) \tag{16}$$

Obviously, the principle of detailed balance, Eq. (13), suffices to guarantee that $P_{eq}(\mathbf{X})$ is the steady-state solution of Eq. (16). If all states are mutually accessible, $P(\mathbf{X}, t)$ must relax towards $P_{eq}(\mathbf{X})$ as $t \rightarrow \infty$ irrespective of the initial condition.

This dynamic interpretation is useful in two respects: (i) One can deal to some extent with the dynamics of polymeric systems [36]. For example, a basic model is the Rouse model [37] describing the Brownian motion of a chain in a heat bath. The heat bath is believed to induce locally stochastic changes of the configuration of a chain. In a lattice model, this is qualitatively taken into account by motions of the types shown in Fig. 4a. Since the Rouse model is believed to describe the dynamics of not too long real chains in melts, there is real interest in investigating the dynamic properties of models such as that shown in Fig. 4a, using lattices filled rather densely with many chains. In fact, the related bond fluctuation model [12,13,38,39], see Fig. 5, has been used successfully to study the dynamics of glass-forming melts [40,41], the Rouse to reptation crossover [42–44] etc. (ii) One can understand the magnitude of statistical errors (see below).

The fact that the algorithm in Fig. 4a should be compatible with the Rouse model can be understood from a simple heuristic argument: As a result of only local motion of beads in a chain, we expect that the center of gravity moves a distance of the order of $\mathbf{a}/N$, where $\mathbf{a}$ is a vector connecting two nearest neighbor sites on the lattice, and whose direction is random.

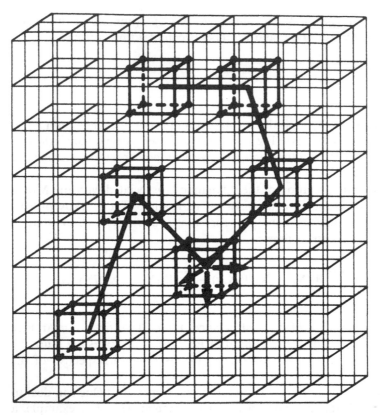

FIG. 5 Schematic illustration of the bond fluctuation model in three dimensions. An effective monomer blocks a cube containing eight lattice sites for occupation by other monomers. The length ℓ of the bonds connecting two neighboring cubes along the chain must be taken from the set $\ell = 2$, $\sqrt{5}$, $\sqrt{6}$, 3, and $\sqrt{10}$ lattice spacings. Chain configurations relax by random diffusive hops of the effective monomers by one lattice spacing in a randomly chosen lattice direction. For the set of bond vectors possible for the bond lengths ℓ, quoted above, the excluded volume constraint that no lattice site can belong to more than one cube automatically ensures that bonds can never cross each other in the course of their random motions. From Deutsch and Binder [13].

These random displacements add up diffusively. If we define the relaxation time τ_N so that after $\tau_N N$ such motions a mean square displacement of the order of the end-to-end distance is reached, we have from Eq. (6)

$$(\mathbf{a}/N)^2 \tau_N N = \langle R^2 \rangle = C_\infty a^2 N^{2\nu} \Longrightarrow \tau_N \propto N^{2\nu+1} \tag{17}$$

Note that the unit of time has been chosen such that each monomer on average attempts to move once, and we have implied that the slowest modes involve long distance properties of the order of the chain's linear dimension. Thus, the chain configuration should be fully relaxed when the center of mass has traveled a distance of the order of $\sqrt{\langle R^2 \rangle}$. This argument implies that the diffusion constant of the chain scales with chain length as

$$\mathcal{D}_N \propto N^{-1} \tag{18}$$

irrespective of excluded volume restrictions.

However, one must be rather careful with such arguments and with the use of algorithms as shown in Fig. 4a in general. For instance, if only end-bond and kink-jump motions are permitted, as in the original Verdier–Stockmayer algorithm [45], Eq. (17) does not hold for the SAW, rather one finds reptation-like behavior $\tau_N \propto N^3$ [46]. As pointed out by Hilhorst and Deutsch [47], the kink-jump move only exchanges neighboring bond vectors in a chain and does not create any new ones—new bond vectors have to diffuse in from the chain ends, explaining that one then finds a reptation-like law [24]. In fact, if one simulated a ring polymer with an algorithm containing only the kink-jump move, the bond vectors contained in the initial configuration would persist throughout the simulation, and clearly good equilibrium could not be established! Thus, when using a new algorithm one has to check carefully whether it has some undesirable conservation laws.

When considering the "slithering snake" algorithm of Fig. 4b, one can argue by an analogous argument as presented above that the center of mass vector moves a distance of $\|\mathbf{R}\|/N$ at each attempted move. If we require again that the mean square displacement is of order $\langle R^2 \rangle$ at time τ_N ($\tau_N N$ attempted moves), we find $(\langle R^2 \rangle/N^2)\tau_N N = \langle R^2 \rangle$, i.e., $\tau_N \propto N$ and $\mathcal{D}_N$ is of order unity. This argument suggests that the slithering snake algorithm is faster by a factor of N compared to any algorithm that updates the monomer positions only locally (kink-jump, etc.).

While such a faster relaxation is desirable when one is just interested in the simulation of static equilibrium properties, the slithering snake algorithm obviously is not a faithful representation of any real polymer dynamics even in a coarse-grained sense. This argument holds for the pivot algorithm (Fig. 4c) *a fortiori:* Due to the "global moves" (i.e., large parts of the chain move together in a single step) the relaxation of long-wavelength properties of the coil is very fast, but the dynamics does not resemble the real dynamics of macromolecules at all.

Nevertheless, the pivot algorithm has one distinctive advantage. It exhibits very good ergodicity properties [5], while slithering snake and

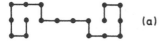 (a)

 (b)

FIG. 6 (a) Example of a SAW on the square lattice that cannot move if the slithering snake algorithm is used. (b) Example of a SAW on the square lattice that cannot move if a combination of the slithering snake algorithm and the Verdier–Stockmayer algorithm is used.

Verdier–Stockmayer type algorithms are manifestly nonergodic [48]. One can show this failure of ergodicity by constructing counterexample configurations that cannot relax at all by the chosen moves (Fig. 6). Such configurations also cannot be reached by the algorithms shown in Fig. 4a and b, and thus represent "pockets" of phase space simply left out completely from the sampling. There is good evidence [2,7,49] that in practice the systematic errors induced by this lack of ergodicity are rather small and are even negligible in comparison with the statistical errors of typical calculations. Nevertheless, in principle the lack of full ergodicity is a serious concern. While the pivot algorithm for single-chain simulations is the method of choice, both because of this problem and because it yields a very fast relaxation, it clearly cannot be used for simulations of dense polymer systems, and therefore algorithms such as those of Fig. 4a and b are still in use.

Irrespective of whether the chosen moves (Fig. 4) resemble the physically occurring dynamics (case (a)) or not (cases (b), (c)), Eqs. (15), (16) always apply and imply that observations of observables A, B are dynamically correlated. A dynamic correlation function, defined as

$$\phi_{AB}(t) = \frac{\langle A(0)B(t)\rangle - \langle A\rangle\langle B\rangle}{\langle AB\rangle - \langle A\rangle\langle B\rangle} \tag{19}$$

is implemented in a Monte Carlo simulation using the estimate (t_{obs} = total observation time)

$$\langle A(0)B(t)\rangle \approx \overline{A(0)B(t)} = \frac{1}{t_{obs} - t - t_0}\int_{t_0}^{t_{obs}-t} A(t)B(t+t)\,dt \tag{20}$$

Even if one is not interested in the study of these correlations per se, they are important because they control the statistical error of static observables. Thus, if we estimate the error δA of an average $\overline{A}$ [Eqs. (14), (15)] as follows (there are n observations of A, i.e., $A(t_\mu) = A_\mu$, $\mu = 1, \ldots, n$)

$$\langle (\delta A)^2 \rangle \approx \overline{(\delta A)^2} = \overline{\left(\frac{1}{n} \sum_{\mu=1}^{n} (A_\mu - \overline{A}) \right)^2} \tag{21}$$

one can to show that [35,3]

$$\langle (\delta A)^2 \rangle \approx \frac{1}{n} [\langle A^2 \rangle - \langle A \rangle^2] \left\{ 1 + 2 \frac{\tau_{AA}}{\delta t} \right\}, \qquad \text{for} \quad t_{\text{obs}} = n \delta t \gg \tau_{AA} \tag{22}$$

where δt is the time interval between subsequent observables included in the sum in Eq. (21) and the relaxation time is defined as follows

$$\tau_{AA} = \int_0^\infty \phi_{AA}(t) \, dt \tag{23}$$

Applying these concepts to the sampling of the end-to-end vector $\mathbf{R}$, for instance, we recognize that statistically independent observations for $\mathbf{R}$ are only obtained if $\delta t > \tau_N$, a time that can be very large for very long chains {cf. Eq. (17)}. In the simulation of models for dense polymer systems with local algorithms of the type of Fig. 2(a) or Fig. 3, the reptation model [24,50] implies that $\tau_N \propto N^3$, and in addition, the prefactor in this law can be rather large, because the acceptance rate of all moves decreases very strongly with increasing density of occupied sites in the lattice. Thus, in practice one has to work with a volume fraction ϕ of occupied lattices that does not exceed 0.8 for the simple SAW model on the simple cubic lattice [7] or 0.6 ($d=3$) or 0.8 ($d=2$) for the bond fluctuation model [12,13]. If one wishes to work with higher volume fractions (or with the fully occupied lattice, $\phi=1$, in the extreme case), no motion at all would be possible with any of the algorithms shown in Figs. 4 and 5. Then the only alternative is to use either "bond-breaking algorithms" [21,22], which have the disadvantage that one does not work with strictly monodisperse chain length, or the "cooperative motion" algorithm (CMA), see the following Chapter 5 by T. Pakula. Neither of these algorithms has any correspondence with the actual chain dynamics, of course.

An intricate problem is how to obtain initial configurations of the system of chains for the dynamic Monte Carlo algorithms. Several recipes have been advocated in the literature. For a system of many short chains, where the linear dimension of the box L is much larger than the length $R_{max} = bN$ of a chain fully stretched to its maximal extension, one can put all chains in this fully stretched state parallel to each other in a lattice direction onto the lattice. Excluded volume interactions can then be trivially respected. The time t_0 necessary to "forget" the initial state that is not characteristic of equilibrium can then simply be measured via the decay of the orientational order of the chains, which is initially present in the system. Although in principle this is a good method, it is clear that for large N and ϕ close to volume fractions that correspond to melt densities (i.e., $\phi \gtrsim 0.5$) the resulting time t_0 is rather large. But the most important drawback is that it cannot be used at all for systems where bN exceeds L (note that in the bond fluctuation model b is between 2 and $\sqrt{10}$ lattice spacings, and simulations up to $N = 2048$ for semidilute solutions using lattice linear dimensions $L \leq 400$ have been carried out [51]). In this case one has the following alternatives: Either the chains in the initial state are put on the lattice with the Rosenbluth–Rosenbluth method or the configurational bias method, or one uses NRRW chains, ignoring excluded volume initially. Then during the initial equilibrium run only moves are accepted that do not increase the number N_v of lattice sites for which the excluded volume constraint is violated, and one has to run the simulation until $N_v = 0$. There is no guarantee, of course, that this latter algorithm works at all for every initial configuration. An alternative would be to replace the hard excluded volume constraint by a soft one, introducing an energy penalty ΔU for every double occupancy of a lattice site, and gradually increase $\Delta U \to \infty$ during the initial equilibration run. Despite the fundamental importance of producing well-equilibrated initial configurations of many-chain (athermal) systems, we are not aware of any systematic comparative evaluation of the efficiency of the various approaches outlined above.

It should also be emphasized that the best method of choosing an initial state for the simulation of a dense polymer system may depend on the physical state of the system that one wishes to simulate. For example, the configurational bias method is a good choice for creating initial configurations of long flexible chains in semidilute solutions [51] or melts [43], but not for melts of stiff chains in a nematic state [14]: In this case, it is much better to start with a monodomain sample of a densely filled lattice of fully stretched chains, remove at random chains such that the desired volume fraction of occupied lattice sites is reached, and then start a relaxation run with slithering snake and local moves. Conversely, if we start from a configuration with no initial nematic order, it already takes a very long time to create small nematic domains out of chain configurations that are initially random-walk like, and relaxing the system further so that the multidomain configuration turns into a nematic monodomain configuration would be prohibitively difficult [14].

As the last point of this section, we briefly comment on several variations of lattice models, such as the SAW (Fig. 4c) or the bond fluctuation model (Fig. 5). What are the respective merits of these models?

The simple SAW has the advantage that it has been under study for more than 40 years, and thus a large body of work exists, to which new studies can be compared. For single short chains, exact enumerations have been carried out, and for some Monte Carlo techniques, e.g., the PERM method, one can proceed to particularly long chains. However, there are some disadvantages: (i) In $d=2$ dimensions, the moves of Fig. 4a do not create new bond vectors, and hence the local algorithm does not yield Rouse dynamics in $d=2$. (ii) Branched chains (stars, polymer networks) cannot be simulated with the dynamic Monte Carlo methods of Fig. 4, since the junction points cannot move. (iii) Since there is a single bond length (one lattice spacing) and only a few bond angles ($0°$ and $90°$ on square or cubic lattices, respectively), there is little variability in the model.

The bond fluctuation model has been used for about 12 years only, but it has numerous advantages: (i) There is a single type of move in the "random hopping" algorithm, namely an effective monomer is moved by a lattice spacing in a randomly chosen lattice direction. It is easy to write a very fast code that executes this move. This move almost always creates new bond vectors, also in $d=2$ dimensions. (ii) If the set of bond vectors is suitably restricted, the noncrossability constraint of bonds is automatically taken into account. (iii) In simulations of branched chains, junction points can move. (iv) Since there exist (in $d=3$ dimensions) 108 bond angles θ, the model in many respects mimics the more realistic behavior of off-lattice models in continuous space. (v) The different choices of bond lengths b allow introduction of an energy function $U(b)$, and this opens the way to modeling the glass transition of polymers [10], or carrying out a mapping of specific polymers to bond fluctuation models with particular choices of potentials $U(b)$, $V(\theta)$ [52]. (vi) While the "random hopping" algorithm of Fig. 5 is not strictly ergodic, the class of configurations that are not sampled is much smaller than that of the algorithms in Fig. 4a, b.

Finally, we also mention models that are in some way intermediate between the SAW and the bond fluctuation model. For example, Shaffer [53,54] has used a model where the monomer takes only a lattice site like the SAW, but the bond length can be 1, $\sqrt{2}$, and $\sqrt{3}$ lattice spacings, and then moves can be allowed where bonds cross each other, junctions can move, etc. This model is useful to separate the effect of excluded volume (site occupancy) from entanglement constraints (noncrossability of chains) in the dynamics [53,54]. Since because of the chain crossability a rather fast relaxation is provided, this "diagonal bond model" is also advocated for the study of ordered phases in models of block copolymers [55].

IV. CONCLUDING REMARKS

Although lattice models of polymers are crudely simplified, they have the big advantage that their study is often much less demanding in computer resources than corresponding off-lattice models. Therefore, a huge activity exists to study the phase behavior of polymer mixtures and block copolymer melts [56,57], while corresponding studies using off-lattice models are scarce. In this context, a variety of specialized techniques have been developed, to extract quantities such as chemical potential, pressure, entropy, etc. from lattice models, see e.g., [58–60]. An important consideration in such problems also is the best choice of statistical ensemble (e.g., semi-grand canonical ensemble for mixtures, etc.) [7,56]. We refer the reader to the more specialized literature for details.

The lattice models of polymers reach their limits when one wants to study phenomena related to hydrodynamic flow. Although study of how chains in polymer brushes are deformed by shear flow has been attempted, by modeling the effect of this simply by assuming a smaller monomer jump rate against the velocity field rather than along it [61], the validity of such "nonequilibrium Monte Carlo" procedures is still uncertain. However, for problems regarding chain configurations in equilibrium, thermodynamics of polymers with various chemical architectures, and even the diffusive relaxation in melts, lattice models still find useful applications.

ACKNOWLEDGMENTS

This chapter has benefited from many fruitful interactions with K. Kremer and W. Paul.

REFERENCES

1. Metropolis, N.; Rosenbluth, A.W.; Rosenbluth, M.N.; Teller, A.H.; Teller, E. J. Chem. Phys. **1953**, *21*, 1087.
2. Kremer, K.; Binder, K. Monte Carlo Simulations of Lattice Models for Macromolecules. Comput. Phys. Rep. **1988**, *7*, 259.
3. Binder, K.; Heermann, D.W.; *Monte Carlo Simulation in Statistical Physics. An Introduction*, Ed.; 3rd Ed, Springer: Berlin, 1997.
4. Milchev, A.; Binder, K.; Heermann, D.W. Z. Phys. **1986**, *B63*, 521.

5. Sokal, A.D. Monte Carlo Methods for the Self-Avoiding Walk. In *Monte Carlo and Molecular Dynamics Simulations in Polymer Science*; Binder, K., Ed.; Oxford University Press: New York, 1995; 47 pp.

6. Flory, P.J. *Principles of Polymer Chemistry*; Cornell University Press: Ithaca, New York, 1953.

7. Sariban, A.; Binder, K. Macromolecules **1988**, *21*, 711.

8. Gibbs, J.H.; Di Marzio. J. Chem. Phys. **1958**, *28*, 373.

9. Gibbs, J.H.; Di Marzio, J.D. J. Chem. Phys. **1958**, *28*, 807.

10. Wolfgardt, M.; Baschnagel, J.; Paul, W.; Binder, K. Phys. Rev. **1996**, *E54*, 1535.

11. Binder, K. In *Computational Methods in Field Theory*; Gausterer, H., Lang, C.B., Eds.; Springer: Berlin-New York, 1992; 59 pp.

12. Carmesin, I.; Kremer, K. Macromolecules **1988**, *21*, 2819.

13. Deutsch, H.-P.; Binder, K. J. Chem. Phys. **1991**, *94*, 2294.

14. Weber, H.; Paul, W.; Binder, K.; Phys. Rev. **1999**, *E59*, 2168.

15. Müller, M.; Binder, K.; Oed, W. J. Chem. Soc. Faraday Trans. **1995**, *91*, 2369.

16. Kron, A.K. Polymer Sci. USSR. **1965**, *7*, 1361.

17. Wall, F.T.; Mandel, F. J. Chem. Phys. **1975**, *63*, 4592.

18. Lal, M. Mol. Phys. **1969**, *17*, 57.

19. Madras, N.; Sokal, A.D. J. Stat. Phys. **1988**, *50*, 109.

20. Siepmann, I. Mol. Phys. **1990**, *70*, 1145.

21. Olaj, O.F.; Lantschbauer, W. Makromol. Chem., Rapid Commun. **1982**, *3*, 847.

22. Mansfield, M.L. J. Chem. Phys. **1982**, *77*, 1554.

23. Mattice, W.L.; Suter, U.W. *Conformational Theory of Large Molecules, the Rotational Isomeric State Model in Macromolecular Systems*; Wiley: New York, 1994.

24. de Gennes, P.G. *Scaling Concepts in Polymer Physics*; Cornell University Press: Ithaca, 1979.

25. Caracciolo, S.; Serena Causo, M.; Pelissetto, A. Phys. Rev. **1998**, *E57*, R1215.

26. Rosenbluth, M.N.; Rosenbluth, A.W. J. Chem. Phys. **1955**, *23*, 356.

27. Batoulis, J.; Kremer, K. J. Phys. **1988**, *A21*, 127.

28. Meirovitch, H. J. Phys. **1982**, *A15*, 735.

29. Alexandrovicz, Z. J. Chem. Phys. **1969**, *51*, 561.

30. Wall, F.T.; Erpenbeck, J.J. J. Chem. Phys. **1959**, *30*, 634.

31. Ohno, K.; Binder, K. J. Stat. Phys. **1991**, *64*, 781.

32. Rapaport, D.C. J. Phys. **1985**, *A18*, 113.

33. Grassberger, P. Phys. Rev. **1997**, *E56*, 3682.

34. des Cloizeaux J.; Jannink, G. *Polymers in Solution: Their Modeling and Structure*; Oxford University Press: Oxford, 1990.

35. Müller-Krumbhaar, H.; Binder, K. J. Stat. Phys. **1973**, *8*, 1.

36. Binder, K.; Paul, W. J. Polym. Sci. **1997**, *B35*, 1.

37. Rouse, P.E. J. Chem. Phys. **1953**, *21*, 1272.

38. Wittmann, H.-P.; Kremer, K. Comp. Phys. Comm. **1990**, *61*, 309.

39. Wittmann, H.-P.; Kremer, K. Comp. Phys. Comm. **1991**, *71*, 343.

40. Paul, W.; Baschnagel, J. Monte Carlo Simulations of the Glass Transition of Polymers. In: *Monte Carlo and Molecular Dynamics Simulations in Polymer Science*; Binder, K., Ed.; Oxford Univeristy Press: New York, 1995; 307 pp.
41. Baschnagel, J.; Dynamic Properties of Polymer Melts above the Glass Transition: Monte Carlo Simulation Results. In *Structure and Properties of Glassy Polymers*, ACS Symposium Series No. 710; Tant, M.R., Hiff, A.J., Eds.; American Chemical Society: Washington, DC, 1998; 53 pp.
42. Paul, W.; Binder, K.; Heermann, D.W.; Kremer, K. J. Phys. **1991**, II (France) *1*, 37.
43. Müller, M.; Wittmer, J.; Barrat, J.-L. Europhys. Lett. **2000**, *52*, 406–412.
44. Kreer, T.; Baschnagel, J.; Müller, M.; Binder, K. Macromolecules **2001**, *34*, 1105–1117.
45. Verdier, P.H.; Stockmayer, W.H. J. Chem. Phys. **1962**, *36*, 227.
46. Verdier, P.H. J. Chem. Phys. **1966**, *45*, 2122.
47. Hilhorst, H.J.; Deutsch, J.M. J. Chem. Phys. **1975**, *63*, 5153.
48. Madras, N.; Sokal, A.D. J. Stat.Phys. **1987**, *47*, 573.
49. Wolfgardt, M.; Baschnagel, J.; Binder, K. J. Phys. **1995**, II (France) 5, 1035.
50. Doi, M.; Edwards, S.F. *Theory of Polymer Dynamics*; Clarendon: Oxford 1986.
51. Müller, M.; Binder, K.; Schäfer, L. Macromolecules **2000**, *33*, 4568.
52. Paul, W.; Binder, K.; Kremer, K.; Heermann, D.W. Macromolecules. **1991**, *24*, 6332.
53. Shaffer, J.S. J. Chem. Phys. **1994**, *101*, 4205.
54. Shaffer, J.S. J. Chem. Phys. **1995**, *103*, 761.
55. Dotera, T.; Hatano, A. J. Chem. Phys. **1996**, *105*, 8413.
56. Binder, K. Monte Carlo Studies of Polymer Blends and Block Copolymer Thermodynamics. In *Monte Carlo and Molecular Dynamics Simulations in Polymer Science*; Binder, K., Ed.; Oxford University Press: New York, 1995; 356 pp.
57. Binder, K.; Müller, M. Monte Carlo Simulations of Block Copolymers. Curr. Opinion Colloid Interface Sci. **2000**, *5*, 315.
58. Dickman, R. J. Chem. Phys. **1987**, *86*, 2246.
59. Müller, M.; Paul, W. J. Chem. Phys. **1994**, *100*, 719.
60. Kumar, S.K.; Szleifer, I.; Panagiotopoulos, A. Phys. Rev. Lett. **1991**, *66*, 2935.
61. Lai, P.-Y.; Binder, K. J. Chem. Phys. **1993**, *98*, 2366.
62. Kinzel, W.; Reents, G. *Physics by Computer*; Springer: Berlin, 1998.

5

Simulations on the Completely Occupied Lattice

TADEUSZ PAKULA   Max-Planck-Institute for Polymer Research, Mainz, Germany, and Technical University of Lodz, Lodz, Poland

I. INTRODUCTION

There is a large number of different methods (algorithms) used for simulation of polymers on the coarse grained molecular scale [1–6]. Models of polymers considered in this range usually disregard the details of the chemical constitution of macromolecules and represent them as assemblies of beads connected by nonbreakable bonds. In order to speed up recognition of neighboring beads, the simplified polymers are often considered to be on lattices with beads occupying lattice sites and bonds coinciding with lines connecting neighboring sites. The methods used for simulation of the lattice polymers can be considered within two groups. The first group includes algorithms that can operate only in systems with a relatively large fraction of lattice sites left free and the second group includes algorithms suitable for lattice systems in which all lattice sites are occupied by molecular elements. Whereas, the systems considered within the first group should be regarded as lattice gases, the systems treated within the second group of methods can be considered as lattice liquids. This reflects the differences in the mechanisms of molecular rearrangements used within these two groups to move the systems through the phase space in order to reach equilibrium. The latter problem concerns the physical nature of molecular rearrangements in dense polymer systems and is related to the microscopic mechanism of motion in molecular liquids. Unfortunately, this is not yet solved entirely. The most popular picture is that a molecule, or a molecular segment in the case of polymers, needs a free space in its neighborhood in order to make a translational step beyond the position usually occupied for quite a long time. Most simulation methods assume this picture and consequently a relatively large portion of the space in the form of free lattice sites has to be left free to allow

147

a reasonable mobility [1–5]. On the other hand, there is only one simulation method that assumes a cooperative nature of molecular rearrangements on the local scale and which does not require such a reserve space to allow the molecular mobility. The method, which uses the mechanism of cooperative rearrangements for polymer systems is the Cooperative Motion Algorithm (CMA) suggested originally in [6] and presented in improved form in subsequent publications [7–10]. A mechanism of this kind has been formulated recently also for low molecular weight liquids, based on assumptions taking into account both a dense packing of molecules interacting strongly due to excluded volume and a condition of preservation of local continuity of the system. The later version of the microscopic mechanism, called the Dynamic Lattice Liquid (DLL) model, has been described in detail [11–14].

The aim of this chapter is to present a background to the simulations of molecular lattice systems on a completely filled lattice (the DLL model) as well as to show some examples of application of the CMA method for simulation of static and dynamic properties of various polymers.

II. THE DYNAMIC LATTICE LIQUID MODEL

Macroscopically, liquids differ from solids by the absence of rigidity and from gases by having a tensile strength. On a microscopic level, this means that the molecules in a liquid can move more easily than in a solid but they remain condensed due to attractive interactions, which are almost negligible in a gaseous phase. Owing to the dense packing of molecules, the dynamic properties of liquids become complex and the relaxations extend over various, usually well distinguishable, time scales. On the short time scale, τ_V, the molecules oscillate around some quasi-fixed positions, being temporarily "caged" by neighbors. It is believed that more extensive translational motions of molecules take place on a much longer time scale, τ_α, due to the breaking down of the cages by cooperative processes. Trajectories of molecules consist, therefore, both of oscillatory components and of occasional longer range translational movements between subsequent quasi-fixed states, as illustrated in Fig. 1a. The macroscopic flow of the liquid is related to the longer time scale (τ_α), in which the arrangement of molecules becomes unstable because each molecule wanders through the material changing neighbors during the translational motion steps.

Although the picture of motion in a liquid shown above is documented by computer simulations of dense Lennard-Jones systems [15,16], it is not quite clear under which conditions single diffusional steps can occur. Theories of transport phenomena in liquids do not consider this problem explicitly [17–22].

(a)

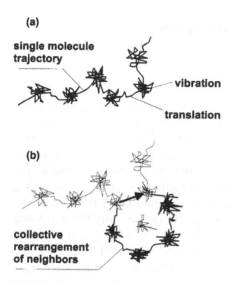

single molecule trajectory

vibration

translation

(b)

collective rearrangement of neighbors

FIG. 1 Schematic illustration of a molecule trajectory in a liquid. (a) A trajectory consisting of vibrations around quasi-localized states and occasional translational steps. (b) Local correlated motions of neighboring molecules contributing to the translational step of the single molecule by a cooperative rearrangement.

The model that has been proposed to answer this question is based on a lattice structure and is called the Dynamic Lattice Liquid model [11,12]. The positions of the molecules are regarded as coinciding with the lattice sites. The lattice is, however, considered only as a coordination skeleton defining the presence of nearest neighbors but not precisely the distances between them. Under the condition of uniform and constant coordination (z), all lattice sites are assumed to be occupied. It is assumed that the system has some excess volume so that molecules have enough space to vibrate around their positions defined by lattice sites but can hardly move over larger distances because all neighboring lattice sites are occupied. The vibrations are assumed to take place with a mean frequency $\nu_V = 1/\tau_V$, related to the short time scale (τ_V). Each large enough displacement of a molecule from the mean position defined by the lattice is considered as an attempt to move to the neighboring lattice site. For simplicity, the attempts are assumed to take place only along the coordination lines but are independent and randomly distributed among z directions. Most of the attempts remain unsuccessful, because it is assumed that all the time the system remains quasi-continuous, which means that no holes of molecular sizes are generated and multiple occupations of lattice sites are excluded (excluded volume condition). The continuity condition of the system, for the vector field of attempted displacements r, can be written as follows

$$\nabla J + \frac{\partial \rho}{\partial t} = \varepsilon \qquad \text{with} \quad \varepsilon \to 0 \tag{1}$$

where J is the current of displacing molecules and ρ is the density and should not deviate considerably from 1 if all lattice sites remain occupied; ε is introduced as a small number allowing density fluctuations but excluding generation of holes. Consequently, most of the attempted displacements have to be compensated by a return to the initial position within the period τ_V. Only those attempts can be successful that coincide in such a way, that along a path including more than two molecules the sum of the displacements is close to zero. In the system considered, only the paths in a form of closed loops can satisfy this condition (Fig. 1b). The probability to find an element taking part in such coincidences will determine the probability of longer range rearrangements and the slower time scale (τ_α).

A determination of this probability reduces to counting the number of self-avoiding "n-circuits" on a given lattice, i.e., to a problem that has already been extensively considered in the literature [23]. The generally accepted result describing the probability to find the self-avoiding circuits is given by

$$p(n) = Bn^{-h}\mu^n \tag{2}$$

where B is a lattice dependent constant, μ plays the role of an effective coordination number (called the "connective constant") of the lattice, and the exponent h is positive and dependent on the dimensionality d of the lattice, but is presumably largely independent of the detailed structure of the lattice. Related theories predict only bounds for the exponent h ($h \geq d/2$). More detailed information is obtained from enumerations performed for various lattices [23].

The model discussed can easily be implemented as a dynamic Monte Carlo algorithm. A system of beads on the fcc lattice is considered. The beads occupy all lattice sites. It is assumed that the beads vibrating with a certain frequency around the lattice sites attempt periodically to change their positions towards nearest neighboring sites. The attempts are represented by a field of unit vectors, assigned to beads and pointing in a direction of attempted motion, chosen randomly. Attempts of all beads are considered simultaneously.

An example of such an assignment of attempted directions of motion is shown in Fig. 2, for a system of beads on a triangular lattice, taken here as a two-dimensional illustration only. From the field of attempts represented in

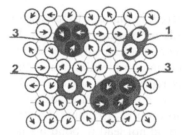

FIG. 2 An illustration of the vector field representing attempts of molecular displacements towards neighboring lattice sites in the lattice liquid model. Shaded areas represent various local situations: (1) unsuccessful attempt when neighboring elements try to move in the opposite direction, (2) unsuccessful attempt because the element in the center will not be replaced by any of the neighbors, and (3) successful attempts, in which each element replaces one of its neighbors. System elements taking part in cooperative translational rearrangements are shown in black.

this way, all vectors that do not contribute to correlated sequences (circuits) satisfying the continuity condition Eq. (1) are set to 0. This concerns, for example, such situations as attempts of displacements in opposite directions (e.g., area 1 in Fig. 2) or attempts of motion starting from lattice sites, towards which any other bead is not trying to move at the same time (e.g., area 2 in Fig. 2). What remains after this operation are vectors contributing to a number of closed loop traces, which are considered as paths of possible rearrangements (e.g., areas 3 in Fig. 2). If the system is considered as athermal, all possible rearrangements found are performed by shifting beads along closed loop traces, each bead to the neighboring lattice site.

The above procedure, consisting of (1) generation of motion attempts, (2) elimination of unsuccessful attempts, and (3) displacing beads within closed loop paths, is considered as a Monte Carlo step and assumed to take place within the time scale τ_V. The procedure is exactly repeated in subsequent time steps always with a new set of randomly chosen directions of attempted displacements.

The system treated in this way can be regarded as provided with the dynamics consisting of local vibrations and occasional diffusional steps resulting from the coincidence of attempts of neighboring elements to displace beyond the occupied positions. Within a longer time interval, this kind of dynamics leads to displacements of individual beads along random walk trajectories with steps distributed randomly in time. Small displacements related to vibrations of beads could be considered explicitly but are neglected here, for simplicity.

Models of macromolecular melts are considered, in analogy to models of simple liquids, as systems of structureless beads occupying lattice sites. In the case of polymers, the beads are connected by bonds to form linear chains or other more complex assemblies regarded as macromolecules. An extension of the algorithm to simulation of systems representing polymer melts appears to be straightforward. The presence of bonds between beads influences only the second step of the described procedure. Assuming that the bonds are not breakable, only attempts that would not lead to bond breaking can be considered as possible. Therefore, the selection of rearrangement possibilities is made under this additional condition, i.e., after having found all possible rearrangement paths, in parallel, as for nonbonded beads, all those ones are rejected that would lead to a disruption of polymer chains. All other steps in the algorithm are identical with those described for systems representing simple liquids. The procedure results in local cooperative conformational rearrangements of polymers, which can take place simultaneously at various places of the system; however, within a single time step one bead can be moved only once and only to a neighboring lattice site. It is a characteristic feature of this algorithm that types of local conformational changes are not precisely specified. All changes of chain conformations that satisfy the assumed conditions of system continuity and nonbreakability of bonds are allowed. This makes the algorithm nonspecific for any type of polymer architecture. During motion, the identities of polymers given by numbers and the sequences of beads within chains are preserved. Local chain moves in this algorithm are probably similar to those in the CMA but some other conformational rearrangements are probably also possible, when two or more displacement loops influence neighboring parts of a chain simultaneously (within a single time step).

Ergodicity has not been shown for any polymer algorithm. But for dimers, it was shown, for instance, for the CMA [24,25]. Therefore, in the DLL algorithm, which leads to the same types of cooperative moves, the ergodicity can be assumed for dimers, as well. The additional types of moves mentioned above can only improve the accessibility of various conformational states.

The requirement of a detailed balance in the athermal polymer system, as considered here, reduces to showing that the transition probabilities between two neighboring states A and B are equal. In this algorithm, two such states are always reversible and are separated by cooperative rearrangements along loops of the same size and form but different motion directions. Because loops consist of vectors that are pointing equally probably at any direction, this condition is satisfied. Moreover, it remains valid for any polymer system because the loops are independent of the structure.

Here, only the athermal version of the DLL model has been described. It has been shown elsewhere [11–13] that the model is able to represent the temperature dependent dynamics of systems as well. This, however, requires additional assumptions concerning free volume distribution and related potential barriers for displacements of individual elements. It has been demonstrated [11–13] that in such a version the model can describe a broad class of temperature dependencies of relaxation rates in supercooled liquids with the variety of behavior ranging between the extremes described by the free volume model on one side and the Arrhenius model on the other side.

III. THE COOPERATIVE MOTION ALGORITHM

Cooperative rearrangements based on displacements of system elements (beads) along closed loops have been introduced originally for dense lattice polymer models within the Cooperative Motion Algorithm (CMA). In this algorithm rearrangements are performed along random self-avoiding closed trajectories generated sequentially in randomly chosen places of the simulated systems. Each attempt to perform a rearrangement consists of a random choice of an element in the system and of searching for a self-avoiding non-reversal random walk in the form of a closed loop, which starts and ends at that element. Subsequent attempts are counted as time. Any distortion in searching loops, such as a reversal attempt or a cross point in the searching trajectory, breaks the attempt and a new attempt is started. If a loop is found, all elements lying on the loop trajectory are translated by one lattice site so that each element replaces its nearest neighbor along the loop. In systems representing polymers, only rearrangements that do not break bonds and do not change sequences of segments along individual chains are accepted. Chains can rearrange only by displacements involving conformational changes. As a result of a large number of such rearrangements, elements of the system move along random walk trajectories. The motion takes place in the system in which all lattice sites are occupied but the system remains continuous and the excluded volume condition is satisfied.

In this algorithm, both the time step definition and the distribution of rearrangement sizes are different than in the previously described DLL model. The time step in the CMA has usually been chosen as corresponding to the number of attempts to perform cooperative rearrangements resulting on average in one attempt per system element (bead). This can be considered as comparable with the time step definition in the DLL algorithm in which each element attempts to move within each time step. This involves, however, a difference in distributions of waiting times. A more distinct difference between these two algorithms is in the distribution of sizes of performed rearrangements, which has only been tested precisely for an athermal non-polymeric system. The distribution of rearrangement sizes within the DLL algorithm is well described by Eq. (2) with parameters $\mu = 0.847$, $h = 3$, and $B = 0.4$, whereas the distribution of rearrangement sizes in the CMA is considerably broader. It has been established that the two distributions differ by the factor n^2. This difference results in much higher efficiency of simulation by means of CMA, achieved however at the expense of some simplifications concerning short time dynamics of the system with respect to that in the DLL model.

On the other hand, it has been proven for melts of linear polymers that switching between DLL and CMA does not influence static properties of the system. The CMA, similarly to the DLL algorithm, involves only conformational changes within polymers which all the time preserve their identities given by the number of elements and by their sequences in individual chains. In the CMA as in the DLL, a variety of conformational rearrangements results from the algorithm and cannot be precisely specified. An important advantage of both DLL and CMA methods with respect to other simulation algorithms of polymers on lattices, consists in allowed moves of chain fragments both along contours of chains and transversally to the chain contours. This probably makes both algorithms ergodic.

The CMA has been applied successfully for simulation of static and dynamic properties of a variety of dense complex polymeric systems. Some examples of such applications are presented in subsequent sections.

IV. EXAMPLES OF APPLICATION

A. Melts of Linear Polymers

Systems representing melts of linear polymers have been simulated using both DLL and CMA methods. Linear polymers of various lengths N have as usual been represented as chains of beads connected by nonbreakable bonds. The face-centered cubic lattice has been used in both cases with chains occupying all lattice sites. Analysis of static properties of such systems has shown that, using both methods, Gaussian chain conformations are generated as indicated by the characteristic scaling laws for chain dimensions represented by the mean square values of the end-to-end distances or by radii of gyration (e.g., [26,27]).

Dynamic properties of such systems have been analyzed using the DLL method in the range of short chains (up to $N = 32$) [11–14] and the CMA for larger chains (up to $N = 800$) [26,27]. In the DLL method the presence of bonds between beads imposes additional limits on rearrangement possibilities, which can be considered as an additional reduction of the connectivity constant of the lattice. This results in a strong effect of the number of bonds and consequently chain length on bead relaxation rates. It has been shown that the chain length dependence of the end-to-end vector relaxation time is much stronger, especially for short chains, than the Rouse model and other simulations predict. On the other hand, when the chain relaxation times are normalized by the segmental mobility described by τ_S they nearly satisfy the N^2 dependence, as expected for the Rouse chains. This suggests that the τ_S vs. N dependence represents the effect of chain length on local dynamics, which in the Rouse model is included in the friction coefficient and is not considered explicitly. It has been tested experimentally that the effects observed in model systems are also seen in real polymer melts of polyisobutylene samples with chain lengths in a range comparable with that of the simulated systems [11,12]. The simulation results used for the comparison with experiments have been obtained for systems at athermal states. This means that the effects of the influence of chain lengths on rates of segmental relaxation are obtained without any additional assumptions, such as, for example, additional free volume effects at chain ends [28]. In the model described, they are only caused by a reduction of the connectivity constant within the lattice involved by introduction of nonbreakable bonds between lattice elements. Nevertheless, the range of changes of segmental relaxation times with the chain length is comparable with the corresponding range of changes observed in experiments.

In application to polymers, the DLL algorithm has many similarities with the CMA [11–14,26]. There is, however, an important difference between them, consisting in a parallel treatment of all system elements in the case of DLL in contrast to the sequential treatment in the CMA. This difference results in differences in distributions of rearrangement sizes and consequently in some specific effects related to the local dynamics. The effects of chain length on the local relaxation rates observed in systems simulated using DLL are, therefore, not seen in systems simulated by means of the CMA. On the other hand the higher computational efficiency of the CMA allows the study of systems with longer chains and within a broader time range.

The dynamical properties of the model systems, with n chains of length N, are usually characterized by the following quantities:

1. the autocorrelation function of a vector representing bond orientation

$$\rho_b(t) = \frac{1}{Nn} \sum_n \sum_i^N (\mathbf{b}_i(t)\mathbf{b}_i(0)) \tag{3}$$

where $\mathbf{b}_i$ are unit vectors representing bond orientation;

2. the autocorrelation function of the end-to-end vector of chains

$$\rho_R(t) = \frac{1}{n}\sum_n \mathbf{R}(0) \cdot \mathbf{R}(t) \tag{4}$$

with end-to-end vectors $\mathbf{R}(0)$ and $\mathbf{R}(t)$ at time $t=0$ and t, respectively; and

3. the mean squared displacements of monomers and of the centers of mass of chains

$$\langle r_m^2 \rangle = \frac{1}{Nn}\sum_n \sum_N [\mathbf{r}_m(t) - r_m(0)]^2 \tag{5}$$

$$\langle r_{cm}^2 \rangle = \frac{1}{n}\sum_n [\mathbf{r}_{cm}(t) - \mathbf{r}_{cm}(0)]^2 \tag{6}$$

where $\mathbf{r}_m(t)$ and $\mathbf{r}_{cm}(t)$ are monomer and chain center of mass coordinates at time t, respectively.

The first two quantities allow determination of relaxation times of corresponding objects in the model (bonds and chains) and the last two allow determination of diffusion constants. Chain length dependencies of the self-diffusion constant and of the relaxation times of bonds and chains are presented in Fig. 3. These results show that the dynamic behavior of

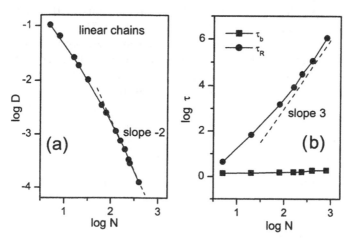

FIG. 3 Chain length dependencies of (a) self-diffusion constants of chains and (b) relaxation times of bonds and end-to-end vectors in model systems of linear chain melts.

simulated systems corresponds well to the behavior of real polymer melts, as detected experimentally. In the simulation, the diffusion constants for long chains reach the well known scaling dependence $D \sim N^{-2}$ and the relaxation times of chains reach the scaling law $\tau \sim N^{\omega}$, with $\omega > 3$, as observed in chain length dependencies of melt viscosities for long chains. Simulation of models with chain lengths much longer than these presented here becomes unrealistic at the moment because of limits imposed by the calculation speed. More detailed analysis of the static and dynamic behavior of polymer melts simulated by means of the CMA has already been presented in other publications [26,27], but this problem still remains not completely understood and further results concerning the mechanism of motion of polymer chains will be published elsewhere. Models considered as polymer melts have been simulated by many other methods [4,29,30], in which, however, a dense packing of molecules under the excluded volume condition is hard to achieve.

B.   Melts of Macromolecules with Complex Topology

The advantages of the CMA—its high efficiency and high flexibility in the representation of complex molecular objects—can be well demonstrated in simulations of melts of stars and microgels. Experimental observations of melts of multiarm polymer stars [31,32] and microgels [33] generate questions concerning the dynamics of molecular rearrangements in such systems. As an illustration of the application of the CMA to such systems, fragmentary results concerning the structure and dynamics of three systems illustrated in Fig. 4 are presented here. All three systems consist of nearly the same number of monomer units ($N = 500$) but the units are joined together to form different molecular objects. In the first case they form a linear chain, in the second a multiarm star (24 arms), and in the third case the linear chain is crosslinked internally by additional bonds (25 per chain) to a kind of microgel particle. The structure and dynamics of such systems in the melt have been studied [34,35]. Figure 4 shows a comparison of mass distributions, $\phi(r)$, in such objects around their centers of mass as well as a comparison of the pair correlation functions, $g(r)$, of the centers of mass of different chains. The first quantity, $\phi(r)$, describes the averaged fraction of space occupied by elements of a given molecule at the distance r from the mass center of that molecule, and the second, $g(r)$, describes the probability that at the distance r from the mass center of one molecule a mass center of another molecule will be found. A considerable difference in the structure of stars and microgels in comparison to linear chains can be seen when these characteristics are considered. Whereas, the linear chains are loosely coiled

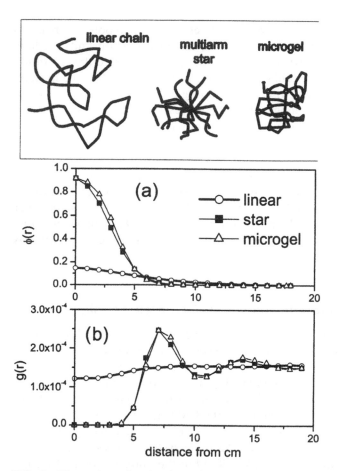

FIG. 4 Illustration of molecular objects of the same total chain length but different topology and a comparison of intrachain segment densities (a) and center of mass pair correlation functions (b) for three computer simulated melts with different molecular objects: linear chains, multiarm stars, and microgels.

and interpenetrate each other, the stars and microgels constitute compact objects with intramolecular density almost reaching the density of the system. This means that the latter molecules do not interpenetrate each other and remain at well correlated distances as indicated by the center of mass pair correlation functions with distinct maxima at preferred intermolecular distances (Fig. 4b).

The dynamics of the three systems is characterized in Fig. 5, where various correlation functions are compared. For the linear chain system the segment position (ρ_s) and the end-to-end vector (ρ_R) [see Eq. (4)]

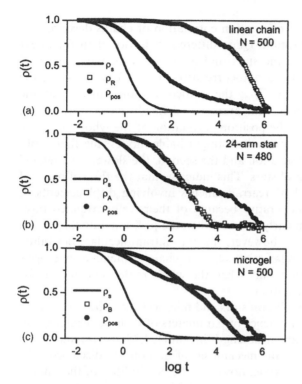

FIG. 5 Dynamical properties of three simulated polymer melts with various molecules: (a) linear chains, (b) stars, and (c) microgels. Various correlation functions are shown: ρ_s — segment position autocorrelation, ρ_R — end-to-end vector autocorrelation for linear chains, ρ_A — arm end-to-end autocorrelation for stars, ρ_B — autocorrelation of microgel chain segments of length $N = 20$, ρ_{pos} — position autocorrelation for all types of molecules.

autocorrelation functions represent the local and the longest relaxation rates, respectively. The third correlation function (ρ_{pos}) describing a correlation of chains with their initial positions (at $t = 0$) characterizes the translational motion. The position correlation of a structural element, a chain segment, or the whole molecule, is defined as

$$\rho_{pos}(t) = \frac{1}{Nn} \sum_n \sum_i^N C_i(0)C_i(t) \tag{7}$$

where $C_i = 1$ at all places occupied by a given structural element and $C_i = 0$ everywhere else. For linear chains, this correlation with the initial position decays much faster than the orientation correlation of chains described by the end-to-end vector relaxation. Quite different behavior than for linear chains is observed both for the stars and microgels with the same local mobility indicated by the bond autocorrelation. In the case of stars, the orientation relaxation of arms [ρ_A, the autocorrelation function of the vector connecting star center with arm end, defined analogously to ρ_R, Eq. (4)] is much faster than the position correlation of the whole star. The position correlation shows two well distinguishable modes, the first probably related to the shape relaxation and the second, the slower one, related to the translational motion of stars. This indicates that the flow of melts of stars should be dominated by rearrangements involving displacements of stars, which have to be cooperative because of their well correlated positions. Microgels behave similarly to stars when the position autocorrelation is considered. They differ, however, in the intramolecular flexibility. Whereas the arms of stars relax relatively fast, chain elements of microgels of corresponding length relax only when the whole molecule can relax by changing position and orientation.

The results concerning multiarm stars and microgels have been presented in detail elsewhere and the influence of parameters of their molecular structure, such as arm number and arm length in stars [34,35] or size and cross-link density in microgels, on the mechanism of motion are discussed. The results presented here demonstrate, however, the possibilities of the simulation method used in studies of structure and dynamics of melts of these complex molecules. The high efficiency of the CMA method has allowed the first simulations of melts of such molecules. Other simulations of dynamic properties of star molecules [36] have been performed for single stars and in a rather narrow time range.

C. Block Copolymers

In order to illustrate the application of the CMA to simulate heterogeneous systems, we present here results concerning properties of a diblock copolymer melt considered in a broad temperature range including both the homogeneous and the microphase separated states. Only symmetric diblock copolymers, of composition $f = 0.5$ of repulsively interacting comonomers A and B, are shown here. The simulation, in this case, allows information about the structure, dynamics, and thermodynamic properties of systems [9,37–40] to be obtained.

In the case of copolymers, two types of partially incompatible monomers (A and B) are considered, characterized by direct interaction parameters ε_{ij}, and often it is assumed that the energy of mixing is given only by the interaction of monomers of different types, i.e., that $\varepsilon_{AA} = \varepsilon_{BB} = 0$ and $\varepsilon_{AB} = 1$. The effective energy of a monomer E_m is given by the sum of ε_{AB} over z nearest neighbors and depends in this way on the local structure. The moving of chain elements alters the local energy because the monomers contact new neighbors. In order to get an equilibrium state at a given temperature the probability of motion is related to the interaction energy of the monomers in the attempted position in each Monte Carlo step. This means that at a given temperature, the Boltzmann factor $p = \exp(-E_{m,final}/k_B T)$ is compared with a random number r, $0 < r < 1$. If $p > r$, the move is performed, and another motion is attempted. Under such conditions, at low temperatures, the different types of monomers tend to separate from each other in order to reduce the number of AB contacts and consequently to reduce the energy. The simulation performed in this way allows information about the structure, dynamics, and thermodynamic properties of systems to be obtained. It is worthwhile to notice that the frequently used Metropolis method of equilibration of systems can lead to a nonrealistic dynamics, therefore it should not be used in cases when the dynamics is of interest.

Examples of temperature dependencies of the thermodynamic quantities recorded during heating of the initially microphase separated system of a symmetric diblock copolymer are shown in Figs. 6a and 6b. The following quantities have been determined: (1) the energy of interaction of a monomer, E_m, determined as the average of interactions of all monomer pairs at a given temperature

$$E_m = \sum_{i=1}^{z} \varepsilon_{kl}(i)/z \tag{8}$$

and (2) the specific heat calculated via the fluctuation-dissipation theorem

$$c_V = \frac{\langle E^2 \rangle - \langle E \rangle^2}{k_B T^2} \tag{9}$$

where the brackets denote averages over energy of subsequent states sampled during simulation of the system at constant temperature. The temperature at which a stepwise change in the energy and the corresponding peak in the specific heat are observed is regarded as the temperature of the order-to-disorder transition, T_{ODT}.

The nature of the transitions corresponding to structural changes in copolymers can be well established from an analysis of distributions of local concentrations, which are directly related to the free energy. An

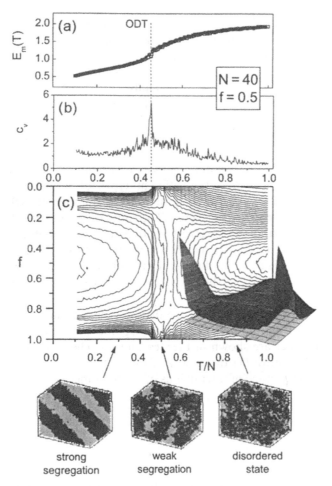

FIG. 6 Temperature dependencies of: (a) the average interaction energy per monomer, (b) the specific heat, and (c) concentration distributions in small volume elements consisting of the nearest neighbors of each chain segment. Characteristic structures corresponding to various temperature ranges are illustrated.

example of such distributions for a symmetric diblock copolymer, in a broad temperature range, is shown in Fig. 6c, by means of contour lines of equal composition probability projected on the composition–temperature plane. Such contour plots reflect many details of the thermodynamics and structure of the system. It is easily seen that, at high temperatures, the system can be considered as homogeneous because locally the most probable concentration corresponds to the nominal composition in the diblock. This is changed at temperatures close to T_{ODT} where at first a plateau and later two maxima corresponding to two coexisting phases are detected. At T_{ODT}, a sudden change transforms the system to a state with well defined micro-phases indicated by the most probable local concentrations corresponding to pure components. These results indicate three characteristic ranges of thermodynamic behavior of the system assigned as (1) disordered, (2) weakly segregated, and (3) strongly segregated regimes appearing with decreasing temperature. Structures of simulated systems corresponding to these regimes are illustrated in Fig. 6 by assuming different colors for different copolymer constituents.

The structure of the simulated block copolymer systems has been characterized in detail [38–40]. Temperature dependencies of various structural parameters have shown that all of them change in a characteristic way in correspondence to T_{ODT}. The microphase separation in the diblock copolymer system is accompanied by chain extension. The chains of the diblock copolymer start to extend at a temperature well above that of the transition to the strongly segregated regime. This extension of chains is related also to an increase of local orientation correlations, which appear well above the transition temperature. On the other hand, the global orientation correlation factor remains zero at temperature above the microphase separation transition and jumps to a finite value at the transition.

In order to get information about dynamic properties of the system various quantities have been monitored with time at equilibrium states corresponding to various temperatures [38–40]: the mean squared displacement of monomers, $\langle r_m^2 \rangle$, the mean squared displacement of the center of mass of chains, $\langle r_{cm}^2 \rangle$, the bond autocorrelation function, $\rho_b(t)$), the end-to-end vector autocorrelation function, $\rho_R(t)$, and the autocorrelation of the end-to-end vector of the block, $\rho_{bl}(t)$. On the basis of these correlation functions, various quantities characterizing the dynamic properties of the systems can be determined, i.e., the diffusion constant of chains and various relaxation times corresponding to considered correlations.

Examples of various correlation functions for the diblock copolymer system at high and at low temperatures are shown in Fig. 7. It has been observed that at high temperatures $(T/N = 1)$, the systems behave like a homogeneous melt. All correlation functions show a single step relaxation. The fastest is the bond relaxation and the slowest is the chain relaxation described by the end-to-end vector autocorrelation function. The relaxation of the block is faster than the whole chain relaxation by a factor of approximately two. Such relations between various relaxation times in the disordered state of the copolymer can be regarded as confirmed experimentally for some real systems, in which the dielectric spectroscopy allows distinction of the different relaxation modes [41]. At low temperatures, drastic changes

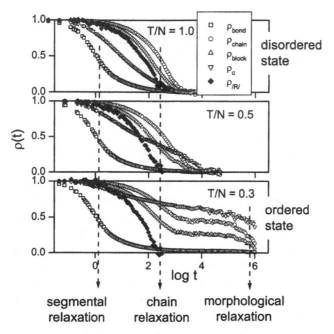

FIG. 7 Various correlation functions determined at various temperatures for the symmetric diblock copolymer melt. The two extreme temperatures correspond to the edges of the temperature range within which the system has been simulated. The high temperature $(T/N = 1.0)$ corresponds to the homogeneous regime and the low temperature $(T/N = 0.3)$ to the strongly segregated limit. The intermediate temperature $(T/N = 0.5)$ is only slightly higher than the temperature of the order-to-disorder transition for this system.

can be noticed for the dynamics of the block copolymer. At temperatures $T/N < 0.45$ (see Fig. 6) the diblock system is in the microphase separated regime and most of the correlation functions determined show bifurcation of the relaxation processes into fast and slow components. The fast components of chain, block, and concentration relaxations are attributed to the almost unchanged in rate, but limited, relaxation of chains when fixed at the A–B interface and the slow components indicate the part of relaxation coupled to the relaxation of the interface within uniformly ordered grains with the lamellar morphology. The concentration relaxation becomes the slowest one in such a state of the system. The dynamic behavior of diblock copolymers is presented in detail and discussed in [38–40], where the spectra of various relaxation modes have been determined in order to compare simulation results with dielectric spectra determined for real copolymer systems in the vicinity of the microphase separation transition [41].

The diffusion in the systems studied has been detected by monitoring in time the mean squared displacements of monomers and centers of mass of chains. Typical results for the diblock copolymer system are shown in Fig. 8. They indicate that the short time displacement rates are not sensitive to temperature but the long time displacements are influenced slightly by the microphase separation. The self-diffusion constants of chains determined at the long time limit are shown in the inset of Fig. 8a, where the effects of the microphase separation in the diblock can be clearly noticed. The slowing down observed at the microphase separation of the system is, however, rather small and indicates a considerable mobility of chains left even when the chains are confined at interfaces. The nature of this mobility has been analyzed by monitoring the correlation between orientation of chain axes and directions of chain displacements (Fig. 8b). It is established that

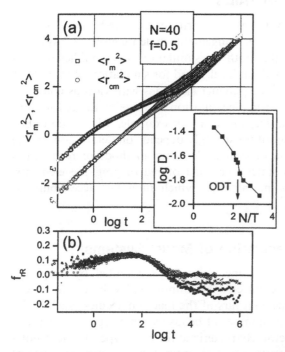

FIG. 8 (a) Mean square displacements of monomers and mean square displacements of chain centers of mass vs. time for the diblock copolymer system at various temperatures. Temperature dependence of the self-diffusion constant of block copolymer chains is shown in the inset. (b) Orientation correlation factor between the end-to-end vector and the center of mass displacement vector of copolymer chains at various temperatures above and below the ODT.

in the phase separated regime the chains diffuse easily in directions parallel to the interface. This is indicated by negative values of the orientation correlation factor and means that the limitations in mobility of diblock chains imposed by the morphology concern only the diffusion through the interface.

Besides the above presented example, the CMA has been applied for simulation of various other copolymer systems with more complex topology [42] of macromolecules and other distributions of comonomers along chains [43,44]. In all cases, the high computational efficiency of the method has enabled detailed information about the structure and dynamics to be obtained, including also the microphase separated states, in which the dynamics becomes considerably slower.

V. IMPLEMENTATION DETAILS

There are essentially several versions of algorithms that are based on similar ideas of performing cooperative rearrangements in dense molecular systems. All versions are equally suitable both for simulation of assemblies of non-bonded beads representing small molecules and for simulation of assemblies of lattice structures, which mimic polymer skeletons with various complexities. Some details essential for implementation of these methods for complex polymers on the fcc lattice, taken as an example, will be described here. The efficiency of an algorithm strongly depends on details concerning methods used for description of system elements, methods of recognition and description of systems states, methods of rearranging the elements, and finally on the programming methods, which allow fast accessibility to large data arrays.

A. Description and Generation of Model Systems

The architecture of complex polymers can be represented by simplified models consisting of beads connected by nonbreakable bonds in a way that corresponds to backbone contours of the macromolecules. Such molecules consist usually of a large number of beads assuming specific positions within a complex bond skeleton characteristic for each type of macromolecule. In this simplified representation of macromolecular structures, sizes of monomers are not distinguishable. With this approximation, the macromolecules can, however, be represented on lattices. The lattice plays the role of a topological skeleton of space and allows fast identification of neighbors. An example of such a representation of a linear macromolecule on the face-centered cubic (fcc) lattice is shown in Fig. 9a.

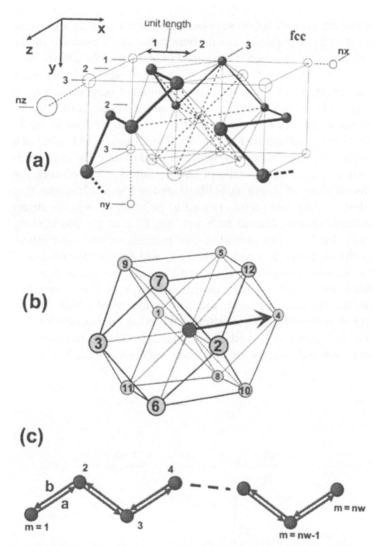

FIG. 9 Illustration of details concerning description of polymer chains and coding directions in the fcc lattice: (a) Fragment of a simplified linear polymer chain on the fcc lattice with beads occupying lattice sites and bonds of constant length connecting only nearest neighbors. The size of the model system is given by numbers nx, ny, and nz of length units in the x, y, and z directions, respectively. (b) The coordination cell of the fcc lattice illustrating the orientation code of the site–site vectors. The vector shown has the direction code $cd = 4$. (c) Description of a linear chain of length nw by means of numbering of beads and using two sets of vectors **a** and **b** pointing in opposite directions for defining of the chain contour.

In a model system with all lattice sites occupied, each bead at a position (x, y, z) is described by a sequence of numbers defining its position in the chain, $m(x, y, z)$, the chain to which it belongs, $k(x, y, z)$, and the connectivity with other beads by means of up to four vectors, $\mathbf{a}(x, y, z)$, $\mathbf{b}(x, y, z)$, $\mathbf{c}(x, y, z)$, and $\mathbf{d}(x, y, z)$, which point towards the bonded neighbors. Vector orientation is given by a code number, cd, assuming values from 1 to 13, which describe the 12 possible orientations of the site–site vectors in the 12 coordinated fcc lattice, as illustrated in Fig. 9b. The value $d = 13$ is used as a code for the vector of length equal to 0, which is used to describe chain ends and single beads representing elements of simple liquid (e.g., solvent). An example of a linear chain of length nw is illustrated in Fig. 9c. It is seen that each bond is described by two vectors (**a** and **b**) pointing in opposite directions. This has the advantage that, at each position, (x, y, z), the connectivity and consequently the local conformation can immediately be established. Table 1 shows the conformation code, $con(\mathbf{b},\mathbf{a})$, used for the description of various angles between the neighboring bonds along the chain contour. Positions of beads are described by the coordinates (x, y, z). It has appeared useful to introduce matrices xnp, ynp, and znp by means of which coordinates (x_n, y_n, z_n) of neighboring lattice sites in direction cd can be found as $x_n = xnp(x, cd)$, $y_n = ynp(y, cd)$, and $z_n = znp(z, cd)$. These matrices can consider system sizes and the type of boundary conditions (e.g., periodic).

TABLE 1   Conformation Code used in the DLL and CMA Simulations

Code (*con*)	Angle between neighboring bonds	Illustration	Range of application
1	180°		
2	120°		Chain interior
3	90°		$1 < m < nw$
4	60°		$\mathbf{a} < 13$ and $\mathbf{b} < 13$
5	—	$\mathbf{b} = 13,\ m = 1$	Ends of chains
6	—	$\mathbf{a} = 13,\ m = nw$	
7	—	$\mathbf{a} = 13,\ \mathbf{b} = 13$; free bead (e.g., solvent)	

The initial states of model systems can generally be obtained in two ways: (1) by generating ordered structures of uniform polymers and subsequent "melting" of the ordered systems or (2) by polymerization in bulk, which can be performed according to various polymerization mechanisms leading in this way to systems with realistic nonuniformities of molecular parameters. The first method has mainly been used in published papers [26,27,34,35,37], whereas the second method has been used recently [45].

B. Implementation of the DLL Model

This is the simplest version of the cooperative algorithm. It has minimum assumptions concerning types of moves and has no limits in complexity of the macromolecular structures. Moreover, it represents probably the most plausible dynamics on the local scale.

An implementation of the DLL model for nonbonded beads is extremely simple. In an athermal case, it consists in periodic repetition of the following sequence of steps: (1) generation of the vectors (one per bead) representing attempts of bead motion to neighboring lattice sites, (2) recognition of attempts forming circuits for which the continuity condition in the simplest form applies, and (3) performing rearrangements by displacing elements along the found circuits, each to a neighboring position. In the non-athermal case and in the case when beads are bounded to more complex structures (macromolecules), additional conditions immobilizing some system elements should be taken into account before the second step [11–14]. One period including the above described steps is considered in this algorithm as a unit of time. The second step of the procedure is the step that controls the computational efficiency. It has been recognized that the vectors of attempted moves form highly branched structures sporadically including closed loops. When the branches, which usually end with situation 2 illustrated in Fig. 2, are declared as nonmobile, only the loops remain. Further details of implementation are dependent on the type of hardware and software used. Large fragments of the algorithm based on this model (steps 1 and 3) can be vectorized, therefore an implementation on a vector computer may be efficient as well.

It is worthwhile to mention that the DLL model can be considered as defining a special purpose parallel computer. A large number of microprocessors, each controlling the logic of a single lattice site, when arranged in an architecture of a spatial, highly coordinated lattice system could constitute a machine, the size of which and consequently the size of the simulated system will be limited less by the computation time than by the cost of corresponding investment. A microprocessor with the single lattice site logic has already been designed [46] and further steps of corresponding development depend on availability of financial support.

C. The CMA (Cooperative Motion Algorithm)

In contrast to the parallel DLL algorithm the CMA, as most other known lattice simulation methods, uses a sequential searching for motion possibilities in randomly chosen fragments of the system. A virtual point, called the "searching point" is introduced to search for the mobile loops. The searching procedure will be described here for a simple, two-dimensional example (Fig. 10) of nonbonded beads on a triangular lattice. This procedure is applied periodically and consists of the following steps: (1) random choice of the initial position of the searching point (1 in Fig. 10), (2) random choice of the direction to one of the nearest neighbors (2 in Fig. 10), (3) moving of a bead from position 2 to 1 (a temporary double occupation defect is created in this way in position 1 and the site denoted as 2 becomes temporarily empty), (4) moving the searching point to position 2, (5) random choice of a new direction (e.g., vector pointing to position 3 in Fig. 10), (6) bead from position 3 moves to position 2, (7) searching point moves to position 3, (8) random choice of a new direction from position 3 (e.g., position 1), and (9) the bead from position 1 moves to position 3. With the last step the loop is closed, each element along the loop has been moved by one lattice site–site distance, and the temporary defects are relaxed. The procedure can be repeated again starting with the random choice of the new searching point position. Not all attempted loops are successful as in the illustrated example. The acceptability depends on further conditions imposed on the type of random walk allowed. Various assumptions here lead to various versions of the algorithm. The following possibilities have been considered: (1) random walk, (2) non-reversal random walk, and (3) self-avoiding walk. The case (3) leads to the dynamics which is the closest to that of the DLL model, whereas the case (2) can be very efficiently used for studies of the static properties of systems or for dynamic effects much slower than the single local rearrangements.

For motion of chains, the searching becomes more complicated because of the demands imposed on the system such as nonbreakability of chains, conservation of sequences of beads along chains, conservation of chain lengths and architecture, etc. There are two possibilities (illustrated in Fig. 11) to move chain parts satisfying the above conditions: (i) the searching path is crossing the chain contour (Fig. 11a) and a displacement of the bead involved is possible just by rotation of bonds connecting that bead to the

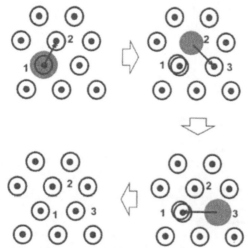

FIG. 10   An example of a simple rearrangement loop as searched using the CMA for a system of nonbonded beads (shown as rings) on a two-dimensional lattice (thick points represent lattice sites). The virtual "searching point" is denoted by means of a gray circle. The initial state is shown in the upper left figure. The searching point is located at position 1 by a random choice and the direction to position 2 is chosen randomly as well. Shifting the bead from position 2 to 1 creates two temporary defects: a double occupancy at position 1 and a vacancy at position 2, as illustrated in the upper right figure. In this state, the searching point is shifted to position 2 and by chance the position 3 is chosen. As the lower right part illustrates the bead from position 3 moves to position 2 and the searching point moves in the opposite direction. Finally, again by chance from position 3, the position 1 is found and the bead located originally at this position is moved to position 3. This closes the loop and relaxes the temporary defects. As a result of such a single rearrangement the state shown in the lower left figure is obtained, in which all lattice sites are occupied but at positions 1, 2, and 3 the elements (beads) changed their positions with respect to the original state replacing each other along the closed loop.

remaining chain structure and (ii) the searching path enters the chain contour and leaves it after a short walk along the chain (Fig. 11b). In the first case, the motion can be performed without extending any bonds, whereas, in the second case, in each step of the searching point, one and only one bond is attempted to be extended but this extension must immediately be relaxed by motion of other beads along the chain as long as the searching point leaves the chain at places with suitable local conformations or through chain ends. The latter forces the searching point to go along the chain contour for some time. In fact, the first possibility can also be considered as a special

(a)

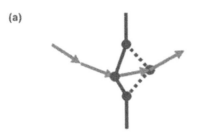

(b)

FIG. 11 Illustration of two types of rearrangements, which are considered in a system of beads bonded to chains: (a) the path of the searching point (gray arrows) crosses the chain contour and the resulting rearrangement takes place by rotation of bonds connecting the moved bead with the remaining parts of the chain and (b) the path of the searching point enters the chain contour, goes locally along the chain, and leaves it after shifting several beads. In both cases, the chain is neither broken nor extended and the sequence of beads along the chain is preserved. The only result is a local conformational change incorporating one or several beads.

case of the second possibility, in which the path of the searching point along the chain is reduced to zero.

A very high efficiency of the implementation of the CMA for polymers has been achieved, when it was recognized that there is a limited number of situations in which the searching point can meet while making a walk through a system containing beads bonded to polymers. A specific situation code, $pat = cr(cd, \mathbf{a}, \mathbf{b})$, is used in order to recognize these situations for various orientations of bonds, various directions of chains, and various local conformations. There are 15 situations distinguished; these are listed in Table 2. From these 15 cases, one corresponds to the case, when propagation of motion of the searching point becomes impossible ($pat = 1$), because the attempted movement would extend two bonds simultaneously. Two cases, $pat = 2$ and $pat = 3$, describe the motion of the searching point along a chain in two different directions. In cases $pat = 4, 5$ and $pat = 10, 11$, the searching point can enter the chain contour through a chain end or through a suitable local conformation, respectively. When $pat = 6, 7$ or $pat = 13, 14$, the searching point leaves the chain contour. The other cases correspond to situations of the type (i) illustrated in Fig. 11a. In all cases,

TABLE 2   The Situation Code $[pat = cr(cd, \mathbf{a}, \mathbf{b})]$ used in the CMA Procedures

Code (*pat*)	Situation before → after	Description
1		Motion impossible—the searching point attempts to stretch two bonds
2		Motion of the searching point along the chain in direction of vector **a**
3		Motion of the searching point along the chain in direction of vector **b**
4		Searching point enters the chain contour in direction of **a** via the end of chain (**b** = 13)
5		Searching point enters the chain contour in direction of **b** via the end of chain (**a** = 13)
6		Searching point leaves the chain contour via the end of chain (**b** = 13)
7		Searching point leaves the chain contour via the end of chain (**a** = 13)
8		Rotation of bond with the chain end (**b** = 13)
9		Rotation of bond with the chain end (**a** = 13)
10		Searching point enters the chain contour along the direction of **b**
11		Searching point enters the chain contour along the direction of **a**
12		Searching point crosses the chain contour
13		Searching point leaves the chain in direction **a**
14		Searching point leaves the chain in direction **b**
15		Searching point meets a single bead

✿—Temporarily free lattice site.

when the searching point is not forced to move along the chain, the direction of motion is chosen randomly. Coding of situations and resulting rearrangements can considerably speed up the simulation. The described codes should be considered as suitable examples. Other, maybe more effective, solutions are certainly possible.

The various versions of the CMA are suitable for implementation and usage on personal computers. With recently available technology, dense polymeric systems of up to one million beads organized in various macromolecular architectures can be simulated in a reasonably inexpensive way.

VI. CONCLUDING REMARKS

It has been demonstrated that a lot of problems concerning the behavior of complex polymer systems can be analyzed successfully by the methods discussed. Difficulties in simulations of dense systems and systems with complex molecular topologies seem to be overcome in the DLL and CMA algorithms. Both can be very efficiently applied to the study of static and dynamic properties of polymer melts with consideration of various topologies of molecules and various interactions between molecular elements.

This success has been achieved by an application of a reasonably simplified dynamic model of cooperative rearrangements, which cuts off details of the dynamics characteristic for shorter time ranges. Such treatment, consisting in a simplification of the dynamics to features characteristic to a given time range, should be considered as analogous to simplifications of the structure in relation to various size scales. Unfortunately, a scheme of dynamic simplifications over a broad time range corresponding to the scheme of structures considered with variable size resolution is generally not as complete as the latter.

The DLL model can be considered as defining a special purpose parallel computer, the development of which could solve the problem of strong size limits for simulated systems caused usually by strong dependencies of computation times on system sizes.

REFERENCES

1. Verdier, P.H.; Stockmeyer, W.H. J. Chem. Phys. **1962**, *36*, 227.
2. Lal, M. Mol. Phys. **1969**, *17*, 57.
3. Wall, F.T.; Mandel, F. J. Chem. Phys. **1975**, *63*, 4592.
4. Carnesin, I.; Kremer, K. Macromolecules **1988**, *21*, 2819.
5. Binder, K.; Kremer, K. Comput. Phys. Rep. **1988**, *7*, 261.
6. Pakula, T. Macromolecules **1987**, *20*, 679.
7. Geyler, S.; Pakula, T.; Reiter, J. J. Chem. Phys. **1990**, *92*, 2676.

8. Reiter, J.; Edling, T.; Pakula, T. J. Chem. Phys. **1990**, *93*, 837.
9. Gauger, A.; Weyersberg, A.; Pakula, T. Macromol. Chem. Theory Simul. **1993**, *2*, 531.
10. Gauger, A.; Pakula, T. Macromolecules **1995**, *28*, 190.
11. Pakula, T.; Teichmann, J. Mat. Res. Soc. Symp. Proc. **1997**, *495*, 211.
12. Teichmann, J. Ph.D. thesis, University of Mainz, 1996.
13. Pakula, T.J. Mol. Liquids **2000**, *86*, 109.
14. Polanowski, P.; Pakula, T. J. Chem. Phys. **2002**, *117*, 4022.
15. Barker, J.A.; Henderson, D. Rev. Mod. Phys. **1976**, *48*, 587.
16. Alder, B.J.; Wainwright, T.E. J. Chem. Phys. **1969**, *31*, 459.
17. Widom, A. Phys. Rev. A **1971**, *3*, 1394.
18. Kubo, R. Rept. Progr. Phys. **1966**, *29*, 255.
19. London, R.E. J. Chem. Phys. **1977**, *66*, 471.
20. Götze, W.; Sjögren, L. Rep. Prog. Phys. **1992**, *55*, 241.
21. Cohen, M.H.; Grest, G.S. Phys. Rev. B **1979**, *20*, 1077.
22. Adam, G.; Gibbs, J.H. J. Chem. Phys. **1965**, *43*, 139.
23. Hughes, B.D. *Random Walks and Random Environments*; Clarendon Press: Oxford, 1995.
24. Reiter, J. Macromolecules **1990**, *23*, 3811.
25. Reiter, J. Physica A **1993**, *196*, 149.
26. Pakula, T.; Geyler, S.; Edling, T.; Boese, D. Rheol. Acta **1996**, *35*, 631.
27. Pakula, T.; Harre, K. Comp. Theor. Polym. Sci. **2000**, *10*, 197.
28. Fox, T.G.; Flory, P.J. J. Appl. Phys. **1950**, *21*, 581.
29. Paul, W.; Binder, K.; Herrmann, D.W.; Kremer, K. J. Chem. Phys. **1991**, *95*, 7726.
30. Kolinski, A.; Skolnick, J.; Yaris, R. J. Chem. Phys. **1987**, *86*, 7164.
31. Fetters, L.J.; Kiss, A.D.; Pearson, D.S.; Quack, G.F.; Vitus, F.J. Macromolecules **1993**, *26*, 647.
32. Roovers, J.; Lin-Lin Zbou, Toporowski, P.M.; van der Zwan, M.; Iatron, H.; Hadjichristidis, N. Macromolecules **1993**, *26*, 4324.
33. Antonietti, M.; Pakula, T.; Bremser, W. Macromolecules **1995**, *28*, 4227.
34. Pakula, T. Comp. Theor. Polym. Sci. **1998**, *8*, 21.
35. Pakula, T.; Vlassopoulos, D.; Fytas, G.; Roovers, J. Macromolecules **1998**, *31*, 8931.
36. Grest, G.S.; Kremer, K.; Milner, S.T.; Witten, T.A. Macromolecules **1989**, *22*, 1904.
37. Pakula, T. J. Comp. Aided Mat. Des. **1996**, *3*, 329.
38. Pakula, T.; Karatasos, K.; Anastasiadis, S.H.; Fytas, G. Macromolecules **1997**, *30*, 8463.
39. Pakula, T.; Floudas, G. In *Block Copolymers*, Balta Calleja, F.J., Roslaniec, Z., Eds.; Marcel Dekker Inc.: New York, 2000; 123–177.
40. Pakula, T. J. Macromol. Sci.—Phys. **1998**, *B37*, 181.
41. Karatasos, K.; Anastasiadis, S.H.; Floudas, G.; Fytas, G.; Pispas, S.; Hadjichristidis, N.; Pakula, T. Macromolecules **1996**, *29*, 1326.

42. Floudas, G.; Pispas, S.; Hadjichristidis, N.; Pakula, T.; Erukhimovich, I. Macromolecules **1996**, *29*, 4142.
43. Vilesov, A.D.; Floudas, G.; Pakula, T.; Melenevskaya, E. Yu.; Birstein, T.M.; Lyatskaya, Y.V. Macromol. Chem. Phys. **1994**, *195*, 2317.
44. Pakula, T.; Matyjaszewski, K. Macromol. Theory Simul. **1996**, *5*, 987.
45. Miller, P.J.; Matyjaszewski, K.; Shukla, N.; Immaraporn, B.; Gelman, A.; Luokala, B.B.; Garoff, S.; Siclovan, T.M.; Kickelbick, G.; Vallant, T.; Hoffmann, H.; Pakula, T. Macromolecules **1999**, *32*, 8716.
46. Polanowski, P. Ph.D. thesis, Technical University of Lodz, 2002.

6

Molecular Dynamics Simulations of Polymers

VAGELIS A. HARMANDARIS and VLASIS G. MAVRANTZAS   Institute of Chemical Engineering and High-Temperature Chemical Processes, and University of Patras, Patras, Greece

I. THE MOLECULAR DYNAMICS TECHNIQUE

Molecular dynamics (MD) is a powerful technique for computing the equilibrium and dynamical properties of classical many-body systems. Over the last fifteen years, owing to the rapid development of computers, polymeric systems have been the subject of intense study with MD simulations [1].

At the heart of this technique is the solution of the classical equations of motion, which are integrated numerically to give information on the positions and velocities of atoms in the system [2–4]. The description of a physical system with the classical equations of motion rather than quantum-mechanically is a satisfactory approximation as long as the spacing $h\nu$ between the successive energy levels described is $h\nu < k_B T$. For a typical system at room temperature this holds for $< \sim 0.6 \times 10^{13}$ Hz, i.e., for motions of time periods of about $t > \sim 1.6 \times 10^{-13}$ sec or 0.16 ps.

A simple flow diagram of a standard MD algorithm is shown in Fig. 1 and includes the following steps:

1. First, a model configuration representing a molecular-level snapshot of the corresponding physical system is chosen or constructed, and is initialized (initial positions, velocities of each particle within the system).

2. Then the total force acting on each particle within the system is computed. For polymer systems such a force has two

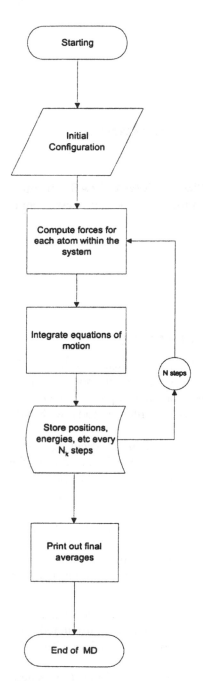

FIG. 1 A simple flow diagram of a standard MD algorithm.

components: intermolecular (from atoms belonging to different polymer chains) and intramolecular (from atoms belonging to the same chain).

3. The integration of the equations of motion follows with an appropriate method. The most popular of these will be described in detail in the next section.

4. Actual measurements are performed (positions, velocities, energies, etc., are stored) after the system has reached equilibration, periodically every N_k steps.

5. After completion of the central loop (N steps), averages of the measured quantities and of the desired properties are calculated and printed.

II. CLASSICAL EQUATIONS OF MOTION

As stated above, at the heart of an MD simulation is the solution of the classical equations of motion. Let us consider a system consisting of N interacting molecules described by a potential energy function V. Let us also denote as q_k and $\dot{q}_k$ the generalized coordinates describing the molecular configuration and their time derivatives, respectively. The classical equations of motion for this system can be formulated in various ways [5]. In the Lagrangian formulation, the trajectory $\mathbf{q}(t)$ $(=q_1(t), q_2(t), \ldots, q_k(t), \ldots)$ satisfies the following set of differential equations:

$$\frac{\partial L}{\partial q_k} = \frac{d}{dt}\left(\frac{\partial L}{\partial \dot{q}_k}\right) \tag{1}$$

where L is the Lagrangian of the system. This is defined in terms of the kinetic energy, K, and potential energy, V, as $L = L(\mathbf{q}, \dot{\mathbf{q}}, t) \equiv K - V$. The generalized momenta p_k conjugate to the generalized coordinates q_k are defined as

$$p_k = \frac{\partial L}{\partial \dot{q}_k} \tag{2}$$

Alternatively, one can adopt the Hamiltonian formalism, which is cast in terms of the generalized coordinates and momenta. These obey Hamilton's equations

$$\dot{q}_k = \frac{\partial H}{\partial p_k}, \qquad \dot{p}_k = -\frac{\partial H}{\partial q_k} \tag{3}$$

where H is the Hamiltonian of the system, defined through the equation

$$H(\mathbf{p}, \mathbf{q}) = \sum_k \dot{q}_k p_k - L \tag{4}$$

If the potential V is independent of velocities and time, then H becomes equal to the total energy of the system: $H(\mathbf{p}, \mathbf{q}) = K(\mathbf{p}) + V(\mathbf{q})$ [5]. In Cartesian coordinates, Hamilton's equations of motion read:

$$\dot{\mathbf{r}}_i \equiv \mathbf{v}_i = \frac{\mathbf{p}_i}{m_i}, \qquad \dot{\mathbf{p}}_i = -\nabla_{\mathbf{r}_i} V \equiv -\frac{\partial V}{\partial \mathbf{r}_i} = \mathbf{F}_i \tag{5}$$

hence

$$m_i \ddot{\mathbf{r}}_i \equiv m_i \dot{\mathbf{q}}_i = \mathbf{F}_i \tag{6}$$

where $\mathbf{F}_i$ is the force acting on atom i. Solving the equations of motion then involves the integration of the $3N$ second-order differential equations (6) (Newton's equations).

The classical equations of motion possess some interesting properties, the most important one being the conservation law. If we assume that K and V do not depend explicitly on time, then it is straightforward to verify that $\dot{H} = dH/dt$ is zero, i.e., the Hamiltonian is a constant of the motion. In actual calculations this conservation law is satisfied if there exist no explicitly time- or velocity-dependent forces acting on the system.

A second important property is that Hamilton's equations of motion are reversible in time. This means that, if we change the signs of all the velocities, we will cause the molecules to retrace their trajectories backwards. The computer-generated trajectories should also possess this property.

There are many different methods for solving ordinary differential equations of the form of Eq. (6). Criteria for the proper choice of an algorithm include the following:

- The algorithm must not require an expensively large number of force evaluations per integration time step. Many common techniques for the solution of ordinary differential equations (such as the fourth-order Runge–Kutta method) become inappropriate, since they do not fulfill this criterion.
- The algorithm should satisfy the energy conservation law. It is also desirable that it be time reversible and conserve volume in phase space (be symplectic).

- The algorithm should permit the use of a large time step dt.
- The algorithm should be fast and require little memory.

Concerning the solution of equations of motion for very long times, it is clear that no algorithm provides an essentially exact solution. But this turns out to be not a serious problem, because the main objective of an MD simulation is not to trace the exact configuration of a system after a long time, but rather to predict thermodynamic properties as time averages and calculate time correlation functions descriptive of the dynamics.

In the following we briefly describe the two most popular families of algorithms used in MD simulations for the solution of classical equations of motion: the higher-order methods and the Verlet algorithms.

A. Higher-Order (Gear) Methods

The basic idea in the higher-order methods is to use information about positions and their first, second, ..., nth time derivatives at time t in order to estimate positions and their first, second, ..., nth time derivatives at time $t + dt$ [2]. If we consider the Taylor expansion of the position vectors of a given particle at time $t + dt$ including terms up to fourth-order we have

$$\mathbf{r}^p(t + dt) = \mathbf{r}(t) + dt\,\mathbf{v}(t) + \frac{dt^2}{2}\ddot{\mathbf{r}}(t) + \frac{dt^3}{6}\dddot{\mathbf{r}}(t) + \frac{dt^4}{24}\ddddot{\mathbf{r}}(t) + \cdots \tag{7}$$

$$\mathbf{v}^p(t + dt) = \mathbf{v}(t) + dt\,\ddot{\mathbf{r}}(t) + \frac{dt^2}{2}\dddot{\mathbf{r}}(t) + \frac{dt^3}{6}\ddddot{\mathbf{r}}(t) + \cdots \tag{8}$$

$$\ddot{\mathbf{r}}^p(t + dt) = \ddot{\mathbf{r}}(t) + dt\,\dddot{\mathbf{r}}(t) + \frac{dt^2}{2}\ddddot{\mathbf{r}}(t) + \cdots \tag{9}$$

$$\dddot{\mathbf{r}}^p(t + dt) = \dddot{\mathbf{r}}(t) + dt\,\ddddot{\mathbf{r}}(t) + \cdots \tag{10}$$

In the above equations, the superscript p is used to denote "predicted" values and the dots, time derivatives. Equations (7)–(10) do not generate classical trajectories, since we have not as yet introduced the equations of motion. To do this we estimate the size of the error incurred by the expansion, $\Delta \mathbf{x}$, by calculating the forces (or, equivalently, the accelerations) at the predicted positions

$$\Delta \mathbf{x} \equiv \ddot{\mathbf{r}}(\mathbf{r}^p(t + dt)) - \ddot{\mathbf{r}}^p(t + dt)$$
$$= -diag(1/m_1, 1/m_2, \ldots, 1/m_N)\nabla_r V(\mathbf{r}^p(t + dt)) - \ddot{\mathbf{r}}^p(t + dt) \tag{11}$$

The error is accounted for and corrected in a "corrector" step, that is

$$\mathbf{r}^c(t + dt) = \mathbf{r}^P(t + dt) + c_0 \Delta \mathbf{x} \tag{12}$$

$$\mathbf{v}^c(t + dt) = \mathbf{v}^P(t + dt) + c_1 \Delta \mathbf{x} \tag{13}$$

$$\ddot{\mathbf{r}}^c(t + dt) = \ddot{\mathbf{r}}^P(t + dt) + c_2 \Delta \mathbf{x} \tag{14}$$

$$\dddot{\mathbf{r}}^c(t + dt) = \dddot{\mathbf{r}}^P(t + dt) + c_3 \Delta \mathbf{x} \tag{15}$$

where c_i, $i = 1, \ldots, n$ are constants. The values of c_i are such that they yield an optimal compromise between desired level of accuracy and algorithm stability [2].

The general scheme of an algorithm based on the predictor-corrector method goes as follows:

1. Predict positions and their first, second, $\ldots$, nth time derivatives at time $t + dt$ using their values at time t.
2. Compute forces using the predicted positions and then the corresponding error $\Delta \mathbf{x}$ from the differences between accelerations as calculated from forces and accelerations as predicted by the prediction scheme.
3. Correct the predicted positions and their first, second, $\ldots$, nth time derivatives guided by $\Delta \mathbf{x}$.

B. Verlet Methods

Algorithms in this family are simple, accurate, and, as we will see below, time reversible. They are the most widely used methods for integrating the classical equations of motion. The initial form of the Verlet equations [3] is obtained by utilizing a Taylor expansion at times $t - dt$ and $t + dt$

$$\mathbf{r}(t + dt) = \mathbf{r}(t) + dt\,\mathbf{v}(t) + \frac{dt^2}{2}\ddot{\mathbf{r}}(t) + \frac{dt^3}{6}\dddot{\mathbf{r}}(t) + \mathcal{O}(dt^4) \tag{16}$$

$$\mathbf{r}(t - dt) = \mathbf{r}(t) - dt\,\mathbf{v}(t) + \frac{dt^2}{2}\ddot{\mathbf{r}}(t) - \frac{dt^3}{6}\dddot{\mathbf{r}}(t) + \mathcal{O}(dt^4) \tag{17}$$

Summing the two equations gives

$$\mathbf{r}(t + dt) = 2\mathbf{r}(t) - \mathbf{r}(t - dt) + dt^2\ddot{\mathbf{r}}(t) + \mathcal{O}(dt^4) \tag{18}$$

with $\ddot{\mathbf{r}}(t)$ calculated from the forces at the current positions.

Two modifications of the Verlet scheme are of wide use. The first is the "leapfrog" algorithm [3] where positions and velocities are not calculated at the same time; velocities are evaluated at half-integer time steps:

$$\mathbf{r}(t + dt) = \mathbf{r}(t) + dt \; \mathbf{v}\left(t + \frac{dt}{2}\right) \tag{19}$$

$$\mathbf{v}\left(t + \frac{dt}{2}\right) = \mathbf{v}\left(t - \frac{dt}{2}\right) + dt \; \ddot{\mathbf{r}}(t) \tag{20}$$

In order to calculate the Hamiltonian H at time t, the velocities at time t are also calculated as averages of the values at $t + dt/2$ and $t - dt/2$:

$$\mathbf{v}(t) = \frac{1}{2}\left(\mathbf{v}\left(t + \frac{dt}{2}\right) + \mathbf{v}\left(t - \frac{dt}{2}\right)\right) \tag{21}$$

The problem of defining the positions and velocities at the same time can be overcome by casting the Verlet algorithm in a different way. This is the velocity-Verlet algorithm [3,6], according to which positions are obtained through the usual Taylor expansion

$$\mathbf{r}(t + dt) = \mathbf{r}(t) + dt \; \mathbf{v}(t) + \frac{dt^2}{2}\ddot{\mathbf{r}}(t) \tag{22}$$

whereas velocities are calculated through

$$\mathbf{v}(t + dt) = \mathbf{v}(t) + \frac{dt}{2}[\ddot{\mathbf{r}}(t) + \ddot{\mathbf{r}}(t + dt)] \tag{23}$$

with all accelerations computed from the forces at the configuration corresponding to the considered time. To see how the velocity-Verlet algorithm is connected to the original Verlet method we note that, by Eq. (22),

$$\mathbf{r}(t + 2dt) = \mathbf{r}(t + dt) + dt \; \mathbf{v}(t + dt) + \frac{dt^2}{2}\ddot{\mathbf{r}}(t + dt) \tag{24}$$

If Eq. (22) is written as

$$\mathbf{r}(t) = \mathbf{r}(t + dt) - dt \; \mathbf{v}(t) - \frac{dt^2}{2}\ddot{\mathbf{r}}(t) \tag{25}$$

then, by addition, we get

$$\mathbf{r}(t+2dt)+\mathbf{r}(t)=2\mathbf{r}(t+dt)+dt[\mathbf{v}(t+dt)-\mathbf{v}(t)]+\frac{dt^2}{2}[\ddot{\mathbf{r}}(t+dt)-\ddot{\mathbf{r}}(t)]$$

(26)

Substitution of Eq. (23) into Eq. (26) gives

$$\mathbf{r}(t+2dt)+\mathbf{r}(t)=2\mathbf{r}(t+dt)+dt^2\,\ddot{\mathbf{r}}(t+dt)$$

(27)

which is indeed the coordinate version of the Verlet algorithm. The calculations involved in one step of the velocity algorithm are schematically shown in [2, figure 3.2, page 80].

A sample code of the velocity-Verlet integrator is shown in Algorithm 1. In this algorithm, N is the total number of atoms in the system and the subroutine *get_forces* calculates the total force on every atom within the system.

Algorithm 1: Velocity-Verlet Integration Method

```
......
do i = 1, N
    r(i) = r(i) + dt * v(i) + dt * dt/2 * F(i)      ! update positions at t + dt
                                                      using velocities and forces at t
    v(i) = v(i) + dt/2 * F(i)                        ! update velocities at t + dt using
                                                      forces at t
end do

call get_forces (F)                                 ! calculate forces at t + dt

do i = 1, N
    v(i) = v(i) + dt/2 * F(i)                        ! update velocities at t + dt
                                                      using forces at t + dt
end do
......
```

In general, higher-order methods are characterized by a much better accuracy than the Verlet algorithms, particularly at small times. Their biggest drawback is that they are not reversible in time, which results in other problems, such as insufficient energy conservation, especially in very long-time MD simulations. On the other hand, the Verlet methods are not essentially exact for small times but their inherent time reversibility guarantees that the energy conservation law is satisfied even for very long times [4]. This feature renders the Verlet methods, and particularly the velocity-Verlet algorithm, the most appropriate ones to use in long atomistic MD simulations.

III. MD IN OTHER STATISTICAL ENSEMBLES

The methods described above address the solution to Newton's equations of motion in the microcanonical (*NVE*) ensemble. In practice, there is usually the need to perform MD simulations under specified conditions of temperature and/or pressure. Thus, in the literature there exist a variety of methodologies for performing MD simulations under isochoric or isothermal conditions [2,3]. Most of these constitute a reformulation of the Lagrangian equations of motion to include the constraints of constant *T* and/or *P*. The most widely used among them is the Nosé–Hoover method.

A. The Nosé–Hoover Thermostat

To constrain temperature, Nosé [7] introduced an additional degree of freedom, *s*, in the Lagrangian. The parameter *s* plays the role of a heat bath whose aim is to damp out temperature deviations from the desired level. This necessitates adding to the total energy an additional potential term of the form

$$V_s = g k_B T \ln s \tag{28}$$

and an additional kinetic energy term of the form

$$K_s = \frac{Q}{2} \left(\frac{\dot{s}}{s} \right)^2 = \frac{p_s^2}{2Q} \tag{29}$$

In the above equations, *g* is the total number of degrees of freedom. In a system with constrained bond lengths, for example, $g = 3 N_{atoms} - N_{bonds} - 3$, with N_{atoms} and N_{bonds} standing for the total numbers of atoms and bonds, respectively; the value of 3 subtracted in calculating *g* takes care of the fact that the total momentum of the simulation box is constrained to be zero by the periodic boundary conditions. *Q* and p_s represent the "effective mass" and momentum, respectively, associated with the new degree of freedom *s*. Equations of motion are derived from the Lagrangian of the extended ensemble, including the degree of freedom *s*. Their final form, according to Hoover's analysis [8], is

$$\dot{\mathbf{r}}_i = \frac{\mathbf{p}_i}{m_i} \tag{30}$$

$$\dot{\mathbf{p}}_i = -\frac{\partial V}{\partial \mathbf{r}_i} - \frac{\dot{s}}{s} \mathbf{p}_i \tag{31}$$

$$\dot{p}_s = \left(\sum_{i=1}^{N} \frac{\mathbf{p}_i^2}{m_i} - g k_B T \right) / Q, \qquad p_s = Q \frac{\dot{s}}{s} \tag{32}$$

An important result in Hoover's analysis is that the set of equations of motion is unique, in the sense that no other equations of the same form can lead to a canonical distribution.

The total Hamiltonian of the system, which should be conserved during the MD simulation, is

$$H_{Nosé-Hoover} = \sum_{i=1}^{N} \frac{\mathbf{p}_i^2}{m_i} + V(\mathbf{r}^N) + gk_BT \ln s + \frac{p_s^2}{2Q} \tag{33}$$

To construct MD simulations under constant P, an analogous reformulation of the Lagrangian was proposed by Andersen [9]. The constant-pressure method of Andersen allows for isotropic changes in the volume of the simulation box. Later, Hoover [8] combined this method with the isothermal MD method described above to provide a set of equations for MD simulations in the NPT ensemble. Parrinello and Rahman [10] extended Andersen's method to allow for changes not only in the size, but also in the shape of the simulation box. This is particularly important in the simulation of solids (e.g., crystalline polymers) since it allows for phase changes in the simulation involving changes in the dimensions and angles of the unit cell.

B. The Berendsen Thermostat—Barostat

Berendsen proposed a simpler way for performing isothermal and/or isobaric MD simulations without the need to use an extended Lagrangian, by coupling the system into a temperature and/or pressure bath [11]. To achieve this, the system is forced to obey the following equations

$$\frac{dT}{dt} = (T - T_{ext})/\tau_T \tag{34}$$

and

$$\frac{dP}{dt} = (P - P_{ext})/\tau_P \tag{35}$$

where T_{ext} and P_{ext} are the desired temperature and pressure values and τ_T and τ_P are time constants characterizing the frequency of the system coupling to temperature and pressure baths. T and P are the instantaneous values of temperature and pressure, calculated from the momenta and configuration of the system [2]. The solution of these equations forces velocities and positions to be scaled at every time step by factors χ_T and χ_P, respectively, with

$$\chi_T = \left(1 + \frac{dt}{\tau_T}\left(\frac{T}{T_{ext}} - 1\right)\right)^{1/2} \tag{36}$$

$$\chi_P = 1 - \beta_T \frac{dt}{\tau_P}(P - P_{ext}) \tag{37}$$

and β_T being the isothermal compressibility of the system.

The method proposed by Berendsen is much simpler and easier to program than that proposed by Nosé and Hoover. It suffers, however, from the fact that the phase-space probability density it defines does not conform to a specific statistical ensemble (e.g., NVT, NPT). Consequently, there exists no Hamiltonian that should be conserved during the MD simulation.

C. MD in the $NTL_x\sigma_{yy}\sigma_{zz}$ Ensemble

To further illustrate how extended ensembles can be designed to conduct MD simulations under various macroscopic constraints, we discuss here the $NTL_x\sigma_{yy}\sigma_{zz}$ ensemble. $NTL_x\sigma_{yy}\sigma_{zz}$ is an appropriate statistical ensemble for the simulation of uniaxial tension experiments on solid polymers [12] or relaxation experiments in uniaxially oriented polymer melts [13]. This ensemble is illustrated in Fig. 2. The quantities that are kept constant during a molecular simulation in this ensemble are the following:

- the total number of atoms in the system N,
- the temperature T,
- the box length in the direction of elongation L_x, and
- the time average values of the two normal stresses σ_{yy} and σ_{zz}.

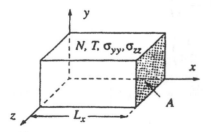

FIG. 2 The $NTL_x\sigma_{yy}\sigma_{zz}$ statistical ensemble.

The $NTL_x\sigma_{yy}\sigma_{zz}$ ensemble can be viewed as a hybrid between the NVT ensemble in the x direction and the isothermal–isobaric (NPT) ensemble in the y and z directions. The temperature T is kept fixed at a prescribed value by employing the Nosé–Hoover thermostat; the latter introduces an additional dynamical variable in the system, the parameter s, for which an evolution equation is derived. Also kept constant during an MD simulation in this ensemble is the box length L_x in the x direction; on the contrary, the box lengths in the other two directions, L_y and L_z, although always kept equal, are allowed to fluctuate. This is achieved by making use in the simulation of an additional dynamical variable, the cross-sectional area $A(=L_yL_z)$ of the simulation cell in the yz plane, which obeys an extra equation of motion involving the instantaneous average normal stress $(\sigma_{yy}+\sigma_{zz})/2$ in the two lateral directions y and z, respectively; $(\sigma_{yy}+\sigma_{zz})/2$ remains constant on average and equal to $-P_{ext}$ throughout the simulation.

The derivation of the equations of motion in the $NTL_x\sigma_{yy}\sigma_{zz}$ ensemble has been carried out in detail by Yang et al. [12], and goes as follows: Consider a system consisting of N atoms with $\mathbf{r}_{ik}$ being the position of atom i belonging to polymer chain k. The bond lengths are kept fixed, with $\mathbf{g}_{ik}$ denoting the constraint forces on atom i. The Lagrangian is written as a function of the "extended" variables $\{\tilde{\mathbf{R}}_k, \mathbf{x}_{ik}, A, s\}$ where $\tilde{\mathbf{R}}_k$ is the scaled (with respect to the box edge lengths) position of the center of mass of every chain k, and $\mathbf{x}_{ik}$ is the position of atom i in chain k measured relative to the chain center of mass. This ensemble is "extended" in the sense that it invokes the additional variables A and s, makes use of a scaled coordinate system, and is formulated with respect to a "virtual" time t'. The equations of motion are derived from the extended Lagrangian by exactly the same procedure as for the other statistical ensembles. The final equations are further recast in terms of real coordinates and real time and have the following form:

$$m_i\ddot{r}_{xik} = F_{xik} + g_{xik} - \frac{\dot{s}}{s}p_{xik} \tag{38}$$

$$m_i\ddot{r}_{yik} = F_{yik} + g_{yik} - \frac{\dot{s}}{s}p_{yik} + \frac{m_iR_{yk}}{2A}\left(\ddot{A} - \frac{\dot{A}^2}{2A}\right) \tag{39}$$

$$m_i\ddot{r}_{zik} = F_{zik} + g_{zik} - \frac{\dot{s}}{s}p_{zik} + \frac{m_iR_{zk}}{2A}\left(\ddot{A} - \frac{\dot{A}^2}{2A}\right) \tag{40}$$

$$Q\ddot{s} = Q\frac{\dot{s}^2}{s} + s\left[\sum_k\sum_i\frac{p_{xik}^2 + p_{yik}^2 + p_{zik}^2}{m_i} - (g+1)k_BT\right] \tag{41}$$

$$W\ddot{A} = W\frac{\dot{s}\dot{A}}{s} + s^2L_x\left[\frac{1}{2}\left((-\sigma_{yy}) + (-\sigma_{zz})\right) - P_{ext}\right] \tag{42}$$

where the forces with two indices indicate center of mass forces, while those with three indices are forces on atoms within a particular polymer chain. $\mathbf{R}_k$ denotes the center of mass of molecule k, while Q and W are inertial constants governing fluctuations in the temperature and the two normal stresses σ_{yy} and σ_{zz}, respectively. The total Hamiltonian of the extended system, derived from the Lagrangian, has the form:

$$H_{NTL_x\sigma_{yy}\sigma_{zz}} = \sum_i \frac{p_i^2}{2m_i} + V(\mathbf{r}) + \frac{Q}{2}\left(\frac{\dot{s}}{s}\right)^2 + (g+1)\frac{\ln s}{\beta} + \frac{W}{2}\left(\frac{\dot{A}}{s}\right)^2 + P_{ext}L_x A$$

(43)

The first term on the right hand side represents the kinetic energy, the second term is the potential energy, and the last four terms are the contributions due to the thermostat and the fluctuating box cross-sectional area in the plane yz. Conservation of $H_{NTL_x\sigma_{yy}\sigma_{zz}}$ is a good test for the simulation.

For the solution of equations of motion, a modification of the velocity-Verlet algorithm proposed by Palmer [14] can be followed.

IV. LIOUVILLE FORMULATION OF EQUATIONS OF MOTION—MULTIPLE TIME STEP ALGORITHMS

In Sections II.A and II.B we presented the most popular algorithms for integrating Newton's equations of motion, some of which are not reversible in time. Recently, Tuckerman et al. [15] and Martyna et al. [16] have shown how one can systematically derive time reversible MD algorithms from the Liouville formulation of classical mechanics.

The Liouville operator L of a system of N degrees of freedom is defined in Cartesian coordinates as

$$iL = \sum_{i=1}^{N}\left[\dot{\mathbf{r}}_i\frac{\partial}{\partial\mathbf{r}_i} + \mathbf{F}_i\frac{\partial}{\partial\mathbf{p}_i}\right]$$

(44)

If we consider the phase-space of a system, $\Gamma = \{\mathbf{r}, \mathbf{p}\}$, the evolution of the system from time 0 to time t, can be found by applying the evolution operator

$$\Gamma(t) = \exp(iLt)\Gamma(0)$$

(45)

The next step is to decompose the evolution operator into two parts such that

$$iL = iL_1 + iL_2 \quad \text{with} \quad iL_1 = \sum_{i=1}^{N} \left[F_i \frac{\partial}{\partial \mathbf{p}_i} \right], \qquad iL_2 = \sum_{i=1}^{N} \left[\dot{\mathbf{r}}_i \frac{\partial}{\partial \mathbf{r}_i} \right] \qquad (46)$$

For this decomposition, a short-time approximation to the evolution operator can be generated via the Trotter theorem [16] as

$$\exp(iLt) = \exp(i(L_1 + L_2)t/P)^P$$
$$= (\exp(iL_1(dt/2)) \exp(iL_2 dt) \exp(iL_1(dt/2)))^P + O(t^3/P^2) \qquad (47)$$

where $dt = t/P$. Thus, the evolution operator becomes

$$\exp(iLdt) = \exp\left(iL_1 \frac{dt}{2} \right) \exp(iL_2 dt) \exp\left(iL_1 \frac{dt}{2} \right) + O(dt^3) \qquad (48)$$

The evolution of the system at time t using the above factorization, Eq. (48), is described through the following scheme [16]

$$\mathbf{r}(dt) = \mathbf{r}(0) + dt\,\mathbf{v}(0) + \frac{dt^2}{2m} \mathbf{F}[\mathbf{r}(0)] \qquad (49)$$

$$\mathbf{v}(dt) = \mathbf{v}(0) + \frac{dt^2}{2m} (\mathbf{F}[\mathbf{r}(0)] + \mathbf{F}[\mathbf{r}(dt)]) \qquad (50)$$

which can be derived using the identity $\exp[a(\partial/\partial g(x))]x = g^{-1}[g(x) + a]$. The result is the well-known velocity-Verlet integration scheme, described before, which is now derived in a different way.

Based on the previous factorization a very efficient algorithm can be developed, through the use of different time steps for integrating the different parts of the Liouville operator. This is the so-called reversible REference System Propagator Algorithm (rRESPA).

A. The rRESPA Algorithm

In the rRESPA algorithm, the above factorization is employed together with an integration of each part of the Liouville operator with a different time step. In addition, the forces $\mathbf{F}$ are also decomposed into fast (short-range) forces $\mathbf{F}^f$, and slow (long-range) forces $\mathbf{F}^s$, according to

$F(r) = F^f(r) + F^s(r)$. The total evolution operator is broken up into $iL = iL_1 + iL_2 + iL_3$ with

$$iL_1 = \sum_{i=1}^{N} \left[F_i^f(\mathbf{r}) \frac{\partial}{\partial \mathbf{p}_i} \right], \quad iL_2 = \sum_{i=1}^{N} \left[\dot{\mathbf{r}}_i \frac{\partial}{\partial \mathbf{r}_i} \right], \quad iL_3 = \sum_{i=1}^{N} \left[F_i^s(\mathbf{r}) \frac{\partial}{\partial \mathbf{p}_i} \right] \quad (51)$$

The heart of the rRESPA algorithm is that the equations of motion are integrated by using two different time steps, it is therefore a Multiple Time Step (MTS) method: the slow modes (slow forces, iL_3) are integrated with a larger time step, Δt, whereas the fast modes (fast forces and velocities, iL_1, iL_2) with a smaller time step, $\delta t (\delta t = \Delta t/n)$. In this case the evolution operator becomes [16]

$$\exp(iL\Delta t) = \exp\left(iL_3 \frac{\Delta t}{2}\right) \left[\exp\left(iL_1 \frac{\delta t}{2}\right) \exp(iL_2 \delta t) \exp\left(iL_1 \frac{\delta t}{2}\right) \right]^n \exp\left(iL_3 \frac{\Delta t}{2}\right)$$
$$+ O(\Delta t^3) \quad (52)$$

The force calculated n times (fast force) is called the reference force. A sample code of an MTS integrator is given in Algorithm 2.

Algorithm 2: rRESPA Integration

```
.......
do i = 1, N
    v(i) = v(i) + Δt/2 * Fˢ(i)              ! update velocities using
                                            ! slow forces at t
end do

do j = 1, n
    do i = 1, N
        v(i) = v(i) + δt/2 * Fᶠ((j−1)δt)    ! update velocities using
                                            ! fast forces at t+(j−1)δt
        r(i) = r(i) + δt * v(i)             ! update positions at t+jδt
    end do
    call fast_forces (Fᶠ)                   ! get fast forces at t+jδt
    do i = 1, N
        v(i) = v(i) + δt/2 * Fᶠ(jδt)
    end do
end do
call slow_forces (Fˢ)                       ! get slow forces at t+Δt
do i = 1, N
    v(i) = v(i) + Δt/2 * Fˢ(i)              ! update velocities using
                                            ! slow forces at t+Δt
end do
.......
```

B.   rRESPA in the *NVT* Ensemble

For MD simulations in *NVT* ensemble, a modification of the rRESPA algorithm has been proposed. The method uses a modification of the Lagrangian of the system based on the Nosé–Hoover approach, described in Section III.A. The difference from the standard rRESPA scheme described before is that now the total Liouville operator is decomposed as

$$iL = iL_1 + iL_2 + iL_3 + iL_{NH} \tag{53}$$

with

$$iL_1 = \sum_{i=1}^{N}\left[\mathbf{F}_i^f(\mathbf{r})\frac{\partial}{\partial\mathbf{p}_i}\right], \qquad iL_2 = \sum_{i=1}^{N}\left[\dot{\mathbf{r}}_i\frac{\partial}{\partial\mathbf{r}_i}\right], \qquad iL_3 = \sum_{i=1}^{N}\left[\mathbf{F}_i^s(\mathbf{r})\frac{\partial}{\partial\mathbf{p}_i}\right] \tag{54}$$

Also,

$$iL_{NH} = -\sum_{i=1}^{N} \upsilon_\xi \mathbf{v}_i \cdot \frac{\partial}{\partial\mathbf{v}_i} + \upsilon_\xi \frac{\partial}{\partial\xi} + G\frac{\partial}{\partial\upsilon_\xi} \tag{55}$$

where

$$G = \left(\sum_{i=1}^{N}\frac{\mathbf{p}_i^2}{m_i} - gk_BT\right)/Q, \qquad \upsilon_\xi = \dot{\xi} \tag{56}$$

and ξ is a transformation of the additional degree of freedom s, $\log s = N\xi$.

Two modifications of the standard RESPA method exist, depending on the application of the extended operator $\exp(iL_{NH}t)$. The first variant of RESPA is useful when the evolution prescribed by the operator $\exp(iL_{NH}t)$ is slow compared to the time scale associated with the reference force. It is formed by writing

$$\exp(iL\Delta t) = \exp\left(iL_{NH}\frac{\Delta t}{2}\right)\exp\left(iL_3\frac{\Delta t}{2}\right)$$
$$\times \left[\exp\left(iL_1\frac{\delta t}{2}\right)\exp(iL_2\delta t)\exp\left(iL_1\frac{\delta t}{2}\right)\right]^n$$
$$\times \exp\left(iL_3\frac{\Delta t}{2}\right)\exp\left(iL_{NH}\frac{\Delta t}{2}\right) + O(\Delta t^3) \tag{57}$$

and is named XO-RESPA (eXtended system Outside-REference System Propagator Algorithm). In general, XO-RESPA can be applied to systems characterized by fast vibrations, as the time scale associated with the extended system variable is usually chosen to be quite slow compared with these motions.

If the motion prescribed by the operator $\exp(iL_{NH}t)$ occurs on the same time scale as that generated by the "fast" forces, then a useful RESPA algorithm includes the application of this operator for the small time step dt. The evolution operator takes then the form

$$\exp(iL\Delta t)$$

$$= \exp\left(iL_{NH}\frac{\delta t}{2}\right)\exp\left(iL_3\frac{\Delta t}{2}\right)\exp\left(-iL_{NH}\frac{\delta t}{2}\right)$$

$$\times\left[\exp\left(iL_{NH}\frac{\delta t}{2}\right)\exp\left(iL_1\frac{\delta t}{2}\right)\exp(iL_2\delta t)\exp\left(iL_1\frac{\delta t}{2}\right)\exp\left(iL_{NH}\frac{\delta t}{2}\right)\right]^n$$

$$\times\exp\left(-iL_{NH}\frac{\delta t}{2}\right)\exp\left(iL_3\frac{\Delta t}{2}\right)\exp\left(iL_{NH}\frac{\delta t}{2}\right)+O(\Delta t^3) \tag{58}$$

The resulting integrator is named XI-RESPA (eXtended system Inside-REference System Propagator Algorithm).

Modifications of the RESPA method for MD simulations in the NPT statistical ensemble have also been formulated in an analogous manner. More details can be found in the original papers [15,16].

V. CONSTRAINT DYNAMICS IN POLYMERIC SYSTEMS

One of the most important considerations in choosing the best algorithm for the solution of the classical equations of motion is, as we saw above, the time step of integration. This should be chosen appreciably shorter than the shortest relevant time scale in the simulation. For long-chain polymeric systems, in particular, where one explicitly simulates the intramolecular dynamics of polymers, this implies that the time step should be shorter than the period of the highest-frequency intramolecular motion. This renders the simulation of long polymers very expensive. One solution to this problem is provided by the MTS algorithm discussed above. Another technique developed to tackle this problem is to treat bonds between atoms, characterized by the highest-frequency intramolecular vibrations, as rigid. The MD equations of motion are then solved under the constraint that bond lengths are kept constant during the simulation. The motion associated with the remaining degrees of freedom is presumably slower, permitting the use of a longer time step in the simulation.

In general, system dynamics should satisfy many constraints (e.g., many bond lengths should be constant) simultaneously. Let us denote the functions describing the constraints by $\sigma_k = 0$ with $\sigma_k = \mathbf{r}_{ij}^2 - d_{ij}^2$, meaning that atoms i and j are held at a fixed distance d_{ij}. A new system Lagrangian is introduced that contains all constraints

$$L^c = L - \sum_k \lambda_k \sigma_k(\mathbf{r}) \tag{59}$$

where k denotes the set of constraints and λ_k the corresponding set of Lagrange multipliers. The equations of motion corresponding to the new Lagrangian are

$$m_i \ddot{\mathbf{r}}_i = \mathbf{F}_i - \sum_k \lambda_k \frac{\partial \sigma_k}{\partial \mathbf{r}_i} = \mathbf{F}_i - \mathbf{g}_i \tag{60}$$

where the second term on the right hand side of Eq. (60) denotes the constraint forces. The question then is how to calculate the set of Lagrange multipliers λ_k. Two methods that have widely been used in our MD simulations are discussed here: The Edberg–Evans–Morriss method and the SHAKE method.

A. The Edberg–Evans–Morriss Algorithm

This algorithm [17] starts by considering a set of a linear system of equations in $\{\lambda_{ij}\}$, which are formulated by taking the second derivatives of the constraint equations in time:

$$\mathbf{r}_{ij}^2 - d_{ij}^2 = 0 \Rightarrow 2\mathbf{r}_{ij} \cdot \dot{\mathbf{r}}_{ij} = 0 \Rightarrow \mathbf{r}_{ij} \cdot \ddot{\mathbf{r}}_{ij} + \left(\dot{\mathbf{r}}_{ij}\right)^2 = 0 \tag{61}$$

One then solves the following set of algebraic and differential equations simultaneously:

$$m_i \ddot{\mathbf{r}}_i = \mathbf{F}_i + \mathbf{g}_i \tag{62}$$

$$\mathbf{g}_i = \sum_k \lambda_k \frac{\partial \sigma_k}{\partial \mathbf{r}_i} \tag{63}$$

$$\mathbf{r}_{ij} \cdot \ddot{\mathbf{r}}_{ij} + \left(\dot{\mathbf{r}}_{ij}\right)^2 = 0 \tag{64}$$

Note that site velocities enter this formulation explicitly. Upon substitution of the site accelerations from Eq. (60) into Eq. (64) one obtains a system of linear equations in $\{\lambda_{ij}\}$; thus, the determination of the λ_{ij}'s reduces to the solution of a linear matrix equation which should be addressed at each time step.

B. The SHAKE–RATTLE Algorithm

The approach described above suffers from the problem that it is computationally expensive, since it requires a matrix inversion at every time step. The problem gets worse with increasing chain length, with the algorithm becoming practically inappropriate for chains of more than about 100 atoms long. Ryckaert et al. [18] developed a simpler scheme, named SHAKE, to satisfy the constraints in this case.

If one considers the classical form of the Verlet algorithm, then in the presence of constraints

$$\mathbf{r}_i^c(t+dt) = \mathbf{r}_i^u(t+dt)_i - \frac{dt^2}{m_i} \sum_k \lambda_k \frac{\partial \sigma_k(t)}{\partial \mathbf{r}_i} \tag{65}$$

where $\mathbf{r}_i^c$ are the constrained and $\mathbf{r}_i^u$ the unconstrained positions. If the constraints are satisfied at time $t+dt$, then $\sigma_k^c(t+dt) = 0$. But if the system moved along the unconstrained trajectory, the constraints would not be satisfied at $t+dt$. In this case, by performing a Taylor expansion around the unconstrained positions, we get

$$\sigma_k^c(t+dt) = \sigma_k^u(t+dt) + \sum_{i=1}^{N} \left(\frac{\partial \sigma_k}{\partial \mathbf{r}_i}\right)_{\mathbf{r}_i^u(t+dt)} \cdot \left[\mathbf{r}_i^c(t+dt) - \mathbf{r}_i^u(t+dt)\right] + \mathrm{O}(dt^4) \tag{66}$$

and by using Eq. (65)

$$\sigma_k^u(t+dt) = \sum_{i=1}^{N} \frac{dt^2}{m_i} \sum_{k'} \lambda_{k'} \left(\frac{\partial \sigma_k}{\partial \mathbf{r}_i}\right) \left(\frac{\partial \sigma_{k'}}{\partial \mathbf{r}_i}\right) \tag{67}$$

The above equation has the structure of a matrix equation

$$\sigma_k^u(t+dt) = dt^2 \mathbf{M}\mathbf{\Lambda} \tag{68}$$

By inverting the matrix, one can solve for the vector $\mathbf{\Lambda}$. However, since the Taylor expansion in Eq. (66) has been truncated, the σ's should be computed at the corrected positions, and the preceding equations should be iterated until convergence is reached.

This procedure is also computationally expensive, because it requires a matrix inversion at every iteration, as does the Edberg–Evans–Morriss algorithm. Ryckaert proposed a new method, SHAKE, where the iterative scheme is not applied to all constraints simultaneously but to each constraint in succession. Thus the need to invert a large matrix is avoided. The key point is that $\mathbf{r}_i^c - \mathbf{r}_i^u$ is approximated as

$$\mathbf{r}_i^c(t + dt) - \mathbf{r}_i^u(t) \approx -\frac{dt^2 \lambda_k}{m_i} \frac{\partial \sigma_k(t)}{\partial \mathbf{r}_i} \tag{69}$$

By inserting the above equation into Eq. (66), one gets

$$\sigma_k^u(t + dt) = dt^2 \lambda_k \sum_{i=1}^{N} \frac{1}{m_i} \frac{\partial \sigma_k(t + dt)}{\partial \mathbf{r}_i} \frac{\partial \sigma_k(t)}{\partial \mathbf{r}_i} \tag{70}$$

from which

$$\lambda_k dt^2 = \frac{\sigma_k^u(t + dt)}{\sum_{i=1}^{N} (1/m_i) (\partial \sigma_k(t + dt)/\partial \mathbf{r}_i)(\partial \sigma_k(t)/\partial \mathbf{r}_i)} \tag{71}$$

In an MD simulation, the constraints are treated in succession during one cycle of the iteration and the process is repeated until all constraints have converged to the desired accuracy. An improvement of the SHAKE algorithm is the RATTLE algorithm, which was proposed by Andersen [19]. In RATTLE, the velocity-Verlet algorithm is employed to integrate the dynamical equations.

As was also stated above, there are several applications of MD simulations in polymer science. An example taken from a recent study of polymer melt viscoelasticity is presented in the following section.

VI. MD APPLICATIONS TO POLYMER MELT VISCOELASTICITY

Despite its simplicity and unquestionable utility, a brute-force application of the atomistic MD technique to polymeric systems is problematic, due to the enormously large computation time needed to track the evolution of such systems for times comparable to their longest relaxation times [1,2]. This is the well-known *problem of long relaxation times*. To overcome this problem, a number of approaches have been proposed over the years. The first is to develop new, more efficient, "clever" algorithms, such as the multiple time step algorithms for the integration of equations of motion described in Section IV, which have allowed extension of the simulation to times almost an order of magnitude longer than what is usually achieved with conventional algorithms. The second is to increase the range of length scales simulated with MD by using a number of processors (nodes) and special parallelization techniques; such techniques are described in detail in the next section.

Alternatively, a hierarchical approach can be adopted, which uses information from many different levels of abstraction, ultimately connecting to the atomistic level, and a combination of different molecular simulation methods and theoretical approaches. Such a methodology can be followed, for example, for the atomistic MD simulation of the viscoelastic properties of polymer melts. The methodology has been described in a number of publications [13,20] and includes two variants. In the first, equilibrium atomistic MD simulations are conducted on model polymer melt configurations preequilibrated with the powerful connectivity-altering end-bridging Monte Carlo (MC) algorithm; the latter algorithm is not subject to the limitations associated with long relaxation times faced by MD. Dynamical as well as viscoelastic properties are then extracted either directly from the MD simulations or indirectly through a mapping of the atomistic trajectories accumulated in the course of the MD simulation onto an analytical coarse-grained model [20–22], such as the Rouse model for unentangled melts or the reptation model for entangled melts. In the second variant, nonequilibrium MD simulations are conducted on model polymer melt configurations which have been preoriented and thoroughly equilibrated with a field-on MC method [23,24] which generates configurations representative of a melt under conditions of steady-state uniaxial elongational flow. The MD tracks the relaxation of the preoriented chains back to equilibrium upon cessation of the flow. In this case, again, the linear viscoelastic properties of the melt are extracted either directly by the simulation or indirectly by utilizing a mapping onto an analytical coarse-grained model [13].

A. Study of Polymer Viscoelasticity Through Equilibrium MD Simulations

In detail, the methodology followed for the prediction of the viscoelastic properties of polymer melts under equilibrium conditions is a three-stage hierarchical approach, whereby the dynamical properties of polymer melts are calculated through the following procedure:

1.   First exhaustive end-bridging Monte Carlo (EBMC) simulations [23,24] are conducted to equilibrate the melts at all length scales. The EBMC algorithm employs moves that modify the connectivity among polymer segments, while preserving a prescribed (narrow) molecular weight distribution. It can thus equilibrate the long-length scale features of a polymer melt orders of magnitude more efficiently than MD or other MC methods, its relative efficiency increasing rapidly with increasing chain length.

2.   Relaxed configurations thus obtained are subjected to equilibrium MD simulations to monitor their evolution in time and extract dynamic properties. During the atomistic MD simulations, a large number of dynamical trajectories are accumulated.

3.   Finally, the trajectories accumulated are mapped onto theoretical mesoscopic (coarse-grained) models to extract the values of the parameters invoked in the mesoscopic model description of the same systems.

With the above methodology, atomistic MD simulations were performed on united-atom model linear polyethylene (PE) melts with molecular length ranging from $N = 24$ up to $N = 250$ in the canonical NVT ensemble ($T = 450K$, $P = 1$ atm). To speed-up the MD simulations the multiple time step rRESPA algorithm, presented in Section IV, was used. The overall simulation time ranged from 100 ns to 300 ns, depending on the chain lengths of the systems studied. Many of the dynamical properties (such as the self-diffusion coefficient D) were calculated directly from the MD simulations. Others, however, such as the zero-shear rate viscosity η_0, required mapping atomistic MD data upon a mesoscopic theoretical model. As such one can choose the Rouse model for relatively short PE melts and the reptation model for the longer-chain melts.

Figure 3 shows the mean square displacement of the chain center of mass, $\langle (R_{cm}(t) - R_{cm}(0))^2 \rangle$, in time for the longer-chain systems, C_{156}, C_{200}, and C_{250}. From the linear part of these curves the self-diffusion coefficient D can be obtained using the Einstein relation,

$$D = \lim_{t \to \infty} \frac{\langle (\mathbf{R}_{cm}(t) - \mathbf{R}_{cm}(0))^2 \rangle}{6t} \tag{72}$$

Figure 4 presents predictions for the self-diffusion coefficient D as a function of mean chain length N. For comparison, also shown in

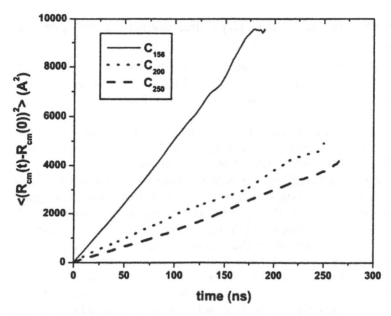

FIG. 3 Mean square displacement of the center of mass for the C_{156} (solid line), C_{200} (dotted line), and C_{250} (dashed line) systems.

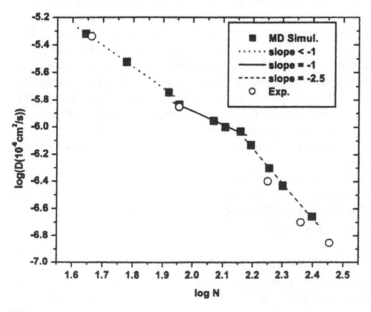

FIG. 4 Predicted and experimental [25] self-diffusion coefficients D vs. chain length N in a log–log plot ($T = 450K$, $P = 1$ atm).

the figure are experimental data [25]. Three distinct regions appear in the figure:

1.   A region of a small-MW, alkane-like behavior ($N < 60$), where D follows a power-law dependence $D \sim M^{-b}$, with $b > 1$. In this regime chain end effects, which can be described through a free volume theory, dominate system dynamics [21].

2.   An intermediate, Rouse-like regime (from $N = 60\text{–}70$ up to $N = 156$) where $b \approx 1$. System dynamics in this regime is found to obey the Rouse model, at least for the overall relaxation of the chain [20].

3.   A long chain-length, reptation-like regime ($N > 156$), where chain diffusivity exhibits a dramatic slow down, $b \approx 2.4$. According to the original formulation of reptation theory, the latter exponent should be 2. Phenomena such as contour length fluctuations (CLF) and constraint release (CR) typically accelerate the escape of the chain from the tube, causing an increase in D and a decrease in η_0 [26]. A recently proposed theory that incorporates CLF and CR phenomena predicts a stronger exponent, between -2.2 and -2.3 [27]. These values agree with recent experimental results for concentrated polymer solutions and melts, which suggest an exponent between -2.2 and -2.4 for a variety of polymer systems [25].

In contrast to D, the prediction of other viscoelastic properties, such as the friction coefficient ζ or the zero-shear rate viscosity η_0, requires that the atomistic MD data be mapped upon a mesoscopic theoretical model. For unentangled polymer melts, such a model is the Rouse model, wherein a chain is envisioned as a set of Brownian particles connected by harmonic springs [25,28]. For entangled polymer melts, a better model that describes more accurately their dynamics is the tube or reptation model [26]. According to this model, the motion of an individual chain is restricted by the surrounding chains within a tube defined by the overall chain contour or primitive path. During the lifetime of this tube, any lateral motion of the chain is quenched.

The Rouse model is formulated in terms of three parameters: the number of beads N, the length of the Kuhn segment b, and the monomeric friction coefficient ζ. The friction coefficient ζ can be calculated directly from the diffusion coefficient D through

$$\zeta = \frac{k_B T}{ND} \tag{73}$$

while the zero-shear rate viscosity η_0 can be calculated from the density ρ, the end-to-end distance $\langle R^2 \rangle$, and the diffusion coefficient D through

$$\eta_0 = \frac{\rho RT \langle R^2 \rangle}{36MD} \tag{74}$$

Reptation theory is formulated in terms of four parameters: N, b, ζ, and the entanglement spacing (or, alternatively, the tube diameter) α. If α were known, ζ and η_0 could be calculated through:

$$\zeta = \frac{k_B T \alpha^2}{3N\langle R^2\rangle D} \tag{75}$$

and

$$\eta_0 = \frac{\rho RT}{36M}\frac{\langle R^2\rangle}{D}\frac{\langle R^2\rangle}{\alpha^2} \tag{76}$$

The calculation of the tube diameter α is a formidable task and can be addressed either through a topological analysis of accumulated polymer melt configurations thoroughly equilibrated with an efficient MC algorithm [29] or by utilizing a geometric mapping of atomistically represented chain configurations onto primitive paths [22,30]. The latter mapping is realized through a projection operation involving a single parameter ξ, which governs the stiffness of the chain in the coarse-grained (primitive path) representation. The parameter ξ is mathematically defined as the ratio of the constants of two types of Hookean springs: The first type connects adjacent beads within the projected primitive path, and the second type connects the projected beads of the primitive path with the corresponding atomistic units [30]. Different values of ξ lead to different parameterizations, i.e., to different primitive paths and, consequently, to different values of the contour length L. Once a value for ξ has been chosen, the primitive path is fully defined which allows the calculation of the tube diameter α through the following equation of reptation theory

$$L\alpha = \langle R^2\rangle \tag{77}$$

To find the proper value of the projection parameter ξ, one can follow a self-consistent scheme based on the mean square displacement of the primitive path points $\phi(s, s; t)$ [22]. $\phi(s, s; t)$ is defined as

$$\phi(s, s; t) \equiv \langle(\mathbf{R}(s, t) - \mathbf{R}(s, 0))^2\rangle \tag{78}$$

where $\mathbf{R}(s, t)$ is the position vector of the primitive segment at contour length s at time t, and $\mathbf{R}(s, 0)$ is the position vector of the primitive segment at contour length s at time 0. According to reptation theory

$$\phi(s, s; t) = 6Dt + \sum_{p=1}^{\infty}\frac{4\langle R^2\rangle}{p^2\pi^2}\cos\left(\frac{p\pi s}{L}\right)\left[1 - \exp\left(-\frac{tp^2}{\tau_d}\right)\right] \tag{79}$$

where the sum is over all normal modes p and τd denotes the longest relaxation or disengagement time. For small times $(t < \tau_d)$, $\phi(s, s; t)$ is dominated by the terms with large p and the above equation becomes

$$\phi(s, s; t) = 6Dt + \int_0^\infty dp \frac{4La}{p^2\pi^2} \frac{1}{2}\left(1 - \exp\left(-\frac{tp^2}{\tau_d}\right)\right)$$

$$= 6Dt + 2\left(\frac{3}{\pi}\langle R^2\rangle D\right)^{1/2} t^{1/2} \tag{80}$$

Equation (80) offers a nice way of mapping atomistic MD trajectories uniquely onto the reptation model, through a self-consistent calculation of the parameter ξ. First, a value of ξ is chosen and the mapping from the atomistic chain onto its primitive path is carried out by following the procedure described by Kröger [30]. Then, Eq. (78) is used to calculate $\phi(s, s; t)$ for the primitive path points, averaged over all s values. For times $t < \tau_d$, the resulting curve is compared to that obtained from Eq. (80), using the values of $\langle R^2\rangle$ and D (long-time diffusivity of the centers of mass) calculated directly from the atomistic MD simulations. The procedure is repeated until convergence is achieved, that is until a ξ value is found for which the two curves coincide. This mapping is performed self-consistently, without any additional adjustable parameters or any experimental input. It allows a reliable estimation of the tube diameter α, by utilizing atomistically collected MD data only for times shorter than τ_d. Thus, the total duration of the MD simulations required is governed solely by the time needed reliably to calculate the center-of-mass diffusion coefficient D. With the values of $\langle R^2\rangle$, D, and α, those of ζ and η_0 can be calculated using Eqs. (75) and (76).

With the above procedure the tube diameter α was calculated to be $\alpha \sim 60\,\text{Å}$ for the longer-chain systems C_{200} and C_{250}, whereas for the shorter systems, $N < 200$, no proper value of the parameter ξ could be identified [22].

Figure 5 shows results for the monomeric friction factor ζ as a function of mean chain length N, over the entire range of molecular lengths studied. Filled squares depict results obtained by mapping atomistic MD data onto the Rouse model, whereas open circles depict results obtained from mapping the atomistic MD data onto the reptation model. According to its definition, ζ should be independent of chain length, its value determined solely by the chemical constitution of the melt. The figure shows clearly that, at around C_{156}, a change in the mechanism of the dynamics takes place, which cannot be accommodated by the Rouse model unless a chain-length dependent ζ is assumed. In contrast, in this regime ($N > 156$), the reptation model provides a consistent description of the system dynamics characterized by a constant $(0.4 \times 10^{-9}\,\text{dyn s/cm})$ chain-length independent ζ value per methylene or methyl segment.

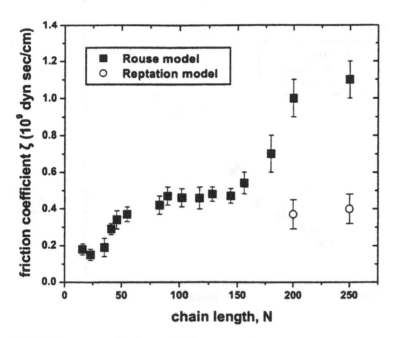

FIG. 5 Monomer friction coefficient ζ vs. chain length N, obtained from mapping the atomistic MD data onto the Rouse model (squares) or the reptation model (circles).

Figure 6 presents the zero-shear rate viscosity η_0 as a function of molecular weight for all systems studied here. For systems of chain length less than C_{156} the Rouse model, Eq. (74), was used, whereas for the longer systems the reptation equation, Eq. (76), was used. Also reported in the figure are experimental η_0 values from measurements [22] conducted in a series of linear monodisperse PE melts. The η_0 predictions from the reptation model were obtained using the value of $\alpha = 60\,\text{Å}$ for the entanglement spacing. The agreement of the simulation results with the experimental ones is remarkable.

B. Study of Polymer Viscoelasticity Through Nonequilibrium MD Simulations—Simulation of the Stress Relaxation Experiment

An alternative way to learn about viscoelastic properties of polymer melts in the linear regime is to conduct MD simulations of preoriented polymer melt

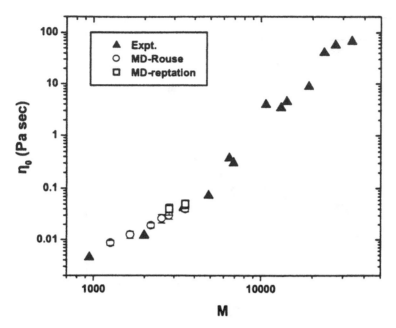

FIG. 6 Zero-shear rate viscosity η_0 vs. molecular weight M, obtained from the MD simulation and the Rouse model for small M (circles) or the reptation model for high M (squares). Also shown in the figure are experimentally obtained η_0 values (triangles).

configurations generated by a field-on MC algorithm. It involves three stages:

1.   First, a coarse-grained description of the polymer melt is invoked through the definition of the conformation tensor, **c**, which is a global descriptor of the long-length scale conformation of polymer chains. The conformation tensor **c** is defined as the second moment tensor of the end-to-end distance vector of a polymer chain reduced by one-third of the unperturbed end-to-end distance and averaged over all chains in the system:

$$\mathbf{c}(t) = 3\left\langle \frac{\mathbf{R}(t)\mathbf{R}(t)}{\langle R^2 \rangle_0} \right\rangle \tag{81}$$

In the above equation, **R** stands for the end-to-end vector of a macromolecule and $\langle R^2 \rangle_0$ is the mean-squared magnitude of that vector in the equilibrium, quiescent state, where chains are unperturbed to an excellent approximation. With the above definition, a series of detailed atomistic MC simulations can be initiated on model melt systems at various values of the orienting thermodynamic field α_{xx}, starting from the zero value ($\alpha_{xx}=0$, equilibrium, quiescent, field-free state) that drive the tensor **c** away from equilibrium [23,24].

2. In the second stage, the isothermal relaxation of these configurations to thermodynamic equilibrium is monitored, keeping their dimension along the x direction constant and the average normal pressure in the y and z directions equal to the atmospheric pressure. The experiment simulated is that of stress relaxation upon cessation of a steady-state uniaxial elongational flow. The MD simulation takes place in the $NTL_x\sigma_{yy}\sigma_{zz}$ statistical ensemble discussed in Section III. The macroscopic variables kept constant in this ensemble are the typical macroscopic constraints encountered in the process of fiber spinning at the end of the spinning operation, when the fibers are under constant extension and the stress σ_{xx} in the direction of pulling is allowed to relax from its initial value to the equilibrium, field-free value, equal to $-P_{ext}$ [i.e., $\sigma_{xx}(t \rightarrow \infty)=-P_{ext}$]. In addition to monitoring the temporal evolution of the stress component $\sigma_{xx}(t)$ during the $NTL_x\sigma_{yy}\sigma_{zz}$ MD simulation, also recorded is the evolution of certain ensemble-averaged descriptors of the chain long-length scale configuration. These descriptors include the diagonal components of the chain conformation tensor (c_{xx}, c_{yy}, and c_{zz}) and the chain mean square end-to-end distance $\langle R^2 \rangle$.

3. The third stage includes the development of expressions describing analytically the time evolution of these quantities by solving the Rouse model under the initial and boundary conditions corresponding to our atomistic computer experiment.

Results are presented here from averaging over about 100 $NTL_x\sigma_{yy}\sigma_{zz}$ MD trajectories for each stress relaxation experiment, initiated at ensembles of strained configurations of two PE melt systems: a 32-chain C_{24} and a 40-chain C_{78} PE melt.

Figures 7a and 7b show the time evolution of the diagonal components c_{xx}, c_{yy}, and c_{zz} of the conformation tensor for the C_{24} and C_{78} melts, respectively. For both systems, the initial value of c_{xx} is significantly higher than 1, whereas those of c_{yy} and c_{zz} are a little less than 1, indicative of the oriented conformations induced by the imposed steady-state elongational structure of flow field α_{xx}. As time evolves, c_{xx} decreases whereas c_{yy} and c_{zz} increase continuously, approaching the steady-state, field-free value of 1, indicative of fully equilibrated, isotropic structures in the absence of any deforming or orienting field.

Figure 8 shows the time evolution of the stress tensor component σ_{xx} for the C_{24} PE melt systems studied. The stress tensor is calculated in two ways.

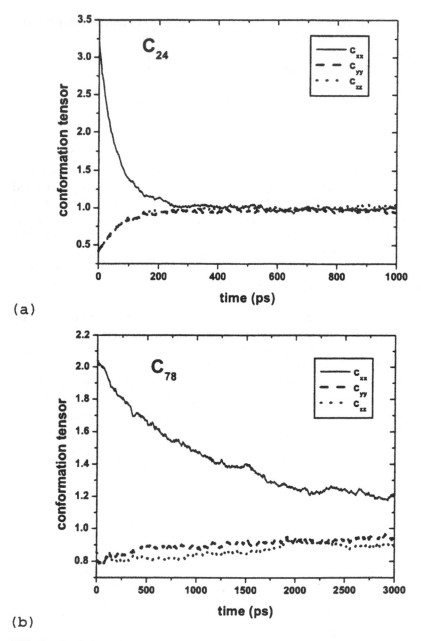

(a)

(b)

FIG. 7 Evolution of the diagonal components c_{xx}, c_{yy} and c_{zz} of the conformation tensor $\mathbf{c}$ with time t for (a) the C_{24} and (b) the C_{78} PE melt systems. Results are averaged over all $NTL_x\sigma_{yy}\sigma_{zz}$ trajectories ($T = 450$K, $P_{ext} = 1$ atm).

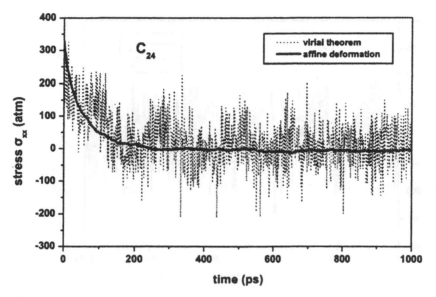

FIG. 8 Evolution of the component σ_{xx} of the stress tensor with time t for the C_{24} system. The results at every time t have been obtained either by applying the virial theorem and averaging over all dynamical trajectories (broken line) or by using a thermodynamic expression based on the free energy as a function of the conformation tensor (thick solid line) ($T = 450K$, $P_{ext} = 1$ atm).

The first (thin broken line) tracks the evolution of σ_{xx} as obtained from applying the molecular virial theorem on the relaxing configurations and averaging over all dynamical trajectories. The second way (thick solid line) uses the Helmholtz energy function and an affine deformation assumption for chain ends [23,24]. According to the latter approach, the stress tensor at every time t is calculated from the ensemble-averaged values of mass density ρ, conformation tensor c_{xx}, and partial derivative of the Helmholtz energy function with respect to c_{xx} at time t, through

$$\sigma_{xx}(t) = -P_{ext} + 2\frac{R}{M}T\rho(t)c_{xx}\left[\frac{\partial(A/N_{ch})}{\partial c_{xx}}\bigg|_{T,\,\rho,\,c_{|xx|}}\right]_{c_{xx}=c_{xx}(t)} \qquad (82)$$

where P_{ext} denotes the equilibrium (atmospheric) pressure and M the number average MW of the system. This approach tracks the evolution of σ_{xx} as obtained from applying the thermodynamic stress equation, Eq. (82),

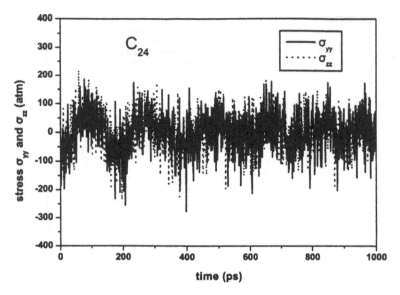

FIG. 9 Evolution of the components σ_{yy} and σ_{zz} of the stress tensor with time t for the C_{24} system ($T = 450\mathrm{K}$, $P_{ext} = 1$ atm).

based on the current values of c_{xx} and $\partial(A/N_{ch})/\partial c_{xx}$, the latter taken from the melt elasticity simulations presented in [23,24]. As expected, in the estimates based on the virial theorem are subject to much higher statistical uncertainty, owing to the fluctuations in the instantaneous configurations. Clearly, averaging over many configurations is needed in order to improve the statistical quality of the virial theorem results. Apart from high-frequency noise, the virial theorem results display an oscillatory character. When the noise and oscillations are smoothed out, the ensemble-averaged stress $\sigma_{xx}(t)$ from the virial theorem is in very good agreement with the thermodynamic estimate obtained from c_{xx} and $\partial(A/N_{ch})/\partial c_{xx}$. This is an important result, as it opens up the possibility of calculating stress with high precision directly from ensemble average conformational properties, based on a free energy function accumulated via efficient MC runs. The transverse components σ_{yy} and σ_{zz} are displayed in Fig. 9. Both σ_{yy} and σ_{zz} fluctuate continuously around the constant value $-P_{atm}$, as required by the macroscopic restrictions placed on the $NTL_x\sigma_{yy}\sigma_{zz}$ ensemble. The plots for the C_{78} system are similar. Relaxation times extracted from these stress relaxation computer experiments are identical to those determined from equilibrium MD simulations (Section VI.A).

VII. PARALLEL MD SIMULATIONS OF POLYMER SYSTEMS

In the previous sections, we presented two different approaches for addressing the problem of long relaxation times, which plagues the conventional MD method: the multiple time step algorithm and a hierarchical methodology that leads to the prediction of the dynamical and rheological properties of polymer melts by mapping simulation data onto analytical theories. Enhanced MD simulation algorithms can further be developed by resorting to special parallelization techniques that allow sharing the total simulation load over a number of processors or nodes. The key idea in all these techniques is to compute forces on each atom or molecule and to perform corresponding velocity/position updates independently but simultaneously for all atoms. It is also desired that the force computations be evenly divided across the processors so as to achieve the maximum parallelism.

MD simulations of polymer systems, in particular, require computation of two kinds of interactions: *bonded* forces (bond length stretching, bond angle bending, torsional) and *nonbonded* forces (van der Waals and Coulombic). Parallel techniques developed [31–33] include the *atom-decomposition* (or *replicated-data*) method, the *force-decomposition* method, and the *spatial (domain)-decomposition* method. The three methods differ only in the way atom coordinates are distributed among the processors to perform the necessary computations. Although all methods scale optimally with respect to computation, their different data layouts incur different interprocessor communication costs which affect the overall performance of each method.

A. Parallel MD Algorithms

Here we will focus on a discussion of algorithms for parallelizing MD simulations with short-range interactions where the nonbonded forces are truncated, so that each atom interacts only with other atoms within a specified cutoff distance. More accurate MD models with long-range forces are more expensive to deal with. Thus, special techniques are required for parallelizing MD simulations with long-range interactions.

1. Atom-Decomposition (Replicated-Data) Method

The most commonly used technique for parallelizing MD simulations of molecular systems is the *replicated-data* (RD) method [33]. In the literature there are numerous parallel algorithms and simulations that have been

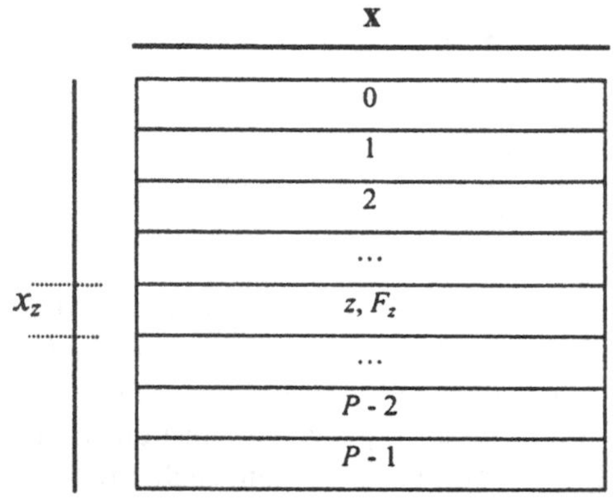

FIG. 10 Division of the force matrix among P processors in the atom-decomposition method.

developed based on this approach [34]. The key idea of this method is that each processor is assigned a subset of atoms and updates their positions and velocities during the entire simulation, regardless of where they move in the physical domain.

Let us consider a polymeric system with a total number of atoms N, where both intramolecular and intermolecular interactions are present. The system is simulated in a parallel system of P processors. We first define $\mathbf{x}$ and $\mathbf{f}$ as vectors of length N, which store the position and total force on each atom, respectively. We also define the force matrix $\mathbf{F}$, of dimensions $N \times N$, with F_{ij} denoting the force on atom i due to atom j. In the RD method each processor is assigned a group of N/P atoms at the beginning of the simulation. Each processor is also assigned a subblock of the force matrix $\mathbf{F}$ which consists of N/P rows of the matrix, as shown in Fig. 10. If z indexes the processors from 0 to $P - 1$, then processor z computes forces in the F_z subblock of rows. It also is assigned the corresponding position and force subvectors of length N/P denoted as $\mathbf{x}_z$ and $\mathbf{f}_z$. The computation of the nonbonded force F_{ij} requires only the two atom positions, x_i and x_j. But to compute all forces in F_z, processor P_z will need the positions of many atoms owned by other processors. In Fig. 10 this is represented by having the horizontal vector $\mathbf{x}$ at the top of the figure span all the columns of $\mathbf{F}$, implying that each processor must store a copy of the *entire* $\mathbf{x}$ vector, hence the name replicated-data. This also implies that, at each time step, each processor must receive updated atom positions from all other processors and send the positions of its own atoms to all other processors, an operation called *all-to-all* communication.

A single time step of an RD algorithm comprises the following steps:

1. First every processor z computes its own part of the nonbonded forces F_z. In MD simulations this is typically done using neighbor lists to calculate nonbonded interactions only with the neighboring atoms. In an analogous manner with the serial algorithm, each processor would construct lists for its subblock F_z once every few time steps. To take advantage of Newton's third law (that $F_{ij} = -F_{ji}$), each processor also stores a copy of the entire force vector **f**. As each pairwise nonbonded force between atoms i and j is calculated, the force component is summed for atom i and negated for atom j. Next, the bonded forces are computed. Since each processor z knows the positions of all atoms, it can compute the nonbonded forces for its subvector x_z and sum the resulting forces into its local copy of **f**. Calculation of both bonded and nonbonded forces scales as N/P, i.e., with the number of nonbonded interactions computed by each processor.

2. In this step, the local force vectors are summed across all processors in such a way that each processor ends up with the total force on each of its N/P atoms. This is the subvector f_z. This force summation is a parallel communication over all processors, an operation known as *fold* [35]. In the literature there are various algorithms that have been developed for optimizing this operation. The key characteristic is that each processor must receive N/P values from every other processor to sum the total force on its atoms. This requires total communication of P times N/P; i.e., the fold operation scales as N.

3. Once each processor has the total force on its subvector x_z in step (3), it can update the positions and the velocities of each atom (*integration step*) with no communication at all. Thus, this operation scales as N/P.

4. Finally the updated positions of each processor $\mathbf{x}_x$ should be shared among all P processors. Each processor must send N/P positions to every other processor. This operation is known as *expand* [35] and scales as N.

A crucial aspect in any parallel algorithm is the issue of load-balance. This concerns the amount of work performed by each processor during the entire simulation, which ideally should be the same for all processors. As we saw before, the RD algorithm divides the MD force computation (the most time consuming part in a typical MD simulation) and integration evenly across all processors. This means that steps (1) and (3) scale optimally as N/P. Load-balance will be good so long as each processor's subset of atoms interacts with roughly the same number of neighbor atoms. This usually occurs naturally if the atom density is uniform across the simulation domain (e.g., bulk simulations). In a different case (e.g., polymer chains at interfaces or adsorbed polymers) load-balance can be achieved by randomizing the order of the atoms at the start of the simulation or by adjusting the size of the subset of each processor dynamically during the simulation to tune the load-balance; these are called dynamical load-balancing techniques [36].

In summary, the RD algorithm divides the force computation evenly across all processors. At the same time, its simplicity makes it easy to implement in existing codes. However, the algorithm requires global communication in steps (2) and (4), as each processor must acquire information from all other processors. This communication scales as N, independently of the number of processors P. Practically this limits the number of processors that can be used effectively.

2. Force-Decomposition Method

The next parallel MD algorithm discussed here is based on a block-decomposition of the force matrix rather than the row-wise decomposition used in the RD algorithm. The partitioning of the force matrix $\mathbf{F}$ is shown in Fig. 11 and the algorithm is called the *force-decomposition* (FD) algorithm [37]. The method has its origin in block-decomposition of matrices, which is commonly encountered in linear algebra algorithms for parallel machines.

The block-decomposition, shown in Fig. 11, is actually applied on a permuted force matrix $\mathbf{F}'$, which is formed by rearranging the columns of the original $\mathbf{F}$ in a particular way. The (ij) element of $\mathbf{F}$ is the force acting on atom i in vector $\mathbf{x}$ due to atom j in the permuted vector $\mathbf{x}'$. Now the F'_z subblock owned by each processor z is of size $(N/P^{1/2}) \times (N/P^{1/2})$. As shown

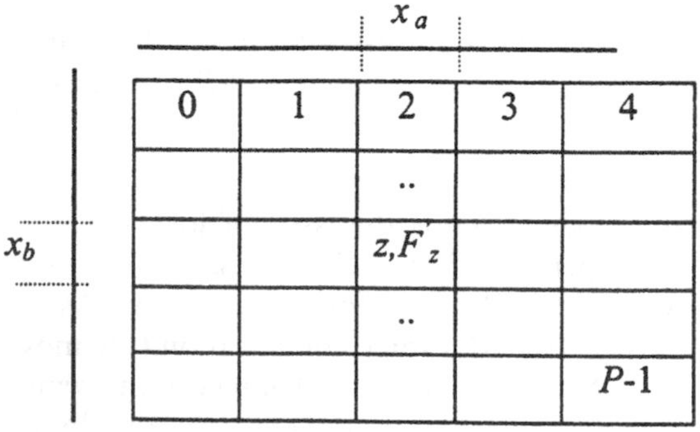

FIG. 11 The division of the force matrix among P processors in the force-decomposition method.

in the figure, to compute the nonbonded forces in F'_z, processor z must know one $N/P^{1/2}$-length piece of each of the $\mathbf{x}$ and $\mathbf{x}'$ vectors, i.e., the subvectors x_a and x'_b. As these elements are computed they will be accumulated into the corresponding subblocks f_a and f'_b. The subscripts a and b each run from 0 to $P^{1/2}$ and reference the row and the column position occupied by processor z.

As in the RD algorithm, each processor has updated copies of the atom positions x_a and x'_b needed at the beginning of the time step. A single time step of the FD algorithm consists of the following steps:

1. The first step is the same as that of the RD algorithm. First the nonbonded forces F'_z are computed. The result is summed into both f_a and f'_b. Next, each processor computes a fraction N/P of the bonded interactions. A critical point here is that in a preprocessing step of the run, we should guarantee that each processor knows all the atom positions needed for the bonded (intramolecular) interactions. This step again scales as N/P.

2. Step (2) is also the same as that of the RD algorithm. The key difference is that now the total force on atom i is the sum of the elements in row i of the force matrix minus the sum of the elements in column i', where i' is the permuted position of column i. Thus this step performs a fold of f_a (f'_b) within each row (column) of processors to sum-up these contributions. The important point is that now the vector f_a (f'_b) being folded is only of length $(N/P^{1/2})$ and only the $P^{1/2}$ elements in one row (column) are participating in the fold. Thus, this operation scales as $N/P^{1/2}$ instead of N as in the RD algorithm. Finally, the two contributions are jointed to yield the total forces f_z (f'_z) on the atoms owned by processor P_z.

3. The processor can now perform the integration for its own atoms, as in the RD case. This operation scales as N/P.

4. Step (4) shares these updated positions with the processors that need them for the next time step. As with the fold operation the processors in each row (column) expand their x_a (x'_b) subvectors within the row (column) so that each acquires the entire x_a (x'_b). This operation scales again as $N/P^{1/2}$ instead of N, as in the RD algorithm.

The FD algorithm, as the RD algorithm, divides the force computation evenly across all processors. The key point is that the communication and memory costs in steps (2) and (4) scale as $N/P^{1/2}$, rather than as N as in the RD algorithm. When large number of processors are used, this can be very important. At the same time, although more steps are needed, the FD algorithm retains the overall simplicity and structure of the RD method.

3.   Domain-Decomposition Method

The third parallel method discussed here for MD simulations of systems with short-range interactions is domain-decomposition (DD) method [31,38–40]. In this method the physical simulation box is divided into small 3D boxes, one for each processor. The partitioning of a simulation box of length L in a DD algorithm is shown in Fig. 12 (2D projection). Each processor z owns a sub-box labeled B_z with edge length L_z ($L_z = L/P$) and updates the positions of all atoms within its own box, x_z, at each time step. Atoms are reassigned to new processors as they move through the physical domain. In order to compute the forces on its atoms, a processor must know the positions of atoms in nearby boxes (processors), y_z. Thus the communication required is local in contrast to global in the AD and FD algorithms. As it computes the force f_z on its atoms, the processor also computes the components of forces f_z^n on the nearby atoms.

A single time step of a DD algorithm consists of the following steps:

1. For each processor P_z, the first step concerns the calculation of bonded and nonbonded forces for atoms within box B_z. This step scales with the numbers of atoms N/P per processor.
2. In step (2) the forces g_z are shared with the processors owning neighboring boxes. The received forces are summed with the previously computed f_z to create the total force on the atoms owned by the processor. The amount of data exchanged in this operation (and consequently the scaling of this step) is a function of the force cutoff distance and box length.
3. After computing f_z, the atomic positions x_z are updated. This operation also scales as N/P.

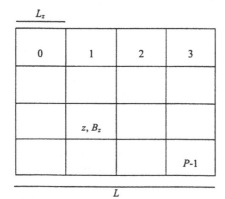

FIG. 12   Partitioning of the simulation domain in a DD algorithm.

4. Next the updated positions are communicated to processors owning neighboring boxes so that all processors can update their y_z list of nearby atoms.
5. Finally, periodically, atoms that have left box B_z are moved into the appropriate new processor.

The scaling of steps (1) and (3) in the DD algorithm is again the optimal N/P. The communication cost involved in steps (2), (4), and (5) is more complicated. It proves to be dependent on the relative value of the cutoff distance r_c in comparison with the subdomain edge length L_z [31]. More specifically, if $L_z > r_c$ the scaling goes as the surface-to-volume ratio $(N/P)^{2/3}$. If $L_z \approx r_c$, then the communication scales as N/P. In practice, however, in the MD simulations of polymer systems there are several obstacles to minimizing communication costs in the DD algorithm.

*   If Coulombic interactions are present then, because of the $1/r$ dependence, long cutoffs should be used. Thus $L_z < r_c$ and extra communication is needed in steps (2) and (4). Data from many neighboring boxes must be exchanged and the communication operation scales as r_c^3. Special techniques such as Ewald summation, particle–particle, or particle–mesh methods can be implemented in parallel [41].
*   As atoms move to new processors in step (5) the molecular connectivity information should be exchanged and updated between processors. This requires extra communication cost, depending on the type of the bonded interactions.
*   If macromolecular systems are simulated uniformly in a simulation domain, then all boxes have a roughly equal number of atoms (and surrounding atoms); load-balance occurs. This will not be the case if the physical domain is nonuniform (e.g., for polymers in vacuum or with surrounding solvent). In this case it is not trivial to divide the simulation domain so that each processor has an equal number of atoms. Sophisticated load-balancing algorithms have been developed [36] to partition an irregular or nonuniformly dense physical domain, but the result is subdomains which are irregular in shape, or connected in an irregular fashion to their neighboring boxes. In both cases the communication operation between processors becomes more costly. If the physical atom density changes with time, then the subject of load-balance becomes more problematic. Dynamical load-balancing schemes are again needed, which require additional computational time and data transfer.

In general, the DD algorithm is more difficult to integrate to existing serial codes than the RD and FD ones. This fact, coupled with the specific problems for macromolecular systems, has made DD implementations less common than RD in the simulations of polymer systems. For simulations with very large systems, however, and in terms of optimal communication scaling, DD is more advantageous. It allows distribution of the computational work over a large number of processors, and this explains why it is nowadays preferred in commercial MD packages such as LAMMPS [41,42], NAMD [43], and AMBER.

B. Efficiency—Examples

Examples of applications of the above techniques can be found in various articles that describe parallel MD applications for systems consisting of many thousands (or even millions) of atoms performed on a large number of processors [31,40]. Here we focus on the implementation of the parallel algorithm for the united-atom polyethylene melt MD simulations described in the previous section [44].

A critical point in any parallel implementation of an existing serial code is the percentage of the code that can be parallelized. To measure the performance of parallel implementations of existing sequential algorithms, the *speed-up* parameter S is used. This is defined as the ratio of the execution time of the serial (sequential) algorithm on a single processor to the execution time of the parallel algorithm running on P processors

$$S(P) = \frac{\tau_s}{\tau_p} \tag{83}$$

where τ_s and τ_p denote the execution time of the algorithms on one and P processors, respectively. For a fully parallel code running on P processors, $\tau_p = \tau_s/P$ and $S(P) = P$ (linear speed-up).

A typical example where parallel implementation of atomistic MD algorithms can substantially speed-up code execution is in $NTL_x\sigma_{yy}\sigma_{zz}$ simulations described in Section VI.B (stress relaxation simulations). In this case the simulated systems are independent and parallelism is straightforward to achieve by assigning each relaxing configuration to a different node (processor). Since no data communication between different systems is required in this case, excellent speed-up should be achieved.

This has been verified with trial runs on a Cray T3E 900 machine at the Edinburgh Parallel Computing Center (EPCC) using the standard MPI approach [45]. Figure 13 presents a graph of the speed-up of the (parallel) code for a number of simulations, each one being executed with a model system containing 40 chains of C_{78} PE melt. As was expected, the speed-up is practically perfect (linear).

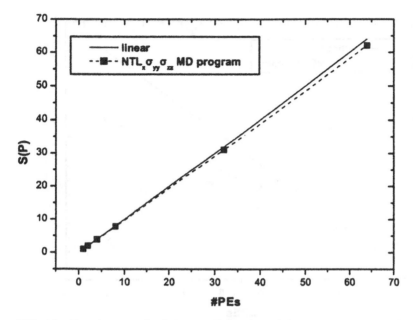

FIG. 13 Speed-up graph of the parallelization of the $NTL_x\sigma_{yy}\sigma_{zz}$ MD simulation runs and the optimal (linear) speed-up.

Figure 14 presents the corresponding speed-up graph for a system containing 24 chains of a C_{1000} PE melt (a total of 24,000 atoms). The runs were executed on a shared memory machine (Sun Enterprise 3000/3500 cluster at EPCC) using OpenMP Fortran [46], which is suitable for developing parallel MD applications in machines with shared memory architecture. Results are shown from parallel runs with different numbers of processors, ranging from 1 to 8 (dotted line) [44]. For comparison, the optimal (linear) speed-up is also shown (solid line). A speed-up of 5.5 is easily reached by using just eight processors.

C. Parallel Tempering

A trend in modern MD (and MC) simulations is to enhance system equilibration at low temperatures by the use of novel parallel techniques. One such technique which has recently attracted considerable attention in different variations is "parallel tempering" (PT) [47,48]. PT was introduced in the context of spin glass simulations, but the real efficiency of the method was demonstrated in a variety of cases, such as in the study of the conformational properties of complex biological molecules [49,50]. Sugita

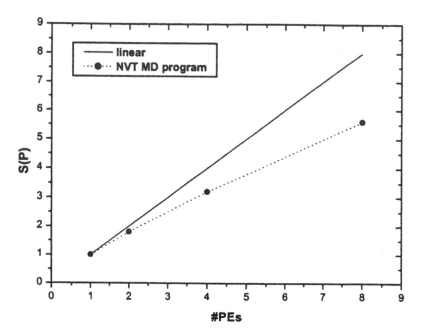

FIG. 14 Speed-up graph of the parallelization of the *NVT* MD simulation runs (dotted line). Also shown is the optimal speed-up (solid line).

and Okamoto [51] used a similar method called "replica exchange molecular dynamics" to simulate protein folding and overcome the multiple-minima problem. They also used multiple-histogram reweighting techniques to calculate thermodynamic quantities in a wide temperature range. Recently, Yamamoto and Kob [52] used the same method to equilibrate a two-component Lennard-Jones mixture in its supercooled state. They found that the replica-exchange MC method is 10–100 times more efficient than the usual canonical molecular dynamics simulation.

PT has also been used successfully to perform ergodic simulations with Lennard-Jones clusters in the canonical and microcanonical ensembles [53,54]. Using simulated tempering as well as PT, Irbäck and Sandelin studied the phase behavior of single homopolymers in a simple hydrophobic/hydrophilic off-lattice model [55]. Yan and de Pablo [56] used multidimensional PT in the context of an expanded grand canonical ensemble to simulate polymer solutions and blends on a cubic lattice. They indicated that the new algorithm, which results from the combination of a biased, open ensemble and PT, performs more efficiently than previously available techniques. In the context of atomistic simulations PT has been employed in a recent study by Bedrov and Smith [57] who report parallel tempering molecular dynamics simulations of atomistic 1,4 polybutadiene polymer melts, in the 323–473K temperature domain at atmospheric pressure. It has also been employed in the context of MC strategies in order to access the regime of low temperatures in the simulations of cis-1,4 polyisoprene melts [58]. In both of these works, it was shown that for a polymer melt well above the glass transition temperature, PT ensures a thorough sampling of configuration space, which is much more effective than afforded by conventional MD or MC simulation methods.

In PT, not a single system but an extended ensemble of n systems, labeled as $i = 1, \ldots, n$, is considered. Each such system is viewed as a copy of the original system, equilibrated, however, at a different temperature T_i, $i = 1, \ldots, n$, with $T_1 < T_2 < \cdots < T_n$. The partition function of this extended ensemble is given by

$$Q = \prod_{i=1}^{n} Q_i \tag{84}$$

where Q_i denotes the individual partition function of the ith system in its relevant statistical ensemble. The strategy of simultaneously equilibrating not *a single* but *a number* of systems at different temperatures accelerates the equilibration of the lowest temperature systems by accepting configurations from the higher temperature systems, for which the rate of the system equilibration is higher. Thus, in the PT method, configurations are swapped between systems being equilibrated at adjacent temperatures. The acceptance probability of a swapping move between system configurations i and $j = i + 1$ within a parallel tempering series of NPT simulations is given by:

$$\begin{aligned} acc[(i,j) \rightarrow (j,i)] &= \min[1, \exp\{\beta_i(U_j + P_j V_j) - \beta_j(U_i + P_i V_i) \\ &\quad + \beta_j(U_i + P_i V_i) + \beta_j(U_j + P_j V_j)\}] \\ &= \min[1, \exp\{\Delta\beta\Delta(U + PV)\}] \end{aligned} \tag{85}$$

with U, P, and V symbolizing the potential energy function, pressure, and volume, respectively [58].

For swapping to occur between two systems, it is important that their $U + PV$ "instantaneous enthalpy" histograms overlap, since only then can there be a nonzero probability to exchange configurations. In this case, the success rate of configuration swapping can increase by performing the simulation with smaller-size systems characterized by larger $(U + PV)$ fluctuations or with systems whose temperatures are not too wide apart.

PT is ideal to implement in parallel architectures by assigning each system to a single node, and by using standard MPI libraries [45] for inter-node communication.

REFERENCES

1. Binder, K., Ed. *Monte Carlo and Molecular Dynamics Simulations in Polymer Science*; Oxford University Press: New York, 1995.
2. Allen, M.P.; Tildesley, D.J. *Computer Simulation of Liquids*; Oxford University Press: Oxford, 1987.
3. Frenkel, D.; Smith, B. *Understanding Molecular Simulations: From Algorithms to Applications*; Academic Press: New York, 1996.
4. Ciccoti, G.; Hoover, W.G. In *Molecular Dynamics Simulations of Statistical Mechanics Systems*, Proceedings of the 97th International "Enrico Fermi" School of Physics, North Holland, Amsterdam, 1986.
5. Goldstein, H. *Classical Mechanics*; Addison-Wesley: Reading, MA, 1980.
6. Swope, W.C.; Andersen, H.C.; Berens, P.H.; Wilson, K.R. J. Chem. Phys. **1982**, *76*, 637–649.
7. Nosé, S. Mol. Phys. **1984**, *52*, 255–268.
8. Hoover, W.G. Phys. Rev. A **1985**, *31*, 1695–1697.
9. Andersen, H.C. J. Chem. Phys. **1980**, *72*, 2384–2393.
10. Parrinello, M.; Rahman, A. Phys. Rev. Lett. **1980**, *45*, 1196–1199.
11. Berendsen, H.J.C.; Postma, J. P.M.; van Gunsteren, W.F.; DiNola, A.; Haak, J.R. J. Chem. Phys. **1984**, *81*, 3684–3690.
12. Yang, L.; Srolovitz, D.J.; Yee, A.F. J. Chem. Phys. **1997**, *107*, 4396–4407.
13. Harmandaris, V.A.; Mavrantzas, V.G.; Theodorou, D.N. Macromolecules **2000**, *33*, 8062–8076.
14. Palmer, B.J. J. Comp. Phys. **1993**, *104*, 470–472.
15. Tuckerman, M.; Berne, B.J.; Martyna, G.J. J. Chem. Phys. **1992**, *97*, 1990–2001.
16. Martyna, G.J.; Tuckerman, M.E.; Tobias, D.J.; Klein, M.L. Mol. Phys. **1996**, *87*, 1117–1157.
17. Edberg, R.; Evans, D.J.; Morriss, G.P. J. Chem. Phys. **1986**, *84*, 6933–6939.
18. Ryckaert, J.P.; Ciccoti, G.; Berendsen, H.J.C. J. Comp. Phys. **1977**, *23*, 327–341.
19. Andersen, H.C. J. Comp. Phys. **1983**, *52*, 24–34.
20. Harmandaris, V.A.; Mavrantzas, V.G.; Theodorou, D.N. Macromolecules **1998**, *31*, 7934–7943.
21. Harmandaris, V.A.; Doxastakis, M.; Mavrantzas, V.G.; Theodorou, D.N. J. Chem. Phys. **2002**, *116*, 436–446.
22. Harmandaris, V.A.; Mavrantzas, V.G.; Theodorou, D.N.; Kröger, M.; Ramírez, J.; Öttinger, H.C.; Vlassopoulos, D. Macromolecules **2003**, *36*,1376–1387.
23. Mavrantzas, V.G.; Theodorou, D.N. Macromolecules **1998**, *31*, 6310–6332.
24. Mavrantzas, V.G.; Öttinger, H.C. Macromolecules **2002**, *35*, 960–975.

25. Doi, M.; Edwards, S.F. *The Theory of Polymer Dynamics*; Clarendon: Oxford, 1986.

26. Milner, S.T.; McLeish, T.C.B. Phys. Rev. Lett. **1998**, *81*, 725–728.

27. Lodge, T.P.; Phys. Rev. Lett. **1999**, *83*, 3218–3221.

28. Ferry, J.D. *Viscoelastic Properties of Polymers*; J. Wiley & Sons: New York, 1980.

29. Uhlherr, A.; Doxastakis, M.; Mavrantzas, V.G.; Theodorou, D.N.; Leak, S.J.; Adam, N.E.; Nyberg, P.E. Atomic structure of a high polymer melt. Europhys. Lett. **2002**, *57*, 506–511.

30. Kröger, M.; Ramírez, J.; Öttinger, H.C. Polymer **2002**, *43*, 477–487.

31. Plimpton, S. J. Comp. Phys. **1995**, *117*, 1–19.

32. Gupta, S. Comp. Phys. Comm. **1992**, *70*, 243–270.

33. Smith, W. Comp. Phys. Comm. **1991**, *62*, 229–248.

34. Smith, W.; Forester, T.R. Comp. Phys. Comm. **1994**, *79*, 52–62.

35. Fox, G.C.; Johnson, M.A.; Lyzenga, G.A.; Otto, S.W.; Salmon, J.K.; Walker, D.W. *Solving Problems on Concurrent Processors*; Prentice Hall: Englewood Cliffs, NJ, 1988; Vol. 1.

36. Deng, Y.; Peiers, R.; Rivera, C. J. Comp. Phys. **2000**, *161*, 250–263.

37. Hendrickson, B.; Plimpton, S. J. Parr. Distr. Comp. **1995**, *27*, 15–25.

38. Clark, T.W.; Hanxleden, R.V.; McCammon, J.A.; Scott, L.R. In Proc. Scalable High Performance Computing Conference–94, IEEE Computer Society Press: Los Alamitos, CA, 1994; 95–102 pp.

39. Brown, D.; Clarke, J.H.R.; Okuda, M.; Yamazaki, T. Comp. Phys. Commun. **1994**, *83*, 1–13.

40. Pütz, M.; Kolb, A. Comp. Phys. Commun. **1998**, *113*, 145–167.

41. Plimpton, S.J.; Pollock, R.; Stevens, M. *Particle-Mesh Ewald and rRESPA for Parallel Molecular Dynamics Simulations*, In Proceedings of the Eighth SIAM Conference on Parallel Processing for Scientific Computing, SIAM, Minneapolis, MN, 1997.

42. http://www.cs.sandia.gov/~sjplimp/lammps.html

43. http://www.ks.uiuc.edu/Research/namd

44. Unpublished data collected at the Edinburgh Parallel Computing Centre (EPCC) under the TRACS programme during the period 10 January 2000–20 February 2000.

45. http://www.mpi-forum.org/

46. http://www.openmp.org/

47. Newman, M.E.J.; Barkema, G.T. *Monte Carlo in Statistical Physics*; Clarendon: Oxford, 1999.

48. Geyer, C.J. *Computing Science and Statistics*, Proceedings of the 23rd Symposium on the Interface, Interface Foundation of North America: Fairfax Station, VA, 1991, p. 156.

49. Hansmann, U.H.E. Chem. Phys. Lett. **1997**, *281*, 140–150.

50. Wu, M.G.; Deem, M.W. Mol. Phys. **1999**, *97*, 559–580.

51. Sugita, Y.; Okamoto, Y. Chem. Phys. Lett. **1999**, *314*, 141–151.

52. Yamamoto, R.; Kob, W. Phys. Rev. E **2000**, *61*, 5473–5476.

53. Neirotti, J.P.; Calvo, F.; Freeman, D.L.; Doll, J.D. J. Chem. Phys. **2000**, *112*, 10340–10349.

54. Calvo, F.; Neirotti, J.P.; Freeman, D.L.; Doll, J.D. J. Chem. Phys. **2000**, *112*, 10350–10357.

55. Irbäck, R.; Sandelin, E. J. Chem. Phys. **1999**, *110*, 12256–12262.

56. Yan, Q.; de Pablo, J.J. J. Chem. Phys. **2000**, *113*, 1276–1282.

57. Bedrov, D.; Smith, G.D. J. Chem. Phys. **2001**, *115*, 1121–1124.

58. Doxastakis, M.; Mavrantzas, V.G.; Theodorou, D.N. J. Chem. Phys. **2001**, *115*, 11352–11361.

7

Configurational Bias Techniques for Simulation of Complex Fluids

T. S. JAIN and J. J. DE PABLO   University of Wisconsin–Madison, Madison, Wisconsin, U.S.A.

I. INTRODUCTION

Monte Carlo methods offer a useful alternative to Molecular Dynamics techniques for the study of the equilibrium structure and properties, including phase behavior, of complex fluids. This is especially true of systems that exhibit a broad spectrum of characteristic relaxation times; in such systems, the computational demands required to generate a long trajectory using Molecular Dynamics methods can be prohibitively large. In a fluid consisting of long chain molecules, for example, Monte Carlo techniques can now be used with confidence to determine thermodynamic properties, provided appropriate techniques are employed.

A generic algorithm for carrying out a Monte Carlo simulation constructs a weighted walk in the configuration space of the system, and samples different states according to their equilibrium probability distribution function. This is typically achieved by proposing a small, random change (about some original state) in one or more degrees of freedom of the system, which is subsequently accepted or rejected according to specific probability criteria [1] pertaining to the statistical ensemble of interest. In a canonical ensemble, for example, random trial displacements of the particles' positions would be accepted according to

$$P_{acc} = \min[1, \exp(-\beta \Delta U)] \tag{1}$$

where ΔU is the change in energy in going from one state of the system to another, and β is the inverse of $k_B T$, with k_B being Boltzmann's constant and T the temperature.

For many complex systems, "blind," or naive random displacements of the particles' coordinates can lead to inefficient algorithms. Poor sampling can in turn yield simulated properties that are not representative of the system under study. Configurational bias techniques provide a fairly general means of *biasing* trial moves in such a way as to guide the system to favorable states, thereby sampling configuration space more effectively.

We begin this chapter by providing a few illustrative examples and by discussing some of the typical problems that can be encountered in a simulation. We then explain, in general terms, how an arbitrary bias can be incorporated in a Monte Carlo simulation by introducing the concept of detailed balance and that of rejection algorithms. We then present a number of different techniques, all based on biased sampling ideas, that can facilitate considerably the simulation of complex systems. This is done in the context of specific examples taken from our own recent work and that of others. Here we note that the goal of this chapter is not to provide a review of configurational bias techniques, but rather to introduce these methods to the reader and to present a broad view of how these ideas can be used in a wide variety of systems to improve sampling efficiency.

II. SHORTCOMINGS OF METROPOLIS SAMPLING

As mentioned above, the main drawback of a naive, Metropolis-sampling based simulation is that trial moves are proposed at random, without ever taking into account the specific nature of the system. Unless the random trial moves are small, most of them will be rejected. If the trial displacements are small, trial moves are more likely to be accepted, but successive configurations end up being highly correlated; the averages obtained from the resulting trajectories may not provide an accurate description of the system as a whole. This problem can be particularly acute in molecular systems, where topological constraints (e.g., as in long polymeric molecules) cause most random displacements of an entire molecule to be rejected. In the case of polar systems (e.g., water), a molecule in a new, trial position (and orientation) must not only conform to its new geometrical environment but, if the move is to have a reasonable probability of being accepted, the energetic contributions arising from polar or electrostatic effects must be comparable to those for the original state.

The above examples serve to illustrate that for complex systems, or even for simple systems at elevated densities, a naive random sampling of configuration space does not provide an efficient simulation algorithm. In order for the system to "evolve" to different states, it will have to be guided out of local energy minima by biasing proposed trial moves. These biases will in general be system specific; they can be designed in such a way as to influence the position of a molecule, its orientation, or its internal configuration. Whatever the bias is, however, it must be subsequently removed in order to generate the desired, correct distribution of states dictated by the ensemble of interest. The next section discusses how this can be achieved.

III. DETAILED BALANCE AND CONFIGURATIONAL BIAS

The condition of detailed balance ensures that the states of the system are sampled according to the correct probability distribution in the asymptotic limit, i.e., in the limit of a large number of proposed transitions. The condition of detailed balance [2] simply ensures that the probability of going from one state of the system to another is the same in the forward and reverse directions. To introduce the concept of detailed balance, we denote the overall state of a system (e.g., the coordinates of all of the particles for the canonical, NVT ensemble or the coordinates of all of the particles and the volume of the system for the NPT ensemble) collectively by X. This probability can be constructed as the product of two probabilities: the probability of being in a particular state X, and the conditional probability of moving from that state to another state Y. This can be written as

$$K(X|Y)f(X) = K(Y|X)f(Y) \tag{2}$$

where $f(X)$ is the probability of being in state X and $K(X|Y)$ is the probability of moving to Y given that the system is at X. The latter probability can be further decomposed into the product of two factors: the probability of proposing the transition, $T(X|Y)$, and that of accepting the transition, $A(X|Y)$. With this separation, Eq. (2) becomes:

$$A(X|Y)T(X|Y)f(X) = A(Y|X)T(Y|X)f(Y) \tag{3}$$

The way in which the moves are proposed is arbitrary; any function $T(X|Y)$ can be used as long as it is properly normalized, i.e.,

$$\int T(X|Y)\,dY = 1 \tag{4}$$

In order to satisfy detailed balance, the function $A(X|Y)$ must have a specific form. In the prescription of Metropolis et al. [1], moves are accepted with probability

$$A(X|Y) = \min\left(1, \frac{T(Y|X)f(Y)}{T(X|Y)f(X)}\right) \tag{5}$$

where "$\min(x, y)$" is used to denote the smaller of its two arguments. It can be shown that in the limit of a large number of trial moves, the sequence of states generated by accepting trial moves according to Eq. (5) follows the desired probability distribution function f.

In the particular case of a naive algorithm, i.e., the case in which displacements are proposed at random, one can see that the transition probabilities $T(X|Y)$ and $T(Y|X)$ are the same. For a canonical, NVT ensemble, the function f takes the form

$$f(X) = \frac{1}{Q}\exp(-\beta U(X)) \tag{6}$$

where Q is the partition function of the system. For the canonical ensemble Eq. (5) therefore reduces to Eq. (1).

In a Configurational Bias algorithm, a bias is introduced in the forward or reverse direction; as a result of the bias, the forward and reverse transition probabilities are no longer equal to each other. This leads to additional flexibility in the construction of a trial move, and can result in considerable improvements of sampling efficiency. This bias is removed in the acceptance criteria according to Eq. (5), thereby leading to sampling according to the correct, original equilibrium probability distribution f.

A few illustrative examples on the use of Eq. (5) are discussed in the next section. In each case, the correct acceptance criteria are arrived at by establishing the form of the functions $T(X|Y)$ and $T(Y|X)$.

IV. CASE STUDIES

A. Orientational Configurational Bias

Orientational bias moves are useful for the study of fluids that exhibit highly directional forces. A prime example of such a fluid is water, where strong hydrogen bonding interactions lead to well defined, relatively stable low-energy molecular structures. Cracknell et al. [3] were among the first to use orientational bias for the study of water. A typical orientational bias algorithm consists of the following steps:

1. Displace the center of mass of the molecule at random to a new position
2. If the potential energy can be separated into orientation-dependent and -independent parts, calculate the orientation independent part, U'_{new}.
3. Generate a set of k orientations $O^F = \{o_1, o_2, \ldots, o_k\}$ about the new center of mass for the forward move. Calculate the orientation dependent potential energy $U^o(l)$ for each orientation in O^F. Note that henceforth, superscripts F and R will be used to denote corresponding quantities for the forward and reverse moves respectively.
4. Evaluate the so-called "Rosenbluth" weight for the new position as follows

$$R_W^F = \sum_{l=1}^{k} \exp(-\beta U^o(l)) \tag{7}$$

5. Select one orientation j, from the set O^F, according to the probability distribution

$$p(j) = \frac{\exp(-\beta U^o(j))}{R_W^F} \tag{8}$$

This implies that

$$T(X|Y) = p(j) \tag{9}$$

6. For the reverse move construct a *new* set of $k-1$ orientations $O^R = \{o_1, o_2, \ldots, o_{k-1}\}$ and calculate the orientation dependent energy of the molecule in its *old*, or original orientation $U^o(k)$. Calculate the Rosenbluth weight for the old position:

$$R_W^R = \sum_{l=1}^{k-1} \exp(-\beta U^o(l)) + \exp(-\beta U^o(k)) \tag{10}$$

For the reverse move we have

$$T(Y|X) = \frac{\exp(-\beta U^o(k))}{R_W^R} \tag{11}$$

7. This leads to the following acceptance criteria for this move

$$A(X|Y) = \min\left(1, \frac{R_W^F}{R_W^R}\exp\left(-\beta(U'_{new} - U'_{old})\right)\right) \tag{12}$$

>   where subscripts *old* and *new* are used to denote quantities evaluated
>   in the original (old) and the trial (new) configurations, respectively.

The above algorithm can be applied to both lattice and continuum
simulations. One important difference, however, must be noted. In the
case of a lattice, all possible trial orientations of the molecule can be
considered explicitly; the sets O^F and O^R are therefore the same (on a
cubic lattice, a simple dimer or a bond can adopt an orientation along the
coordinate axes, and $n = 6$). In a continuum, a molecule can adopt one of
an infinite number of trial orientations. By restricting the trial orientations
to the finite sets O^F and O^R, the probability of generating these sets also
enters the transition matrix along with the probability of choosing
a direction from the probability distribution (which itself is a function
of these sets [4]). The acceptance criteria are therefore also a function of
these two sets:

$$A(X|Y, O^F, O^R) = \min\left[1, \frac{T(Y|X)P(O^R)f(Y)}{T(X|Y)P(O^F)f(X)}\right] \tag{13}$$

The *a priori* probabilities, $P(O^F)$ and $P(O^R)$, of generating these sets of trial
orientations are the same and drop out of the final expression. In order to
satisfy detailed balance, the condition of "super-detailed balance" is
imposed, i.e., detailed balance should be obeyed for every pair of sets
chosen. Hence the acceptance criteria of Eq. (12) generate the true equilibrium distribution.

1. Continuum Example: Simulation of Water Clay Hydrates

Bulk water has been the subject of a large number of simulation studies.
Water in confined environments has been studied less extensively and, more
importantly to the goals of this chapter, it represents a case where
orientational bias moves are particularly helpful.

The question at hand is to determine the structure of water in the
presence of geometrical constraints, at elevated densities. For concreteness,
we consider an example taken from our studies of clay hydrates. At small
length scales, these hydrates can be viewed as a stack of closely spaced disk-
like particles, having a thickness of a few nanometers. The disks are

negatively charged, and electroneutrality is maintained by the presence of compensating cations such as sodium or calcium. The interstitial spaces between different disks (also called galleries) are occupied by water; depending on pressure, temperature, and ionic strength, more or less water can enter the galleries and swell the hydrate. Configurational bias is helpful because the presence of charged species and strong confinement restrict considerably the freedom of water molecules.

Figure 1 depicts a typical configuration of a hydrated sodium montmorillonite clay [5]. The structure and partial charges of the model

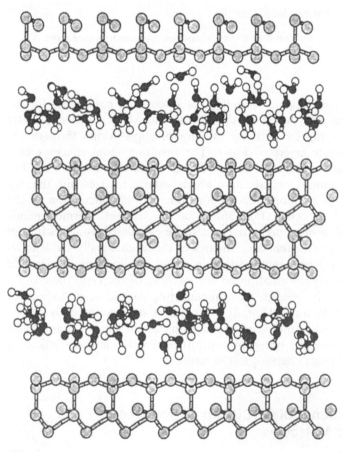

FIG. 1 Representative equilibrium configuration of a system comprising two clay sheets, twelve sodium ions, and interlayer water. Sodium ions are not shown for clarity.

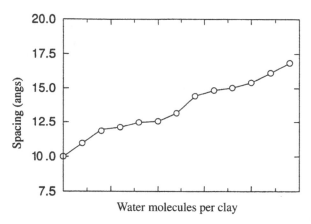

FIG. 2 Swelling curve for two clay sheets of sodium montmorillonite from $NP_{zz}T$ simulations.

clay were taken from Skipper et al. [6]. The interaction parameters for the clay–water system were based on the TIP4P and MCY models of water [7–9]. Simulations suggest that the swelling of clays occurs through the formation of discrete layers of water. The resulting basal spacings as a function of number of water molecules (Fig. 2) agree with experiment. The use of a grand canonical formalism, in which water is inserted into the clay galleries via a configurational bias algorithm, permits calculation of the water content of the clay (Fig. 3b) as a function of thermodynamic conditions. The disjoining pressure can also be calculated as a function of basal distance to provide reliable estimates of the mechanically stable basal spacings (Fig. 3a), which is important in many applications of clays. In addition to providing information on the swelling of the clay, these simulations reveal that, in general, the compensating cations (e.g., sodium or calcium) are fully hydrated (Fig. 4). Arriving at such conclusions on the basis of conventional Monte Carlo simulations would require considerable computational resources; using an orientational bias increases efficiency considerably and lends credence to the results.

B. Configurational Bias (CB) for Articulated or Polymeric Molecules

Polymer molecules can be studied at various levels of detail, from highly coarse-grained lattice representations to fully atomistic models. Whatever the level of detail, however, the connectivity of a complex molecule imposes

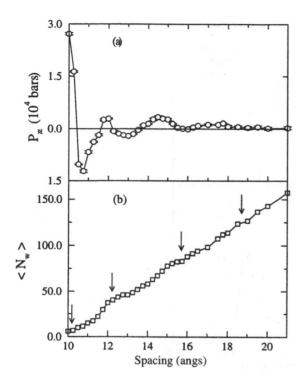

FIG. 3 Grand canonical ensemble simulation results at $T = 300\,K$ and chemical potential $\beta\mu = -17.408$, which is the value that corresponds to TIP4P water in the bulk (at $T = 300\,K$ and $P = 1$ bar). (a) Normal component of the pressure tensor. (b) Average number of water molecules per clay.

severe constraints on the types of trial moves that can be used to generate distinct configurations of the system. At liquid-like densities, it is clear that a random displacement of the center of mass of a molecule (while maintaining the internal configuration) would lead to overlaps with other molecules, and a rejection of the trial move.

This problem can be partially alleviated by conducting highly localized trial moves. On a simple cubic lattice, for example, the conformation of a linear polymeric molecule can be explored by resorting to "local" trial moves such as "kink-jumps," "end-rotations," or "crankshafts"[10]. These moves typically displace one or two sites of the chain. The so-called "reptation" [11] moves involve displacing an entire chain in a slithering, snake-like motion along its contour. In a continuum, simple random moves of individual sites can be used to sample the configuration space of the

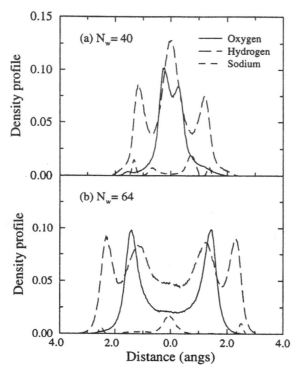

FIG. 4 Distribution of oxygens, hydrogens, and sodium ions inferred from an $NP_{zz}T$ simulation at $P_{zz} = 1$ bar and $T = 300K$, using two clay sheets: (a) $N_w = 40$, (b) $N_w = 64$.

system. These moves, however, are generally insufficient to achieve appreciable relaxation of the intramolecular conformation of long chain molecules [12]. Configurational bias ideas can be used to probe energetically favorable configurations over length scales of several sites, thereby accelerating the relaxation of configurations.

In its simplest form, a configurational bias trial move for a linear polymer chain involves cutting off a terminal part of the chain, and regrowing the end sites using an energetic bias (Fig. 5). Such a move is inspired by the seminal work of Rosenbluth and Rosenbluth [13]. In the canonical ensemble, the algorithm for this move is as follows [14,15]:

1. Select one end of the chain of length L at random and remove n interaction sites.
2. Use i to denote the current site to be added back to the chain.
3. Calculate the energy of site i, in k trial positions.

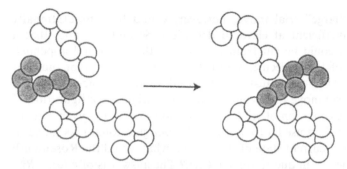

FIG. 5 Illustration of configurational bias move.

4. Select one of the trial positions, j, from the probability distribution given by

$$p_i^{(j)} = \frac{e^{-\beta U_i^{(j)}}}{\sum_{l=1}^{k} e^{-\beta U_i^{(l)}}} \tag{14}$$

5. After all the sites have been grown calculate the Rosenbluth weight of the chain as follows:

$$R_W^F = \prod_{i=L-n+1}^{L} \sum_{l=1}^{k} e^{-\beta U_i^{(l)}} \tag{15}$$

The construction of the move was such that

$$T(X|Y) = \prod_{i=L-n+1}^{L} p_i^{(j)} \tag{16}$$

6. Similarly, to construct the reverse move, a *new* set of $k-1$ orientations is generated about the old positions, and the Rosenbluth factor of the original configuration of the molecule, R_W^R, is calculated.
7. Substitution into Eq. (5) gives acceptance criteria of the form:

$$A(X|Y) = \min\left(1, \frac{R_W^F}{R_W^R}\right) \tag{17}$$

A number of remarks are in order. First, we note that there is no restriction on the number of sites that can be cut and regrown for this move. In fact, an entire chain can be removed and regrown at another point if so desired (such a "large" trial move, however, would be computationally demanding and inefficient at elevated densities). Second, we note that a piece of the chain could be cut from one side of the chain and appended to the other end of the molecule. Third, it is important to point out that to arrive at Eq. (17) care must be exercised in the construction of a hypothetical, reverse move that would lead to the original configuration of the molecule (the calculation of R_W^R). The question to ask at a given state of the system is: If we had used a (hypothetical) configurational bias move to arrive at the precise, current state of the system, what would the Rosenbluth weight of the molecule in question have been? The answer is of course R_W^R, but the concept of conducting a hypothetical process to recreate the original state of the molecule can be difficult to grasp.

The algorithm outlined above is equally applicable to chains on a lattice or chains in a continuum. In the particular case of models having strong intramolecular interactions (e.g., bond-stretching or bond-bending potential-energy functions), minor alterations of this algorithm can be implemented to improve sampling. The k trial orientations for i can be generated using the Boltzmann distribution arising from the intramolecular potential [16]

$$p_i^{(l)} = Ce^{-\beta U_i^{bonded(l)}} \qquad l = 1, 2, \ldots, k \tag{18}$$

where C is a normalization constant. For each of these trial positions, the *external* Boltzmann factor arising from intermolecular interactions, external fields, and intramolecular nonbonded interactions is calculated. The subsequent procedure to be followed is the same as above, but the Rosenbluth weight is calculated from the sum of the *external* Boltzmann factors. For the reverse move, a *new* set of $k-1$ orientations is generated and the Rosenbluth weight is calculated. The arguments of super-detailed balance presented for the Orientational Bias algorithm in the continuum are applicable to the growth of each site. The acceptance criteria reduce to Eq. (17).

1. Expanded Grand Canonical Ensemble Simulation of Polymer Chains Using Configurational Bias

In the case of long polymer chains, the insertion or deletion of entire molecules required for grand canonical ensemble simulations is difficult, even when configurational bias techniques are employed. In that case it is beneficial to implement configurational bias moves in the context of an expanded ensemble formalism [17], which essentially allows one to create or delete a smaller number of sites of a molecule (as opposed to an entire chain) in each trial move, thereby increasing the likelihood of acceptance. The combination of an expanded grand canonical technique with configurational bias can lead to highly effective insertion schemes, even for long chain molecules at elevated densities [18].

Formally, an expanded grand canonical ensemble can be defined through the partition function:

$$\Omega = \sum_{N=0}^{\infty} \sum_{y=1}^{M} Q(N, y, V, T) \exp(\psi_y) \exp(\beta\mu N) \tag{19}$$

where μ is the chemical potential, and Q is the canonical partition function for a system of volume V, temperature T, and N full molecules and one tagged molecule in state y. States 1 and M correspond to a fully decoupled and a fully coupled molecule, respectively. The ψ_y's are arbitrary weighting factors whose relevance is discussed below. The probability of finding the system in a configuration described by a particular set of V, T, N, y is given by

$$p(y) = \frac{1}{\Omega} \exp(\beta\mu N) \exp(-\beta U(N, y)) \exp(\psi_y) \tag{20}$$

The ensemble average probability of finding the system in a state given by N, y is

$$\langle p(y) \rangle = \frac{Q(N, V, T, y)}{\Omega} \exp(\beta\mu N) \exp(\psi_y) \tag{21}$$

The goal of the algorithm is to sample fluctuations of the number of molecules effectively. If the molecules of interest consist of L sites, this can be done by allowing a tagged chain to fluctuate in length; at the two extremes of the spectrum these fluctuations can cause the tagged chain to comprise L interaction sites and become a real, full chain, or they can cause the tagged chain to have zero sites and disappear. In order for these sequential insertion or deletion processes to be effective, the probability with which terminal states of the tagged chain are visited should be appreciable. This can be controlled by judicious assignment of the weighting factors appearing in Eq. (19). The ratio of the probabilities for two states x and y of the system is given by

$$\frac{\langle p(x) \rangle}{\langle p(y) \rangle} = \exp(\psi_x - \psi_y) \left[\frac{Q(N, V, T, x)}{Q(N, V, T, y)} \right] \tag{22}$$

The logarithm of the ratio of the partition functions, under identical values of N, V, and T, is equal to the difference in the chemical potential for a transition from state x to state y. In order to avoid a "bottleneck" at any one state, we can prescribe that the probabilities of visiting different states of the expanded ensemble be equal, which leads to

$$\psi_x - \psi_y = -\beta\Delta\mu^{ex}(x \rightarrow y) = \beta\mu^{ex}(x, N) - \beta\mu^{ex}(y, N) \tag{23}$$

where $\mu^{ex}(y, N)$ denotes the excess chemical potential of a system having N regular molecules and a tagged chain in state y.

A natural choice for the weights ψ_y is therefore:

$$\psi_y = \beta\mu^{ex}(y, N) = \alpha_y\beta\mu^{ex}(N) = \alpha_y\left[\beta\mu - \ln\left(\frac{\Lambda^3 N_y}{V}\right)\right] \tag{24}$$

where $\mu^{ex}(N)$ is the chemical potential required to insert an entire chain in a system with N full chains, and N_y is the number of chains in the system including the tagged chain if it is of nonzero length. The physical significance of the weights ψ_y is now apparent: They are chemical potentials for particular states of the growing chains. The prefactors α_y should satisfy $\alpha_1 = 0$ and $\alpha_M = 1$. These factors can be approximated as $\alpha_y = L_y/L$ for homonuclear, monodisperse polymer chains, where L_y is the length of the polymer molecule in state y. This approximation is based on the result that, in a simple polymer melt, the incremental chemical potential (chemical potential required to insert one additional site in the chain) of a polymer chain is relatively insensitive to chain length [19,20]. In other cases, they can be determined from a preliminary run so that eventually each state is uniformly sampled. If the number of states of the system is reduced to two, then the method reduces to the simple Grand Canonical method.

The algorithm for performing a trial move in an expanded grand canonical ensemble is as follows:

1. Given that the system is in state x, propose a move to a neighboring state y.
2. If the transition involves
    a. Adding n sites to the chain, then calculate the Rosenbluth weight of the added sites as discussed previously.
    b. Removing n sites from the chain, then calculate the Rosenbluth weight of the existing sites.
3. Normalize the Rosenbluth weight by dividing by n^k, where k is the number of trial positions proposed.
4. The modified acceptance criteria for the move are

$$A(X|Y) = \min\left[1, R_W^\Delta \exp(\psi_y - \psi_x)\right] \tag{25}$$

where $\Delta = +1$ if the move involves a growth process and $\Delta = -1$ if the move involves a deletion.

The choice of intermediate states for the expanded ensemble can vary for different systems. In certain cases, the interaction of the entire tagged molecule with the system can be "turned on" through a gradual progression from an ideal noninteracting state to a fully interacting one [21–23], characterized by the parameter y.

2. Adsorption of Hard-Core Flexible Chain Polymers in a Slit-Like Pore

As with the previous example concerning simulations of clay hydrates, a striking demonstration of the efficiency of an expanded ensemble/configurational bias algorithm is provided by the study of long chain molecules confined to narrow slit pores. One of the goals of that work would be to investigate how polymeric molecules can segregate between a bulk solution or melt and a porous system. The use of EGCMC can alleviate the problems associated with the insertion and relaxation of molecules in such a porous system.

Simulations of hard-core flexible tetramers, hexadecamers, and 100-mers have been performed by Escobedo and de Pablo [18] to study the equilibrium segregation of large molecules between a pore and bulk fluid. The width of the pore was kept constant at 5σ. The pore walls were impenetrable. Figure 6 shows some of the results. It can be seen that, in spite of the fact that the packing fractions in the pore are relatively high (above 0.4), EGCMC is able to provide reliable estimates of the concentration of long chains in the pore. For small chains (tetramers), simple molecular dynamics simulations can be used to follow the actual diffusion of the molecules into the pore, and in that case good agreement is found between EGCMC and MD [24].

3. Critical Behavior in Polymer Solutions

The EGCMC formalism has been used extensively to investigate the phase behavior of chain molecules. One recent application which is particularly instructive is concerned with the critical scalings of large-molecular weight polymer solutions [25]. Polymer solutions (e.g., polystyrene in methylcyclohexane) often exhibit an upper critical solution temperature, whose precise location depends on molecular weight [26]; the nature of that dependence is of interest for both theoretical and practical considerations. Using EGCMC simulations of lattice polymers, it has now been possible to show that the scaling of the critical density with molecular weight follows a scaling relation of the form $\rho_c \sim MW^\nu$, with $\nu = 0.5$ in the limit of high molecular weights. Figure 7a shows simulated and experimental phase diagrams for several

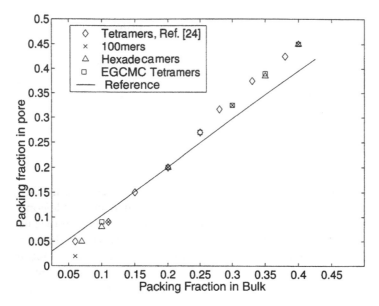

FIG. 6 Simulated adsorption isotherms for hard-sphere chains of 4, 16, and 100 sites in a slit-like pore. Reference refers to [24].

polystyrene solutions. It can be seen that simulations are consistent with experiment, and that a good mapping of the actual data onto a cubic lattice model can be achieved. The chain lengths examined in simulations are found to be comparable with, and in many cases longer than, those considered in experiments. Most experimental studies had indicated that $\nu = 0.38$; simulations of long chains were necessary to show that a gradual crossover occurs from 0.38 to 0.5 (Fig. 7b), and that it occurs at large molecular weights (polystyrene of about $MW \approx 1{,}000{,}000$). These findings are consistent with theoretical arguments [27], and serve to illustrate that even for large molecules Monte Carlo simulations can be a useful complement to experimental data, provided appropriate techniques are employed.

C. Topological Configurational Bias

The methods described so far for study of polymeric molecules rely on the presence of end sites. For long chain molecules or for cyclic or ring molecules, they are of little use. It turns out that in such cases it is also possible to construct configurational-bias based algorithms in which the source of the bias is not only the potential energy of individual sites in the new, trial position, but also a connectivity constraint. Topological configurational bias moves are similar in spirit, but different in implementation, to concerted rotation (CONROT) moves [28,29], which are not elaborated upon here.

In a "topological" configurational bias move, an arbitrary number of inner sites of a chain are excised and regrown by taking into account energetic biases, as well as geometric (topology) considerations that ensure that connectivity is preserved. Detailed balance is satisfied by determining the number of fixed-end random walks that exist between two sites of the molecule. While different schemes may vary in the way in which the number of random walks, $N^{(j)}$, is estimated, the general idea remains the same. We begin by discussing the implementation on a lattice, and then describe several more recent methods suitable for a continuum.

1. Lattice Case

An algorithm to perform such a move on a *cubic lattice* [30] could be:

1. A piece of the chain consisting of n sites is cut starting from site s of the chain. The sites to be regrown are numbered from 1 to n.
2. The current site to be regrown is labelled i.
3. Each of the possible k trial orientations for site i are visited and the following quantity is calculated:

$$N^{(l)}(i, n+1)e^{-\beta U_i^{(l)}} \tag{26}$$

where $N^{(l)}(i, n+1)$ is the number of random walks (this quantity is discussed below) from site i in trial position l to site $n+1$, and $U_i^{(l)}$ is the energy of site i in trial position l.

4. A position, j, is selected for site i from the following probability distribution:

$$p_i^{(j)} = \frac{N^{(j)}(i, n+1)e^{-\beta U_i^{(j)}}}{\sum_{l=1}^{k} N^{(l)}(i, n+1)e^{-\beta U_i^{(l)}}} \tag{27}$$

5. The modified Rosenbluth weight and Random Walk weight for the chain are constructed as follows:

$$R_W^F = \prod_{i=1}^{n} \sum_{l=1}^{k} N^{(l)}(i, n+1)e^{-\beta U_i^{(l)}} \tag{28}$$

$$G_{RW}^F = \prod_{i=1}^{n} N^{(j)}(i, n+1) \tag{29}$$

6. To estimate the weight of the original configuration, the same procedure is repeated for the original positions of the sites that were excised.

7. The modified acceptance criteria for this move then take the form

$$A(X|Y) = \min\left(1, \frac{R_W^F G_{RW}^R}{R_W^R G_{RW}^F}\right) \tag{30}$$

In the case of a lattice model, the expression for $N^{(j)}$ is given by [30]:

$$N^{(j)}(i, s+n) = \sum_{\bar{x}=0}^{N^- - \bar{y}} \sum_{\bar{y}=0}^{N^-} \frac{N_s!}{\bar{x}!(\bar{x}+\Delta x)!\bar{y}!(\bar{y}+\Delta y)!(N^- - \bar{x} - \bar{y})!(N^+ - \bar{x} - \bar{y})!} \tag{31}$$

$$N^- = \frac{(N_s - \Delta x - \Delta y - \Delta z)}{2} \tag{32}$$

$$N^+ = \frac{(N_s - \Delta x - \Delta y + \Delta z)}{2} \tag{33}$$

where $\Delta x = x - \bar{x}$, $\Delta y = y - \bar{y}$, $\Delta z = z - \bar{z}$ are the difference in the coordinates between site i and site $n+1$, N_s is the number of steps, $n - i + 1$, along the chain between sites i and $n+1$, and x, y, z are the number of steps along the positive coordinate axes and $\bar{x}, \bar{y}, \bar{z}$ are the steps along the negative coordinate axes.

In order to assess the performance of this algorithm, we have studied the decay of the bond autocorrelation function as a function of CPU time. The bond autocorrelation function is defined by

$$b(t) = \frac{\langle \vec{B}(t) \cdot \vec{B}(0) \rangle}{\langle \vec{B}(0) \cdot \vec{B}(0) \rangle} \tag{34}$$

where $\vec{B}(t)$ denotes the vector along a bond of the molecule at time t. Figure 8 shows the bond autocorrelation function corresponding to a simulation for cyclic chains consisting of 100 sites, in which only local moves (kink-jump and crankshaft) were employed, and one in which topological configurational bias moves were implemented. One can clearly see that the decay in the latter case is much faster, thereby resulting in a much more efficient simulation algorithm. In the case of ring molecules, care must be exercised in the selection of the number of sites to be cut and regrown. This is because the topological nature of the molecule can change (e.g., from unknotted to knotted) if a large number of sites are selected. In the case of linear chains this restriction is irrelevant and, in principle, any number

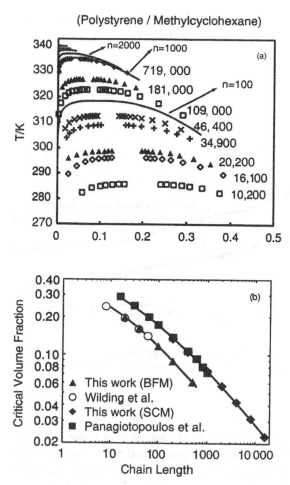

FIG. 7 (a) Phase diagram for solutions of polystyrene in methylcyclohexane. The lines correspond to results of expanded ensemble simulations, and the symbols refer to experimental data [26]. (b) Scaling of the critical volume fraction with chain length. The squares and the circles correspond to literature simulation data [49,50]; the diamonds and the triangles correspond to expanded ensemble simulations for a simple cubic lattice and for a bond fluctuation model, respectively [25].

of sites can be used in an individual trial move. It turns out that an optimum number of sites can be identified because the computational expense of the trial move increases with the number of cut sites. This point is illustrated by Fig. 9, which depicts the decay of the bond autocorrelation function for linear chains of 500 sites on a lattice. For a fixed amount of CPU time, the

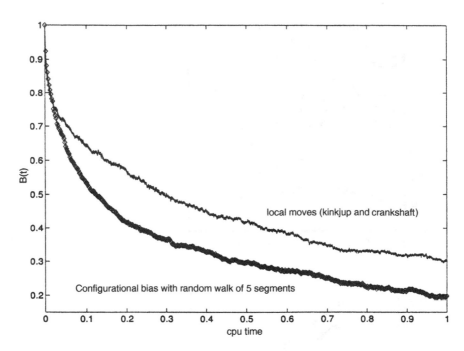

FIG. 8 Rate of decay of the bond autocorrelation function $(b(t))$ for cyclic (ring) chains of 100 sites, using local moves and topological configurational bias moves, as a function of CPU time.

decay of the bond autocorrelation function is plotted as a function of the number of sites used in the topological configurational bias move. As indicated above, there is a clear optimum number of sites for the move. These simulations correspond to thin polymer films at relatively high volume fractions ($\phi \approx 0.95$); they are of relevance for the study of glass transition phenomena in polymeric films, where the presence of free interfaces introduces pronounced departures from bulk behavior [31]. Recently, configurational-bias techniques have also been combined with a density of states formalism to calculate thermodynamic properties over a wide range of temperature from a single simulation [32].

2. Continuum Case

In the continuum case, various strategies can be used to achieve closure. In the case of Extended continuum configurational bias (ECCB) Monte Carlo [33], a simple geometric constraint is employed to ensure that the chain is correctly closed. Since the geometric constraint does not take intramolecular

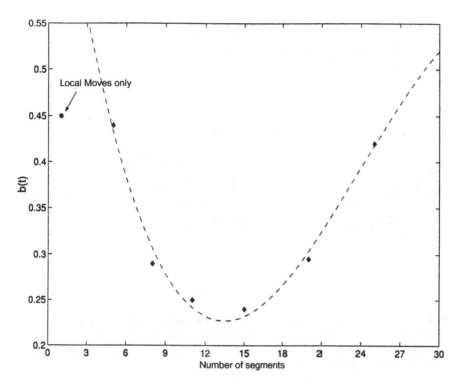

FIG. 9 Decay of the bond autocorrelation function ($b(t)$) for linear chains of length 500 as a function of number of sites cut and regrown, for a given amount of CPU time.

interactions into account, the ECCB method is efficient for molecules with relatively flexible backbones. This method is described in some detail below.

1. Select a chain at random to be moved.
2. Select two sites, a and b, at random within the chain such that the number of sites between them (including the two chosen sites), n, is less than or equal to some specified maximum N_{max}. The number of sites to be cut and regrown for a particular move is n.
3. If one of the sites happens to be an end site, then the usual configurational bias move is performed.
   a. If $n = 1$, then a simple crankshaft move is carried out.
   b. If the current site to be appended is i, then lower and upper bounds exist for the angles, θ_{min} and θ_{max}. These can be determined from the line joining sites i and $i-1$ with the line joining

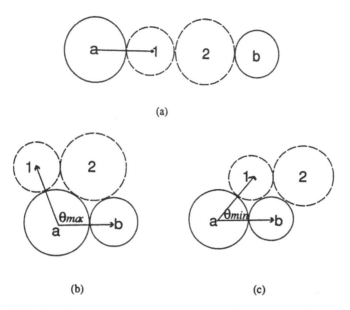

(a)

(b)   (c)

FIG. 10 Limiting cases for the bounds of the θ angle for the two regrowing sites (1 and 2) involved in an extended configurational bias move (ECCB). (a) Neighbor sites a and b in the configuration of maximum separation, (b) and (c) neighbor sites a and b in the configuration of closest approach.

        $i-1$ and b, where b is the first uncut site on the side of the cut opposite to the growing end (see Fig. 10).

4. Once the bounds for θ have been established, several orientations of the bond vector $\mathbf{b}_{i,\,i-1}$ are sampled according to the usual configurational bias method, and one is selected from the probability distribution given by Eq. (14).

5. The same procedure is repeated for the reverse move. There is a temporary change in degrees of freedom used to generate the new configuration because of the restricted sampling of θ (Fig. 11). In order to correct for this, the Jacobian determinants must be incorporated into the acceptance criteria. The acceptance criteria for this move become

$$A(X|Y) = \min\left(1, \frac{R_W^F J^F}{R_W^R J^R}\right) \tag{35}$$

where J^F and J^R are the Jacobian determinants for the forward and reverse moves, respectively. For a detailed explanation on the calculation of the bound of θ and the Jacobians readers are referred to [33].

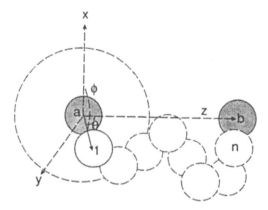

FIG. 11 The entire spherical sampling space is not accessible for selection of the trial orientation of site 1, because the θ angle must lie within the bounds $[\theta_{min}, \theta_{max}]$ (see Fig. 10). Note that there are no bounds on the spherical angle ϕ for the bond vector.

This algorithm can also be extended to simulation of branched chains and cross-linked structures [34]. In some cases, it is useful to target or restrict ECCB moves to a particular region of space. If such a region is spherical, for example, sites within that sphere can be regrown into new positions by using ECCB (if they are inner sites) or configurational bias (if they are terminal or end sites). This approach has been referred to as the HOLE method in the literature (Fig. 12). Figure 13 shows the bond and end-to-end relaxation functions for a system of 16-mer hard spheres.

Another strategy for topological bias moves involves the use of an analytical probability distribution function (for ideal, non-self-avoiding random walks) to bias the choice of trial orientations [4,35]. Again, this algorithm is applicable to relatively flexible molecules.

For molecules with strong intramolecular interactions, inner sites can be regrown using the concerted rotation (CONROT) method [28,29]. This method is particularly useful for molecular models that exhibit rigid constraints (e.g., rigid bonds or rigid angles). It involves cutting a trimer from the inner sites of the chain. Before the trimer is regrown, either one or both of the sites adjacent to the ends of the trimer are displaced. The problem of regrowing the trimer is handled by solving a system of simultaneous equations [28,29,36].

Configurational Bias methods have now also been used to simulate molecules with strong intramolecular potentials [37,38]; in that case, it is necessary to incorporate such potentials in the biasing factors used to generate trial orientations. If the trial orientations are proposed at random, the efficiency of the trial move is low, due to high intramolecular energies

chain 2

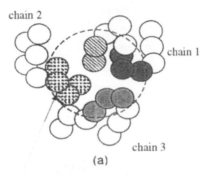

chain 1

chain 3

(a)

HOLE Boundary

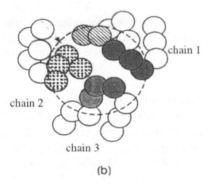

chain 1

chain 2

chain 3

(b)

FIG. 12 Illustration of the HOLE method. (a) Local configuration before the move; (b) configuration after the move.

for a significant number of the proposed positions. The use of intramolecular potentials along with the probability distribution for distances between sites has been successful in biasing the random walk and achieving closure [37]. A preliminary run is necessary in order to generate a probability distribution for site separations as a function of site spacing along the chain. A brief outline of the implementation as proposed by Escobedo and Chen is provided below:

1. A chain is selected at random, and part of the chain consisting of n sites is cut, starting from site s of the chain. The sites to be regrown are numbered from 1 to n.
2. Assume that the current site to be regrown is i.
3. Select k trial orientations for i, using the Boltzmann distribution arising from the intramolecular potential

$$p_i^{(l)} = Ce^{-\beta U_i^{bonded(l)}} \qquad l = 1, 2, \ldots, k \tag{36}$$

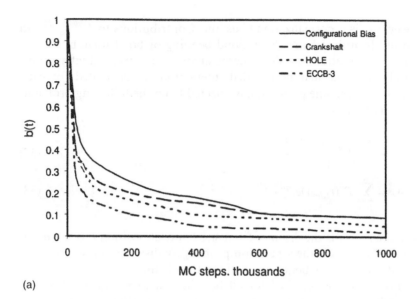

(a)

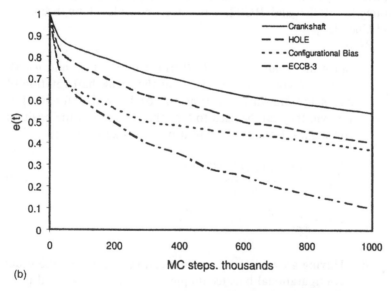

(b)

FIG. 13 Correlation functions for crankshaft, Configurational bias, ECCB-3, and HOLE moves for a system of 16-mer hard spheres at $\eta = 0.3$. (a) Bond autocorrelation function ($b(t)$); (b) end-to-end autocorrelation function ($e(t)$).

where C is a normalization constant. Contributions to $U_i^{bonded(l)}$ can result from bond-stretching, bond-bending or bond-torsion.

4. The contribution to a site's energy arising from external interactions and intramolecular nonbonded interactions is calculated for each trial position; one position, j, is selected from the following probability distribution:

$$p_i^{(j)} = \frac{P_i^o(r_{j,n+1})e^{-\beta U_i^{ext}(j)}}{w_i} \tag{37}$$

$$w_i = \sum_{l=1}^{k} P_i^o(r_{l,n+1})e^{-\beta U_i^{ext}(l)} \tag{38}$$

where $r_{l,n+1}$ is the distance from site i in trial position l to site $n+1$, and $P_i^o(r_{l,n+1})$ is the separation probability distribution for two sites that are $n-i+1$ bonds apart along the chain.

5. This process is repeated for all but the last two sites, $n-1$ and n, which are grown using a *look-ahead* strategy (Fig. 14). Two such strategies, RCB1 and RCB2, have been considered by Escobedo and Chen [37].

    a. RCB1

        i. Since the bending angle, θ_n (formed by $n-1$, n, and $n+1$), and the Jacobian factor, J_{n-1}, of the crankshaft rotation of site n are known as soon as the trial position for $n-1$ is known, they can be used to bias the selection of the position for site $n-1$ out of k_{n-1} trial positions according to

$$p_{n-1}^{(j)} = \frac{e^{-\beta U_{n-1}^{ext}(j)}e^{-\beta U^{bend}(\theta_n^{(j)})}J_{n-1}^{(j)}}{w_{n-1}} \tag{39}$$

$$w_{n-1} = \sum_{l=1}^{k_{n-1}} e^{-\beta U_{n-1}^{ext}(l)}e^{-\beta U^{bend}(\theta_n^{(l)})}J_{n-1}^{(l)} \tag{40}$$

        ii. Having grown site $n-1$, site n can be selected by the usual configurational bias technique out of a set of k_n trial positions according to Eq. (14).

        iii. After all the sites have been grown, the Rosenbluth weight for the entire chain is constructed:

$$R_W^F = \prod_{i=1}^{n} w_i \tag{41}$$

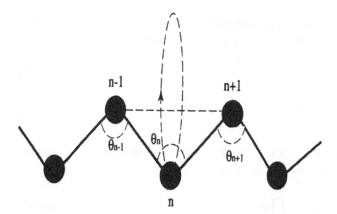

FIG. 14 Illustration and notation of the reconstruction process of the last two sites ($n = 1$ and n) of an ECCB move for a chain that exhibits stiff intramolecular energy functions [37].

iv. The modified acceptance criteria for this move are

$$A(X|Y) = \min\left[1, \frac{R_W^F\left(\prod_{i=1}^{n-2} P_i^o(r_{jn+1})\right)^R}{R_W^R\left(\prod_{i=1}^{n-2} P_i^o(r_{jn+1})\right)^F}\right] \qquad (42)$$

b. RCB2

i. In this variation, instead of selecting one trial position for $n - 1$, a total of m final trial positions are selected from mk_{n-1} trial choices, using the following probability distribution:

$$p_{n-1,t}^{(j)} = \frac{e^{-\beta U^{bend}(\theta_{n,t}^{(j)})} J_{n-1,t}^{(j)}}{w_{n-1,t}} \qquad t = 1, 2, \ldots, m \qquad (43)$$

$$w_{n-1,t} = \sum_{l=1}^{k_{n-1}} e^{-\beta U^{bend}(\theta_{n,t}^{(l)})} J_{n-1,t}^{(l)} \qquad (44)$$

where subscript index t runs over the m sets of trial positions.

ii. For each final trial position of $n - 1$, $j_{n-1,t}$, one trial position of n is chosen from k_n trial choices, just as in (a). This gives a set of Rosenbluth weights, $w_{n,t}$, and a set of final trial orientations, $j_{n,t}$.

iii. Out of the m position pairs of sites $n-1$ and n, one pair, j, is finally selected from the probability distribution:

$$p_{n-1,n}^{(j)} = \frac{e^{-\beta U_{n-1}^{ext}(j)} e^{-\beta U_n^{ext}(j)} w_{n-1,j} w_{n,j}}{w_{n,n-1}} \tag{45}$$

$$w_{n,n-1} = \sum_{t=1}^{m} e^{-\beta U_{n-1}^{ext}(t)} e^{-\beta U_n^{ext}(t)} w_{n-1,t} w_{n,t} \tag{46}$$

iv. After all the sites have been grown, the Rosenbluth weight for the entire chain is constructed:

$$R_W^F = w_{n,n-1} \prod_{i=1}^{n-2} w_i \tag{47}$$

v. The acceptance criteria are the same as given by Eq. (42).

3. Simulation of Linear and Cyclic Alkanes Using Configurational Bias Approach

Escobedo and Chen [37] have used this technique to study linear and cyclic alkanes (using a united-atom force field [39]). In order to test the efficiency of these algorithms, a model system consisting of one isolated linear alkane molecule at $T = 400K$ was simulated. A "half-chain end-to-end vector" autocorrelation function was used as a measure of the relaxation of the chain. This function was measured in the simulation as a function of number of sites regrowing in every RCB2 move (Fig. 15). The dependence of the autocorrelation function on the choice of $P_i^o(r_{l,n+1})$ was considered (Fig. 16). A comparison between RCB1, RCB2, and simple crankshaft moves was also considered (Fig. 17). One can see that the improvement over the simple crankshaft move is significant for both the RCB1 and RCB2 moves. The optimum number of sites to be regrown depends on the system conditions, but in general the improvements tend to be large for the first sites that are excised, and they taper off gradually.

D. Parallel Tempering and Configurational Bias

A problem that pervades simulations of complex fluids, particularly at elevated densities and low temperatures (e.g., near the glass transition temperature), is that the system gets trapped in local energy minima from which it is unable to escape. In such cases Configurational Bias moves are often insufficient, and additional "tricks" are necessary to improve sampling. One class of methods that is particularly helpful and easy to implement is provided by "Parallel Tempering" techniques.

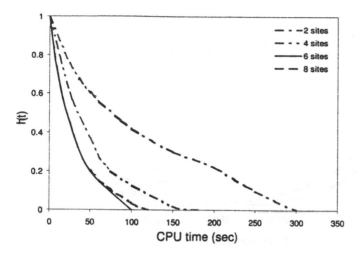

FIG. 15 Half-chain end-to-end correlation function ($h(t)$) for RCB2, regrowing exclusively 2, 4, 6, or 8 sites in a system consisting of one isolated C_{30} chain at 400K.

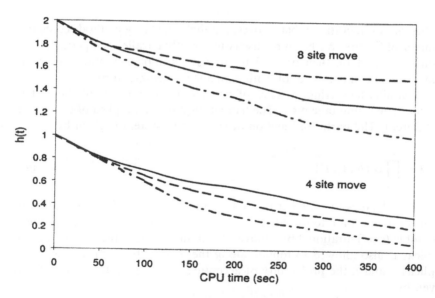

FIG. 16 Half-chain end-to-end correlation function ($h(t)$) for RCB1 for an isolated C_{30} molecule at 250K for 4 and 8 site moves. Three different forms of $P_i^o(r_{l,n+1})$ are used: the Continuous Unperturbed Chain (CUC) $P_i^o(r_{l,n+1})$ (dash-dot line), the uniform $P_i^o(r_{l,n+1})$ implied by conventional ECCB moves (dashed line), and $P_i^o(r_{l,n+1})$ for a fully flexible Guassian chain (solid line).

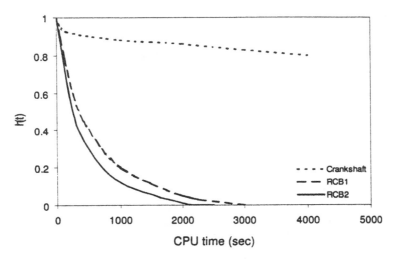

FIG. 17 Half-chain end-to-end correlation function ($h(t)$) for crankshaft, RCB1, and RCB2 moves for an isolated linear C_{70} molecule at 400K. The number of sites regrown ranged from 2 to 10.

Just as configurational bias moves, parallel tempering methods come in a number of flavors. We begin by discussing briefly Parallel Tempering in one dimension, and we then proceed to discuss its implementation in several dimensions or in combination with expanded ensemble ideas.

In parallel tempering, a series of noninteracting replicas of the system are simulated simultaneously. The temperature of each replica of the system is different. The partition function of the overall system is given by

$$Q = \prod_{i=1}^{N_r} Q_i(N, V, T_i) \tag{48}$$

where N_r denotes the number of replicas and T_i denotes the temperature of replica i. In addition to the equilibration moves carried out in the canonical ensemble (configurational bias moves, local moves, etc.), the identity of two replicas is also allowed to mutate along the T axis (Fig. 18). For any two replicas, i and j, the probability of accepting such a mutation (or swap) is given by

$$A(X|Y) = \min[1, \exp(\Delta \beta_{i,j} \Delta U_{i,j})] \tag{49}$$

where $\Delta \beta_{i,j}$ is the difference in the inverse temperatures of the two replicas and $\Delta U_{i,j}$ is the difference in the potential energies. It can be seen from Eq. (49) that the trial move will have a reasonable chance of being accepted only if the energy distributions for the two replicas overlap to some extent.

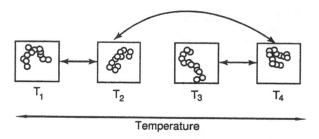

Temperature

FIG. 18 Illustration of Parallel Tempering Monte Carlo.

A preliminary run may therefore be necessary to determine appropriate temperature spacings between replicas. Parallel tempering in temperature has been used to study simple glasses and isolated biological molecules [40–43]. It has also been used in conjunction with topological configurational bias to study linear and cyclic alkanes [37].

1. Multidimensional and Hyperparallel Parallel Tempering

In many cases Parallel Tempering in temperature does not provide significant improvements over a conventional, one-replica simulation. It turns out that the performance of parallel tempering can be improved considerably by considering several dimensions, e.g., replicas having different T and μ or different T and P. We refer to such methods as Multidimensional Parallel Tempering. In a grand canonical ensemble, for example, different replicas i can have different values of chemical potential μ_i and temperature T_i; configuration swaps between any two replicas are accepted with probability:

$$P_{acc} = \min[1, \exp(\Delta\beta_{i,j}\Delta U_{i,j} - \Delta N_{i,j}\Delta(\beta\mu)_{ij})] \tag{50}$$

where $\Delta\beta_{i,j} = \beta_i - \beta_j$, $\Delta U_{i,j} = U_i - U_j$, $\Delta N_{i,j} = N_i - N_j$ and $\Delta(\beta\mu)_{i,j} = (\beta\mu)_i - (\beta\mu)_j$. Multidimensional Parallel Tempering has been employed successfully to study the phase behavior of electrolytes at low temperatures [44,45]. It has also been used to investigate the behavior of glass-forming disaccharides near the glass transition [46].

In the particular case of polymeric molecules, it is also possible to combine multidimensional parallel tempering and expanded ensemble ideas into a powerful technique, which we call hyperparallel tempering (HPTMC) [47]. In that method, a mutation along the number of sites of a tagged chain is also considered (see Section IV.B.1); Fig. 19 provides a schematic representation of the algorithm. Two of the axes refer to replicas having different chemical potentials or different temperatures. A third axis or dimension is used to denote the fact that, in each replica of the system,

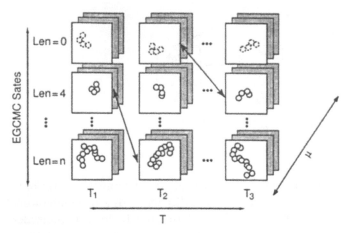

FIG. 19 Schematic representation of the implementation of hyperparallel Monte Carlo. Each box in the figure represents a distinct replica of the system; these replicas are simulated simultaneously in a single run. In addition to traditional Monte Carlo trial moves within each replica, distinct replicas can (1) change their state variables in the expanded dimension and (2) exchange or swap configuration with each other, thereby visiting different values of T and μ.

one can have N_i full molecules and one tagged molecule of length y (recall Section IV.B.1). When two replicas exchange configuration, they also swap a tagged chain. Within each of the replicas, however, the tagged chain is allowed to grow or shrink according to a conventional expanded ensemble formalism.

This technique has been applied to the study of phase behavior of long polymeric molecules. Figure 20 compares the performance of various simulation techniques to that of Hyperparallel Tempering Monte Carlo. It can be seen that the decay of the end-to-end autocorrelation function for HPTMC is considerably faster than that for conventional canonical or grand canonical simulations. However, it is important to point out that much of the performance of that method is due to the efficiency of the underlying moves, namely the expanded ensemble configurational bias gradual growth or deletion of sections of the molecules.

V. FUTURE DIRECTIONS

Configurational bias trial moves offer significant improvements in efficiency over conventional, blind trial moves. As discussed throughout this chapter, local random moves usually provide small displacements of the system, whereas biased moves can efficiently displace several sites per move while

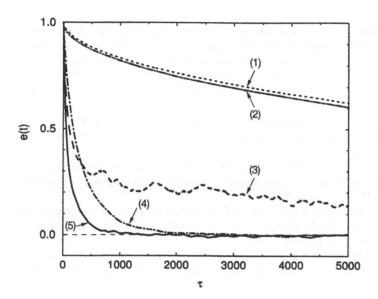

FIG. 20 End-to-end vector autocorrelation function ($e(t)$) for polymer chains obtained by different simulation methods: (1) short dashed line: canonical ensemble; (2) dashed-dotted-dotted line: grand canonical ensemble (3) dashed line: multi-dimensional parallel tempering; (4) dashed-dotted line expanded grand canonical ensemble; (5) solid line: hyperparallel tempering.

preserving a high acceptance rate. As illustrated by the variety of systems and moves discussed here, configurational bias techniques offer considerable flexibility; new implementations will surely be proposed as researchers continue to explore increasingly complex systems.

In the particular context of polymeric molecules, recent developments with end-bridging and double-bridging moves [48] indicate that a powerful algorithm could be arrived at by using topological configurational bias to cut simultaneously several chains, and to reconnect them in different ways, thereby achieving considerable molecular relaxation in just a few moves. The details of such moves remain to be worked out, particularly for the more complex case of highly branched molecules. But it is a direction in which we are likely to see interesting new developments in the near future.

REFERENCES

1. Metropolis, N.; Rosenbluth, A.W.; Rosenbluth, M.N.; Teller, A.H.; Teller, E. Equation of state calculations by fast computing machines. J. Chem. Phys. **1953**, *21*, 1087–1092.

2. Kalos, M.H.; Whitlock, P.A. *Monte Carlo Methods*; Wiley: New York, 1986.

3. Cracknell, R.F.; Nicholson, D.; Parsonage, N.G.; Evans, H. Rotational insertion bias: a novel method for simulating dense phases of structured particles, with particular application to water. Mol. Phys. **1990**, *71* (5), 931–943.

4. Frenkel, D.; Smit, B. *Understanding Molecular Simulation*; Academic: New York, 1996.

5. Chávez-Páez, M.; Van Workum, K.; de Pablo, L.; de Pablo, J.J. Monte Carlo simulations of wyoming sodium montmorillonite hydrates. J. Chem. Phys. **2001**, *114* (3), 1405–1413.

6. Skipper, N.T.; Chang, F.R.C.; Sposito, G. Monte-Carlo simulation of interlayer molecular-structure in swelling clay-minerals. 1. methodology. Clay Clay Miner. **1995**, *43* (3), 285–293.

7. Boek, E.S.; Convey, P.V.; Skipper, N.T. Molecular modeling of clay hydration: a study of hysteresis loops in the swelling curves of sodium montmorillonites. Langmuir **1995**, *11* (7), 4629–4631.

8. Jorgensen, W.L.; Chandrasekhar, J.; Madura, J.D. Comparison of simple potential functions for simulating liquid water. J. Chem. Phys. **1983**, *79* (2), 926–935.

9. Lie, G.C.; Clementi, E.; Yoshimine, M. Study of the structure of molecular complexes. XIII. Monte Carlo simulation of liquid water with a configuration interaction pair potential. J. Chem. Phys. **1976**, *64* (6), 2314–2323.

10. Lax, M.; Brender, K. Monte Carlo studies of lattice polymer dynamics. J. Chem. Phys. **1977**, *67* (4), 1785–1787.

11. Mandel, F. Macromolecular dimensions obtained by an efficient Monte Carlo method: the mean square end-to-end separation. J. Chem. Phys. **1979**, *70* (8), 3984–3988.

12. de Pablo, J.J.; Yan, Q.L.; Escobedo, F.A. Simulation of phase transitions in fluids. Annu. Rev. Phys. Chem. **1999**, *50*, 377–411.

13. Rosenbluth, M.N.; Rosenbluth, A.W. Monte Carlo calculation of the average extension of molecular chains. J. Chem. Phys. **1955**, *23* (2), 356–359.

14. de Pablo, J.J.; Laso, M.; Suter, U.W. Simulation of polyethylene above and below the melting-point. J. Chem. Phys. **1992**, *96* (3), 2395–2403.

15. Siepmann, J.I.; Frenkel, D. Configurational bias Monte-Carlo—A new sampling scheme for flexible chains. Mol. Phys. **1992**, *75* (1), 59–70.

16. Smit, B.; Karaborni, S.; Siepmann, J.I. Computer simulations of vapour-liquid phase equilibria of n-alkanes. J. Chem. Phys. **1995**, *102* (5), 2126–2140.

17. Lyubartsev, A.P.; Martinovski, A.A.; Shevkunov, S.V.; Vorontsov-Velyaminov, P.N. New approach to Monte Carlo calculation of the free energy: Method of expanded ensembles. J. Chem. Phys. **1991**, *96* (3), 1776–1783.

18. Depablo, J.J.; Escobedo, F.A. Expanded grand canonical and Gibbs ensemble Monte Carlo simulation of polymers. J. Chem. Phys. **1996**, *105* (10), 4391–4394.

19. Kumar, S.K.; Szleifer, I.; Panagiotopoulos, A.Z. Determination of the chemical-potentials of polymeric systems from Monte-Carlo simulations. Phys. Rev. Lett. **1991**, *66* (23), 2395–2398.

20. de Pablo, J.J.; Escobedo, F.A. Chemical-potential and equations of state of hard-core chain molecules. J. Chem. Phys. **1995**, *103* (9), 1946–1956.
21. Attard, P. Simulation of the chemical potential and the cavity free energy of dense hard-sphere fluids. J. Chem. Phys. **1993**, *98* (3), 2225–2231.
22. Wilding, N.B.; Muller, M. Accurate measurements of the chemical-potential of polymeric systems by Monte-Carlo simulation. J. Chem. Phys. **1994**, *101* (5), 4324–4330.
23. Rutledge, G.C.; Khare, A.A. Chemical potential of aromatic compounds in pure n-alkanes using expanded ensemble Monte Carlo simulations. J. Phys. Chem. B **2000**, *104* (15), 3639–3644.
24. Yethiraj, A.; Hall, C.K. Monte-Carlo simulation of the equilibrium partitioning of chain fluids between a bulk and a pore. Mol. Phys. **1991**, *73* (3), 503–515.
25. Yan, Q.L.; de Pablo, J.J. Critical behavior of lattice polymers studied by Monte Carlo simulations. J. Chem. Phys. **2000**, *113* (14), 5954–5957.
26. Dobashi, T.; Nakata, M.; Kaneko, M. Coexistence curve of polystyrene in methylcyclohexane. 1. Range of simple scaling and critical exponents. J. Chem. Phys. **1980**, *72* (12), 6685–6697.
27. Frauenkron, H.; Grassberger, P. Critical unmixing of polymer solutions. J. Chem. Phys. **1997**, *107* (22), 9599–9608.
28. Dodd, L.R.; Boone, T.D.; Theodorou, D.N. A concerted rotation algorithm for atomistic Monte-Carlo simulation of polymer melts and glasses. Mol. Phys. **1993**, *78* (4), 961–996.
29. Pant, P.V.K.; Theodorou, D.N. Variable connectivity method for the atomistic Monte-Carlo simulation of polydisperse polymer melts. Macromolecules **1995**, *28* (21), 7224–7234.
30. Dijsktra, M.; Frenkel, D.; Hansen, J.P. Phase separation in binary hard-core mixtures. J. Chem. Phys. **1994**, *101* (4), 3179–3189.
31. Jain, T.S.; de Pablo, J.J. Monte Carlo simulation of free-standing polymer films near the glass transition temperature. Macromolecules **2002**, *35* (6), 2167–2176.
32. Jain, T.S.; de Pablo, J.J. A biased Monte Carlo technique for calculation of the density of states of polymer films. J. Chem. Phys. **2002**, *116* (16), 7238–7243.
33. Escobedo, F.A.; de Pablo, J.J. Extended continuum configurational bias Monte Carlo methods for simulation of flexible molecules. J. Chem. Phys. **1994**, *102* (6), 2636–2652.
34. Escobedo, F.A.; de Pablo, J.J. Molecular simulation of polymeric networks and gels: phase behavior and swelling. Phys. Rep. **1999**, *318* (3), 86–112.
35. Vendruscolo, M. Modified configurational bias Monte Carlo method for simulation of polymer systems. J. Chem. Phys. **1997**, *106* (7), 2970–2976.
36. Wu, M.G.; Deem, M.W. Analytical rebridging Monte Carlo: application to cis/trans isomerization in proline-containing, cyclic peptides. J. Chem. Phys. **1999**, *111* (14), 6625–6632.
37. Escobedo, F.A.; Chen, Z. A configurational-bias approach for the simulation of inner sections of linear and cyclic molecules. J. Chem. Phys. **2000**, *113* (24), 11382–11392.

38. Siepmann, J.I.; Wick, C.D. Self-adapting fixed-end-point configurational-bias Monte Carlo method for the regrowth of interior segments of chain molecules with strong intramolecular interactions. Macromolecules **2000**, *33* (19), 7207–7218.

39. Nath, S.K.; Escobedo, F.A.; de Pablo, J.J. On the simulation of vapor-liquid equilibria for alkanes. J. Chem. Phys. **1998**, *108* (23), 9905–9911.

40. Hansmann, U.H.E. Parallel tempering algorithm for conformational studies of biological molecules. Chem. Phys. Lett. **1997**, *281* (1–3), 140–150.

41. Deem, M.W.; Wu, M.G. Efficient Monte Carlo methods for cyclic peptides. Mol. Phys. **1999**, *97* (4), 559–580.

42. Okamoto, Y. Tackling the multiple-minima problem in protein folding by Monte Carlo simulated annealing and generalized-ensemble algorithms. Int. J. Mod. Phys. C **1999**, *10* (8), 1571–1582.

43. Marinari, G.; Parisi, E.; Ricci-Tersenghi, F.; Zuliani, F. The use of optimized Monte Carlo methods for studying spin glasses. J. Phys. A-Math. Gen. **2001**, *34* (3), 383–390.

44. Yan, Q.L.; de Pablo, J.J. Phase equilibria of size-asymmetric primitive model electrolytes. Phys. Rev. Lett. **2001**, *86* (10), 2054–2057.

45. Yan, Q.L.; de Pablo, J.J. Phase equilibria and clustering in size-asymmetric primitive model electrolytes. J. Chem. Phys. **2001**, *114* (4), 1727–1731.

46. Ekdawi-Sever, N.C.; Conrad, P.B.; de Pablo, J.J. Molecular simulation of sucrose solutions near the glass transition temperature. J. Phys. Chem. A **2001**, *105* (4), 734–742.

47. Yan, Q.L.; de Pablo, J.J. Hyperparallel tempering Monte Carlo simulation of polymeric systems. J. Chem. Phys. **2000**, *113* (3), 1276–1282.

48. Theodorou, D.N. Variable-connectivity Monte Carlo algorithms for the atomistic simulation of long-chain polymer systems. *Bridging Time Scales: Molecular Simulations for the Next Decades*, Nielaba, P., Mareschal, M., Cicotti, G., Eds.; Springer Verlag: Berlin, 2002, 67–127.

49. Panagiotopoulos, A.Z.; Wang, V. Phase equilibria of lattice polymers from histogram reweighting Monte Carlo simulations. Macromolecules **1998**, *31* (3), 912–918.

50. Wilding, N.B.; Muller, M.; Binder, K. Chain length dependence of the polymer-solvent critical point parameters. J. Chem. Phys. **1996**, *105* (2), 802–809.

8

Molecular Simulations
of Charged Polymers

ANDREY V. DOBRYNIN   University of Connecticut, Storrs,
Connecticut, U.S.A.

I. INTRODUCTION

Considerable theoretical and experimental work during the past half century
has been devoted to charged polymers [1–9]—macromolecules with ion-
izable groups. Under appropriate conditions, such as in aqueous solutions,
these groups dissociate, leaving ions on chains and counterions in solutions.
If the charges on the polymers are all positive or all negative, these polymers
are called polyelectrolytes. Common polyelectrolytes are polyacrylic and
methacrylic acids and their salts, cellulose derivatives, sulfonated polystyrene, DNA and other polyacids and polybases. If after dissociation of the
charged groups the polymers carry both positive and negative charges, they
are called polyampholytes. Examples of polyampholytes include proteins,
for example gelatin, and synthetic copolymers made of monomers with
acidic and basic groups. If these groups are weak acids or bases, the net
charge of polyampholytes can be changed by varying the *pH* of aqueous
solutions and at high charge asymmetry these polymers demonstrate
polyelectrolyte-like behavior.

Despite these extensive efforts we are still far away from a complete
understanding of the behavior of polymeric systems with electrostatic
interactions. The main factor hindering our progress is the long-range
nature of the Coulombic forces between charged species. At this stage
computer simulations have proven to be a valuable tool for the elucidation
of structural and physical properties of charged systems as well as for
verification of old and creation of new theoretical models. The explosive

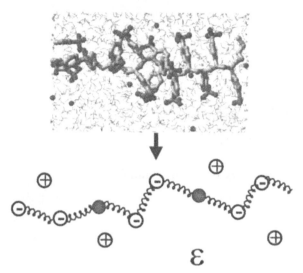

FIG. 1 Illustration of the mapping procedure of sodium polystyrene sulfonate (NaPSS) in water where the repeat units of the NaPSS chain are replaced by spherical monomers connected by springs and the solvent molecules are represented by a continuum with dielectric constant ε.

growth of computer power over the last few years has led to development of large scale simulation techniques whose goal is to reproduce and simulate processes on the molecular level. It is possible now to simulate thousands of atoms over nanosecond time scales. But such detailed molecular simulations are still impractical for the simulation of real polymeric systems, for which typical time scales to reach thermal equilibrium vary from microseconds to seconds.

The coarse-grained models of polymers (see [10] for a review) allow the extension of molecular simulations beyond nanosecond time scales by leaving aside the atomistic details of the solvent and polymers and concentrating on their macroscopic properties. In this approach the monomers are not chemical monomers, but rather groups of chemical units, and solvent molecules are represented by a continuum with macroscopic physical properties. Figure 1 shows an example of such mapping by presenting the snapshots of sodium polystyrene sulfonate in water and its representation as a coarse-grained chain with counterions in a continuum with macroscopic physical properties. Of course, the challenging part is to relate the parameters of the coarse-grained models with the real atomistic ones by calculating the parameters of intermolecular and intramolecular coarse-grained potentials [10].

II. COMPUTER SIMULATIONS OF SINGLE CHAIN PROPERTIES

A. Models and Methods

We begin our discussion by considering the conformation of a single chain in solution under the influence of the long-range electrostatic and short-range U_{sh} ($|\mathbf{r}_i - \mathbf{r}_j|$) monomer–monomer interactions. The potential energy $U(\{\mathbf{r}_i\})$ of the charged polymer chain of the degree of polymerization N in a solvent with dielectric permittivity ε with monomers located at positions $\mathbf{r}_1, \mathbf{r}_2, \ldots, \mathbf{r}_N$ and carrying the charges $eq_1, eq_2, \ldots, eq_N$ is*

$$U(\{\mathbf{r}_i\}) = \sum_{bond} U_{bond}(\{\mathbf{r}_i\})$$

$$+ \sum_i \sum_{i<j} \left(k_B T l_B \frac{q_i q_j}{|\mathbf{r}_i - \mathbf{r}_j|} \exp(-\kappa |\mathbf{r}_i - \mathbf{r}_j|) + U_{sh}(|\mathbf{r}_i - \mathbf{r}_j|) \right) \quad (1)$$

where $U_{bond}(\{\mathbf{r}_i\})$ is the bond potential describing the effect of the connectivity of monomers into the polymer chain, k_B is the Boltzmann constant, T is the absolute temperature, and l_B is the Bjerrum length ($l_B = e^2/\varepsilon k_B T$,[†] the length scale at which the electrostatic interaction between two elementary charges e in the medium with dielectric permittivity ε is of the order of the thermal energy $k_B T$). The electrostatic interactions between free ions and charged monomers are not explicitly included in the chain potential energy; instead, their effect is treated through the dependence of the inverse Debye–Huckel screening length κ on the electrolyte concentration

$$\kappa^2 = 4\pi l_B \sum_s q_s^2 c_s \quad (2)$$

where c_s is the concentration of small ions of type s and q_s is their valence.

As discussed in previous chapters, the choice of the bond $U_{bond}(\{\mathbf{r}_i\})$ and short-range $U_{sh}(|\mathbf{r}_i - \mathbf{r}_j|)$ potential varies from simulation to simulation. Off-lattice models, for example, have used the harmonic-spring potential, the FENE (finitely extendable, nonlinear elastic) potential, the rigid bond with fixed valence angles, and the freely-jointed chain model to represent the bonding interaction between adjacent monomers. For the short-range

---

*CGS units are used in this chapter, unless specifically indicated.
[†]In SI units it is $l_B = e^2/(4\pi\varepsilon\varepsilon_0 k_B T)$.

monomer–monomer interactions the Lennard-Jones and hard-core type potentials are commonly used in simulations of charged polymers.

To sample the chain configurational space one can perform canonical Monte Carlo (MC) [11–16] or molecular dynamics (MD) simulations [15–17]; see also the discussions in previous chapters in this book. The Monte Carlo method involves generation of successive "trial" chain configurations which are then accepted or rejected with the correct statistical weight. The trial configurations are generated by randomly modifying the monomer positions. For example, a new chain conformation may be obtained by *reptation* [18,19] and *pivot* [20–22] moves or by *local* moves randomly displacing a single monomer. In reptation or the "slithering snake" algorithm one chooses an end of the chain at random and transfers a monomer from this end to other end. Applied to an N-bead chain this method leads to a statistically independent configuration after N^2 attempts to exchange the ends of a chain. In the pivot algorithm a new chain configuration is obtained by rotating a part of the chain around a randomly selected bond by a randomly selected angle. The pivot moves are very radical: only after a few accepted moves may a chain reach an essentially new configuration. (For lattice models of a chain the moves have to be modified to preserve the lattice structure.)

To decide whether or not a trial conformation should be accepted or rejected, the change in energy, ΔU, associated with the move is calculated, and the move is accepted or rejected according to the Metropolis algorithm with probability

$$acc(old \rightarrow new) = \min\left[1, \exp\left(-\frac{\Delta U}{k_B T}\right)\right] \tag{3}$$

If the new conformation is rejected then the current configuration is recounted in the averaging process.

The choice of one or another type of move or combination of moves is dictated by the most efficient sampling of the configurational space of the chain during the course of simulations. The art is to find their optimal combination. Below we give a theoretical estimate of the computational efficiency for different types of moves defined as the CPU time required to generate a new statistically independent chain configuration. The less computationally efficient are the local moves, when a new chain configuration is generated by random displacement of a single monomer. This MC scheme describes the Verdier–Stockmayer dynamics [25] of a polymer chain, for which a chain renews its configuration after N^2 successful moves per monomer. Since the monomers are chosen at random, each monomer moves on average after N elementary steps. This leads to an increase of relaxation time τ with the number of monomers on a chain N as $\tau \simeq N^3$. After each move one has to recalculate the electrostatic interactions with $N-1$ monomers, assuming that each monomer is charged and Coulomb interactions are unscreened. The total CPU time required for chain relaxation will grow with the number of monomers N as $\tau_{flip} \simeq \tau N \simeq N^4$. For pivot moves with acceptance probability p_{acc}, the new configuration is reached after $\tau \simeq p_{acc}^{-1}$ attempts. After each pivot move, a part of a chain with, say, n monomers is displaced with respect to the remaining $N-n$ monomers. And $(N-n)\,n$ new pairs are created for which the electrostatic interactions have to be reevaluated. Since the number of new pairs cannot exceed $N^2/4$, the CPU time for pivot moves cannot increase faster than $\tau_{pivot} \leq N^2/p_{acc}$. It is important to point out that the efficiency of the pivot moves is strongly dependent on the acceptance probability p_{acc}. In a system with a high acceptance rate, $p_{acc} \simeq 1$, the CPU time may grow as N^2. However, for low acceptance probability it can grow faster than N^3. For reptation moves the CPU time grows as $\tau_{rep} \simeq N^3$, because a chain renews its configuration after N^2 steps and at each step one has to recalculate the electrostatic interactions with $N-1$ monomers. Examples of computational efficiency of different types of moves for a polyelectrolyte chain are shown in Fig. 2.

Molecular dynamics simulations [16,17] are based on numerical integration of the equations of motion of the system. During the course of simulation the system moves in phase space generating new chain

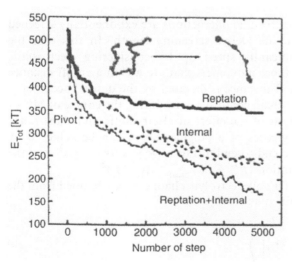

FIG. 2 Variation of the total energy of the system E_{tot} versus the number of MC steps for different types of moves. (From Ref. [69].)

conformations along its physical trajectory determined by the equations of motion. The equation of motion of the ith monomer with mass m is

$$m\ddot{\mathbf{r}}_i(t) = -\nabla_i U\big(\{\mathbf{r}_j(t)\}\big) - m\Gamma\dot{\mathbf{r}}_i(t) + \mathbf{F}_i(t) \tag{4}$$

where the first term describes the deterministic forces between monomers and the last two terms implicitly take into account the effect of the solvent by coupling the system to a Langevin thermostat [23,24] which maintains a constant average temperature of the system. In Eq. (4) the parameter Γ is the friction coefficient and $\mathbf{F}_i(t)$ is a random force with zero average value $\langle\mathbf{F}_i(t)\rangle$ and delta-functional time correlations

$$\langle\mathbf{F}_i(t)\mathbf{F}_j(t')\rangle = 6k_B Tm\Gamma\delta_{ij}\delta(t - t') \tag{5}$$

The MD simulation coupled with the Langevin thermostat simulates Rouse dynamics of a polymer chain. The Rouse relaxation time scales with the number of monomers on a chain as N^2 and it is necessary to perform at least cN^2 (where constant c depends on the value of the integration time step Δt) integrations of the equation of motion for a chain to completely renew its configuration. During each time step, Δt, $N(N-1)/2$ calculations of forces between monomers are performed. The CPU time required to do cN^2 integrations of the equations of motion will grow with the number of monomers on a chain as $\tau_{MD} \simeq N^4$. Thus, the computational efficiency of MD simulation has the same N dependence as a MC simulation with only local moves.

In a salt solution the electrostatic interactions are exponentially screened over distances larger than the Debye screening length. In this case the simulations may be significantly speed up by considering electrostatic interactions between only those monomers that are within a cutoff distance r_{cut}. The choice of the cutoff distance is dictated by the desired accuracy.

After generating a set of chain conformations during the course of MC or MD simulations, the analysis of the effect of electrostatic and short-range interactions on global properties of a polymer chain can be achieved by looking at the mean square end-to-end distance $\langle R_e^2\rangle = \langle(\mathbf{r}_N - \mathbf{r}_1)^2\rangle$ and the mean square radius of gyration $\langle R_g^2\rangle = N^{-2}\sum_{i<j}\langle(\mathbf{r}_i - \mathbf{r}_j)^2\rangle$.

The internal structure of a polyelectrolyte chain can be obtained from the chain structure factor, defined as

$$S(q) = \left\langle N^{-1}\left|\sum_{j=1}^{N}\exp(i\mathbf{q}\cdot\mathbf{r}_j)\right|^2\right\rangle \tag{6}$$

where $\mathbf{q}$ is the wave vector.

B. Polyelectrolyte Chain in θ and Good Solvents

1. Chain Conformation in Dilute Salt-Free Solutions

In a θ-solvent the short-range interactions cancel to zero and only bonded and electrostatic potentials are present in the chain potential energy $U(\{\mathbf{r}_i\})$, Eq. (1). In spite of its simplicity, this model already contains two important ingredients responsible for the unique behavior of polyelectrolytes—chain connectivity and the long-range electrostatic interactions. In salt-free solutions ($\kappa \to 0$) the size L_0 of a polyelectrolyte chain with fN charged monomers can be estimated from a simple Flory argument by balancing the chain elastic energy $k_B T L_0^2/(b^2 N)$ (here b is the bond length) and electrostatic energy $k_B T l_B f^2 N^2 \ln(L_0)/L_0$; the logarithm in the electrostatic energy is accounting for chain elongation. This leads to the well known result that in salt-free solutions the chain size

$$L_0 \approx b\left(uf^2\right)^{1/3} N \ln^{1/3}(N) \tag{7}$$

is proportional to the number of monomers N with weak logarithmic correction. In Eq. (7) u is the ratio of the Bjerrum length l_B to the bond length b. The parameter uf^2 is sometimes called the coupling parameter. In fact, polyelectrolytes are classified according to the strength of this parameter. Polyelectrolyte chains with small values of the coupling parameter $uf^2 \ll 1$ are referred to as weakly charged polyelectrolytes, while ones with $uf^2 \gtrsim 1$ are considered to be strongly charged. The larger the value of the coupling parameter uf^2, the more elongated the polyelectrolyte chain is.

To better understand the different length scales involved in the problem, it is useful to introduce the concept of the electrostatic blob [26–31]. The conformation of a chain inside the electrostatic blob is almost unperturbed by electrostatic interactions, with the number of monomers in it being $g_e \approx (D_e/b)^2$ in a θ-solvent. The size of the electrostatic blob D_e containing g_e monomers can be found by comparison of the electrostatic energy of a blob $e^2 g_e^2 f^2/\varepsilon D_e$ with the thermal energy $k_B T$. This leads to the electrostatic blob size

$$D_e \approx b\left(uf^2\right)^{-1/3} \tag{8}$$

and the number of monomers in it

$$g_e \approx \left(uf^2\right)^{-2/3} \tag{9}$$

For weakly charged polyelectrolytes ($uf^2 \ll 1$) there is more than one monomer in an electrostatic blob ($g_e > 1$). At length scales larger than the electrostatic blob, electrostatic interactions are much stronger than the thermal energy $k_B T$ and the chain configuration is that of a fully extended array of N/g_e electrostatic blobs of length L_0

$$L_0 \approx \frac{N}{g_e} D_e \approx bN \left(uf^2\right)^{1/3} \tag{10}$$

However, for a polyelectrolyte chain to be elongated, the number of monomers in a chain N should be larger than the number of monomers in an electrostatic blob g_e $[N > (uf^2)^{-2/3}]$. Thus, short chains with number of monomers N smaller than $N_G \approx (uf^2)^{-2/3}$ will still be Gaussian.

Of course, this simple scaling picture does not account for the nonuniform stretching of a polyelectrolyte chain. In reality the chain is more strongly stretched in the middle than at the ends. Logarithmic corrections to the chain size may be obtained by allowing the blob size to vary along the chain.

The generalization of this scaling approach to the case of the good solvent is straightforward by replacing the relation between the blob size D_e and the number of monomers in it g_e from random to self-avoiding random walk statistics [26–31].

The conformational properties of an isolated polyelectrolyte chain in a salt-free solution have been investigated by lattice Monte Carlo simulations [32–35], and by off-lattice Monte Carlo simulation of a chain made of hard spheres connected by rigid bonds with fixed valance angles [36], of a freely-jointed chain [37], and of a bead-spring chain [38–42]. These simulations support the trivial scaling prediction that the chain crosses over from a coil to a rod-like conformation with increasing the strength of the electrostatic interactions (increasing value of the coupling parameter uf^2). This was done either by adding ionized groups to a chain or by increasing the Bjerrum length l_B.

Quantitative analysis of the scaling model of a polyelectrolyte chain as a chain of electrostatic blobs was done by Higgs and Orland [37] and by Barrat and Boyer [38]. Figure 3 shows the mean square end-to-end distance as a function of the degree of polymerization N for different values of the coupling parameter uf^2 (in references [37,38] the parameter f was kept constant and equal to unity). For each curve there is a crossover between two regimes. For small N the electrostatic interactions between charges have little or no effect at all on the conformation of the polyelectrolyte chain and the chain remains Gaussian $\langle R_e^2 \rangle \simeq N$. For large N the electrostatic interactions dominate and the chain adopts an elongated conformation with size $\langle R_e^2 \rangle \simeq N^2$.

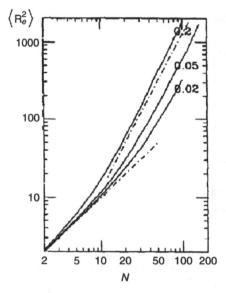

FIG. 3 Mean square end-to-end distance $\langle R_e^2 \rangle$ as a function of number of monomers N for a polyelectrolyte chain in a θ-solvent. Curves are for three different values of the parameter $u = 0.02$, 0.05, and 0.2. The dashed–dotted lines have gradients 1 and 2. (From Ref. [37].)

In Fig. 4 the dependence of the mean square end-to-end distance $\langle R_e^2 \rangle$ is shown as a function of the coupling parameter uf^2 for several values of N. As expected from the blob model, the chain size $\langle R_e^2 \rangle$ demonstrates only weak dependence on the coupling parameter for small values of uf^2 and increases with uf^2 for larger values of the coupling parameter. For large values of the coupling parameter the increase in the mean square end-to-end distance $\langle R_e^2 \rangle$ follows the scaling law with the exponent 2/3 [see Eq. (10)].

In their simulation Barrat and Boyer [38] gave a quantitative definition of the electrostatic blob. Using the model of the chain under tension they established the quantitative relation between the number of monomers n in a section and its size $R(n)$. The electrostatic blob size D_e was directly related to the prefactor of n^2 in the expression for the mean square section size $\langle R(n)^2 \rangle = nb^2 + n^2b^2/g_e$. To verify the dependence of the blob size on the value of the coupling parameter uf^2 Barrat and Boyer also calculated the ratio $D_e(uf^2)^{-1/3}/b$. This ratio varies between 1.45 and 0.92 when the value of the coupling parameter uf^2 changes between 0.01 and 1, providing reasonable agreement with the blob model [26–31].

Systematic MC studies of the chain size dependence on its degree of polymerization, performed by different groups [38–42], have shown that the

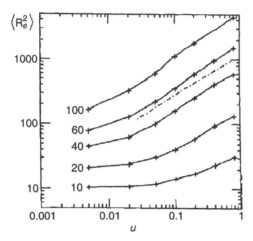

FIG. 4 Scaling of $\langle R_e^2 \rangle$ with the parameter u for a polyelectrolyte in a θ-solvent. Chain lengths between 10 and 100 are shown. The dashed–dotted line shows the theoretical prediction $\langle R_e^2 \rangle \simeq u^{2/3}$. (From Ref. [37].)

chain size grows faster than linearly, $L_0 \simeq N(\ln N)^{1/3}$, with the degree of polymerization N. Figure 5 shows the dependence of the normalized root-mean-square end-to-end distance $\sqrt{\langle R_e^2 \rangle}/N$ on the parameter $(uf^2)^{1/3}(\ln(N/e))^{1/3}$ for the chains with N ranging from 20 to 2000. The nature of this deviation from the scaling law, $L_0 \simeq N$, is due to nonuniform (logarithmic) stretching of the polyelectrolyte chain.

2. Effects of Added Salt on Chain Conformation and Electrostatic Persistence Length

When a finite concentration of salt is present in a solution the electrostatic interactions between charged monomers are screened by salt ions and fall off exponentially with distance. However, at distances smaller than the Debye screening length κ^{-1} the charges still interact through a bare Coulomb potential. A polyelectrolyte chain will not feel the presence of the salt until the Debye screening length is larger than the chain size L_0. On the other hand, at very high salt concentrations such that the Debye screening length κ^{-1} is smaller than the electrostatic blob size D_e, the electrostatic interactions can be viewed as short-range ones with the effective monomeric second virial coefficient $v_{el} \approx f^2 l_B \kappa^{-2}$. At these salt concentrations a polyelectrolyte chain has the same structure as a neutral polymer in a good solvent with chain size depending on the Debye screening length and degree of polymerization as $L \simeq \kappa^{-2/5} N^{3/5}$. But the question is what

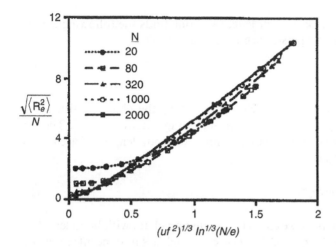

FIG. 5 Root-mean-square end-to-end distance $\sqrt{\langle R_e^2 \rangle}$ divided by N vs. $(uf^2)^{1/3}(\ln N/e)^{1/3}$, for chains with $N = 20, 80, 320, 1000,$ and 2000 (filled circles, open squares, filled triangles, open circles, and filled squares, respectively). (From Ref. [39].)

conformation is adopted by the polyelectrolyte chain at intermediate salt concentrations $D_e < \kappa^{-1} < L_0$.

It was first shown independently by Odijk [43] and by Skolnick and Fixman [44] (OSF) that for intrinsically stiff chains, the intrachain electrostatic repulsion can induce additional chain stiffening beyond the Debye screening length κ^{-1}. To understand this result let us consider the variation in the electrostatic energy between two consecutive chain sections of length κ^{-1} due to bending by small angle θ. This energy variation is

$$\delta E_\theta \approx k_B T l_B Z_\kappa^2 \kappa \theta^2 \qquad (11)$$

where Z_κ is a charge carried by a section of size κ^{-1}, which for weakly charged polyelectrolytes is $Z_\kappa \approx f g_e \kappa^{-1}/D_e$. The typical mean square value of the bending angle θ caused by the thermal fluctuations ($\delta E_\theta \approx k_B T$) is $\langle \theta^2 \rangle \approx D_e \kappa$. The bending of each pair of sections along a polyelectrolyte chain can be considered to be statistically independent, because electrostatic interactions are exponentially weak at these length scales. A chain section containing l units of size κ^{-1} will deviate from its initial direction by an angle $\langle \theta^2(l) \rangle \approx l \langle \theta^2 \rangle \approx l D_e \kappa$ and forgets its initial orientation after $1/D_e \kappa$ "steps" when the value of $\langle \theta^2(l) \rangle$ becomes of the order of unity. Multiplying the number of sections $1/D_e\kappa$ by their length κ^{-1} one finds the following expression for the electrostatic persistence length [27,45–47]

$$L_e \approx \kappa^{-2}/D_e \tag{12}$$

Indeed for salt concentrations such that $\kappa^{-1} > D_e$ the chain is stiffened at the length scales larger than the Debye screening length $L_e > \kappa^{-1}$. The size of a chain with persistence length L_e and a contour length of a chain of electrostatic blobs L_0 is

$$L^2 \approx L_e L_0 \approx \kappa^{-2} N/g_e \tag{13}$$

and crossover from a rod-like chain to a Gaussian chain with bond length L_e occurs at $\kappa^{-1} \approx D_e\sqrt{N/g_e} \approx bN^{1/2}$. For even higher salt concentrations the number of the persistence segments increases, making it more probable for interactions between segments separated by larger distances along the chain contour to occur. The second virial coefficient υ for these interactions can be estimated as that between two rods of length L_e and thickness κ^{-1} ($\upsilon \approx L_e^2\kappa^{-1}$). The excluded volume effect becomes important, when the interaction parameter z of a polyelectrolyte chain

$$z \approx \left(\frac{L_0}{L_e}\right)^{1/2}\frac{\upsilon}{L_e^3} \approx \kappa^{-2}D_e bN^{1/2} \tag{14}$$

becomes larger than unity. In this range of salt concentrations ($\kappa^{-1} < \sqrt{D_e bN^{1/2}}$) the chain size scales as $L \propto \kappa^{-3/5}N^{3/5}$. This regime continues until the Debye radius is larger than the electrostatic blob size D_e.

However, the predictions of this model are still challenged in the literature. Computer simulations of weakly charged polyelectrolyte chains [38,58–61] and some experiments [6,48–54] as well as analytical calculations [55–57] indicate that the exponent for the dependence of the electrostatic persistence length L_e on the Debye screening length is closer to 1 rather than to 2 or even shows sublinear [62] dependence.

The most complete analysis to date of the electrostatic persistence length dependence on the Debye screening length was performed by the Kremer group [62]. Micka and Kremer have performed hybrid MC and MD simulations of a polyelectrolyte chain with Debye–Hückel interaction bonded by harmonic springs. For different bond lengths ($b = 2, 4, 8, 10, 16$) and chain lengths ($N = 16, 32, 64, 128, 256, 512$) they investigated the dependence of the chain dimensions and electrostatic persistence length on the Debye screening length in the interval $0.001 \leq \kappa^{-1} \leq 0.48$. The data for the mean square end-to-end distance can be collapsed into universal curves reasonably well by either assuming linear or quadratic dependence of the electrostatic persistence length L_e on the Debye radius κ^{-1}. Both of these plots show crossover from a rod-like chain ($\langle R_e^2 \rangle \simeq N^2$) to a chain with excluded volume interactions ($\langle R_e^2 \rangle \simeq N^{6/5}$) as the value of the Debye radius decreases. In order to differentiate between the models, the persistence length of the chain was calculated from the bond angle correlation function $G(n)$. This function is defined by the scalar product of two normalized bond vectors $\mathbf{b}(k)$ and $\mathbf{b}(k+n)$

$$G(n) = \langle \mathbf{b}(k) \cdot \mathbf{b}(k+n) \rangle \tag{15}$$

where the brackets $\langle\rangle$ denote the ensemble average over all chain conformations. The averaging procedure was improved by moving reference point k along the chain. The persistence length was estimated from the exponential decay of the function $G(n)$. This method lead to sublinear dependence of the electrostatic persistence length on the Debye radius (see Fig. 6). Similar sublinear dependence was obtained by analyzing the chain structure factor $S(q)$.

The effect of internal chain stiffness on the dependence of the electrostatic persistence length on the Debye radius was studied in [63,64].

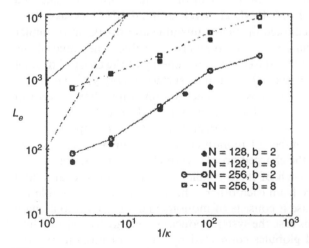

FIG. 6 Persistence length L_e dependence on the Debye screening length κ^{-1}. The two lines in the upper left corner indicate the slopes predicted by OSF (dashed) and the variational method (solid). (From Ref. [62].)

It was shown that with increasing internal chain stiffness the effective exponent y for $L_e \simeq \kappa^{-y}$ crosses over from a value of one toward two as the internal stiffness of a chain increases. The quadratic dependence of the electrostatic persistence length on the Debye radius for the discrete Kratky–Porod model of the polyelectrolyte chain was recently obtained in [65]. It seems that the concept of electrostatic persistence length works better for intrinsically stiff chains rather than for flexible ones. Further computer simulations are required to exactly pinpoint the reason for its failure for weakly charged flexible polyelectrolytes.

C. Polyelectrolyte Chain in a Poor Solvent

In a poor solvent, the chains have a negative second virial coefficient, corresponding to an effective attraction between monomers. This attraction causes a neutral polymer chain without charged groups to collapse into dense spherical globules in order to maximize the number of favorable monomer–monomer contacts. The size R of such a globule scales with the number of monomers on a chain as $N^{1/3}$ [31]. By charging the polymer chain one forces the polymeric globule to change its shape. The shape of a polyelectrolyte in a poor solvent is determined by the interplay between long-range electrostatic and short-range attractive interactions [68,71,72]. It is interesting to point out that the problem of the shape of a charged globule bears similarities with the classical problem of the instability of a charged droplet, considered by Lord Rayleigh over a hundred years ago [66]. Rayleigh showed that a charged droplet is unstable and breaks into smaller droplets if the electric charge exceeds some critical value. The value of the critical charge is controlled by the electrostatic energy of the charged droplet of size R and carrying charge Q, Q^2/R, and its surface energy γR^2, where γ is the surface tension. Balancing those two energies one finds that the critical charge Q_{crit} scales with the size of a droplet as $R^{3/2}$. [For a charged globule with size $R \simeq bN^{1/3}$ the value of the critical charge Q_{crit} is proportional to the square root of the number of monomers on a chain N ($Q_{crit} \approx \sqrt{N}$).] The equilibrium state of the charged droplet with $Q > Q_{crit}$ is a set of smaller droplets with charge on each of them smaller than the critical charge and placed at infinite distance from each other. This final state is impossible for a charged polymer because it consists of monomers connected into a chain by chemical bonds. In this case the system reduces its energy by splitting into a set of smaller charged globules connected by strings of monomers—the necklace globule. The number of monomers m_b in each bead is determined by Rayleigh's stability condition—the surface energy of a bead γD_b^2 is of the order of its electrostatic energy $k_B T l_B f^2 m_b^2 / D_b$. For dense beads with

$\gamma \approx k_B T / b^2$ and $D_b \approx b m_b^{1/3}$ this results in the number of monomers in a bead $m_b \approx 1/u f^2$. Those beads are separated by strings whose length l_{str} is obtained by equating the surface energy of a string $\gamma b l_{str}$ with the electrostatic repulsion between two consecutive beads $k_B T l_B f^2 m_b^2 / l_{str}$. The string length l_{str} is proportional to the critical charge on a bead $f m_b$ ($l_{str} \approx b / \sqrt{u f^2}$).

The qualitative features of an abrupt conformational transition in a polyelectrolyte chain in a poor solvent were first studied by Hooper et al. [67] and by Higgs and Orland [37] by performing lattice Monte Carlo simulations of a polyelectrolyte chain with attractive segment–segment interactions. These authors find that with increasing charge, the chain undergoes an abrupt transition from collapsed to extended conformations. As the strength of the segment–segment attraction decreases, the transition becomes less pronounced [67].

The details of the transition and conformations of the polyelectrolyte chain above the transition were studied by Dobrynin, Rubinstein, and Obukhov [68]. Performing Monte Carlo simulations of a freely-jointed uniformly charged polyelectrolyte chain with fractional charge on each monomer f, they have shown that the critical charge Q_{crit} on a chain at which the charged globule becomes unstable is proportional to $\sqrt{N}$ (see Fig. 7). For charge on the chain Q above the critical value Q_{crit}, the polyelectrolyte chain first assumes a dumbbell configuration (see Fig. 8). At still higher charge, the polymer forms a necklace with three beads joined by two strings (see Fig. 8). These simulations have shown that in fact there is a

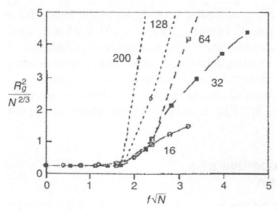

FIG. 7 Reduced mean square radius of gyration $\langle R_g^2 \rangle / N^{2/3}$ as a function of reduced valence $f N^{1/2}$ for chains with degrees of polymerization $N = 16, 32, 64, 128$, and 200.

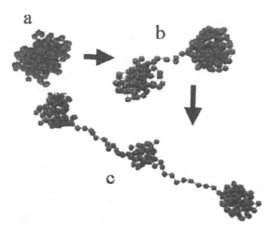

FIG. 8 Typical configurations of a freely-jointed uniformly charged chain with $N = 200$ interacting via Coulomb and Lennard-Jones potentials (with $\varepsilon_{LJ} = 1.5$ and $u = 2$) for three different charge densities: (a) spherical globule for $f = 0$; (b) dumbbell for $f = 0.125$; (c) necklace with three beads for $f = 0.15$.

cascade of transitions between necklaces with different numbers of beads as charge on the chain increases.

The effect of solvent quality and salt concentrations on the cascade of transitions between different pearl-necklace structures was investigated by Monte Carlo [69] and by molecular dynamics [70] simulations. In these simulations the effect of the salt has been taken into account through the Debye–Huckel potential. In [69] the authors have investigated the full domain of stability of pearl-necklace globules in the solvent quality/salt concentration plane. They found necklaces with up to twelve beads for a polyelectrolyte chain with degree of polymerization $N = 200$. With increasing salt concentration the necklace transforms into a cylindrical globule and at very high salt concentrations, when electrostatic interactions are completely screened, a polyelectrolyte globule takes once again a spherical shape.

These results of computer simulations are in a good qualitative agreement with theoretical models [68,71,72] of a polyelectrolyte chain in a poor solvent.

D. Conformational Properties of a Polyampholyte Chain

Polyampholytes are charged polymers with both positively and negatively charged monomers. In this case the Coulombic interactions between charges are not only repulsive, as in the case of polyelectrolytes, but are also attractive in nature. The overall size and shape of polyampholytes is determined by the balance of four factors [73–78].

1. Chain entropy tends to keep the polymer configuration as close to the Gaussian statistics as possible.
2. Fluctuation-induced attractions between charges (similar to those observed in electrolyte solutions [79]) tend to collapse the chain into a globule.
3. Excluded volume interactions between monomers stabilize the size of globule.
4. If the chain has a nonzero total charge (either positive or negative) the overall Coulomb repulsion between excess charges tends to elongate the chain.

The relative importance of these factors depends on the number of positive N_+ and negative N_- charges on the chain, on the degree of polymerization N, and on the ratio of the Bjerrum length l_B to the bond size b, defining the strength of the electrostatic interactions.

The effect of the overall excess charge $\delta Q = e|\sum q_i| = e|N_+ - N_-|$ on the ensemble averaged properties of polyampholyte chains was the central point of lattice Monte Carlo simulations of polyampholytes [80–84], of off-lattice Monte Carlo simulations of a bead-spring model of a polyampholyte chain [85], and of the MD simulations [86,87]. These simulations have shown that nearly symmetric random polyampholytes with small charge asymmetry, δQ smaller than $\sqrt{N_+ + N_-}$, collapse into a globule as the temperature decreases. This collapse is caused by fluctuation-induced attraction between charged monomers. Assuming that the charges of both signs are distributed with average concentration $c_{ch} \simeq (N_+ + N_-)/R_0^3$ within the volume of a polymer coil, the fluctuation-induced attraction energy $W_{att} \simeq -k_B T(l_B c_{ch})^{3/2} R_0^3$ [73–75] of a Gaussian polyampholyte chain with size $R_0 \approx b\sqrt{N}$. A polyampholyte chain collapses when the fluctuation-induced attraction energy W_{att} becomes stronger than the thermal energy $k_B T$. This happens for a value of the interaction parameter u larger than $\sqrt{N}/(N_+ + N_-)$.

Polyampholytes with charge imbalance, δQ larger than $\sqrt{N_+ + N_-}$, form a necklace globule at low temperatures. There is a striking similarity between the instability of a polyampholyte globule with excess charge $\delta Q \approx \sqrt{N_+ + N_-}$ and the necklace instability of a polyelectrolyte chain in a poor solvent [68]. But for polyampholytes the factors responsible for the shape of a globule are all electrostatic in nature. The main difference between polyampholytes and uniformly charged polyelectrolytes is the randomness in the charge sequence. For polyampholytes the structure of the necklace is predetermined by the initial charge distribution. Monte Carlo studies by Kantor and Kardar [82,83] showed that the necklace may consist of a few almost neutral globules connected by charged necks, or even of one big neutral globule with a tail sticking out of it. The necklace instability was also observed in MC simulation of circular polyampholytes by Lee and Obukhov [88].

The effect of charge sequences on the collapse of polyampholytes was studied by lattice MC simulations [89,90]. Polyampholytes with alternating distribution of charges behave like polymers with short-range interactions and can be characterized by an effective second virial coefficient [91]. The collapse transition of these polyampholytes is similar to the coil–globule transition of a neutral polymer [89]. A qualitatively different picture of the collapse transition was discovered for diblock polyampholytes. The collapse of these polymers happens in two stages. First, at high temperatures, a zipping transition occurs that corresponds to the formation of dipolar pairs between oppositely charged monomers. Second, at lower temperatures this preassembled zipped structure undergoes an ordinary coil–globule transition. The possibility of a freezing transition in random polyampholyte globules was investigated in computer simulations [92].

The influence of a uniform external electric field E on the conformation and dynamics of polyampholyte chains has been studied by the MD simulations of Soddemann et al. [87]. These simulations have shown that the polyampholyte globule becomes unstable above the critical external electric field E_{c1} and breaks up forming a random necklace structure. A polyampholyte collapses back into a globule as the strength of the external electric field E is lowered below E_{c2}. The strength of the electric field E_{c2} at which a polyampholyte collapses back into a globule is weaker than that required to break up a globule E_{c1} ($E_{c2} < E_{c1}$). There is a hysteresis in chain size dependence on the external electric field. This hysteresis is a manifestation of the coexistence of two states of a chain—collapsed and elongated—separated by a barrier.

III. SIMULATION METHODS FOR SOLUTIONS OF CHARGED POLYMERS

Molecular simulations of polyelectrolyte solutions at finite concentrations with explicit counterions and salt ions require special handling of the Coulombic interaction between the charges. These simulations are commonly performed under periodic boundary conditions to ensure that small sample surface effects are suppressed and that the results obtained for systems with various numbers of particles can be considered as approximations of properties of the hypothetical infinite system. The earliest computer simulations of polyelectrolyte solutions [93–97] were done by using the minimum image convention for electrostatic interactions, which includes Coulombic interactions only between charges within the first periodic image. The minimum image convention is usually sufficient for short-range interactions, but the long-range nature of Coulombic interactions causes a pair of charges to interact far beyond their first periodic image. The Ewald summation method [98] allows us to overcome this problem and properly account for contributions from all periodic images. We begin this section by describing methods such as the Ewald summation method, particle—particle particle—mesh method (P^3M), particle—mesh Ewald method, and Fast Multipole Method. These methods have become the standard tool for evaluation of long-range electrostatic interactions in simulations of molecular systems [102,103].

A. Lattice-Sum Methods for Calculation of Electrostatic Interactions

1. Ewald Summation

Consider an electroneutral system of N_p charges $eq_1, eq_2, \ldots, eq_{N_p}$ such that $q_1 + q_2 + \cdots + q_{N_p} = 0$, located at positions $\mathbf{r}_1, \mathbf{r}_2, \ldots, \mathbf{r}_{N_p}$ within the unit cell of size L. For simplicity we will use a cubic unit cell with primitive vectors $\mathbf{a}_\alpha$, $\alpha = 1, 2, 3$ forming the edges of the unit cell, however the methods described below are valid for any Bravais lattice. [For the cubic lattice these vectors are $\mathbf{a}_1 = (L, 0, 0)$, $\mathbf{a}_2 = (0, L, 0)$ and $\mathbf{a}_3 = (0, 0, L)$.] The charges $\{eq_j\}$ interact according to Coulomb's law with each other and with their infinite number of replicas in periodic boundary conditions (see Fig. 9). The point

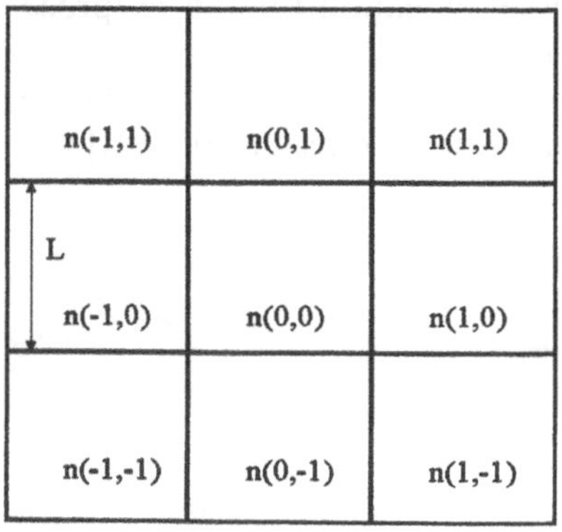

n(-1,1)	n(0,1)	n(1,1)
L n(-1,0)	n(0,0)	n(1,0)
n(-1,-1)	n(0,-1)	n(1,-1)

FIG. 9 2D periodic system of 3×3 periodic lattice built from unit cells.

charge eq_k located at point $\mathbf{r}_k$ interacts with all other charges eq_j at positions $\mathbf{r}_j$ ($k \neq j$) within the cell as well as with all of their periodic images located at $\mathbf{r}_j + n_1\mathbf{a}_1 + n_2\mathbf{a}_2 + n_3\mathbf{a}_3$ where n_1, n_2, n_3 are all integer numbers between $-M$ and M. The electrostatic energy of the unit cell interacting with an infinite number of periodic images can be written as follows

$$\frac{U_{el}(\{\mathbf{r}_j\})}{k_B T} = \lim_{M \to \infty} \frac{l_B}{2} \sum_{\mathbf{n}}^{*} \sum_{k,j} \frac{q_k q_j}{|\mathbf{r}_{kj} + \mathbf{n}|} \tag{16}$$

where we introduced $\mathbf{r}_{kj} = \mathbf{r}_k - \mathbf{r}_j$. The first sum in Eq. (16) is over all vectors $\mathbf{n} = n_1\mathbf{a}_1 + n_2\mathbf{a}_2 + n_3\mathbf{a}_3$, and the asterisk indicates that the terms with $k=j$ for $\mathbf{n}=0$ are excluded from the summation.

The sum (16) is only slowly and conditionally convergent—the result depends on the order in which the terms are added. The original method for summation of the lattice sum was introduced by Ewald [98], who replaced the sum in Eq. (16) by the sum of two absolutely convergent series—a direct sum in Cartesian space and a reciprocal sum in Fourier space. There is a simple physical interpretation of this decomposition of the lattice sum (see Fig. 10). In the direct sum each charge in the system q_i is viewed as being surrounded by a Gaussian charge distribution of the opposite sign $[\rho(\mathbf{r}) = -q_i\beta^3 \exp(-\beta^2\mathbf{r}^2)/\pi^{3/2}$, where the parameter β determines the width of the distribution], such that the net charge of this cloud exactly cancels q_i. The electrostatic interaction between screened charges is a rapidly decaying function of distance, and is effectively reduced to short-range interactions in the direct space. To counteract this diffuse cloud around each charge, a second Gaussian charge distribution of the same sign as the original charge is added for each point charge. This second distribution varies smoothly in space and acts as a source term in a Poisson equation for the reciprocal potential with periodic boundary conditions.

However, these most commonly used Ewald transformation formulas are correct for centrosymmetric crystal structures but give results [98–101] that

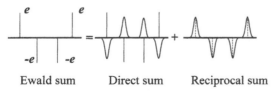

Ewald sum   Direct sum   Reciprocal sum

FIG. 10   The Ewald sum components of a one-dimensional point charge system. The vertical lines are unit charges.

disagree with the explicit evaluation of sum (16) in the case where the unit cell has a net dipole moment [104–107]. The discrepancy is due to uncritical use of the Fourier transformation converting the conditionally convergent lattice sum in direct Cartesian space into a sum in reciprocal Fourier space.

The evaluation of the lattice sum (16) is cumbersome and details of this calculation are given in the Appendix. The final expression for the electrostatic energy of the cell interacting with all its periodic images (16) can be represented as a sum of four terms: sums in the direct U_{dir} ($\{\mathbf{r}_j\}$) and reciprocal U_{rec} ($\{\mathbf{r}_j\}$) space, self-energy term, and shape term that depends on the net dipole moment of the cell

$$\frac{U_{el}\big(\{\mathbf{r}_j\}\big)}{k_B T} = \frac{l_B}{2} \sum_{\mathbf{n}} \sum_{k,j}^{*} q_k q_j \frac{\mathrm{erfc}\big(|\mathbf{r}_{kj} + \mathbf{n}|\beta\big)}{|\mathbf{r}_{kj} + \mathbf{n}|}$$

$$+ \frac{l_B}{2\pi L} \sum_{\mathbf{m} \neq 0} \frac{\exp\big(-\pi^2 \mathbf{m}^2 / L^2 \beta^2\big)}{\mathbf{m}^2} S(\mathbf{m}) S(-\mathbf{m})$$

$$- \frac{\beta l_B}{\sqrt{\pi}} \sum_{j} q_j^2 + \frac{2\pi l_B}{3L^3} \left| \sum_{j} q_j \mathbf{r}_j \right|^2 \tag{17}$$

where $\mathrm{erfc}(x) = 2/\sqrt{\pi} \int_x^{\infty} dt \exp(-t^2)$ is the complementary error function which tends to zero as $x \to \infty$ and $S(\mathbf{m}) = \sum_j q_j \exp(i2\pi\mathbf{m} \cdot \mathbf{r}_j / L)$ is the charge structure factor.

The Coulombic forces on charge k are obtained by differentiating the electrostatic energy of the system $U_{el}(\{\mathbf{r}_j\})$ with respect to $\mathbf{r}_k$. The resulting force

$$\mathbf{F}_k = \mathbf{F}_k^{dir} + \mathbf{F}_k^{rec} + \mathbf{F}_k^{shape} \tag{18}$$

has three contributions: the real space part

$$\mathbf{F}_k^{dir} = k_B T l_B q_k \sum_j q_j \sum_{\mathbf{n}}^{*} \left(\frac{2\beta}{\sqrt{\pi}} \exp\big(-|\mathbf{r}_{kj} + \mathbf{n}|^2 \beta^2\big) + \frac{\mathrm{erfc}\big(|\mathbf{r}_{kj} + \mathbf{n}|\beta\big)}{|\mathbf{r}_{kj} + \mathbf{n}|} \right)$$

$$\times \frac{\mathbf{r}_{kj} + \mathbf{n}}{|\mathbf{r}_{kj} + \mathbf{n}|^2} \tag{19}$$

the reciprocal (Fourier) space part

$$\mathbf{F}_k^{rec} = k_B T \frac{2l_B q_k}{L^2} \sum_{\mathbf{m} \neq 0} \frac{i\mathbf{m} \exp\big(-\pi^2 \mathbf{m}^2 / L^2 \beta^2\big)}{\mathbf{m}^2} \exp\left(\frac{2\pi i \mathbf{m} \cdot \mathbf{r}_k}{L} \right) S(-\mathbf{m}) \tag{20}$$

and the dipolar contribution

$$\mathbf{F}_k^{shape} = -k_B T \frac{4\pi l_B q_k}{3L^3} \sum_j q_j \mathbf{r}_j \tag{21}$$

The value of the parameter β controls the rate of convergence of the direct $U_{dir}(\{\mathbf{r}_j\})$ and reciprocal $U_{rec}(\{\mathbf{r}_j\})$ sums. Increasing β causes the direct sum $U_{dir}(\{r_j\})$ to converge more rapidly, at the expense of slower convergence in the reciprocal sum $U_{rec}(\{\mathbf{r}_j\})$. The usual method is to take a value of β such that only one term with $\mathbf{n} = 0$ is necessary in the direct sum. Thus, the first sum reduces to the normal minimal image conversion. To achieve the same accuracy in the reciprocal sum, a subset of m_0 **m**-vectors has to be used. Usually, in practice the parameter β is set to $5/L$ and it is required to use on the order of 100–200 wave vectors for reasonable convergence of the reciprocal sum. This choice of parameter leads to a linear increase of the computational time for evaluation of the reciprocal sum with the number of particles in the system N_p. The major computational overhead comes from evaluation of the direct sum, which requires evaluation of $N_p^2/2$ error functions. For large systems, $N_p > 10^4$, this method becomes computationally very inefficient.

A more efficient algorithm for evaluation of the lattice sum was designed by Perram et al. [108]. In this algorithm each side of the unit cell is divided into K equal segments, each of length L/K. This leads to tessellation of the original cell into $M_B = K^3$ identical cubic subboxes, each containing on average N_p/M_B particles. If we apply the minimum image convention for each subbox for evaluation of the direct space contribution to the lattice sum, such that the interactions between particles separated by distances larger than L/K ($\beta \simeq K/L$) are neglected, the number of operations required for this evaluation will decrease with the number of subboxes M_B as $M_B(N_p/M_B)^2 = N_p^2/M_B$. To maintain the same accuracy in evaluation of the reciprocal sum, the number of wave vectors in the reciprocal space should be proportional to the volume of a sphere of size β. (The contributions from the vectors with modulus $|\mathbf{m}| > \beta L \simeq K$ are exponentially suppressed.) The number of operations involved in the evaluation of the reciprocal sum will be of the order of $N_p K^3 = N_p M_B$ and will increase with the number of boxes M_B. By determining the number of boxes for optimal computational efficiency of this method one finds that $M_B \simeq \sqrt{N_p}$. If the boxing is implemented with the optimum value $\sqrt{N_p}$ the computer time for lattice sum evaluation increases only as $N_p^{3/2}$. This is a considerable improvement over the N^2 rate for traditional methods with fixed value of β.

Before closing this section, let us comment on the parallel implementation of the Perram et al. algorithm [108]. In practice the computation of the direct and reciprocal space contributions to the Ewald sum are done separately. Since the direct sum represents short-ranged interactions between particles, the parallelization of the summation in the direct sum is handled by three-dimensional spacial domain decomposition based upon blocks of chaining cells, considering the interactions only between neighboring cells. The computation of the sum in the reciprocal (Fourier) space is ideal for parallel implementation. The required charge structure factor $S(\mathbf{m})$ is rewritten as follows

$$S(\mathbf{m}) = \sum_{j} q_j \exp(i2\pi\mathbf{m} \cdot \mathbf{r}_j/L) = \sum_{P} \sum_{j \in P} q_j \exp(i2\pi\mathbf{m} \cdot \mathbf{r}_j/L) = \sum_{P} S_P(\mathbf{m})$$

(22)

where P denotes processors. Particles are distributed to the processors and the partial charge structure factor $S_P(\mathbf{m})$ is computed on each processor for all $\mathbf{m}$ vectors. These partial charge structure factors $S_P(\mathbf{m})$ are then summed across the processors to obtain the total charge structure factor $S(\mathbf{m})$. Computation of the reciprocal forces, where summation of factors $2\pi i\mathbf{m}S(\mathbf{m})$ is involved, does not require further computer communications. The factors $2\pi i\mathbf{m}S_P(\mathbf{m})$ can be summed on each processor for the whole range of $\mathbf{m}$ vectors.

2. Particle Mesh Ewald Method (PME)

The particle mesh Ewald (PME) method [109–113] is based on approximations that turn the structure factor $S(\mathbf{m})$ into discrete Fourier transforms, making the evaluation of the reciprocal sum more computationally efficient. This is achieved by substituting the real charge distribution by a weighted charge distribution on a regular three-dimensional grid.

Consider a grid of length L with K subintervals along L with the lattice spacing of the grid being $h = L/K$. Let us rescale all lengths in the system by the grid spacing h, such that $\mathbf{u} = \mathbf{r}/h$. Due to periodic boundary conditions the rescaled coordinates $\mathbf{u}$ vary in the interval $0 \le u_\alpha < K$ for $\alpha = 1, 2, 3$. After rescaling, the charge structure factor is

$$S(\mathbf{m}) = \sum_{j} q_j \exp\left(i\frac{2\pi\mathbf{m} \cdot \mathbf{u}_j}{K}\right)$$

(23)

In the PME algorithm, the complex exponentials

$$\exp\left(i\frac{2\pi\mathbf{m} \cdot \mathbf{u}}{K}\right) = \exp\left(i\frac{2\pi m_1 u_1}{K}\right) \exp\left(i\frac{2\pi m_2 u_2}{K}\right) \exp\left(i\frac{2\pi m_3 u_3}{K}\right)$$

(24)

are replaced by the linear combination of their values at the nearby grid points. For example, let $[u_\alpha]$ denote the integer part of u_α. Then, using linear interpolation, we can approximate the individual exponents in the r.h.s. of Eq. (24) by their linear combination at the nearest grid points $[u_\alpha]$ and $[u_\alpha] + 1$

$$\exp\left(i\frac{2\pi m_\alpha u_\alpha}{K}\right) \approx (1 - (u_\alpha - [u_\alpha]))\exp\left(i\frac{2\pi m_\alpha}{K}[u_\alpha]\right)$$
$$+ (u_\alpha - [u_\alpha])\exp\left(i\frac{2\pi m_\alpha}{K}[u_\alpha] + 1\right) \tag{25}$$

Similar interpolation can be done for each term in the Eq. (24), with $\alpha = 1, 2, 3$. Let us define $W_2(x)$, the linear hat function given by $W_2(x) = 1 - |x|$ for $|x| \leq 1$, and $W_2(x) = 0$ for $|x| > 1$. Then Eq. (25) can be rewritten as

$$\exp\left(i\frac{2\pi m_\alpha u_\alpha}{K}\right) \approx \sum_{k=-\infty}^{\infty} W_2(u_\alpha - k)\exp\left(i\frac{2\pi m_\alpha}{K}k\right) \tag{26}$$

The sum in the last equation is finite, because the function $W_2(x)$ is nonzero only within the interval $|x| \leq 1$. The accuracy of the approximation can be improved by including more grid points in the interpolation scheme.

In the original implementation of the PME algorithm [109,111], Lagrangian weight functions $W_{2p}(x)$ [116] were used for interpolation of the complex exponents, by using values at $2p$ points in the interval $|x| \leq p$. Unfortunately, these functions $W_{2p}(x)$ are only piecewise differentiable, so the approximate reciprocal lattice sum cannot be differentiated to arrive at the reciprocal part of the Coulomb forces. Thus, the forces were interpolated as well.

Instead of the Lagrangian weight functions $W_{2p}(x)$, the Cardinal B-splines $M_n(x)$ [117,118] were utilized in later versions of the PME method [110,112,113]. These weight functions are continuously differentiable and allow the forces to be obtained from analytical differentiation of the approximation of the reciprocal lattice sum. The Cardinal B-spline of the second order $M_2(x)$ gives the linear hat function $M_2(x) = 1 - |x - 1|$ in the interval $0 \leq x \leq 2$ and $M_2(x) = 0$ for $x < 0$ and $x > 2$. The nth order B-spline satisfies the following properties:

1. $M_n(x) > 0$ for $0 \leq x \leq n$ and $M_n(x) = 0$ for $x < 0$ and $x > n$.
2. $M_n(x) = M_n(n - x)$.
3. $\sum_{j=-\infty}^{\infty} M_n(x - j) = 1$.

4. $M_n(x) = (x/(n-1))M_{n-1}(x) + ((n-x)/(n-1))M_{n-1}(x-1).$
5. $(d/dx)M_n(x) = M_{n-1}(x) - M_{n-1}(x-1).$

The interpolation of the complex exponential, called the Euler exponential spline, has a simple solution that, for even n, has the form

$$\exp\left(i\frac{2\pi m_\alpha u_\alpha}{K}\right) \approx b_\alpha(m_\alpha) \sum_{k=-\infty}^{\infty} M_n(u_\alpha - k)\exp\left(i\frac{2\pi m_\alpha}{K}k\right) \tag{27}$$

where the coefficient $b_\alpha(m_\alpha)$ is given by

$$b_\alpha(m_\alpha) = \exp\left(i\frac{2\pi m_\alpha}{K}(n-1)\right)\left[\sum_{k=0}^{n-2} M_n(k+1)\exp\left(i\frac{2\pi m_\alpha}{K}k\right)\right]^{-1} \tag{28}$$

This approximation for the complex exponential in the charge structure factor leads to the following approximation for $S(\mathbf{m})$

$$S(\mathbf{m}) \approx \sum_{j=1}^{N_P} q_j \prod_{\alpha=1}^{3} b_\alpha(m_\alpha) \sum_{k_\alpha=-\infty}^{\infty} M_n(u_{j,\alpha} - k_\alpha)\exp\left(i\frac{2\pi m_\alpha}{K}k_\alpha\right)$$

$$= \prod_{\alpha=1}^{3} b_\alpha(m_\alpha) \sum_{\mathbf{k}\in V_K} Q(\mathbf{k})\exp\left(i\frac{2\pi}{K}\mathbf{m}\cdot\mathbf{k}\right)$$

$$= b_1(m_1)b_2(m_2)b_3(m_3)\widehat{Q}(\mathbf{m}) \tag{29}$$

where $\mathbf{k}$ is the vector with components (k_1, k_2, k_3), $V_K = \{0 \le k_\alpha \le K-1,$ for $\alpha = 1,2,3\}$, and $\widehat{Q}(\mathbf{m})$ is the discrete Fourier transform of the charge array Q, defined as

$$Q(\mathbf{k}) = \sum_{j=1}^{N_P} q_j \prod_{\alpha=1}^{3} \sum_{s_\alpha} M_n(u_{j,\alpha} - k_\alpha - s_\alpha K) \tag{30}$$

where the inner sum is over all integers s_α. Using this spline representation of the charge structure factor $S(\mathbf{m})$, the approximate reciprocal energy can now be rewritten as

$$\frac{U_{rec}(\{u_j\})}{k_B T} \approx \frac{l_B}{2\pi L} \sum_{\mathbf{m}\neq 0} \frac{\exp(-\pi^2 \mathbf{m}^2/L^2\beta^2)}{\mathbf{m}^2} B(\mathbf{m})\widehat{Q}(\mathbf{m})\widehat{Q}(-\mathbf{m}) \tag{31}$$

where we introduce

$$B(\mathbf{m}) = \prod_{\alpha=1}^{3} |b_\alpha(m_\alpha)|^2 \tag{32}$$

The last expression can be simplified further by introducing the reciprocal pair potential

$$\varphi_{rec}(\mathbf{1}) = \frac{1}{\pi} \sum_{\mathbf{m} \neq 0} B(\mathbf{m}) \frac{\exp(-\pi^2 \mathbf{m}^2 / L^2 \beta^2)}{\mathbf{m}^2} \exp\left(i \frac{2\pi}{K} \mathbf{1} \cdot \mathbf{m}\right)$$
$$= \widehat{C}(\mathbf{1}) \tag{33}$$

where C is defined as

$$C(\mathbf{m}) = \begin{cases} \dfrac{1}{\pi} B(\mathbf{m}) \dfrac{\exp(-\pi^2 \mathbf{m}^2 / L^2 \beta^2)}{\mathbf{m}^2}, & \text{for } \mathbf{m} \neq 0 \\ 0, & \text{for } \mathbf{m} = 0 \end{cases} \tag{34}$$

with vector $\mathbf{m}$, defined as $\mathbf{m} = (m_1', m_2', m_3')$, where $m_\alpha' = m_\alpha$ for $0 \leq m_\alpha \leq K/2$ and $m_\alpha' = m_\alpha - K/2$ otherwise. It is important to point out that $C = (\widehat{\varphi_{rec}})^{-1}$, where $(\widehat{\varphi_{rec}})^{-1}$ is the inverse discrete Fourier transform defined as $A(\mathbf{m}) = K^{-3} \sum_{\mathbf{k} \in V_K} \widehat{A}(\mathbf{k}) \exp(-i(2\pi/K)\mathbf{m} \cdot \mathbf{k})$. Using the property of the Fourier transformation $\widehat{A}(-\mathbf{m}) = K^3 A(\mathbf{m})$, after some algebra, the reciprocal lattice sum reduces to

$$\frac{U_{rec}(\{\mathbf{u}_j\})}{k_B T} \approx \frac{l_B}{2L} \sum_{\mathbf{m} \in V_K} Q(\mathbf{m})(\varphi_{rec} * Q)(\mathbf{m}) \tag{35}$$

where $\varphi_{rec} * Q$ denotes the convolution of φ_{rec} and Q defined as

$$\varphi_{rec} * Q(\mathbf{m}) = \sum_{k_1=0}^{K-1} \sum_{k_2=0}^{K-1} \sum_{k_3=0}^{K-1} \varphi_{rec}(m_1 - k_1, m_2 - k_2, m_3 - k_3) Q(k_1, k_2, k_3) \tag{36}$$

To obtain the reciprocal part of the Coulomb forces, Eq. (35) has to be differentiated with respect to $\mathbf{r}_j$.

$$\mathbf{F}_j^{rec} = -\frac{\partial U_{rec}(\{\mathbf{u}_j\})}{\partial \mathbf{r}_j} = -\frac{k_B T l_B}{L} \sum_{\mathbf{m} \in V_K} \frac{\partial Q(\mathbf{m})}{\partial \mathbf{r}_j} (\varphi_{rec} * Q)(\mathbf{m}) \tag{37}$$

The numerical implementation of this algorithm involves the following steps [110,111]:

1. Determine B-spline coefficients $b_\alpha(m_\alpha)$.
2. Fill the charge grid array Q, using coefficients $M_n(u_{j,\alpha} - k)$ for $j = 1, \ldots, N_p$, $\alpha = 1, 2, 3$, and $j = 1, \ldots, n$, computed from scaled fractional coordinates of the particles $u_{j,\alpha}$. The cost of this step is $O(Nn^3)$, where n is the order of Cardinal B-spline.
3. Calculate the inverse 3DFFT of the charge grid array Q. The cost of this step is $O(K^3 \log K^3)$. For dense systems K^3 is of the order of N.
4. The approximate expression of U_{rec} ($\{\mathbf{u}_j\}$) is computed by using Eq. (35). At the same time the transformed Q array is overwritten by the product of itself with the arrays C and B.
5. Forward 3DFFT on the new Q array to evaluate the convolution $\psi_{rec}{}^*Q$. The cost of this step is again $O(K^3 \log K^3)$.
6. Generate forces by using convolution and Eq. (37). The cost of this step is $O(N)$.

Thus, the overall cost of the PME algorithm is $O(N\log N)$. Direct comparison of the Ewald [108] and PME methods has shown that the PME method is significantly faster than the Ewald method for $N_p > 10^4$ [110,111].

The PME method with Cardinal B-spline interpolation scheme is included as a standard routine in the AMBER [115] and DL_POLY [114] simulation packages.

3. Particle–Particle Particle–Mesh Method (P³M)

The key to the P³M method lies in splitting the interparticle Coulombic forces into a smoothly varying long-range component and a short-range component that is nonzero only for particles whose separation is less than the cutoff distance [119–124]. These two forces must be properly matched for the total force to be correctly reproduced. The total short-range force on the particle is computed by the direct summation of the particle–particle (PP) forces, while the smoothly varying long-range component is approximated by a particle–mesh (PM) force calculation. A key feature of the PM calculation is the use of the FFT convolution method to solve for the electrostatic potential at a given particle distribution.

The traditional Ewald representation of the Coulombic lattice sum as a sum of the two fast converging series is a perfect example of such splitting into the short-range and smoothly varying long-range components. In the Ewald sum the direct sum is due to the point charge and Gaussian counter-ion cloud and is short ranged, while the reciprocal sum is due to the Gaussian charge cloud and is a smooth function, with its Fourier transform rapidly convergent. Below we outline the general steps of particle–mesh calculations of the long-range component.

In the P^3M method the Gaussian charge clouds surrounding each charge within the unit cells are substituted by finite range weight functions $W_n(x)$ [123,124] that interpolate the original charge density over the n grid points, with spacing between them $h=L/K$.

$$Q(\mathbf{r}) = \sum_j \frac{q_j \beta^3}{\pi^{3/2}} \exp\left(-\beta^2 (\mathbf{r} - \mathbf{r}_j)^2\right) \Rightarrow Q_g(\mathbf{k}) = \sum_j q_j W_n(\mathbf{u}_j - \mathbf{k}) \qquad (38)$$

where $\mathbf{k}$ is the radius vector of the grid point and vector $\mathbf{k}=(k_1,k_2,k_3)$ with all $0 \le k_\alpha \le K-1$ being integers. As before, we rescale all lengths in the system by the grid spacing h, such that $\mathbf{u}_j=\mathbf{r}_j/h$. Due to periodic boundary conditions, the rescaled coordinates $\mathbf{u}$ vary in the interval $0 \le u_\alpha < K$ for $\alpha = 1,2,3$.

The simplest one-dimensional weight function $W_1(x)$ [$W_1(x)=1$, for $|x| \le 1/2$ and $W_1(x)=0$, for $|x| > 1/2$] assigns the charge densities to the nearest grid point. The full 3D weight function is a product of weight functions in the x, y, and z directions. Higher order weight functions are convolutions of $W_1(x)$, $W_n(x)= W_1 * W_{n-1}(x)= \int_{-\infty}^{\infty} W_1(x-x_1)W_{n-1}(x_1)\,dx_1$, and span n grid points in each direction. One can also use the Cardinal B-splines, introduced for interpolation of the exponentials in the PME method, as the weight functions for charge assignment. The higher order Cardinal B-splines can also be generated by convolutions of the hat function $M_1(x)$ [$M_1(x)=1$, for $0 \le x \le 1$ and $M_1(x)=0$ otherwise]. In fact, the charge assignment functions $W_n(x)$ and $M_n(x)$ are identical up to translation.

Since the original charge density $Q(\mathbf{u})$ has been replaced by charge density over the grid points $Q_g(\mathbf{k})$, Hockney and Eastwood [119] suggested to minimize the effect of such substitution by replacing the Coulomb Green's function by function $G_n(\mathbf{q})$, which will minimize the mean square error in forces due to the new assignment function W_n and finite size grid errors.

Consider two charges q_1 and q_2 placed at random positions $\mathbf{u}_1$ and $\mathbf{u}_2$ in the unit cell. The mean square error in force due to charge assignment is

$$\Delta = \int \frac{d\mathbf{u}_1}{K^3} \int \frac{d\mathbf{u}_2}{K^3} \left| \mathbf{F}^{EW}(\mathbf{u}_2,\mathbf{u}_1) - \mathbf{F}^{PM}(\mathbf{u}_2,\mathbf{u}_1) \right|^2 \qquad (39)$$

where $\mathbf{F}^{EW}(\mathbf{u}_2, \mathbf{u}_1)$ is the exact reciprocal force and $\mathbf{F}^{PM}(\mathbf{u}_2, \mathbf{u}_1)$ is the approximated force. The optimal Green's function $\widehat{G}_n(\mathbf{q})$ that minimizes the functional (39) is [119,122–124]

$$\widehat{G}_n(\mathbf{q}) = \frac{l_B}{\pi L} \frac{\sum_{\mathbf{b}} \mathbf{q} \cdot (\mathbf{b} + \mathbf{q}) \widehat{A}(\mathbf{q}) \left| \widehat{W}_n(\mathbf{b} + \mathbf{q}) \right|^2}{\mathbf{q}^2 \left[\sum_{\mathbf{b}} \left| \widehat{W}_n(\mathbf{b} + \mathbf{q}) \right|^2 \right]^2} \tag{40}$$

where $\widehat{A}(\mathbf{q}) = 1/\mathbf{q}^2 \exp(-\pi^2\mathbf{q}^2/\beta^2)$ and $\widehat{W}_n(\mathbf{q})$ is the Fourier transform of the 3D assignment function. The sum in Eq. (40) is taken over a Brillouin zone vector $\mathbf{b} = (b_1, b_2, b_3)$ with b_1, b_2, b_3 all integers and the vector $\mathbf{q}$ belongs to the first Brillouin zone with components $-K/2 \leq q_\alpha < K/2$, $\alpha = 1, 2, 3$. Since this expression does not depend on the particle positions, it has to be evaluated only once at the beginning of the simulation.

The usual PM calculations of the long-range forces between particles in the P^3M consist of the following steps (see Fig. 11):

1. Form an effective grid charge density $Q_g(\mathbf{k})$ by assigning the charges over the grid points. This charge density is Fourier transformed by using the forward discrete Fourier transform

$$\widehat{Q}_g(\mathbf{q}) = \sum_{\mathbf{k}} Q_g(\mathbf{k}) \exp(i\mathbf{k} \cdot \mathbf{q}) \tag{41}$$

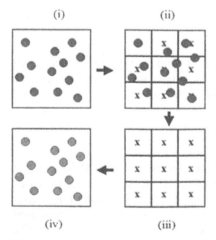

FIG. 11 A 2D schematic of the particle—mesh technique. (i) System of charged particles. (ii) Charges are interpolated on a 2D grid. (iii) Using FFT, to solve for the potential and forces at grid points. (iv) Interpolation of forces back to particles.

where vector $\mathbf{q}$ is the vector with components (q_1, q_2, q_3) with each $q_\alpha = 2\pi m_\alpha / K$ and $0 \le m_\alpha \le K - 1$.

2. Using the modified Green's function $\widehat{G}_n(\mathbf{q})$ and charge distribution $\widehat{Q}_g(\mathbf{q})$ solve for the electrostatic potential $\widehat{\varphi}_g(\mathbf{q})$ over the grid points

$$\widehat{\varphi}_g(\mathbf{q}) = \widehat{G}_n(\mathbf{q})\widehat{Q}_g(\mathbf{q}) \tag{42}$$

and find the Fourier transform of the electric field

$$\widehat{\mathbf{E}}_g(\mathbf{q}) = i\mathbf{q}\widehat{\varphi}(\mathbf{q}) = i\mathbf{q}\widehat{G}_n(\mathbf{q})\widehat{Q}_g(\mathbf{q}) \tag{43}$$

3. Using the inverse Fourier transform find the electric field at the grid points $\mathbf{E}_g(\mathbf{k})$

$$\mathbf{E}_g(\mathbf{k}) = \frac{1}{K^3} \sum_{\mathbf{q}} \widehat{\mathbf{E}}_g(\mathbf{q}) \exp(-i\mathbf{k} \cdot \mathbf{q}) \tag{44}$$

Note, that three inverse three-dimensional Fourier transforms have to be calculated to obtain the electric field distribution over the grid points because $\widehat{\mathbf{E}}_g(\mathbf{q})$ is a vector.

4. Finally, the electric field $\mathbf{E}_g(\mathbf{k})$ is interpolated back to the positions of the particles giving the long-range component of the forces on them

$$\mathbf{F}^{PM}(\mathbf{u}_j) = q_j \sum_{\mathbf{k}} W_n(\mathbf{u}_j - \mathbf{k})\mathbf{E}_g(\mathbf{k}) \tag{45}$$

The interpretation of the back interpolation is very simple, since each charge was replaced by several subcharges located at surrounding grid points. The force acting on each subcharge is given by its charge $q_j W_n(\mathbf{u}_j - k)$ times the electric field at its grid point $\mathbf{E}_g(\mathbf{k})$. Thus, the net force on the original charge q_j is the simple sum of all forces acting on its subcharges.

As one can see from the description of the algorithm for the P³M method, it is very close to the one presented for the PME method in the previous section. The PME differs in the final steps, in which the potential is first interpolated onto the particles and the resulting potential is differentiated with respect to the particle positions to give the electric field and forces. The PME uses half as many FFTs as P³M; it is more efficient where the cost of the FFT becomes an issue, such as in the parallel implementations. The number of FFT transformations in the P³M method can be reduced by first calculating the electrostatic potential over the mesh points by applying the inverse FFT to the electrostatic potential $\varphi_g(\mathbf{q})$. The electric field at the mesh points is then obtained by numerical differentiation of the electrostatic potential at the neighboring grid points. However, this implementation of the P^{3}M is less accurate than ones that use direct differentiation in the Fourier space [125,126]. Extensive comparisons and analysis of both methods can be found in [112,113,125,126].

B. Fast Multipole Method for Ewald Summation

The Fast Multipole Method (FMM) was originally introduced by Greengard and Rokhlin [127–130] for efficient simulation of N_p particles interacting through a Coulomb-like potential confined in a nonperiodic cell. The FMM relies on the standard multipole expansion for the electrostatic potential (forces) by separating the pairwise interactions into two components: one due to nearby particles, computed directly, and another due interaction with distant particles, approximated by their multipole expansions.

Consider a set of point charges q_j ($j=1,\ldots,k$), see Fig. 12, within a sphere of radius R, whose center is at distance $\mathbf{r}$ far away, $r = |\mathbf{r}| > R$, from a test charge located at point Q. The electrostatic potential $\varphi(\mathbf{r})$ at point Q due to all charges q_j at locations (ρ_j, α_j, β_j) is given by the infinite multipole expansion if, for all j, $\rho_j < r$

$$\varphi(\mathbf{r}) = k_B T l_B \sum_{n=0}^{\infty} \sum_{m=-n}^{n} \frac{M_n^m}{r^{n+1}} Y_n^m(\theta, \varphi) \tag{46}$$

where M_n^m are multipole moments, defined as

$$M_n^m = \sum_{j=1}^{k} q_j \rho_j^n Y_n^{-m}(\alpha_j, \beta_j) \tag{47}$$

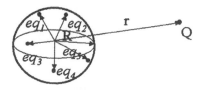

FIG. 12 The multipole expansion method.

where $Y_n^m(\theta, \varphi)$ is the spherical harmonic polynomial [131]. If we approximate this expansion by first $p+1$ terms, the truncation error ϵ is bounded by $(r/R)^{-p}$ [127,128]. Thus, fixing a precision ϵ, the number of terms in the multipole expansion is estimated to be $p = -\log_\gamma \epsilon$, where $\gamma = r/R$.

There is a duality inherent in the situation depicted in Fig. 12; if the location of test charge Q and a system of charges q_j ($j = 1, \ldots, k$) were interchanged, we may write another expansion that represents the potential due to particles outside the sphere of radius r and is correct as long as $\rho_j > r$,

$$\varphi(\mathbf{r}) = k_B T l_B \sum_{n=0}^{\infty} \sum_{m=-n}^{n} r^n L_n^m Y_n^m(\theta, \varphi) \tag{48}$$

where L_n^m are coefficients of the local expansion defined as

$$L_n^m = \sum_{j=1}^{k} q_j \rho_j^{-n-1} Y_n^{-m}(\alpha_j, \beta_j) \tag{49}$$

To illustrate the method, let us consider a square simulation box with sides of length one centered about the origin of the coordinate system and containing N_p particles inside. Fixing the precision ϵ, we choose $r/R = 2$, leading to the number of terms in the multipole expansion to be $p \approx -\log_2 \epsilon$, and specify that no interactions be computed for clusters of particles which are not separated by distances $r < 2R$. In order to impose such a condition, we introduce a hierarchy of meshes which refine the computational box into smaller and smaller regions (Fig. 13). Mesh level 0 is equivalent to the entire box, while mesh level $l+1$ is obtained from level l by dividing each box into four equal parts. The number of distinct boxes at mesh level l is equal to 4^l. A tree structure is imposed on this mesh hierarchy; the four boxes obtained on level $l+1$ by subdivision of the box on level l are considered to be its children. Now consider some box S at level n. It is not justifiable to apply the multipole expansion to nearest neighbor boxes, because particles in

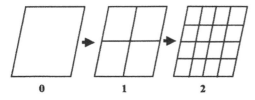

FIG. 13   Hierarchical subdivision of the full simulation space (2D case) into children, grandchildren, etc.

neighboring cells might be within the cutoff distance $2R_n$ for this level. Nearest neighbor cells are defined as cells sharing an edge or a corner (in 3D also a face) with a given cell. The interactions between charges in cell S with its nearest cells are taken by summing the electrostatic interactions directly. We apply this approximation at each step to at least next nearest neighbors and skip squares which lie in the regions that have been treated at previous levels. Therefore, the boxes with which the particles in box S interact at the present level, are (1) those that are not nearest neighbors of S and (2) those whose parents were a nearest neighbor of the parent of S at level $n-1$. Figure 14 shows the squares that are in the interaction list of cell S.

The realization of the Fast Multipole Method involves six steps.

1. The computational cell is subdivided into a hierarchy of meshes. (For a three-dimensional cubic cell the root of the hierarchy is the simulation box itself, which is divided into 8 cubes. Each of these cubes is then subdivided and so on for a prescribed number of subdivisions, l, so that at the finest level there are 8^l cubes.)

2. Compute the multipole moments about the center of each cell at the finest level of hierarchy.

3. *Upward pass.* Sweep up from the smallest cells to the largest cell to obtain the multipole moments M_n^m for cells at each subdivision level. This is done by using the translation operators that allow us to obtain the multipole moment of the parent cell from multipole moments of children cells (see for details [123,128,132]).

4. *Downward pass.* Sweep down from the largest (single) cell to cells at the next hierarchy level to obtain the local expansion coefficients L_n^m by multipole-to-local and local-to-local translations. This local

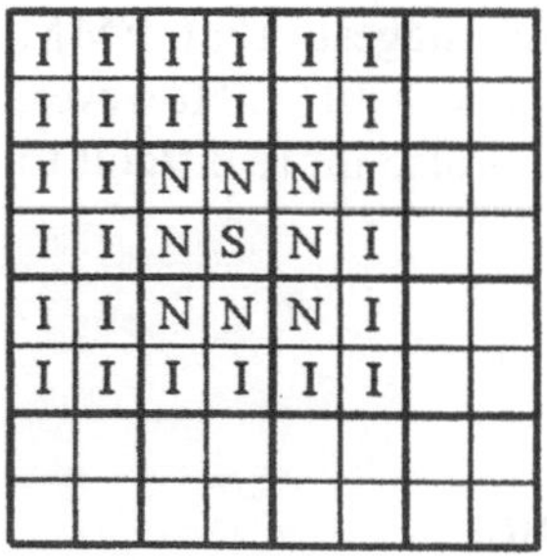

I	I	I	I	I	I		
I	I	I	I	I	I		
I	I	N	N	N	I		
I	I	N	S	N	I		
I	I	N	N	N	I		
I	I	I	I	I	I		

FIG. 14 Interaction list of cell S at level n. The squares at level n are separated by thin lines, their parents at level $n-1$ by heavy lines. The cells in the interaction list are labeled by **I**. The nearest neighbor cells are labeled by **N**.

expansion describes the field due to all particles in the system that are not contained in the current cubic box, its nearest neighbors and second nearest neighbors.

5. Evaluate the potential and fields for each particle using the local expansion coefficients on the finest level.

6. Add the contributions from other charges in the same cell and near cells that are not included in the multipole expansion by direct summation.

This method was later generalized to systems with nonuniform charge distribution by using an adaptive grid [133].

The FMM was extended by Schmidt and Lee [134,135] to systems with periodic boundary conditions. This approach combines both FMM and Ewald techniques. First, the FMM is initiated as in the finite case, calculating the multipole expansion of all cells at all refinement levels. The level zero expansion then contains the multipole expansion for all particles in the original simulation cell. All of its periodic images have the same multipole expansions about their centers. The FMM requires the local expansion of the potential from periodic images except the 26 nearest neighbors of the original simulation cell. This local expansion is obtained by using the Ewald sum formulation for multipole-to-local translation for image cells. After that, the algorithm continues in its downward pass as for the finite system.

Pollock and Glosli [123] have performed a timing comparison of Ewald, P^3M, and FMM algorithms by calculating the forces and potential energies of random periodic configurations of $N_p = 512, 1000, 5000, 10,000$, and $20,000$ charges (see Fig. 15). They concluded that P^3M and FMM are more efficient than Ewald summation, but P^3M is roughly four times faster than the FMM algorithm for all N_p used in this study, despite the superior asymptotic scaling of the FMM $O(N_p)$ vs. the $O(N_p \log N_p)$ for P^3M. (An extrapolation based on the scaling of both methods suggests that FMM becomes faster at some unphysically large system size $N_p > 10^{60}$.) However, the speed is not the only disadvantage of the FMM; it is also more diffcult to code than P^3M.

IV. POLYELECTROLYTE SOLUTIONS

A. Polyelectrolytes in Good and θ Solvents

In very dilute salt-free solutions the interaction between polyelectrolyte chains and counterions is weaker than the contribution of the configurational entropy of counterions, and counterions are thus distributed almost

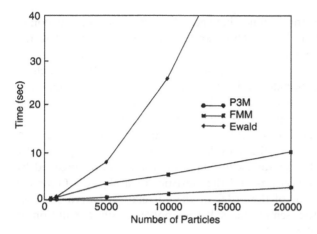

FIG. 15 Timing for complete force calculation of various size systems using standard Ewald, fast multipole method, and P³M. (From Ref. [124].)

homogeneously in the solution with only a small fraction being near the polyelectrolyte chains. The conformation of a chain is then determined by the chain elasticity and by the electrostatic repulsion between charged monomers (see Section II). As a result, chains adopt elongated conformations with size $L \approx bN(uf^2)^{1/3}$ [see Eq. (10)]. With increasing polymer concentration c, more and more counterions are drawn toward polyelectrolyte chains and, for sufficiently strong electrostatic interactions, can even be localized on them [3,136]. This phenomenon is called counterion or Manning condensation [3,136]. Localization of counterions leads to effective renormalization of the total charge on a chain $Q_{eff}=f_{eff}N<fN$. A simple estimate of the effective charge Q_{eff} can be achieved by equating the chemical potential of a counterion in a solution far away from a chain $\mu \approx k_BT \ln (cb^3 f_{eff})$ with that of a counterion localized within the chain volume LD_e^2, $\mu \approx k_BT \ln((f - f_{eff})(uf_{eff}^2)^{1/3}) - k_BTu^{2/3}f_{eff}^{1/3}$ [3,137–140]. This leads to the effective fraction of the charged monomers on the chain

$$f_{eff} \approx f\left(1 - \frac{cb^3}{(uf^2)^{1/3}} \exp\left(u^{2/3}f^{1/3}\right)\right) \tag{50}$$

being a decreasing function of polymer concentration c. The parameter $u^{2/3}f^{1/3}$ is the Manning parameter which determines the onset of counterion condensation. Thus, the chains shrink, $L \approx bN(uf_{eff}^2)^{1/3}$, as polymer concentration increases. However, for small values of the Manning parameter $(u^{2/3}f^{1/3} \ll 1)$ the corrections to the chain size are small and can be neglected.

Chains overlap when the distance between them $R_{cm} \approx (N/c)^{1/3}$ becomes of the order of their size $L \sim N$. This happens at polymer concentration $c^* \sim N^{-2}$. In semidilute solution, $c > c^*$, polyelectrolytes form a temporary network with mesh size ξ. At length scales smaller than ξ, the intrachain electrostatic repulsion dominates and sections of the chain with g_ξ monomer are strongly elongated such that $\xi \sim g_\xi$. Pure geometric arguments can be used to obtain the concentration dependence of the correlation length ξ, by imposing the close packing condition for chain sections of size ξ, $c \approx g_\xi / \xi^3 \sim \xi^{-2}$. This leads to the correlation length $\xi \sim c^{-1/2}$ [26,27,29]. At length scales larger than ξ the electrostatic interactions are screened by the other chains and counterions and the statistics of a chain is Gaussian with effective bond lengths of the order of the correlation length ξ. Thus, the chain size R in the semidilute salt-free polyelectrolyte solution is $R \sim \xi\sqrt{N/g_\xi} \sim N^{1/2}c^{-1/4}$ [26,29].

Polyelectrolyte solutions have a number of properties remarkably different from solutions of uncharged polymers. In particular:

1. There is a well pronounced peak in the scattering function of a homogeneous polyelectrolyte solution, whose position shifts with polymer concentration [1,4,6,7,26,27,29,141–145].
2. At low salt concentrations the main contribution to the osmotic pressure comes from counterions [1,4,6,7,26,27,29].

Extensive molecular dynamics simulations of dilute and semidilute polyelectrolyte solutions of chains with degree of polymerization N ranging from 16 up to 300 were recently performed by Stevens and Kremer [146–148] and by Liao et al. [149]. In these simulations the long-range electrostatic interactions were taken into account by the Ewald summation method, including interactions with all periodic images of the system. Stevens and Kremer [146–148] have used a spherical approximation of Adams and Dubey [150] for the Ewald sum while Liao et al. [149] have applied the PME method [110]. In addition to Coulombic interactions, all particles, including monomers and counterions, interacted via a shifted Lennard-Jones potential with cutoff $r_{cut} = 2^{1/6}\sigma$

$$U_{LJ}(r) = \begin{cases} 4\varepsilon_{LJ}\left[\left(\dfrac{\sigma}{r}\right)^{12} - \left(\dfrac{\sigma}{r}\right)^6 + \dfrac{1}{4}\right], & \text{for } r \leq r_{cut} \\ 0, & \text{for } r > r_{cut} \end{cases} \tag{51}$$

The polymers were modeled as bead-spring chains. The attractive part of the bond potential is described by the FENE potential

$$U_{FENE}(r) = -\frac{1}{2}kr_0^2 \ln\left(1 - \frac{r^2}{r_0^2}\right) \tag{52}$$

with the maximum extent $r_0 = 2\sigma$ and the spring constant $k = 7\varepsilon_{LJ}/\sigma^2$. The repulsive part of the bond potential was described by the shifted Lennard-Jones potential. The strength of the electrostatic interactions in the simulations was controlled by the value of the Bjerrum length. In the simulations of Stevens and Kremer [146–148] most of the results were obtained for the Bjerrum length l_B equal to 0.833σ, which corresponds to the case of weakly charged polyelectrolytes with the value of the coupling parameter $u < 1$, while in the Liao et al. [149] simulations the Bjerrum length l_B was 3σ, corresponding to strongly charged polyelectrolytes.

Figure 16 shows the dependence of the osmotic pressure of salt-free polyelectrolyte solutions. Through almost the entire concentration range considered, the osmotic pressure is proportional to the polymer concentration, supporting that it is controlled by the osmotic pressure of counterions, both above and below the overlap concentration. There appears to be a weak chain length dependence of the osmotic pressure for short chains. However, this N-dependence is consistent with a $1/N$ correction to the osmotic pressure due to chain translational entropy. The deviation from linear dependence of the osmotic pressure π occurs around polymer concentration $c \approx 0.07\sigma^{-3}$, which is above the overlap concentration for all samples. At very high polymer concentrations, where electrostatic interactions are almost completely screened by counterions and by charges on the

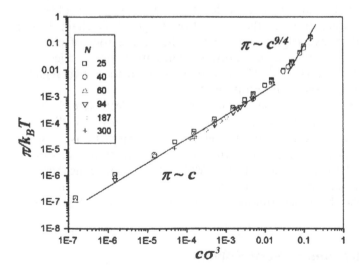

FIG. 16 The osmotic pressure is plotted as a function of the monomer density for various chain lengths.

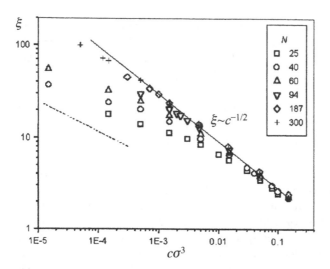

FIG. 17 Plot of the correlation length ξ as a function of density for different chain lengths. The slopes are -0.5 and -0.33 for the solid and dashed lines, respectively.

chains, the osmotic pressure π is proportional to $c^{9/4}$, and the scaling law for neutral polymers in a good solvent is recovered.

The correlation length ξ of the solution (see Fig. 17) exhibits two regimes as observed in experiment [1,6] and predicted by the theory [26,29,143–145]. Above overlap concentrations the correlation length ξ is chain length independent and is inversely proportional to the square root of polymer concentrations ($\xi \simeq c^{-1/2}$). At low polymer concentrations ξ scales with polymer concentration as $c^{-1/3}$. The crossover between these two regimes occurs around the overlap concentration c^*.

The effect of added salt on the conformation of polyelectrolyte chains in dilute and semidilute solutions was investigated by Stevens and Plimpton [151]. At high salt concentrations the electrostatic interactions between charged monomers are screened and the chain conformations are similar to those observed for good solvent, $R \simeq N^{0.588}$. As the salt concentration decreases, chains become more elongated, and finally at very low salt concentrations the chain size saturates at its salt-free value $R \simeq N$ [146,148].

B. Polyelectrolytes in Poor Solvent

Molecular dynamics simulations of partially charged polyelectrolytes in poor solvent conditions were performed by the Kremer group [153,154] for

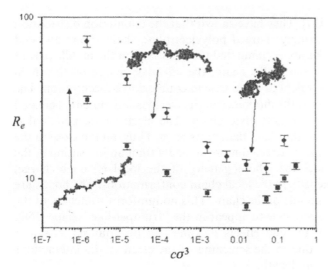

FIG. 18 Dependence of the end-to-end distance for polyelectrolyte chains with number of monomers $N = 94$ (filled circles) and $N = 187$ (filled squares) in poor solvent. Insets are snapshots of chain conformations.

chains with degree of polymerization $N = 94$ and by Liao et al. [155] for $N = 94, 187$. In these simulations, only every third monomer carried an electric charge and poor solvent conditions were imposed by setting the cutoff radius r_{cut} for monomer–monomer interactions to 2.5σ. Figure 18 shows the dependence of chain size on polymer concentration [153–155]. At low polymer concentrations the polyelectrolyte chains form necklaces of beads connected by strings. As the polymer concentration increases, the fraction of the condensed counterions on the chain increases and chains shrink by decreasing the length of the strings and the number of beads on the chain. Eventually, at higher polymer concentrations, polymer chains interpenetrate, leading to a concentrated polyelectrolyte solution. In this range of polymer concentrations, the chain size is observed to increase back towards its Gaussian value. The nonmonotonic dependence of the chain size on polymer concentration is in qualitative agreement with theoretical predictions [156].

C. Counterion Distribution and Condensation in Dilute Polyelectrolyte Solutions

The effect of counterion distributions in the dilute polyelectrolyte solution was the central subject of the molecular dynamics simulations performed by

Limbach and Holm [157]. They have shown that the counterion distribution around a quenched strongly charged polyelectrolyte chain shows an "end effect." The counterions are accumulated preferentially in the middle part of the chain. Toward the end of the chain, the effective charge on the chain increases, which is equivalent to a decrease in counterion concentration. The reason for this behavior is the difference in the electrostatic potential created by the charges of the polyelectrolyte chain. The electrostatic potential is stronger in the middle of the chain than at its ends. Thus, an ion close to the middle part of the chain is attracted more strongly than an ion sitting at the chain end [158]. Besides the inhomogeneity of the local effective charge, there is also inhomogeneity in the local chain conformations. The bonds are stretched more in the middle of the chain. This nonuniform stretching of the chain causes the polyelectrolyte to appear in the "trumpet-like" shape [158]. As one would expect, the end effects become less pronounced as salt is added to the solution, due to the screening of the electrostatic interactions over the Debye screening length.

For polyelectrolytes in a poor solvent, the situation is more peculiar. In this case the end effects are smeared out over the end beads. There is a sharp decrease in the effective charge in the region of the first strings. The modulation in accumulation of the effective charge in the middle of the chain repeats the necklace structure.

Condensation of counterions was studied by molecular dynamics simulations [148,159]. These simulations have shown that in dilute polyelectrolyte solutions at constant polymer concentrations the chain size depends nonmonotonically on the Bjerrum length l_B. First, the chain size increases with increasing Bjerrum length l_B, which is due to increasing strength of the intrachain electrostatic repulsion between charged monomers. This will continue until the Bjerrum length l_B reaches the crossover value l_B^*. Above this value the counterion condensation kicks in, reducing the effective charge on the chain, weakening intrachain electrostatic repulsions, and chains begin to shrink [139,140,160].

D. How Good Is the Debye–Hückel Approximation?

Almost all theoretical works considering polyelectrolytes [7,26,27,29, 141–145] have treated counterions and salt ions at the Debye–Hückel (DH) level by preaveraging the electrostatic interactions between charged monomers over the small ions' degrees of freedom. In this approximation the presence of counterions and salt ions leads to an effective screening of the electrostatic interaction between charged monomers, which interact via a screened electrostatic (Yukawa) potential. Of course, such an approximation is only correct when the presence of polymer chains will only slightly perturb the uniform distribution of the small ions in the system. This approximation should also fail at high polymer and salt concentrations, when excluded volume effects start to control the ionic atmosphere around polyelectrolyte chains.

To investigate the effect of the Debye–Hückel approximation on the solution properties, Stevens and Kremer [152] performed molecular dynamics simulations of salt-free solutions of bead-spring polyelectrolyte chains in which the presence of counterions was treated via a screened Coulomb potential, and compared the results with their simulations with explicit counterions [146,148]. To elucidate the effect of the Debye–Hückel approximation, the dependence of the mean square end-to-end distance, $\langle R_e^2 \rangle$, osmotic pressure, and chain structure factor on polymer concentration was examined. Stevens and Kremer found that $\langle R_e^2 \rangle$ tends to be larger at low densities for DH simulations and is smaller at higher densities. However, the difference in $\langle R_e^2 \rangle$ between DH simulations and simulations with explicit counterions is within 10%. This trend seems to be a generic feature for all N in their simulations. The functional form and density dependence of the chain structure factor are very close in both simulations. The most severe Debye–Hückel approximation affects the dependence of the osmotic pressure on polymer concentration. It appears that in the DH simulations not only is the magnitude of the osmotic pressure incorrect, but also the concentration dependence is wrong.

The effect of the Bjerrum length l_B (strength of the Coulombic interactions) on the applicability of the Debye–Hückel approximation was investigated by Stevens and Plimpton [151]. They found that this approximation works well for weak Coulomb interactions or for the Bjerrum lengths l_B smaller than σ (see Fig. 19). From this figure we see that the Debye–Hückel approximation breaks down for $l_B > \sigma$. In this range of parameters the chain size monotonically increases in the DH simulations, while for simulations with full Coulomb interactions the chain shrinks as the strength of these interactions increases. This discrepancy should not be too surprising, because, for $l_B > \sigma$, the Coulomb attraction of counterions to a polyelectrolyte chain in its vicinity becomes stronger than the thermal energy $k_B T$, leading to effective localization (condensation) of counterions near the polymer chain [3,136,139,140,160]. This condensation reduces the charge on the chain forcing it to contract [139,140,160].

The Debye–Hückel approximation can be improved by considering the full Poisson–Boltzmann approach to electrostatic interactions, in which the counterion condensation phenomena are included implicitly. Tests of the Poisson–Boltzmann approach to a cell model of rigid polyelectrolytes [161,162] were done in references [163–165] by performing molecular

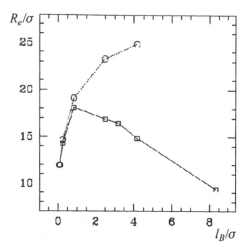

FIG. 19 Plot of the square-root mean square end-to-end distance as a function of the Bjerrum length l_B for DH simulation (open circles) and Coulomb simulation (open squares). For these simulations $N = 32$ and $c = 0.001\sigma^{-3}$. (From Ref. [151].)

simulations of this model. While the agreement between theory and simulation results was excellent in the case of monovalent counterions and weakly charged rods, it deteriorated with increase in the strength of the electrostatic interactions and, in particular, with increase in the counterion valence. At high concentrations of divalent counterions, computer simulations show charge oscillations which the theoretical model is unable to reproduce.

E. Bundle Formation in Polyelectrolyte Solutions

Can like-charged polyions ever attract each other? The answer to this question is counterintuitive and has significant implications. It is a well established fact that many bacteriophages use multivalent cations to package their DNA into compact and ordered forms in vivo. This phenomena is known as DNA condensation [166,167]. Moreover, it turns out that DNA is not the only polyelectrolyte that is able to do so. Other stiff polyelectrolytes such as F-actin [168], tobacco mosaic virus, and the bacteriophage fd [169] are also able to form laterally ordered aggregates (bundles) in the presence of multivalent counterions. Such a variety of systems strongly suggests an electrostatic origin for the bundle formation.

Recent Brownian dynamics simulations of systems of two infinitely long similarly charged rods in the presence of their counterions [170] have supported the electrostatic origin of aggregation by showing that the attraction between similarly charged polyions is indeed mediated by counterions. As the strength of the electrostatic interaction increases (or temperature of the system decreases) fluctuations in the densities of condensed counterions along the axes of the rods become strongly anticorrelated (lock–key structure) leading to effective attraction between rods at small separations [171–173] (see [174] for a review).

The bundle formation in systems of stiff polyelectrolytes in the presence of divalent counterions was simulated by Stevens [175,176]. Figure 20 shows the ordering of stiff polyelectrolytes into a network of bundles. The number of divalent counterions condensed on the chain is extremely large and clearly drives the aggregation. The bundling is due to attraction between two chains caused by correlated counterion fluctuations [171,172]. The precise

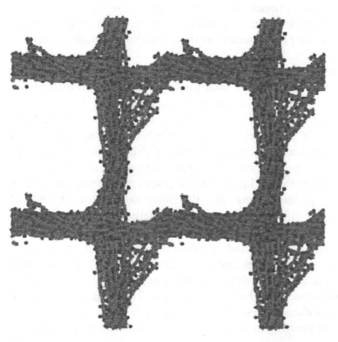

FIG. 20 Bundle configuration of polyelectrolyte chains with $N = 64$ and number of chains in simulation cell $M = 16$. The figure shows four periodic images in order to exhibit the connecting chains between bundles; there are only two bundles in the simulation cell. The counterions are dark spheres and the monomers are light spheres. (From Ref. [175].)

morphology of the bundle depends on chain stiffness, length, and concentration. The bundling occurs for stiff long chains, and when the chain length decreases, the bundling decreases as well. The same happens when the chain stiffness decreases. After the bundle is assembled the chains can only reptate within the bundle, which leads to significant slowing down of the relaxation of the system. It is therefore possible that within achievable simulation times, this bundle network is metastable.

V. WHAT IS NEXT?

Our understanding of polyelectrolytes in solutions has significantly progressed during the last ten years. The properties of polyelectrolyte chains in dilute solution seem to be understood reasonably well. The only problem that remains to be resolved here is the electrostatic persistence length of flexible polyelectrolytes. Computer simulations of semidilute polyelectrolyte solutions provide reasonable qualitative agreement with existing experimental and theoretical works. We now better understand counterion condensation phenomena and their relation to chain attraction and clustering.

However, a lot of interesting work remains to be done as attention turns to the dynamics of polyelectrolyte solutions. All current simulations of polyelectrolytes neglect the hydrodynamic interactions [177]. The inclusion of these interactions into molecular simulations will allow us to perform simulations of "realistic" dilute and semidilute polyelectrolyte solutions. It also will help us to answer the question of how hydrodynamic interactions are really screened in polyelectrolyte solutions. The answer to this question is extremely important for checking the assumptions of existing theoretical models. Hydrodynamic interactions are also long range, thus the use of Ewald summation [178–180], PME, and P^3M methods can significantly speed up the simulations. Work in this direction has just begun.

Another area which has attracted significant interest over the last few years is polyelectrolytes at surfaces and interfaces [8,181–184]. The progress in this area has implications for different areas of science and technology such as colloids, biophysics, biochemistry, medical science, pharmacy, food processing, water purification, etc. Some simulation work [185–191] has already appeared, suggesting a broad spectrum of interesting physical phenomena.

Our list of future directions cannot be complete without mentioning polyelectrolyte gels, polyelectrolyte brushes [192–195] (polyelectrolytes grafted to a surface), and complexation of polyelectrolytes with colloids [196–200], dendrimers [201], surfactants, and proteins. We anticipate serious simulation studies of these systems in the near future that will have tremendous practical applications.

APPENDIX

To perform direct evaluation of the lattice sum in Eq. (16) we first introduce the integral representation of

$$\frac{1}{|\mathbf{r}|} = \frac{2}{\sqrt{\pi}}\left[\int_0^\beta dt \exp(-|\mathbf{r}|^2 t^2) + \int_\beta^\infty dt \exp(-|\mathbf{r}|^2 t^2)\right] \tag{53}$$

With the help of this integral representation we can split the electrostatic energy of the unit cell $U(\{\mathbf{r}_j\})$ into two parts

$$U(\{\mathbf{r}_j\}) = U_{dir}(\{\mathbf{r}_j\}) + U_2(\{\mathbf{r}_j\})$$

where we defined the direct sum

$$\frac{U_{dir}(\{\mathbf{r}_j\})}{k_B T} = \lim_{M \to \infty} \frac{l_B}{\sqrt{\pi}} \sum_{k,j} q_k q_j \sum_{\mathbf{n}}^* \int_\beta^\infty \exp\left(-|\mathbf{r}_{kj} + \mathbf{n}|^2 t^2\right) dt \tag{54}$$

$$= \frac{l_B}{2} \sum_{\mathbf{n}}^* \sum_{k,j} q_k q_j \frac{\operatorname{erfc}(|\mathbf{r}_k - \mathbf{r}_j + \mathbf{n}|\beta)}{|\mathbf{r}_k - \mathbf{r}_j + \mathbf{n}|}$$

and a sum

$$\frac{U_2(\{\mathbf{r}_j\})}{k_B T} = \frac{l_B}{\sqrt{\pi}} \sum_{k,j} q_k q_j \sum_{\mathbf{n}}^* \int_0^\beta dt \exp\left(-|\mathbf{r}_{kj} + \mathbf{n}|^2 t^2\right) \tag{55}$$

In Eq. (54) $\operatorname{erfc}(x) = 2/\sqrt{\pi}\int_x^\infty dt \exp(-t^2)$ is the complementary error function. The direct sum $U_{dir}(\{\mathbf{r}_j\})$ is a rapidly convergent series and we may take the limit $M \to \infty$ without further ado. The second lattice sum in (55) requires more careful attention. This sum is diffcult to evaluate as it is, so we use the identity

$$\exp\left(-|\mathbf{r}_{kj} + \mathbf{n}|^2 t^2\right) = \frac{1}{\pi^{3/2} t^3} \int d^3\mathbf{u} \exp\left(-\frac{\mathbf{u}^2}{t^2} + 2i\mathbf{r}_{kj} \cdot \mathbf{u} + 2i\mathbf{n} \cdot \mathbf{u}\right) \tag{56}$$

to reduce it to three finite geometric series. After substitution of the integral (56) into Eq. (55), the integration over t may be performed immediately by using the substitution $s = t^{-1}$. This leads to the following expression for the lattice sum

$$\frac{U_2(\{\mathbf{r}_j\})}{k_BT} = \lim_{M\to\infty} \frac{l_B}{2\pi^2} \sum_{k,j} q_k q_j \int \frac{d^3\mathbf{u}}{\mathbf{u}^2} \exp(-\mathbf{u}^2/\beta^2 + 2i\mathbf{r}_{kj}\cdot\mathbf{u}) \sum_{\mathbf{n}} \exp(2i\mathbf{n}\cdot\mathbf{u})$$
$$- \frac{\beta l_B}{\sqrt{\pi}} \sum_j q_j^2$$

In the last equation the second term in the r.h.s. is accounting for the self-energy of the charges with $\mathbf{n}=0$ and $k=j$. The main contribution to the sum comes from the regions where $\exp(2i\mathbf{n}\cdot\mathbf{u})=1$ for which $\mathbf{u}=\mathbf{G}= m_1\mathbf{b}_1 + m_2\mathbf{b}_2 + m_3\mathbf{b}_3$ is a vector of the reciprocal lattice, $\mathbf{b}_\alpha$, $\alpha = 1, 2, 3$ are primitive vectors of the reciprocal lattice, and m_1, m_2, m_3 are integer numbers. The vectors $\mathbf{b}_\alpha$ have the property $\mathbf{b}_\alpha\mathbf{a}_\beta = \pi\delta_{\alpha\beta}$. [For the cubic lattice these vectors are $\mathbf{b}_1 = (\pi/L, 0, 0)$, $\mathbf{b}_2 = (0, \pi/L, 0)$, and $\mathbf{b}_3 = (0, 0, \pi/L)$.] This suggests dividing the integration region over $\mathbf{u}$ into a sum of integrals over the volume of a reciprocal lattice unit cell (the first Brillouin zone) $V_G = \{-\pi/2L < v_\alpha < \pi/2L$, for $\alpha = 1, 2, 3\}$. To do so we make the substitution $\mathbf{u} = \mathbf{G} + \mathbf{v}$ where $\mathbf{v} \in V_G$. This leads to

$$\frac{U_{rec}(\{\mathbf{r}_j\})}{k_BT} = \lim_{M\to\infty} \frac{l_B}{2} \sum_{k,j} q_k q_j \left(\sum_{G\neq 0} \exp(-\mathbf{G}^2/\beta^2 + 2i\mathbf{r}_{kj}\cdot\mathbf{G})\mathbf{I}_G(\mathbf{r}_{kj}) + \mathbf{I}_0(\mathbf{r}_{kj}) \right) \tag{58}$$

where we have defined

$$\mathbf{I}_G(\mathbf{r}_{kj}) = \frac{1}{\pi^2} \int_{V_G} d^3\mathbf{v} g_G(\mathbf{v}, \mathbf{r}_{kj}) \sum_{\mathbf{n}} \exp(2i\mathbf{n}\cdot\mathbf{v}) \tag{59}$$

and

$$g_G(\mathbf{v}, \mathbf{r}_{ij}) = (\mathbf{v} + \mathbf{G})^{-2}\exp(-\mathbf{v}^2/\beta^2 - 2\mathbf{v}\cdot\mathbf{G} + 2i\mathbf{r}_{ij}\cdot\mathbf{v})$$

The lattice sum in the integral $\mathbf{I}_G(\mathbf{r}_{kj})$ may be easily evaluated

$$\sum_{\mathbf{n}} \exp(2i\mathbf{n}\cdot\mathbf{v}) = \prod_{\alpha=1}^{3} \sum_{n_\alpha=-M}^{M} \exp(2in_\alpha L v_\alpha) = \prod_{\alpha=1}^{3} \frac{\sin((2M+1)Lv_\alpha)}{\sin(Lv_\alpha)} \tag{60}$$

For any $\mathbf{G} \neq 0$ the integral $I_G(\mathbf{r}_{ij})$ may be written as follows

$$I_G(\mathbf{r}_{kj}) = g_G(0, \mathbf{r}_{kj}) \frac{1}{\pi^2} \int_{V_G} d^3\mathbf{v} \prod_{\alpha=1}^{3} \frac{\sin((2M+1)Lv_\alpha)}{\sin(Lv_\alpha)}$$

$$+ \frac{1}{\pi^2} \int_{V_G} d^3\mathbf{v} \left(g_G(\mathbf{v}, \mathbf{r}_{kj}) - g_G(0, \mathbf{r}_{kj})\right) \prod_{\alpha=1}^{3} \frac{\sin((2M+1)Lv_\alpha)}{\sin(Lv_\alpha)} \quad (61)$$

The first integral is equal to π^3/L^3 and the first term is equal to $\pi/L^3\mathbf{G}^2$. The second integral may be estimated by integrating by parts with the use of $\sin((2M+1)Lv_\alpha) = -((2M+1)L)^{-1} d \cos((2M+1)Lv_\alpha)/dv_\alpha$. The leading terms are equal to zero because $\cos((2M+1)\pi/2) = 0$. Integration by parts again gives terms that are $O(((2M+1)L)^{-2})$ and can be neglected in the limit $M \to \infty$. Collecting all terms together we obtain the final expression for the sum in the reciprocal space with $\mathbf{G} \neq 0$

$$\frac{U_{rec}(\{\mathbf{r}_j\})}{k_B T} = \frac{\pi l_B}{2L^3} \sum_{\mathbf{G} \neq 0} \frac{\exp(-\mathbf{G}^2/\beta^2)}{\mathbf{G}^2} S(\mathbf{G})S(-\mathbf{G}) \quad (62)$$

$$= \frac{l_B}{2\pi L} \sum_{\mathbf{m} \neq 0} \frac{\exp(-\pi^2\mathbf{m}^2/L^2\beta^2)}{\mathbf{m}^2} S(\mathbf{m})S(-\mathbf{m})$$

where we defined the charge structure factor

$$S(\mathbf{G}) = \sum_j q_j \exp(i2\mathbf{G}\mathbf{r}_j) = \sum_j q_j \exp\left(i2\pi \frac{\mathbf{m} \cdot \mathbf{r}_j}{L}\right) \quad (63)$$

where vector $\mathbf{m} = (m_1, m_2, m_3)$ with all integers $m_\alpha \in [-\infty, \infty]$ for $\alpha = 1, 2, 3$. Now let us evaluate $I_0(\mathbf{r}_{kj})$. After some algebra it can be written as follows

$$I_0(\mathbf{r}_{kj}) = I_0(0) + \frac{1}{\pi^2} \int_{V_G} d^3\mathbf{v} \frac{(\exp(2i\mathbf{r}_{kj} \cdot \mathbf{v}) - 1)}{\mathbf{v}^2} \prod_{\alpha=1}^{3} \frac{\sin((2M+1)Lv_\alpha)}{\sin(Lv_\alpha)}$$

$$+ \frac{1}{\pi^2} \int_{V_G} d^3\mathbf{v} \left(\frac{\exp(-\mathbf{v}^2/\beta^2) - 1}{\mathbf{v}^2}\right)$$

$$\times (\exp(2i\mathbf{r}_{ij} \cdot \mathbf{v}) - 1) \prod_{\alpha=1}^{3} \frac{\sin((2M+1)Lv_\alpha)}{\sin(Lv_\alpha)} \quad (64)$$

The function $I_0(0)$ is independent of $\mathbf{r}_{kj}$ and its contribution to the lattice sum cancels for electroneutral systems. The first integral can be estimated by expanding $\exp(2i\mathbf{r}_{ij}\cdot\mathbf{v})$ in powers of $2i\mathbf{r}_{kj}\cdot\mathbf{v}$. All the odd powers in this series do not contribute to the integral due to symmetry of the V_G. Thus the expansion starts with the term $-2(\mathbf{r}_{kj}\cdot\mathbf{v})^2$. This is the only term in which we are interested, because all other terms in the expansion are $O(((2M+1)L)^{-2})$ and can be omitted in the limit $M \to \infty$. The symmetry of the integral allows us to change $(\mathbf{r}_{kj}\cdot\mathbf{v})^2 \, \mathbf{v}^2$ by $\mathbf{r}_{kj}^2/3$ and to perform the integration over $\mathbf{v}$ obtaining π^3/L^3. The second integral is finite and can be evaluated by integrating by parts leading to an asymptotic expansion in inverse powers of $(2M+1)L$. This expansion starts with the term of the order of $O(((2M+1)L)^{-2})$ and once again can be neglected in the limit $M \to \infty$. The final result for this part of the lattice sum is

$$\frac{U_{shape}(\{\mathbf{r}_j\})}{k_B T} = -\frac{2\pi l_B}{3L^3}\sum_{k,j} q_k q_j \mathbf{r}_{kj}^2 = \frac{2\pi l_B}{3L^3}\left|\sum_j q_j \mathbf{r}_j\right|^2 \tag{65}$$

Combining all terms together, we can write the final expression for the electrostatic energy of the cell as a sum of four terms sums in the direct and reciprocal space, self-energy term, and shape term that depends on the net dipole moment of the cell

$$\frac{U_{el}(\{\mathbf{r}_j\})}{k_B T} = \frac{l_B}{2}\sum_{\mathbf{n}}^{*}\sum_{k,j} q_k q_j \frac{\operatorname{erfc}(|\mathbf{r}_{kj}+\mathbf{n}|\beta)}{|\mathbf{r}_{kj}+\mathbf{n}|}$$
$$+ \frac{l_B}{2\pi L}\sum_{m\neq 0} \frac{\exp(-\pi^2 \mathbf{m}^2/L^2\beta^2)}{\mathbf{m}^2} S(\mathbf{m})S(-\mathbf{m})$$
$$- \frac{\beta l_B}{\sqrt{\pi}}\sum_j q_j^2 + \frac{2\pi l_B}{3L^3}\left|\sum_j q_j \mathbf{r}_j\right|^2 \tag{66}$$

REFERENCES

1. *Polyelectrolytes*; Hara, M., Ed.; Marcel Dekker: New York, 1993.
2. Tanford, C. *Physical Chemistry of Macromolecules*; Wiley: New York, 1961.
3. Oosawa, F. *Polyelectrolytes*; Marcel Dekker: New York, 1971.
4. Mandel, M. In *Encyclopedia of Polymer Science and Engineering*, Wiley: New York, 1988.
5. Schmitz, K.S. *Macroions in Solution and Colloidal Suspension*; VCH: Weinheim, 1993.
6. Forster, S.; Schmidt, M. Adv. Polym. Sci. **1995**, *120*, 51–133.
7. Barrat, J.L.; Joanny, J.F. Adv. Chem. Phys. **1996**, *XCIV*, 1–66.
8. *Colloid-Polymer Interactions: From Fundamentals to Practice*; Dubin, P.L., Farinato, R.S., Eds.; Wiley-Interscience: New York, 1999.

9. *Physical Chemistry of Polyelectrolytes*; Radeva, T., Ed.; Marcel Dekker: New York, 2001.
10. Baschanagel, J.; Binder, K.; Doruker, P.; Gusev, A.A.; Hahn, O.; Kremer, K.; Mattice, W.L.; Muller-Plathe, F.; Murat, M.; Paul, W.; Santos, S.; Suter, U.W.; Tries, V. Adv. Polym. Sci. **2000**, *152*, 41–156.
11. Application of the Monte Carlo methods in statistical physics, *Topics in Current Physics*, Binder, K., Ed.; Springer-Verlag: New York, 1987; Vol. 36.
12. The Monte Carlo methods in condensed matter physics. *Topics in Applied Physics*, Binder, K., Ed.; Springer-Verlag: New York, 1995; Vol. 71.
13. Binder, K.; Heermann, D.W. An introduction. *Monte Carlo Simulation in Statistical Physics*, 3rd Ed.; Springer Verlag: Berlin, 1997.
14. Newman, M.E.J.; Barkema, G.T. *Monte Carlo Methods in Statistical Physics*; Oxford University Press: New York, 1999.
15. Allen, M.P.; Tildesley, D.L. *Computer Simulation of Liquids*; Oxford University Press: New York, 1989.
16. Frenkel, D.; Smit, B. *Understanding Molecular Simulations: From Algorithms to Applications*; Academic Press: San Diego, 1996.
17. Rapaport, D.C. *The Art of Molecular Dynamics Simulations*; Cambridge University Press: New York, 1995.
18. Wall, F.T.; Mandel, F. J. Chem. Phys. **1975**, *63*, 4592–4595.
19. Mandel, F. J. Chem. Phys. **1979**, *70*, 3984–3988.
20. Lal, L. Mol. Phys. **1969**, *17*, 57.
21. McDonald, B.; Jan, N.; Hunter, D.L.; Steinitz, M.O. J. Phys. A **1985**, *18*, 2627.
22. Madras, N.; Socal, A.D. J. Stat. Phys. **1988**, *50*, 109.
23. Kremer, K.; Grest, G. J. Chem. Phys. **1990**, *92*, 5057–5086.
24. Schneider, J.; Hess, W.; Klein, R. J. Phys. A **1985**, *18*, 1221.
25. Verdier, P.H.; Stockmayer, W.H. J. Chem. Phys. **1962**, *36*, 227.
26. de Gennes, P.G.; Pincus, P.; Velasco, R.M.; Brochard, F. J. Phys. (Paris) **1976**, *37*, 1461–1473.
27. Khokhlov, A.R.; Khachaturian, K.A. Polymer **1982**, *23*, 1742–1750.
28. Rubinstein, M.; Colby, R.H.; Dobrynin, A.V. Phys. Rev. Lett. **1994**, *73*, 2776–2780.
29. Dobrynin, A.V.; Colby, R.H.; Rubinstein, M. Macromolecules **1995**, *28*, 1859–1871.
30. de Gennes, P.G. *Scaling Concepts in Polymer Physics*; Cornell University Press: Ithaca, 1979.
31. Yu Grosberg, A.; Khokhlov, A.R. *Statistical Physics of Macromolecules*; AIP Press: New York, 1994.
32. Brender, C. J. Chem. Phys. **1990**, *92*, 4468–4472.
33. Brender, C. J. Chem. Phys. **1991**, *94*, 3213–3221.
34. Brender, C.; Danino, M. J. Chem. Phys. **1992**, *97*, 2119–2125.
35. Hooper, H.H.; Blanch, H.W.; Prausnitz, J.M. Macromolecules **1990**, *23*, 4820–4829.
36. Christos, G.A.; Carnie, S.L. J. Chem. Phys. **1989**, *91*, 439–453.
37. Higgs, P.G.; Orland, H. J. Chem. Phys. **1991**, *95*, 4506–4518.

38. Barrat, J.L.; Boyer, D. J. Phys. II (France) **1993**, *3*, 343–356.
39. Ullner, M.; Jonsson, B.; Widmark, P.O. J. Chem. Phys. **1994**, *100*, 3365–3366.
40. Jonsson, B.; Petersen, C.; Soderberg, B. J. Phys. Chem. **1995**, *99*, 1251–1266.
41. Peterson, C.; Sommelius, O.; Soderberg, B. J. Chem. Phys. **1996**, *105*, 5233–5241.
42. Migliorini, G.; Rostiashvili, V.G.; Vilgis, T.A. Eur. Phys. J. E. **2001**, *4*, 475–487.
43. Odijk, T. J. Polym. Sci. Polym. Phys. Ed. **1977**, *15*, 477–483.
44. Skolnick, J.; Fixman, M. Macromolecules **1977**, *10*, 944–948.
45. Li, H.; Witten, T.A. Macromolecules **1995**, *28*, 5921–5927.
46. Netz, R.R.; Orland, H. Eur. Phys. J. B. **1999**, *8*, 81–98.
47. Ha, B.Y.; Thirumalai, D. J. Chem. Phys. **1999**, *110*, 7533–7541.
48. Tricot, M. Macromolecules **1984**, *17*, 1698–1703.
49. Reed, C.E.; Reed, W.F. J. Chem. Phys. **1992**, *97*, 7766–7776.
50. Forster, S.; Schmidt, M.; Antonietti, M. J. Phys. Chem. **1992**, *96*, 4008–4014.
51. deNooy, A.E.J.; Besemer, A.C.; van Bekkum, H.; van Dijk, J.; Smit, J.A.M. Macromolecules **1996**, *29*, 6541–6547.
52. Nishida, K.; Urakawa, H.; Kaji, K.; Gabrys, B.; Higgins, J.S. Polymer **1997**, *38*, 6083–6085.
53. Tanahatoe, J.J. J. Phys. Chem. B **1997**, *101*, 10442–10445.
54. Beer, M.; Schmidt, M.; Muthukumar, M. Macromolecules **1997**, *30*, 8375–8385.
55. Barrat, J.L.; Joanny, J.F. Europhys. Lett. **1993**, *24*, 333–338.
56. Bratko, D.; Dawson, K.A. J. Chem. Phys. **1993**, *99*, 5352–5364.
57. Ha, B.Y.; Thirumalai, D. Macromolecules **1995**, *28*, 577–581.
58. Reed, C.E.; Reed, W.F. J. Chem. Phys. **1991**, *94*, 8479–8486.
59. Seidel, C. Ber. Bunse-Ges. Phys. Chem. **1996**, *100*, 757–763.
60. Schafer, H.; Seidel, C. Macromolecules **1997**, *30*, 6658–6661.
61. Ullner, M.; Jonsson, B.; Peterson, C.; Sommelius, O.; Soderberg, B. J. Chem. Phys. **1997**, *107*, 1279–1287.
62. Micka, U.; Kremer, K. Phys. Rev. E **1996**, *54*, 2653–2662.
63. Micka, U.; Kremer, K. J. Phys. Cond. Matter **1996**, *8*, 9463–9470.
64. Micka, U.; Kremer, K. Europhys. Lett. **1997**, *38*, 279–284.
65. Cannavacciuolo, L.; Sommer, C.; Pedersen, J.S.; Schurtenberger, P. Phys. Rev. E **2000**, *62*, 5409–5419.
66. Lord Rayleigh. Philos. Mag. **1882**, *14*, 182.
67. Hooper, H.H.; Beltran, S.; Sassi, A.P.; Blanch, H.W.; Prausnitz, J.M. J. Chem. Phys. **1990**, *93*, 2715–2723.
68. Dobrynin, A.V.; Rubinstein, M.; Obukhov, S.P. Macromolecules **1996**, *29*, 2974–2979.
69. Chodanowski, P.; Stoll, S. J. Chem. Phys. **1999**, *111*, 6069–6081.
70. Lyulin, A.V.; Dunweg, B.; Borisov, O.V.; Darinskii, A.A. Macromolecules **1999**, *32*, 3264–3278.
71. Solis, F.J.; Olvera de la Cruz, M. Macromolecules **1998**, *31*, 5502–5506.

72. Balazs, A.C.; Singh, C.; Zhulina, E.; Pickett, G.; Chern, S.S.; Lyatskaya, Y. Prog. Surface Sci. **1997**, *55*, 181–269.
73. Edwards, S.F.; King, P.R.; Pincus, P. Ferroelectrics **1980**, *30*, 3–6.
74. Higgs, P.G.; Joanny, J-F. J. Chem. Phys. **1991**, *94*, 1543–1558.
75. Dobrynin, A.V.; Rubinstein, M. J. Phys. II (France) **1995**, *5*, 677–695.
76. Gutin, A.; Shakhnovich, E. Phys. Rev. E **1994**, *50*, R3322–R3325.
77. Quan, C.; Kholodenko, A. J. Chem. Phys. **1988**, *89*, 5273–5281.
78. Bratko, D.; Chakraborty, A.K. J. Phys. Chem. **1996**, *100*, 1164–1173.
79. McQuarrie, D.A. *Statistical Mechanics*; Harper & Row: New York, 1976; Chapter 15.
80. Kantor, Y.; Kardar, M.; Li, H. Phys. Rev. Lett. **1992**, *69*, 61–64.
81. Kantor, Y.; Kardar, M.; Li, H. Phys. Rev. E **1994**, *49*, 1383–1392.
82. Kantor, Y.; Kardar, M. Europhys. Lett. **1994**, *27*, 643–648.
83. Kantor, Y.; Kardar, M. Phys. Rev. E **1995**, *51*, 1299–1312.
84. Kantor, Y.; Kardar, M. Phys. Rev. E **1995**, *52*, 835–846.
85. Yamakov, V.; Milchev, A.; Limbach, H.J.; Dunweg, B.; Everaers, R. Phys. Rev. Lett. **2000**, *85*, 4305–4308.
86. Tanaka, M.; Yu Grosberg, A.; Pande, V.S.; Tanaka, T. Phys. Rev. E **1997**, *56*, 5798–5808.
87. Soddemann, T.; Schiessel, H.; Bluman, A. Phys. Rev. E **1998**, *57*, 2081–2090.
88. Lee, N.; Obukhov, S.P. Eur. Phys. J. B **1998**, *1*, 371–376.
89. Victor, J.M.; Imbert, J.B. Europhys. Lett. **1993**, *24*, 189–195.
90. Imbert, J.B.; Victor, J.M.; Tsunekawa, N.; Hiwatary, Y. Phys. Lett. **1999**, *258*, 92–98.
91. Wittmer, J.; Johner, A.; Joanny, J-F. Europhys. Lett. **1993**, *24*, 263–269.
92. Pande, V.S.; Yu Grosberg, A.; Joerg, C.; Kardar, M.; Tanaka, T. Phys. Rev. Lett. **1996**, *77*, 3565–3569.
93. Brender, C.; Lax, M.; Windwer, S. J. Chem. Phys. **1981**, *74*, 2576–2581.
94. Brender, C.; Lax, M. J. Chem. Phys. **1984**, *80*, 886–892.
95. Christos, G.A.; Carnie, S.L. Chem. Phys. Lett. **1990**, *172*, 249–253.
96. Woodward, C.; Jonsson, B. Chem. Phys. **1991**, *155*, 207–213.
97. Valleau, J. Chem. Phys. **1989**, *129*, 163.
98. Ewald, P.P. Ann. Phys. **1921**, *64*, 253–287.
99. Evjen, H.M. Phys. Rev. **1932**, *39*, 675.
100. Born, M.; Huang, K. *Dynamical Theory of Crystal Lattices*; Oxford University Press: London, 1954.
101. Kittel, C. *Introduction to Solid State Physics*, 5th Ed.; John Wiley & Sons: New York, 1976.
102. Toukmaji, A.Y.; Board, J.A. Jr. Comp. Phys. Comm. **1996**, *95*, 73–92.
103. Sagui, C.; Dardin, T.A. Annu. Rev. Biophys. Biomol. Struct. **1999**, *28*, 155–179.
104. Redlack, A.; Grinday, J. J. Phys. Chem. Solids **1975**, *36*, 73–82.
105. De Leeuw, S.W.; Perram, J.W.; Smith, E.R. Proc. R. Soc. London **1980**, *A373*, 27–56.
106. Smith, E.R. Proc. R. Soc. London **1981**, *A375*, 475–505.

107. Deem, M.W.; Newsam, J.M.; Sinha, S.K. J. Phys. Chem. **1990**, *94*, 8356–8359.
108. Perram, J.W.; Petersen, H.G.; De Leeuw, S.W. Molecular Physics **1988**, *65*, 875–893.
109. Dardin, T.; York, D.; Pedersen, L. J. Chem. Phys. **1993**, *98*, 10089–10092.
110. Essmann, U.; Perera, L.; Berkowitz, M.L.; Dardin, T.; Lee, H.; Pedersen, L.G. J. Chem. Phys. **1995**, *103*, 8577–8593.
111. Petersen, H.G. J. Chem. Phys. **1995**, *103*, 3668–3680.
112. Saqui, C.; Dardin, T.A. In *Simulation and Theory of Electrostatic Interactions in Solutions*, AIP Conference Proceedings, Prat, L.P., Hammer, G., Eds.; Vol. 492, 104–113.
113. Dardin, T.A.; Toukmaji, A.; Pedersen, L.G. J. Chim. Phys. **1997**, *94*, 1346–1364.
114. Kholmurodov, K.; Smith, W.; Yasuoka, K.; Dardin, T.; Ebisazaki, T. J. Comp. Chem. **2000**, *21*, 1187–1191.
115. Crowley, M.F.; Dardin, T.A.; Cheatham, T.E.; Deerfield, D.W. J. Supercomputing **1997**, *11*, 255–278.
116. Press, W.H.; Teukolsky, S.A.; Vetterling, W.T.; Flannery, B.P. Numerical Recipies in C. *The Art of Scientific Computing*; 2nd Ed.; Cambridge University Press: New York, 1992.
117. Chui, C.K. *An Introduction to Wavelets*; Academic Press: San Diego, 1992.
118. Schoenberg, I.J. *Cardinal Spline Interpolation*; SIAM: Philadelphia, 1973.
119. Hockney, R.W.; Eastwood, J.W. *Computer Simulations Using Particles*; McGraw-Hill: New York, 1981.
120. Luty, B.A.; Davis, M.E.; Tirony, I.G.; van Gunsteren, W.F. Mol. Simul. **1994**, *14*, 11–20.
121. Shimada, J.; Kaneko, H.; Takada, T. J. Comp. Chem. **1993**, *14*, 867–878.
122. Ferrell, R.; Bertschinger, E. Int. J. Mod. Phys. C **1994**, *5*, 933–956.
123. Pollock, E.L.; Glosli, J. Comp. Phys. Comm. **1996**, *95*, 93–110.
124. Pollock, E.L. In *Simulation and Theory of Electrostatic Interactions in Solutions*, AIP Conferences Proceedings, Prat, L.P., Hammer, G., Eds.; Vol. 492, 146–158.
125. Deserno, M.; Holm, C. J. Chem. Phys. **1998**, *109*, 7678–7693.
126. Deserno, M.; Holm, C. J. Chem. Phys. **1998**, *109*, 7694–7701.
127. Greengard, L.; Rokhlin, V. J. Comp. Phys. **1987**, *73*, 325–348.
128. Greengard, L.; Rokhlin, V. Lecture Notes in Mathematics. **1988**, *1360*, 121–141.
129. Greengard, L.; Rokhlin, V. Chem. Scrip. **1989**, *29A*, 139–144.
130. Greengard, L. Science **1994**, *265*, 909–914.
131. Jackson, J.D. *Classical Electrodynamics*; 2nd Ed.; Wiley: New York, 1975.
132. Esselink, K. Comp. Phys. Com. **1995**, *87*, 375–395.
133. Carrier, J.; Greengard, L.; Rokhlin, V. SIAM J. Sci. Stat. Comput. **1988**, *9*, 669–686.
134. Schmidt, K.E.; Lee, M.A. J. Stat. Phys. **1991**, *63*, 1223–1235.
135. Schmidt, K.E.; Lee, M.A. J. Stat. Phys. **1997**, *89*, 411–424.
136. Manning, G.S. J. Chem. Phys. **1969**, *51*, 924–933.

137. Khokhlov, A.R. J. Phys. A **1980**, *13*, 979–987.
138. Raphael, E.; Joanny, J.F. Europhys. Lett. **1990**, *13*, 623–628.
139. Schiessel, H.; Pincus, P. Macromolecules **1998**, *31*, 7953–7959.
140. Schiessel, H. Macromolecules **1999**, *32*, 5673–5680.
141. Muthukumar, M. J. Chem. Phys. **1996**, *105*, 5183–5199.
142. Muthukumar, M. J. Chem. Phys. **1997**, *107*, 2619–2635.
143. Donley, J.P.; Rudnik, J.; Liu, A.J. Macromolecules **1997**, *30*, 1188–1193.
144. Yethiraj, A. Phys. Rev. Lett. **1998**, *78*, 3789–3792.
145. Yethiraj, A. J. Chem. Phys. **1998**, *108*, 1184–1192.
146. Stevens, M.J.; Kremer, K. Phys. Rev. Lett. **1993**, *71*, 2228–2231.
147. Stevens, M.J.; Kremer, K. Macromolecules **1993**, *26*, 4717–4719.
148. Stevens, M.J.; Kremer, K. J. Chem. Phys. **1995**, *103*, 1669–1690.
149. Liao, Q.; Dobrynin, A.V.; Rubinstein, M. Macromolecules **2003**, *36*, 3386–3398; 3399–3410.
150. Adams, D.; Dubey, G. J. Comp. Phys. **1987**, *72*, 156–168.
151. Stevens, M.J.; Plimpton, S.J. Eur. Phys. J. B **1998**, *2*, 341–345.
152. Stevens, M.J.; Kremer, K. J. Phys. II France **1996**, *6*, 1607–1613.
153. Micka, U.; Holm, C.; Kremer, K. Langmuir **1999**, *15*, 4033–4044.
154. Micka, U.; Kremer, K. Europhys. Lett. **2000**, *49*, 189–195.
155. Liao, Q.; Dobrynin, A.V.; Rubinstein, M. Macromolecules 2004 (to be published).
156. Dobrynin, A.V.; Rubinstein, M. Macromolecules **2001**, *34*, 1964–1972.
157. Limbach, H.J.; Holm, C. J. Chem. Phys. **2001**, *114*, 9674–9682.
158. Castelnovo, M.; Sens, P.; Joanny, J-F. Eur. Phys. J. E **2000**, *1*, 115–125.
159. Winkler, R.G.; Gold, R.G.; Reineker, P. Phys. Rev. Lett. **1998**, *80*, 3731–3734.
160. Brilliantov, N.V.; Kuznetsov, D.V.; Klein, R. Phys. Rev. Lett. **1998**, *81*, 1433–1436.
161. Fuoss, R.M.; Katchalsky, A.; Lifson, S. Proc. Natl. Acad. Sci. USA. **1951**, *37*, 579.
162. Alfrey, T.; Berg, P.; Morawetz, H.J. J. Polym. Sci. **1951**, *7*, 543–547.
163. Das, T.; Bratko, D.; Bhuiyan, L.B.; Outhwaite, C.W. J. Phys. Chem. **1995**, *99*, 410.
164. Deserno, M.; Holm, C.; May, S. Macromolecules **2000**, *33*, 199–206.
165. Deserno, M.; Holm, C.; Kremer, K. In *Physical Chemistry of Polyelectrolytes*, Radeva, T., Ed.; Marcel Dekker: New York, 2001; 59–110.
166. Bloomfield, V.A. Biopolymers **1991**, *31*, 1471–1481.
167. Bloomfield, V.A. *Current Opinion in Structural Biology* **1996**, *6*, 334–341.
168. Tang, J.X.; Janmey, P.A. J. Biol. Chem. **1996**, *271*, 8556–8563.
169. Tang, J.X.; Wong, S.E.; Tran, P.T.; Janmey, P.A. Ber. der. Buns. Ges. Phys. Chem. Chem. Phys. **1996**, *100*, 796–806.
170. Gronbech-Jensen, N.; Mashl, R.J.; Bruinsma, R.F.; Gelbart, W.M. Phys. Rev. Lett. **1997**, *78*, 2477–2480.
171. Shklovskii, B.I. Phys. Rev. Lett. **1999**, *82*, 3268–3271.
172. Solis, F.J.; de la Cruz, M.O. Phys. Rev. E **1999**, *60*, 4496–4499.
173. Levin, Y.; Arenzon, J.J.; Stilck, J.F. Phys. Rev. Lett. **1999**, *83*, 2680.

174. Ha, B.Y.; Liu, A.J. In *Physical Chemistry of Polyelectrolytes*; Radeva, T., Ed.; Marcel Dekker: New York, 2001; 163–181.
175. Stevens, M.J. Phys. Rev. Lett. **1999**, *82*, 101–103.
176. Stevens, M.J. Biophys. J. **2001**, *80*, 130–139.
177. Doi, M.; Edwards, S.F. *The Theory of Polymer Dynamics*; Oxford University Press: New York, 1986.
178. Beenakker, C.W.J. J. Chem. Phys. **1986**, *85*, 1581.
179. Brady, J.F.; Phillips, R.J.; Lester, J.C.; Bossis, G. J. Fluid Mech. **1988**, *195*, 257.
180. Hase, K.R.; Powel, R.L. J. Phys. of Fluid **2001**, *13*, 32–44.
181. Fleer, G.J.; Cohen Stuart, M.A.; Scheutjens, J.M.H.M.; Gasgove, T.; Vincent, B. *Polymers at Interfaces*; Chapman and Hall: London, 1993.
182. Bajpai, A.K. Prog. Polym. Sci. **1997**, *22*, 523–564.
183. Kawaguchi, M.; Takahashi, A. Adv. Coll. Inter. Sci. **1992**, *37*, 219–317.
184. Decher, G.; Eckle, M.; Schmitt, J.; Struth, B. Curr. Opin. in Coll. Inter. Sci. **1998**, *3*, 32–40.
185. Beltran, S.; Hooper, H.H.; Blanch, H.W.; Prausnitz, J.M. Macromolecules **1991**, *24*, 3178–3184.
186. Muthukumar, M. J. Chem. Phys. **1995**, *103*, 4723–4731.
187. Mashl, R.J.; Gronbech-Jensen, N. J. Chem. Phys. **1998**, *109*, 4617–4623.
188. Kong, C.Y.; Muthukumar, M. J. Chem. Phys. **1998**, *109*, 1522–1527.
189. Mashl, R.J.; Gronbech-Jensen, N.; Fitzsimmons, M.R.; Lutt, M.; Li, D.Q. J. Chem. Phys. **1999**, *110*, 2219–2295.
190. Yamakov, V.; Milchev, A.; Borisov, O.; Dunweg, B. J. Phys. Cond. Matt. **1999**, *11*, 9907–9923.
191. Ellis, M.; Kong, C.Y.; Muthukumar, M. J. Chem. Phys. **2000**, *112*, 8723–8729.
192. Csajka, F.S.; van der Linden, C.C.; Seidel, C. Macromol. Symp. **1999**, *146*, 243–249.
193. Csajka, F.S.; Seidel, C. Macromolecules **2000**, *33*, 2728–2739.
194. Ennis, J.; Sjostrom, L.; Akesson, T.; Jonsson, B. Langmuir **2000**, *16*, 7116–7125.
195. Csajka, F.S.; Netz, R.R.; Seidel, C.; Joanny, J.F. Eur. Phys. J. E **2001**, *4*, 505–513.
196. von Goeler, F. J. Chem. Phys. **1994**, *100*, 7796–7803.
197. Wallin, T.; Linse, P. J. Chem. Phys. **1998**, *109*, 5089–5100.
198. Chodanowski, P.; Stoll, S. Macromolecules **2001**, *34*, 2320–2328.
199. Mateescu, E.M.; Jeppesen, C.; Pincus, P. Eur. Phys. Lett. **1999**, *46*, 493–498.
200. Kunze, K.K.; Netz, R.R. Phys. Rev. Lett. **2000**, *85*, 4398–4392.
201. Welch, P.; Muthukumar, M. Macromolecules **2000**, *33*, 6159–6167.

9

Gibbs Ensemble and Histogram Reweighting Grand Canonical Monte Carlo Methods

ATHANASSIOS Z. PANAGIOTOPOULOS Princeton University, Princeton, New Jersey, U.S.A.

I. INTRODUCTION

The phase behavior of fluids is of central importance to many technological and scientific fields, for example in designing separations for the chemical and pharmaceutical industries or in understanding fundamental processes in living systems. A large body of experimental information has been gathered over the years (e.g., see [1]), and significant efforts have been made to understand the phenomenology of the transitions and to obtain empirical- and theoretical-based models that can be used to correlate and extend the range of experimental data. Experimental measurements are time-consuming and expensive. For multicomponent mixtures, measurements are available only for a limited number of temperatures, pressures, and compositions. Empirical models are only valid over the range of conditions for which experimental data have been used to obtain the model parameters. Even theoretical-based models have limited predictive abilities for conditions and systems different from the ones for which they have been tested against using experimental data [2].

Molecular-based simulations are an increasingly important alternative to experimental measurements and theoretical techniques for obtaining properties of fluids and materials. The focus of the present chapter is on simulations of phase equilibrium properties of fluids. Classical force-field-based simulations start by postulating a functional form for the intermolecular forces in a system. Equilibrium properties can generally be obtained by either Monte Carlo or molecular dynamics methods. Monte Carlo methods are based on generating configurations from the appropriate probability distribution for a statistical mechanical ensemble, while molecular dynamics methods generate configurations by solving Newton's equations of motion. Calculations by simulation of simple structural and energetic properties (such as pair correlation functions, the mean configurational energy, or pressure) are relatively straightforward, but the prediction of the order and precise location of phase transitions is not a simple matter. Phase transitions are collective phenomena that occur over time and length scales that are not directly accessible by molecular dynamics or simple constant-volume Monte Carlo simulations. Until the mid-1980s, obtaining the phase behavior of even a simple one-component system required a major research effort [3]. Methodological developments since then have rendered the determination of phase equilibria by simulation much easier than before. Most of these methodological advances have involved development of novel Monte Carlo algorithms, which are the focus of the present review. In addition, the sustained increases in computing hardware capabilities have greatly expanded the range of systems that can be studied on readily available machines. As a result, the number of simulation studies of both model potentials and realistic systems has dramatically increased.

A number of textbooks, research monographs, and review articles have appeared previously in the area of the present chapter. The book by Allen and Tildesley [4] on computer simulation methods for liquids provides an excellent introduction to molecular dynamics and Monte Carlo methods, but does not cover the major recent methodological advances, since it was published in 1987. The recent book by Frenkel and Smit [5] has comprehensive coverage of molecular simulation methods for fluids, with particular emphasis on algorithms for phase equilibrium calculations. It describes many of the techniques mentioned in the present chapter in significantly more detail than is possible here. The Gibbs ensemble method and its applications have been reviewed in [6–9]. Proceedings of a workshop on simulations of phase transitions [10] and general review articles on simulation methods and their applications (e.g., [11–13]) are also available. The present chapter follows closely an earlier review article by the author [14].

Knowledge of the chemical potential of all components (or the free energy) of a system as a function of temperature, density, and composition is, of course, sufficient to determine the phase behavior. Methods to obtain the chemical potential include thermodynamic integration, a very general technique in which the state of interest is linked via a reversible path to a state of known free energy [5], and the Widom test particle insertion method [15]. The present chapter focuses on methods that were specifically designed for phase equilibrium calculations. The relative precision and accuracy of methods to obtain the chemical potential have been examined in [16]. Applications of direct interfacial simulations, which can be performed by either Monte Carlo or molecular dynamics algorithms, have been reviewed by Rowlinson and Widom [17] and Gubbins [18].

The plan of this chapter is as follows. Section II deals with the Gibbs ensemble Monte Carlo method, which is based on simultaneous calculations in two regions representing equilibrium phases, coupled indirectly via particle transfers and volume changes. The method is now commonly used for obtaining phase equilibria of fluids, because of its simplicity and speed. A single Gibbs ensemble simulation gives a point on the phase envelope of a multicomponent system. A number of other methods designed for direct calculations of phase equilibria are described in Section III. The $NPT+$ test particle method (Section III.A) is based on chemical potential calculations. The method has roughly the same range of applicability and limitations as the Gibbs ensemble, but requires multiple simulations per coexistence point. Gibbs–Duhem integration (Section III.B) does not require particle insertions and removals and is applicable to transitions involving solids. It needs to start, however, from a point on the phase envelope determined by one of the other techniques. Pseudo-ensembles (Section III.C) provide significant flexibility in determinations of phase equilibria under different external constraints and can be implemented in combination with the Gibbs ensemble or Gibbs–Duhem integrations. Histogram reweighting methods (Section IV) provide the free energy and phase behavior with excellent accuracy and can be used in the vicinity of critical points. The majority of simulation methods for calculations of phase transitions rely on particle transfers, which become impractical for dense systems or multisegment molecules. A number of methods have been developed for improving the efficiency of particle transfers and have been instrumental in enabling calculations for realistic potential models. Configurational-bias sampling techniques that perform "smart" insertions at favorable locations are described in Section V.A. Expanded ensembles are based on gradual transfers of parts of molecules and are described in Section V.B. The last part of this review (Section VI) describes applications of simulations to calculations of the phase behavior of polymeric systems. The chapter concludes with a discussion of the relative strengths and weaknesses of the methods discussed and provides some suggestions for possible future research directions in the field.

II. GIBBS ENSEMBLE MONTE CARLO

The Gibbs Ensemble Monte Carlo simulation methodology [19–21] enables direct simulations of phase equilibria in fluids. A schematic diagram of the technique is shown in Fig. 1. Let us consider a macroscopic system with two phases coexisting at equilibrium. Gibbs ensemble simulations are performed in two separate microscopic regions, each within periodic boundary conditions (denoted by the dashed lines in Fig. 1). The thermodynamic requirements for phase coexistence are that each region should be in internal equilibrium, and that temperature, pressure, and the chemical potentials of all components should be the same in the two regions. System temperature in Monte Carlo simulations is specified in advance. The remaining three conditions are satisfied by performing three types of Monte Carlo moves— displacements of particles within each region (to satisfy internal equilibrium), fluctuations in the volume of the two regions (to satisfy equality of pressures), and transfers of particles between regions (to satisfy equality of chemical potentials of all components).

The acceptance criteria for the Gibbs ensemble were originally derived from fluctuation theory [19]. An approximation was implicitly made in the derivation that resulted in a difference in the acceptance criterion for particle transfers proportional to $1/N$ relative to the exact expressions given subsequently [20]. A full development of the statistical mechanics of the

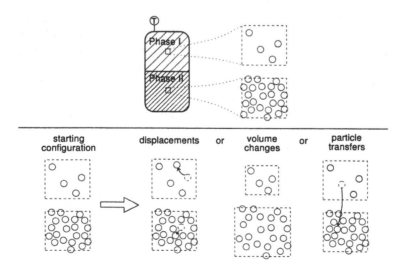

FIG. 1 Schematic diagram of the Gibbs ensemble Monte Carlo simulation methodology.

ensemble was given by Smit et al. [21] and Smit and Frenkel [22], which we follow here. A one-component system at constant temperature T, total volume V, and total number of particles N is divided into two regions, with volumes V_I and $V_{II} = V - V_I$, and number of particles N_I and $N_{II} = N - N_I$. The probability density of states, $\wp(N_I, V_I; N, V, T)$, is proportional to

$$\wp(N_I, V_I; N, V, T)$$
$$\propto \frac{N!}{N_I! N_{II}!} \exp[N_I \ln V_I + N_{II} \ln V_{II} - \beta U_I(N_I) - \beta U_{II}(N_{II})] \qquad (1)$$

Smit et al. [21] used a free energy minimization procedure to show that for a system with a first-order phase transition, the two regions in a Gibbs ensemble simulation are expected to reach the correct equilibrium densities.

The acceptance criteria for the three types of moves can be immediately obtained from Eq. (1). For a displacement step internal to one of the regions, the probability of acceptance is the same as for conventional constant-NVT simulations,

$$\wp_{move} = \min[1, \exp(-\beta \Delta U)] \qquad (2)$$

where ΔU is the configurational energy change resulting from the displacement. For a volume change step during which the volume of region I is increased by ΔV with a corresponding decrease of the volume of region II,

$$\wp_{volume} = \min\left[1, \exp\left(-\beta \Delta U_I - \beta \Delta U_{II} + N_I \ln \frac{V_I + \Delta V}{V_I} + N_{II} \ln \frac{V_{II} - \Delta V}{V_{II}}\right)\right]$$
$$(3)$$

Equation (3) implies that sampling is performed uniformly in the volume itself. The acceptance criterion for particle transfers, written here for transfer from region II to region I is

$$\wp_{transfer} = \min\left[1, \frac{N_{II} V_I}{(N_I + 1) V_{II}} \exp(-\beta \Delta U_I - \beta \Delta U_{II})\right] \qquad (4)$$

Equation (4) can be readily generalized to multicomponent systems. The only difference is that the numbers of particles of species j in each of the two regions, $N_{I,j}$ and $N_{II,j}$, replace N_I and N_{II} respectively. In simulations of multicomponent systems dilute in one component, it is possible that the

number of particles of a species in one of the two regions becomes zero after
a successful transfer out of that region. Equation (4) in this case is taken to
imply that the probability of transfer out of an empty region is zero.

The acceptance rules to this point are for a simulation in which the total
system is at constant number of molecules, temperature, and volume. For
pure component systems, the phase rule requires that only one intensive
variable (in this case system temperature) can be independently specified
when two phases coexist. The vapor pressure is obtained from the simulation. By contrast, for multicomponent systems pressure can be specified
in advance, with the total system being considered at constant NPT. The
probability density for this case, $\wp(N_I, V_I; N, P, T)$ is proportional to

$$\wp(N_I, V_I; N, P, T)$$
$$\propto \frac{N!}{N_I! N_{II}!} \exp\left[N_I \ln V_I + N_{II} \ln V_{II} - \beta U_I(N_I) - \beta U_{II}(N_{II}) - \beta P(V_I + V_{II}) \right]$$

(5)

and the only change necessary in the algorithm is that the volume changes in
the two regions are now made independently. The acceptance criterion
for a volume change step in which the volume of region I is changed by
ΔV, while the other region remains unchanged is then

$$\wp_{volume} = \min\left[1, \exp\left(-\beta \Delta U_I + N_I \ln \frac{V_I + \Delta V}{V_I} - \beta P \Delta V \right) \right]$$

(6)

An interesting extension of the original methodology was proposed by
Lopes and Tildesley [23], to allow the study of more than two phases at
equilibrium. The extension is based on setting up a simulation with as many
boxes as the maximum number of phases expected to be present. Kristof and
Liszi [24,25] have proposed an implementation of the Gibbs ensemble in
which the total enthalpy, pressure, and number of particles in the total
system are kept constant. Molecular dynamics versions of the Gibbs
ensemble algorithm are also available [26–28], and also in the following
Chapter 10 by M. Kotelyanskii and R. Hentschke.

The physical reason for the ability of the Gibbs ensemble to converge to a
state that contains phases at their equilibrium density in the corresponding
boxes, rather than a mixture of the two phases in each box, is the free energy
cost for creating and maintaining an interface. Essentially, in the Gibbs
ensemble, one trades off the directness of the interfacial simulation
approach with the (slight) inconvenience of setting up and maintaining
two or more boxes for the equilibrium phases. However, much smaller
system sizes can be used relative to interfacial simulations, and the

simulations are generally stable, except at the immediate vicinity of critical points.

Near critical points, Gibbs ensemble simulations become unstable because the free energy penalty for creating an interface becomes small. In a detailed study of the behavior of Gibbs ensemble simulations near critical points, Valleau [29] concluded that "it is only with extreme care ... that reliable information on critical parameters or the shapes of coexistence curves may be obtained from Gibbs ensemble simulations." The cause of the problems is that near critical points finite-size effects are present, and there is no mechanism for controlling system size of each individual region in the Gibbs ensemble. A better approach for dealing with systems near critical points is provided by the histogram methods described in Section IV. The finite-size critical behavior of the Gibbs ensemble has been examined by Bruce [30], Mon and Binder [31], and Panagiotopoulos [32]. The "standard" procedure for obtaining critical points from Gibbs ensemble simulations is to fit subcritical coexistence data to universal scaling laws. This approach has a weak theoretical foundation, since the universal scaling laws are only guaranteed to be valid in the immediate vicinity of the critical point, where simulations give the wrong (classical) behavior due to the truncation of the correlation length at the edge of the simulation box. In many cases, however, the resulting critical points are in reasonable agreement with more accurate results obtained from finite-size scaling methods (Section IV.C).

In summary, the Gibbs ensemble Monte Carlo methodology provides a direct and efficient route to the phase coexistence properties of fluids, for calculations of moderate accuracy. The method has become a standard tool for the simulation community, as evidenced by the large number of applications. Histogram reweighting techniques (Section IV) have the potential for higher accuracy, especially if equilibria at a large number of state points are to be determined. Histogram methods are also inherently better at determining critical points. In its original form, the Gibbs ensemble method is not practical for multisegment or strongly interacting systems, but development of configurational-bias sampling methods described in Section V.A has overcome this limitation.

III. THE *NPT* + TEST PARTICLE METHOD, GIBBS–DUHEM INTEGRATION AND PSEUDO-ENSEMBLES

A. The *NPT* + Test Particle Method

The NPT + test particle method [33,34] is based on calculations of the chemical potentials for a number of state points. A phase coexistence point is determined at the intersection of the vapor and liquid branches of the chemical potential vs. pressure diagram. The Widom test particle method [15] or any other suitable method [16] can be used to obtain the chemical potentials. Corrections to the chemical potential of the liquid and vapor phases can be made, using standard thermodynamic relationships, for deviations between the pressure at which the calculations were made and the actual coexistence pressure. Extrapolations with respect to temperature are also possible [35].

In contrast to the Gibbs ensemble, a number of simulations are required per coexistence point, but the number can be quite small, especially for vapor–liquid equilibrium calculations away from the critical point. For example, for a one-component system near the triple point, the density of the dense liquid can be obtained from a single NPT simulation at zero pressure. The chemical potential of the liquid, in turn, determines the density of the (near-ideal) vapor phase so that only one simulation is required. The method has been extended to mixtures [36,37]. Significantly lower statistical uncertainties were obtained in [37] compared to earlier Gibbs ensemble calculations of the same Lennard-Jones binary mixtures, but the NPT + test particle method calculations were based on longer simulations.

The NPT + test particle method shares many characteristics with the histogram reweighting methods discussed in Section IV. In particular, histogram reweighting methods also obtain the chemical potentials and pressures of the coexisting phase from a series of simulations. The corrections to the chemical potentials for changes in pressure [34] and temperature [35] are similar to the concept of reweighting of combined histograms from grand canonical simulations to new densities and temperatures.

Spyriouni et al. [38,39] have presented a powerful method (called "SPECS") for calculations of polymer phase behavior related to the NPT + test particle method. The method of Spyriouni et al. targets the calculation of the phase behavior of long-chain systems for which the test particle method for calculation of chemical potentials fails. For sufficiently long chains, even configurational-bias sampling methods discussed in Section V.A become impractical. For binary mixtures of a low-molecular weight solvent (species 1) and a polymer (species 2), two parallel simulations are performed in the (μ_1, N_2, P, T) ensemble at conditions near the expected coexistence curve. The chemical potential of component 2 is determined through the "chain increment" technique [40]. Iterative calculations at corrected values of the chemical potential of the solvent are performed until the chemical potential of the polymer in the two phases is equal. For the special case of a dilute solution, estimates of the chemical potentials of the solvent and polymer for compositions different from the original simulation conditions can be made using standard thermodynamic relations and the number of required iterations is significantly reduced.

B. Gibbs–Duhem Integration

Most methods for determination of phase equilibria by simulation rely on particle insertions to equilibrate or determine the chemical potentials of the components. Methods that rely on insertions experience severe difficulties for dense or highly structured phases. If a point on the coexistence curve is known (e.g., from Gibbs ensemble simulations), the remarkable method of Kofke [41,42] enables the calculation of a complete phase diagram from a series of constant-pressure simulations that do not involve any transfers of particles. For one-component systems, the method is based on integration of the Clausius–Clapeyron equation over temperature,

$$\left(\frac{dP}{d\beta}\right)_{sat} = -\frac{\Delta H}{\beta \Delta V} \tag{7}$$

where *sat* indicates that the equation holds on the saturation line, and ΔH is the difference in enthalpy between the two coexisting phases. The right hand side of Eq. (7) involves only "mechanical" quantities that can be simply determined in the course of a standard Monte Carlo or molecular dynamics simulation. From the known point on the coexistence curve, a change in temperature is chosen, and the saturation pressure at the new temperature is predicted from Eq. (7). Two independent simulations for the corresponding phases are performed at the new temperature, with gradual changes of the pressure as the simulations proceed to take into account the enthalpies and densities at the new temperature as they are being calculated.

Questions related to propagation of errors and numerical stability of the method have been addressed in [42] and [43]. Errors in initial conditions resulting from uncertainties in the coexistence densities can propagate and increase with distance from the starting point when the integration path is towards the critical point [43]. Near critical points, the method suffers from instability of a different nature. Because of the small free energy barrier for conversion of one phase into the other, even if the coexistence pressure is set properly, the identity of each phase is hard to maintain and large fluctuations in density are likely. The solution to this last problem is to borrow an idea from the Gibbs ensemble and couple the volume changes of the two regions [42]. Extensions of the method to calculations of three-phase coexistence lines are presented in [44] and to multicomponent systems in [43]. Unfortunately, for multicomponent systems the Gibbs–Duhem integration method cannot avoid particle transfers—however, it avoids transfers for one component, typically the one that is the hardest to transfer. The method and its applications have been recently reviewed [45].

In some cases, in particular lattice and polymeric systems, volume change moves may be hard to perform, but particle insertions and deletions may be relatively easy, especially when using configurational-bias methods. Escobedo and de Pablo [46,47] proposed a modification of the Gibbs–Duhem approach that is based on the expression

$$\left(\frac{d(\beta\mu)}{d\beta}\right)_{sat} = -\frac{\Delta(\rho u)}{\Delta\rho}, \tag{8}$$

where ρ is the density ($=N/V$) and u the energy per particle. This method was applied to continuous-phase polymeric systems in [46] and to lattice models in [50].

The Gibbs–Duhem integration method excels in calculations of solid–fluid coexistence [48,49], for which other methods described in this chapter are not applicable. An extension of the method that assumes that the initial free energy difference between the two phases is known in advance, rather than requiring it to be zero, has been proposed by Meijer and El Azhar [51]. The procedure has been used in [51] to determine the coexistence lines of a hard-core Yukawa model for charge-stabilized colloids.

C. Pseudo-Ensembles

The Gibbs–Duhem integration method represents a succesful combination of numerical methods and molecular simulations. Taking this concept even further, Mehta and Kofke [52] proposed a "pseudo-grand canonical ensemble" method in which a system maintains a constant number of particles and temperature, but has a fluctuating volume to ensure that, at the final density, the imposed value of the chemical potential is matched. The formalism still requires that estimates of the chemical potential be made during the simulation. The main advantage of the approach over more traditional grand canonical ensemble methods is that it provides additional flexibility with respect to the method to be used for determination of the chemical potential. For example, the "chain increment" method [40] for chain molecules, which cannot be combined with grand canonical simulations, can be used for the chemical potential evaluations in a pseudo-grand canonical simulation (as in [38]).

The same "pseudo-ensemble" concept has been used by Camp and Allen [53] to obtain a "pseudo–Gibbs" method in which particle transfers are substituted by volume fluctuations of the two phases. The volume fluctuations are unrelated to the ones required for pressure equality [Eq. (3)] but are instead designed to correct imbalances in the chemical potentials of some of the components detected, for example, by test particle insertions.

While the main driving force in [52] and [53] was to avoid direct particle transfers, Escobedo and de Pablo [47] designed a "pseudo-NPT" method to avoid direct volume fluctuations which may be inefficient for polymeric systems, especially on lattices. Escobedo [54] extended the concept for bubble-point and dew-point calculations in a "pseudo-Gibbs" method and proposed extensions of the Gibbs–Duhem integration techniques for tracing coexistence lines in multicomponent systems [55].

IV. HISTOGRAM REWEIGHTING GRAND CANONICAL MONTE CARLO

Early in the history of development of simulation methods it was realized that a single calculation can, in principle, be used to obtain information on the properties of a system for a range of state conditions [56–58]. However, the practical application of this concept was severely limited by the performance of computers available at the time. In more recent years, several groups have confirmed the usefulness of this concept, first in the context of simulations of spin systems [59–61] and later for continuous-space fluids [62,63,65–67]. In the following subsections, we give a pedagogical review of histogram reweighting methods for grand canonical Monte Carlo (GCMC) simulations as applied to one- and multicomponent systems. In addition, the determination of critical parameters from histogram data is briefly reviewed.

A. One-Component Systems

A GCMC simulation for a one-component system is performed as follows. The simulation cell has a fixed volume V, and is placed under periodic boundary conditions. The inverse temperature, $\beta = 1/k_B T$ and the chemical potential, μ, are specified as input parameters to the simulation. Histogram reweighting requires collection of data for the probability $f(N,E)$ of occurrence of N particles in the simulation cell with total configurational energy in the vicinity of E. This probability distribution function follows the relationship

$$f(N, E) = \frac{\Omega(N, V, E)\exp(-\beta E + \beta\mu N)}{\Xi(\mu, V, \beta)} \tag{9}$$

where $\Omega\,(N,V,E)$ is the microcanonical partition function (density of states) and $\Xi(\mu,V,\beta)$ is the grand partition function. Neither Ω nor Ξ are known at this stage, but Ξ is a constant for a run at given conditions. Since the left hand side of Eq. (9) can be easily measured in a simulation, an estimate for Ω and its corresponding thermodynamic function, the entropy $S(N,V,E)$, can be obtained by a simple transformation of Eq. (9):

$$S(N, V, E)/k_B = \ln \Omega(N, V, E) = \ln f(N, E) + \beta E - \beta\mu N + C \qquad (10)$$

C is a run-specific constant. Equation (10) is meaningful only over the range of densities and energies covered in a simulation. If two runs at different chemical potentials and temperatures have a region of overlap in the space of (N,E) sampled, then the entropy functions can be "merged" by requiring that the functions are identical in the region of overlap. To illustrate this concept, we make a one-dimensional projection of Eq. (9) to obtain

$$f(N) = \frac{Q(N, V, \beta)\exp(\beta\mu N)}{\Xi(\mu, V, \beta)} \qquad (11)$$

Histograms for two runs at different chemical potentials are presented in Fig. 2. There is a range of N over which the two runs overlap. Figure 3 shows the function $\ln f(N) - \beta\mu N$ for the data of Fig. 2. From elementary statistical mechanics, this function is related to the Helmholtz energy,

$$\beta A(N, V, \beta) = -\ln Q(N, V, \beta) = \ln f(N) - \beta\mu N + C \qquad (12)$$

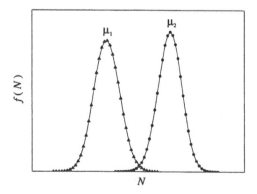

FIG. 2 Schematic diagram of the probability $f(N)$ of occurrence of N particles for two GCMC runs of a pure component system at the same volume V and temperature T, but different chemical potentials, μ_1 and μ_2, respectively. (From [14], ©2000 IOP Publishing Ltd.)

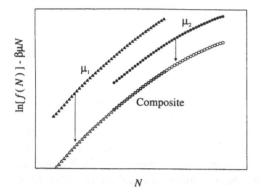

FIG. 3 The function $\ln[f(N)] - \beta\mu N$ for the data of Fig. 2. The figure shows the raw curves for μ_1 and μ_2 as well as a "composite" curve formed by shifting the data by the amount indicated by the arrows. (From [14], ©2000 IOP Publishing Ltd.)

Figure 3 shows the raw curves for μ_1 and μ_2 as well as a "composite" curve formed by shifting data for the two runs by the amount indicated by the arrows. The combined curve provides information over the combined range of particle numbers, N, covered by the two runs. Note that by keeping one-dimensional histograms for N we are restricted to combining runs of the same temparature, while the more general form [Eq. (10)] allows combination of runs at different temperatures.

Simulation data are subject to statistical (sampling) uncertainties, which are particularly pronounced near the extremes of particle numbers and energies visited during a run. When data from multiple runs are combined as shown in Fig. 3, the question arises of how to determine the optimal amount by which to shift the raw data in order to obtain a global free energy function. Ferrenberg and Swendsen [68] provided a solution to this problem by minimizing the differences between predicted and observed histograms. In this approach, it is assumed that multiple overlapping runs, $i = 1, 2, \ldots, R$ are available for a given system. The composite probability, $\wp(N, E; \mu, \beta)$, of observing N particles and energy E, if one takes into account all runs and assumes that they have the same statistical efficiency, is

$$\wp(N, E; \mu, \beta) = \frac{\sum_{i=1}^{R} f_i(N, E) \exp[-\beta E + \beta \mu N]}{\sum_{i=1}^{R} K_i \exp[-\beta_i E + \beta_i \mu_i N - C_i]} \tag{13}$$

where K_i is the total number of observations $[K_i = \sum_{N,E} f_i(N, E)]$ for run i. The constants C_i (also known as "weights") are obtained by iteration from the relationship

$$\exp(C_i) = \sum_E \sum_N \wp(N, E; \mu_i, \beta_i) \tag{14}$$

Given an initial guess for the set of weights C_i, Eqs. (13) and (14) can be iterated until convergence. When many histograms are to be combined this convergence of the Ferrenberg–Swendsen weights can take a long time. Once this has been achieved, however, all thermodynamic quantities for the system over the range of densities and energies covered by the histograms can be obtained. For example, the mean configurational energy $U(\mu, \beta)$ is

$$\langle U \rangle_{\mu, \beta} = \sum_E \sum_N \wp(N, E; \mu, \beta) \times E \tag{15}$$

and the mean density $\rho(\mu, \beta)$

$$\langle \rho \rangle_{\mu, \beta} = \frac{1}{V} \sum_E \sum_N \wp(N, E; \mu, \beta) \times N \tag{16}$$

The pressure of a system can be obtained from the following expression. If the conditions for run 1 are (μ_1, V, β_1) and for run 2 (μ_2, V, β_2), then

$$C_2 - C_1 = \ln\frac{\Xi(\mu_2, V, \beta_2)}{\Xi(\mu_1, V, \beta_1)} = \beta_2 P_2 V - \beta_1 P_1 V \tag{17}$$

where P is the pressure, since $\ln\Xi = \beta P V$. Equation (17) can be used to obtain the absolute value of the pressure for one of the two runs, provided that the absolute pressure can be estimated for the other run. Typically, this is done by performing simulations for low-density states for which the system follows the ideal-gas equation of state, $PV = Nk_BT$.

Up to this point, we assumed that a system exists in a one-phase region over the range of densities and energies sampled. If a phase transition exists, then the system, *in principle*, should sample states on either side of the phase transition, resulting in histograms with multiple peaks. This is illustrated in Fig. 4, in which actual simulation data (from a single run) are plotted for a simple cubic lattice homopolymer system [63] at a slightly subcritical temperature. There are two states sampled by the run, one at low and one at high particle numbers, corresponding to the gas and liquid states. The conditions for phase coexistence are equality of temperature, chemical potential, and pressure—the first two are satisfied by construction. From

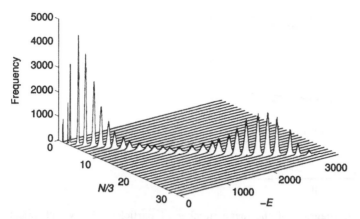

FIG. 4 Frequency of observation of states vs. energy, E, and number of particles, N, for a homopolymer of chain length $r = 8$ and coordination number $z = 6$ on a $10 \times 10 \times 10$ simple cubic lattice. Conditions, following the notation of [63] are $T^* = 11.5$, $\mu^* = -60.4$. In order to reduce clutter, data are plotted only for every third particle. (From [14], ©2000 IOP Publishing Ltd.)

Eq. (17), the integral under the probability distribution function is proportional to the pressure. In the case of two distinct phases, the integrals should be calculated separately under the liquid and gas peaks. The condition of equality of pressures can be satisfied by reweighting the data until this condition is met. In Section IV.C, we discuss how near-critical histogram data can be used to obtain precise estimates of the critical parameters for a transition.

In the absence of phase transitions or at temperatures near a critical point, the values of all observable quantities (such as the histograms of energy and density) are independent of initial conditions, since free energy barriers for transitions between states are small or nonexistent. However, at lower temperatures, free energy barriers for nucleation of new phases become increasingly larger. The states sampled at a given temperature and chemical potential depend on initial conditions, a phenomenon known as hysteresis. This is illustrated schematically in Fig. 5. For a supercritical isotherm, $T > T_c$, the mean value of the density is a continuous function of the chemical potential, and the same value is obtained for given conditions, irrespective of the starting configuration. By contrast, for a subcritical isotherm, when the runs are started from a low-density state, at some value of the chemical potential, a discontinuous "jump" to a state of higher density is observed. The exact location of the jump depends on the initial state and the specific mix of Monte Carlo moves used to change the configuration of the system. When simulations are started in a high-density

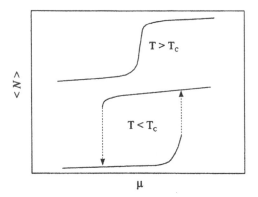

FIG. 5 Schematic diagram of the mean number of particles, $\langle N \rangle$, vs. chemical potential, μ, for a subcritical and a supercritical isotherm of a one-component fluid. The curve for the supercritical isotherm has been shifted up for clarity. (From [14], ©2000 IOP Publishing Ltd.)

state, the system remains on the high-density branch of the isotherm until some value of the chemical potential that is lower than the chemical potential of the jump from low- to high-density states.

The histogram reweighting method can be applied to systems with large free energy barriers for transitions between states, provided that care is taken to link all states of interest via reversible paths. One possibility is to utilize umbrella or multicanonical sampling techniques [61,69] to artificially enhance the frequency with which a simulation samples the intermediate density region [62]. Multicanonical and umbrella sampling require as input an estimate of the free energy in the intermediate density region, which has to be obtained by trial and error. In addition, a significant fraction of simulation time is spent sampling unphysical configurations of intermediate density. An alternative approach is to link states by providing connections through a supercritical path, in a process analogous to thermodynamic integration [5]. This approach is illustrated schematically in Fig. 6. The filled square represents the critical point for a transition, and open squares linked by dashed lines represent tie-lines. Ellipses represent the range of particle numbers and energies sampled by a single simulation. A near-critical simulation samples states on both sides of the coexistence curve, while subcritical simulations are likely to be trapped in (possibly metastable) states on either side. However, as long as there is a continuous path linking all states of interest, the free energies and pressures can be calculated correctly, and an accurate phase envelope can be obtained.

An example of the application of histogram reweighting for determining the phase behavior of a homopolymer model on the simple cubic lattice is

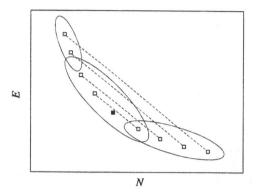

FIG. 6 Schematic diagram of the energy, E, vs. the number of particles, N, for a one-component fluid with a phase transition. Squares linked by dashed lines are coexisting phases joined by tie-lines and the filled square indicates the critical point of the transition. Ellipses represent the range of particle numbers and energies sampled during different GCMC runs. (From [14], ©2000 IOP Publishing Ltd.)

illustrated in Fig. 7. The phase behavior and critical properties of the model for a range of chain lengths have been studied in [63]. The system in this example is for chain length $r = 8$ and coordination number $z = 6$. In this example, we first performed a simulation at reduced temperature $T^* = 11.5$ and chemical potential $\mu^* = -60.4$, for which the raw histogram data are shown in Fig. 4. The resulting average volume fraction for the run is indicated on Fig. 7 by the filled circle at $T^* = 11.5$. The range of volume fractions sampled during the simulation is indicated on Fig. 7 by the arrows originating at the run point. Because this run is near the critical point, a very broad range of particle numbers and thus volume fractions is sampled during this single run. The histogram from this run was then reweighted to lower temperatures and a preliminary phase diagram was obtained. The estimated coexistence chemical potential at $T^* = 9$ was used as input to a new simulation, which sampled states near the saturated liquid line. The same procedure was repeated, now with combined histograms from the first two runs, to obtain an estimate of the coexistence chemical potential at $T^* = 7$. A new simulation was performed to sample the properties of the liquid at that temperature. The total time for the three runs was 10 CPU min on a Pentium III 300 MHz processor. The final result of these three calculations was the phase coexistence lines shown by the thick continuous lines on Fig. 7.

Two general observations can be made in relation to this example. First, it should be pointed out that the histogram reweighting method works much faster on smaller system sizes. As the system size increases, relative

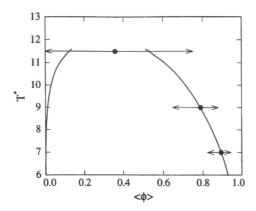

FIG. 7 Phase diagram for a homopolymer of chain length $r = 8$ on a $10 \times 10 \times 10$ simple cubic lattice of coordination number $z = 6$. Filled circles give the reduced temperature, T^*, and mean volume fraction, $\langle \phi \rangle$, of the three runs performed. Arrows from the run points indicate the range of densities sampled for each simulation. The thick continuous line is the estimated phase coexistence curve. (From [14], ©2000 IOP Publishing Ltd.)

fluctutations in the number of particles and energy for a single run at specified conditions decrease as the $1/2$ power of the system volume V. This implies that more simulations are required to obtain overlapping histograms that cover the range of energies and densities of interest. Moreover, the number of Monte Carlo moves required to sample properties increases approximately linearly with system size in order to keep the number of moves per particle constant. The computational cost of each Monte Carlo move is proportional to system size for pairwise additive long-range interactions and independent of system size for short-range interactions. The net effect is that the total computational effort required to obtain a phase diagram at a given accuracy scales as the 1.5 to 2.5 power of system volume, respectively for short- and long-range interactions. Fortunately, away from critical points, the effect of system size on the location of the coexistence curves for first-order transitions is typically small. In this example, calculations on a 15^3 system result in phase coexistence lines practically indistinguishable from the ones shown in Fig. 7. The mean absolute relative differences for the coexistence densities between the small and large systems are 0.1% for the liquid and 1% for the (much lower density) gas, well within the width of the coexistence lines on Fig. 7.

A second observation relates to calculations near critical points. The coexistence lines in Fig. 7 do not extend above a temperature of $T^* = 11.6$ because above that temperature significant overlap exists between the liquid and vapor peaks of the histograms. This overlap renders calculations of the liquid and gas densities imprecise. Larger system sizes suffer less from this effect and can be used to obtain coexistence densities near critical points. As discussed in Section IV.C, a sequence of studies with increasing system size is also required to obtain accurate estimates of critical points.

B. Multicomponent Systems

The histogram reweighting methodology for multicomponent systems [65–67] closely follows the one-component version described above. The probability distribution function for observing N_1 particles of component 1 and N_2 particles of component 2 with configurational energy in the vicinity of E for a GCMC simulation at imposed chemical potentials μ_1 and μ_2, respectively, at inverse temperature β in a box of volume V is

$$f(N_1, N_2, E) = \frac{\Omega(N_1, N_2, V, E)\exp(-\beta E + \beta\mu_1 N_1 + \beta\mu_2 N_2)}{\Xi(\mu_1, \mu_2, V, \beta)} \tag{18}$$

Equations (10) to (17) can be similarly extended to multicomponent systems.

The main complication in the case of multicomponent systems relative to the one-component case is that the dimensionality of the histograms is one plus the number of components, thus making their machine storage and manipulation somewhat more challenging. For example, in the case of one-component systems, it is possible to store the histograms directly as two-dimensional arrays. The memory requirements for storing three-dimensional arrays for a two-component system make it impractical to do so. Instead, lists of observations of particle numbers and energies are periodically stored on disk. It is important to select the frequency of sampling of the histogram information so that only essentially independent configurations are sampled. This implies that sampling is less frequent at high densities for which the acceptance ratio of the insertion and removal steps is lower. Sampling essentially independent configurations also enforces the condition of equal statistical efficiency underlying the Ferrenberg–Swendsen histogram combination equations [(13), (14)].

C. Critical Point Determination

Recent advances in the determination of critical parameters for fluids lacking special symmetries have been based on the concept of mixed-field finite-size scaling and have been reviewed in detail by Wilding [70]. As a critical point is approached, the correlation length ξ grows without bound and eventually exceeds the linear system size L of the simulation box. Singularities and discontinuities that characterize critical behavior in the thermodynamic limit are smeared out and shifted in finite systems. The infinite-volume critical point of a system can, however, be extracted by examining the size dependence of thermodynamic observables, through finite-size scaling theory [71–73]. The finite-size scaling approach proposed by Bruce and Wilding [74,75] accounts for the lack of symmetry between coexisting phases in most continuous-space fluids. Even though some recent work has cast some shadow on its full reliability [76], the approach seems to be quite robust and is easy to apply. For one-component systems, the ordering operator, M, is proportional to a linear combination of the number of particles N and total configurational energy U:

$$M \propto N - sU \tag{19}$$

where s is the field mixing parameter. For multicomponent systems, an extra field mixing parameter appears for each added component; for example for binary systems,

$$M \propto N_1 - sU - qN_2 \tag{20}$$

where q is the field mixing parameter for the number of particles of component 2.

General finite-size scaling arguments predict that the normalized probability distribution for the ordering operator M at criticality, $\wp(M)$, has a universal form. The order parameter distribution for the three-dimensional Ising universality class is shown in Fig. 8 as a continuous line. Also shown in Fig. 8 are data for a homopolymer of chain length $r = 200$ on a $50 \times 50 \times 50$ simple cubic lattice of coordination number $z = 26$ [63]. The data were obtained by histogram reweighting methods, by adjusting the chemical potential, temperature, and field mixing parameter s so as to obtain the best possible fit to the universal distribution. The nonuniversal constant A and the critical value of the ordering operator M_c were chosen so that the data have zero mean and unit variance. Due to finite-size corrections to scaling, the apparent critical temperature, $T_c(L)$, and density, $\rho_c(L)$, deviate from their infinite-system values, $T_c(\infty)$ and $\rho_c(\infty)$. They are expected to follow the scaling relationships with respect to the simulated system size, L:

$$T_c(L) - T_c(\infty) \propto L^{-(\theta+1)/\nu} (\rho_c(L) - \rho_c(\infty)) \propto L^{-(1-\alpha)/\nu} \tag{21}$$

where θ, ν, and α are, respectively, the correction-to-scaling exponent, the correlation length exponent, and the exponent associated with the

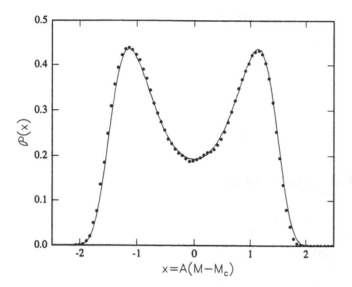

FIG. 8 Ordering operator distribution for the three-dimensional Ising universality class (continuous line) (data are courtesy of N. B. Wilding). Points are for a homopolymer of chain length $r = 200$ on a $50 \times 50 \times 50$ simple cubic lattice of coordination number $z = 26$ [63]. The nonuniversal constant A and the critical value of the ordering operator M_c were chosen so that the data have zero mean and unit variance. (From [14], ©2000 IOP Publishing Ltd.)

heat capacity divergence. For the three-dimensional Ising universality class, the approximate values of these exponents are [77,80] $(\theta, \nu, \alpha) \approx$ $(0.54, 0.629, 0.11)$. Figure 9 demonstrates these scaling relationships for the critical temperature and density of the square-well fluid of range $\lambda = 3$ [81].

D. Thermodynamic and Hamiltonian Scaling

Finally in this section, we would like to mention briefly two methods that are related to histogram reweighting. Thermodynamic scaling techniques proposed by Valleau [82] are based on calculations in the NPT, rather than the grand canonical (μVT) ensemble and provide information for the free energy over a range of volumes, rather than a range of particle numbers. Thermodynamic scaling techniques can also be designed to cover a range of Hamiltonians (potential models) in the Gibbs [83] or grand canonical [84] ensembles. In their Hamiltonian scaling form, the methods are particularly

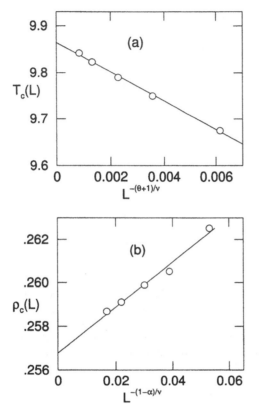

FIG. 9 Critical temperature (a) and density (b) scaling with linear system size for the square-fluid of range $\lambda = 3$. Solid lines represent a least-squares fit to the points. (From [81], ©1999 American Institute of Physics.)

useful for optimizing parameters in intermolecular potential models to reproduce experimental data such as the coexisting densities and vapor pressures. Thermodynamic and Hamiltonian scaling methods require estimates for the free energy of the system as a function of conditions, so that the system can be forced to sample the range of states of interest with roughly uniform probability, as for umbrella sampling Monte Carlo [69].

V. SMART SAMPLING FOR DIFFICULT SYSTEMS

A. Configurational-Bias Sampling

The most common bottleneck in achieving convergence in methods that rely on particle transfers is the prohibitively low acceptance of transfer attempts.

For dense fluid phases, especially for complex, orientation-dependent intermolecular potentials, configurations with "holes" in which an extra particle can be accommodated are highly improbable, and the converse step of removing a particle involves a large cost in energy. Configurational-bias sampling techniques significantly improve sampling efficiency for Gibbs or grand canonical Monte Carlo simulations. The methods have been reviewed in detail in [5,13], the chapter by Frenkel in [10], and in the Chapter 7 by T.S. Jain and J. de Pablo in this book, and will only be covered briefly in the present chapter.

Configurational-bias methods trace their ancestry to biased sampling for lattice polymer configurations proposed by Rosenbluth and Rosenbluth [85]. Development of configurational-bias methods for canonical and grand canonical simulations and for continuous-space models took place in the early 1990s [86–90] and dramatically expanded the range of intermolecular potential models that can be studied by the methods described in the previous sections.

Configurational-bias methods are based on segment-by-segment inser-tions or removals of a multisegment molecule. Several trial directions are attempted for every segment insertion, and a favorable growth direction is preferentially selected for the segment addition. In this way, the acceptance probability of insertions is greatly enhanced. For each segment growth or removal step, a correction factor (often called the "Rosenbluth weight") is calculated. The product of the Rosenbluth weights of all steps is incorporated in the overall acceptance criterion for particle insertions and removals in order to correct for the bias introduced by the nonrandom growth along preferential directions.

B. Expanded Ensembles

Another approach for handling multisegment molecules is based on the concept of expanded ensembles [91–94]. Expanded ensembles for chain molecules construct a series of intermediate states for the molecule of interest, from a noninteracting (phantom) chain to the actual chain with all segments and interactions in place. These intermediate states can be semipenetrable chains of the full length [91,92] or shortened versions of the actual chain [93,94]. Estimates of the free energy of the intermediate states are required to ensure roughly uniform sampling, as for thermodynamic and Hamiltonian scaling methods mentioned in the previous section. The advantage of expanded ensembles over configurational-bias methods is that arbitrarily complex long molecules can be sampled adequately, if sufficient computational effort is invested in constructing good approximations of the free energies of intermediate states.

VI. SOME APPLICATIONS TO POLYMERIC FLUIDS

The methods described in the previous sections enable fast and accurate calculations of the phase behavior of fluids. Their availability has resulted in a veritable explosion in the number of studies of the phase behavior of both simple "toy model" and realistic potentials for fluids in the past decade. Several reviews [6–9,14] have covered applications of such methods to many different systems. Here, we focus on recent applications to polymeric fluids.

Phase equilibria in long-chain lattice polymer models were studied by [63,78,79]. Of particular interest is a recent study by Yan and de Pablo [64] that goes to chains of length 16,000 on the simple cubic lattice. By going to such long chains, the authors were able to illustrate the presence of logarithmic corrections to the exponent characterizing the decrease of critical density with increase in chain length.

Hydrocarbon molecules are ubiquitous in industrial processes and form the building blocks of biological systems. They are nonpolar and consist of a small number of groups, thus making them the logical starting point for potential model development. Siepmann, Karaborni, and Smit [95–97] used configurational-bias Gibbs ensemble simulations to obtain an optimized potential model and the critical properties of the n-alkane homologous series. At the time, there were conflicting experimental data on the dependence of the critical density on chain length, which were resolved with the help of the simulations. Spyriouni et al. [39] have studied the phase behavior of n-hexadecane for the Dodd–Theodorou potential [98] and obtained good agreement for the phase envelope but not for the vapor pressure. Branched alkanes have been studied by [99–102], perfluorinated alkanes by [103], fluoromethanes by [104,105], and α-olephins by [106].

Three accurate united-atom potential sets for n-alkanes have appeared recently. The TRAPPE [107] and NERD models [108] use the Lennard-Jones (12,6) potential to describe nonbonded interactions among methyl and methylene groups, while the model of Errington and Panagiotopoulos [109] uses the exponential-6 functional form. All three reproduce the experimental phase diagrams and critical points. The exponential-6 model is slightly better with respect to representation of the vapor pressures. Figures 10 and 11 illustrate the quality of representation of experimental data for the newer optimized models. Deviations from experimental data for the exponential-6 united-atom model are comparable to those for a recently developed explicit hydrogen model [110].

Alkane mixtures have been studied extensively in recent years. For example, Chen and Siepmann investigated supercritical ethane and n-heptane mixtures and obtained the free energy of transfer for n-pentane and

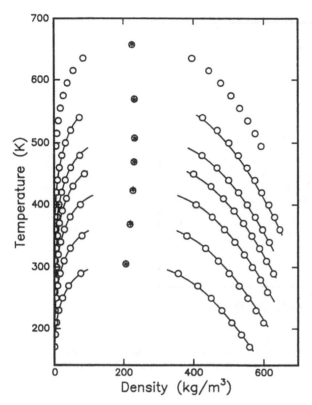

FIG. 10 Phase diagrams of selected *n*-alkanes. The curves from bottom to top are for ethane, propane, butane, pentane, hexane, octane, and dodecane. Circles represent calculations for the model of [109]. Uncertainties are smaller than the size of the symbols. A solid line is used for experimental data and a star for the experimental critical point. (From [109], ©1999 American Chemical Society.)

n-heptane between He and *n*-heptane liquid phases [110]. They have also obtained partition coefficients of alkanes between water and octanol [111]. Delhommelle et al. [112,113] studied mixtures of *n*-alkanes using both a united-atom model [96] and anisotropic united-atom models originated by Toxvaerd [114,115]. Other recent studies of alkane mixtures include [109,116,117]. The solubility of small molecules such as N_2 and methane and their mixtures in polyethylene, including effects of polymer crystallinity, was studied in [118]. Mixtures with α-olephins were studied in [38,106]. In general, excellent agreement between experiment and simulation results is obtained for these nonpolar mixtures, provided that the pure component potentials have been optimized to reproduce the phase envelope and vapor

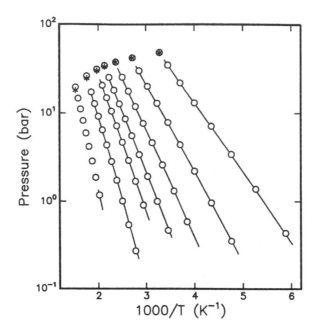

FIG. 11 Vapor pressures of selected *n*-alkanes. The curves from right to left are for ethane, propane, butane, pentane, hexane, octane, and dodecane. Symbols are the same as for Fig. 10. (From [109], ©1999 American Chemical Society.)

pressures of the pure components. No mixture parameters are necessary for the calculations.

VII. CONCLUDING REMARKS

A novice researcher interested in obtaining the phase behavior of a fluid is now faced with a bewildering choice among a number of alternative methods and their variations. In this section, similarities and differences among the methods, their relative performance, and their ease of implementation will be discussed.

Simulations with an explicit interface appear, at first glance, to be relatively simple to implement and perform. Unlike most other methods discussed here, interfacial simulations can also be performed using molecular dynamics codes. However, they provide an inefficient route to the phase coexistence properties. Unless the properties of the interface itself (or the surface tension) are of interest, other methods discussed in the present chapter provide better alternatives.

The majority of recent studies discussed in Section VI have been performed using various implementations of the Gibbs ensemble, often combined with configurational-bias methods to improve sampling for multisegment molecules. The Gibbs method is relatively easy to implement and provides direct information on the properties of coexisting phases from a single simulation. One major weakness of the methodology is that it is not applicable to solid or highly structured phases. For such systems, the only possible choice is the Gibbs–Duhem integration method and its variations. The Gibbs–Duhem method, however, needs to start from a point on the coexistence curve.

The accuracy of the Gibbs ensemble method for a given amount of computer time does not seem to match the accuracy of histogram reweighting methods [117]. Histogram reweighting methods are also inherently better at determining critical points via the finite-size scaling formalism. On the negative side, the effort required to implement histogram combination and reweighting is more than for the Gibbs ensemble. Histogram reweighting methods are also indirect, requiring the assembly of the free energy function of a system from a series of simulations. The efficiency of histogram methods decreases rapidly with increase in system size. Despite these disadvantages, they are probably the most promising for future applications.

The NPT + test particle method shares with histogram reweighting techniques the feature that it proceeds by computation of the chemical potential for a number of state points. Histogram reweighting methods, however, cover a range of densities and temperatures in a single simulation. In addition, data from separate runs can be combined in a systematic way only for histogram reweighting methods. A variation of the NPT + test particle method is the SPECS method for systems of long-chain molecules [38,39]. The SPECS method is a good alternative to expanded-ensemble Gibbs Monte Carlo calculations for cases for which configurational-bias sampling methods become inefficient.

Interesting connections between many of the methods discussed in the present chapter have been pointed out by Escobedo [54,55]. In particular, Escobedo suggests that Gibbs–Duhem integration, pseudo-ensembles, and the NPT + test particle method can be considered as low-order approximations of a histogram reweighting approach.

In the area of applications, an important goal for research in coming years will be to develop a set of pure component group-based potentials and combining rules that can be used for general predictions of both pure component and mixture phase behavior. Early results for realistic mixtures [117] suggest that relatively simple intermolecular potential models can be used to predict the phase behavior of broad classes of binary systems. For mixtures with large differences in polar character of the components, however, present models do not always result in quantitative agreement with experiment. New models that include higher-order interactions such as polarizability may be suitable for this purpose, a hypothesis that will need to be tested in the future.

ACKNOWLEDGEMENTS

Research in the author's group on which this chapter is based has been supported by the Department of Energy (Office of Basic Energy Sciences), the National Science Foundation, the American Chemical Society Petroleum Research Fund, and the Camille and Henry Dreyfus Foundation.

REFERENCES

1. Rowlinson, J.S.; Swinton, F. *Liquids and Liquid Mixtures*, 3rd Ed.; Butterworth Scientific: London, 1982.
2. Prausnitz, J.M.; Lichtenthaler, R.N.; de Azevedo, E.G. *Molecular Thermodynamics of Fluid Phase Equilibria*, 3rd Ed.; Prentice-Hall: New York, 1999.
3. Hansen, J.-P.; Verlet, L. Phys. Rev. **1969**, *184*, 151–161.
4. Allen, M.P.; Tildesley, D.J. *Computer Simulation of Liquids*; Oxford University Press: Oxford, 1987.
5. Frenkel, D.; Smit, B. *Understanding Molecular Simulation*; Academic Press: London, 1996.
6. Panagiotopoulos, A.Z. Molec. Simulation **1992**, *9*, 1–23.
7. Smit. B. In *Computer Simulations in Chemical Physics*, Allen, M.P., Tildesley, D.J., Eds.; NATO ASI Ser C, **1993**, *397*, 173–209.
8. Panagiotopoulos, A.Z. In Kiran, E., Levelt Sengers, J.M.H., Eds.; NATO ASI Ser. E, **1994**, *273*, 411–437.
9. Panagiotopoulos, A.Z. In *Observation, Prediction and Simulation of Phase Transitions in Complex Fluids*. NATO ASI Ser. C **1995**, *460*, 463–501.
10. Baus, M.; Rull, L.F.; Ryckaert, J.-P. Eds. *Observation, Prediction and Simulation of Phase Transitions in Complex Fluids*; NATO ASI Ser. C, **1995**, *460*.
11. Binder, K. Rep. Prog. Phys. **1997**, *60*, 487–559.
12. Deem, M.W. AIChE Journal **1998**, *44*, 2569–2596.
13. Siepmann, J.I. Adv. Chem. Phys. **1999**, *105*, 443–460.
14. Panagiotopoulos, A.Z. J. Phys.: Condens Matter **2000**, *12*, R25–52.
15. Widom, B. J. Chem. Phys. **1963**, *39*, 2808–2812.
16. Kofke, D.A.; Cummings, P.T. Molec. Phys. **1997**, *92*, 973–996.
17. Rowlinson, J.S.; Widom, B. *Molecular Theory of Capillarity*; Clarendon Press: Oxford, Ch. 6, 1982.
18. Gubbins, K.E. Molec. Simulation **1989**, *2*, 223–252.
19. Panagiotopoulos, A.Z. Molec. Phys. **1987**, *61*, 813–826.

20. Panagiotopoulos, A.Z.; Quirke, N.; Stapleton, M.; Tildesley, D.J. Molec. Phys. **1988**, *63*, 527–545.
21. Smit, B.; de Smedt, P.; Frenkel, D. Molec. Phys. **1989**, *68*, 931–950.
22. Smit, B.; Frenkel, D. Molec. Phys. **1989**, *68*, 951–958.
23. Lopes, J.N.C.; Tildesley, D.J. Molec. Phys. **1997**, *92*, 187–195.
24. Kristof, T.; Liszi, J. Molec. Phys. **1997**, *90*, 1031–1034.
25. Kristof, T.; Liszi, J. Molec. Phys. **1998**, *94*, 519–525.
26. Hentschke, R.; Bast, T.; Aydt, E. et al. J. Molec. Model **1996**, *2*, 319–326.
27. Kotelyanskii, M.J.; Hentschke, R. Molec. Simulation **1996**, *17*, 95–112.
28. Baranyai, A.; Cummings, P.T. Molec. Simulation **1996**, *17*, 21–25.
29. Valleau, J.P. J. Chem. Phys. **1998**, *108*, 2962–2966.
30. Bruce, A.D. Phys. Rev. E **1997**, *55*, 2315–2320.
31. Mon, K.K.; Binder, K. J. Chem. Phys. **1992**, *96*, 6989–6995.
32. Panagiotopoulos, A.Z. Int. J. Thermophys. **1989**, *10*, 447–457.
33. Möller, D.; Fischer, J. Molec. Phys. **1990**, *69*, 463–473; erratum in Molec. Phys. **1992**, *75*, 1461–1462.
34. Lotfi, A.; Vrabec, J.; Fischer, J. Molec. Phys. **1992**, *76*, 1319–1333.
35. Boda, D.; Liszi, J.; Szalai, I. Chem. Phys. Lett. **1995**, *235*, 140–145.
36. Vrabec, J.; Fischer, J. Molec. Phys. **1995**, *85*, 781–792.
37. Vrabec, J.; Lotfi, A.; Fischer, J. Fluid Phase Equil. **1995**, *112*, 173–197.
38. Spyriouni, T.; Economou, I.G.; Theodorou, D.N. Phys. Rev. Lett. **1998**, *80*, 4466–4469.
39. Spyriouni, T.; Economou, I.G.; Theodorou, D.N. Macromolecules **1998**, *31*, 1430–1431.
40. Kumar, S.K.; Szleifer, I.; Panagiotopoulos, A.Z. Phys. Rev. Lett. **1991**, *66*, 2935–2938.
41. Kofke, D.A. Molec. Phys. **1993**, *78*, 1331–1336.
42. Kofke, D.A. J. Chem. Phys. **1993**, *98*, 4149–4162.
43. Mehta, M.; Kofke, D.A. Chem. Eng. Sci. **1994**, *49*, 2633–2645.
44. Agrawal, R.; Mehta, M.; Kofke, D.A. Int. J. Thermophys. **1994**, *15*, 1073–1083.
45. Kofke, D.A. Adv. Chem. Phys. **1999**, *105*, 405–441.
46. Escobedo, F.A.; Pablo, J.J. J. Chem. Phys. **1997**, *106*, 9858–9868.
47. Escobedo, F.A.; Pablo, J.J. J. Chem. Phys. **1997**, *106*, 2911–2923.
48. Agrawal, R.; Kofke, D.A. Phys. Rev. Lett. **1995**, *74*, 122–125.
49. Bolhuis, P.; Frenkel, D. J. Chem. Phys. **1997**, *106*, 666–687.
50. Escobedo, F.A.; de Pablo, J.J. Europhys. Lett. **1997**, *40*, 111–117.
51. Meijer, E.J.; El Azhar, F. J. Chem. Phys. **1997**, *106*, 4678–4683.
52. Mehta, M.; Kofke, D.A. Molec. Phys. **1995**, *86*, 139–147.
53. Camp, P.J.; Allen, M.P. Molec. Phys. **1996**, *88*, 1459–1469.
54. Escobedo, F.A. J. Chem. Phys. **1998**, *108*, 8761–8772.
55. Escobedo, F.A. J. Chem. Phys. **1999**, *110*, 11999–12010.
56. McDonald, I.R.; Singer, K. J. Chem. Phys. **1967**, *47*, 4766–4772.
57. Wood, E.E. J. Chem. Phys. **1968**, *48*, 415–434.
58. Card, D.N.; Valleau, J.P. J. Chem. Phys. **1970**, *52*, 6232–6240.

59. Bennett, C.H. J. Comp. Phys. **1976**, *22*, 245–268.
60. Ferrenberg, A.M.; Swendsen, R.H. Phys. Rev. Lett. **1988**, *61*, 2635–2638.
61. Berg, B.A.; Neuhaus, T. Phys. Rev. Lett. **1992**, *68*, 9–12.
62. Wilding, N.B. Phys. Rev. E **1995**, *52*, 602–611.
63. Panagiotopoulos, A.Z.; Wong, V.; Floriano, M.A. Macromolecules **1998**, *31*, 912–918.
64. Yan, Q.L.; de Pablo, J.J. J. Chem. Phys. **2000**, *113*, 5954–5957.
65. Wilding, N.B. Phys. Rev. E **1997**, *55*, 6624–6631.
66. Wilding, N.B.; Schmid, F.; Nielaba, P. Phys. Rev. E **1998**, *58*, 2201–2212.
67. Potoff, J.J.; Panagiotopoulos, A.Z. J. Chem. Phys. **1998**, *109*, 10914–10920.
68. Ferrenberg, A.M.; Swendsen, R.H. Phys. Rev. Lett. **1989**, *63*, 1195–1198.
69. Torrie, G.M.; Valleau, J.P. J. Comp. Phys. **1977**, *23*, 187–199.
70. Wilding, N.B. J. Phys.: Condens Matter **1997**, *9*, 585–612.
71. Ferdinand, A.E.; Fisher, M.E. Phys. Rev. **1969**, *185*, 832.
72. Fisher, M.E.; Barber, M.N. Phys. Rev. Lett. **1972**, *28*, 1516.
73. Privman, V. Ed. *Finite Size Scaling and Numerical Simulation of Statistical Mechanical Systems*, World Scientific: Singapore, 1990.
74. Bruce, A.D.; Wilding, N.B. Phys. Rev. Lett. **1992**, *68*, 193–196.
75. Wilding, N.B.; Bruce, A.D. J. Phys.: Condens Matter **1992**, *4*, 3087–3108.
76. Orkoulas, G.; Fisher, M.E.; Üstün, C. J. Chem. Phys. **2000**, *113*, 7530–7545.
77. Sengers, J.V.; Levelt-Sengers, J.M.H. Ann. Rev. Phys. Chem. **1986**, *37*, 189–222.
78. Yan, Q.; Liu, H.; Hu, Y. Macromolecules **1996**, *29*, 4066–4071.
79. Frauenkron, H.; Grassberger, P. J. Chem. Phys. **1997**, *107*, 9599–9608.
80. Liu, A.J.; Fisher, M.E. Physica A **1989**, *156*, 35.
81. Orkoulas, G.; Panagiotopoulos, A.Z. J. Chem. Phys. **1999**, *110*, 1581–1590.
82. Valleau, J.P. Advances in Chemical Physics **1999**, *105*, 369–404.
83. Kiyohara, K.; Spyriouni, T.; Gubbins, K.E. et al. Molec. Phys. **1996**, *89*, 965–974.
84. Errington, J.R.; Panagiotopoulos, A.Z. J. Chem. Phys. **1998**, *109*, 1093–1100.
85. Rosenbluth, M.N.; Rosenbluth, A.W. J. Chem. Phys. **1955**, *23*, 356–359.
86. Frenkel, D.; Mooij, G.C.A.M.; Smit, B. J. Phys.: Condens Matter **1992**, *4*, 3053–3076.
87. de Pablo, J.J.; Laso, M.; Siepmann, J.I.; Suter, U.W. Molec. Phys. **1993**, *80*, 55–63.
88. Siepmann, J.I.; McDonald, I.R. Molec. Phys. **1993**, *79*, 457–473.
89. Mooij, G.C.A.M.; Frenkel, D.; Smit, B. J. Phys.: Condens Matter **1992**, *4*, L255–L259.
90. Laso, M.; de Pablo, J.J.; Suter, U.W. J. Chem. Phys. **1992**, *97*, 2817–2819.
91. Lyubartsev, A.; Martsinovski, A.; Shevkunov, S. J. Chem. Phys. **1992**, *96*, 1776–1783.
92. Vorontsov-Velyaminov, P.; Broukhno, A.; Kuznetsova, T. J. Phys. Chem. **1996**, *100*, 1153–1158.
93. Wilding, N.B.; Müller, M. J. Chem. Phys. **1994**, *101*, 4324–4330.
94. Escobedo, F.; de Pablo. J.J. J. Chem. Phys. **1996**, *105*, 4391–4394.

95. Siepmann, J.I.; Karaborni, S.; Smit, B. J. Am. Chem. Soc. **1993**, *115*, 6454–6455.
96. Smit, B.; Karaborni, S.; Siepmann, J.I. J. Chem. Phys. **1995**, *102*, 2126–2140.
97. Siepmann, J.I.; Karaborni, S.; Smit, B. Nature **1993**, *365*, 330–332.
98. Dodd, L.R.; Theodorou, D.N. Adv. Pol. Sci. **1994**, *116*, 249–281.
99. Cui, S.T.; Cummings, P.T.; Cochran, H.D. Fluid Phase Equil. **1997**, *141*, 45–61.
100. Siepmann, J.I.; Martin, M.G.; Mundy, C.J.; Klein, M.L. Mol. Phys. **1997**, *90*, 687–693.
101. Martin, M.G.; Siepmann, J.I. J. Phys. Chem. B **1999**, *103*, 4508–4517.
102. Zhuravlev, N.D.; Siepmann, J.I. Fluid Phase Equil. **1997**, *134*, 55–61.
103. Cui, S.T.; Siepmann, J.I.; Cochran, H.D.; Cummings, P.T. Fluid Phase Equil. **1998**, *146*, 51–61.
104. Potter, S.C.; Tildesley, D.J.; Burgess, A.N. et al. Molec. Phys. **1997**, *92*, 825–833.
105. Jedlovszky, P.; Mezei, M. J. Chem. Phys. **1999**, *110*, 2991–3002.
106. Spyriouni, T.; Economou, I.G.; Theodorou, D.N. J. Am. Chem. Soc. **1999**, *121*, 3407–3413.
107. Martin, M.G.; Siepmann, J.I. J. Phys. Chem. B **1998**, *102*, 2569–2577.
108. Nath, S.K.; Escobedo, F.A.; de Pablo, J.J. J. Chem. Phys. **1998**, *108*, 9905–9911.
109. Errington, J.R.; Panagiotopoulos, A.Z. J. Phys. Chem. B **1999**, *103*, 6314.
110. Chen, B.; Siepmann, J.I. J. Phys. Chem. B **1999**, *103*, 5370–5379.
111. Chen, B.; Siepmann, J.I. J. Am. Chem. Soc. **2000**, *122*, 6464–6467.
112. Delhommelle, J.; Boutin, A.; Tavitian, B.; Mackie, A.D.; Fuchs, A.H. Molec. Phys. **1999**, *96*, 1517–1524.
113. Ungerer, P.; Beauvais, C.; Delhommelle, J.; Boutin, A.; Rousseau, B.; Fuchs, A.H. J. Chem. Phys. **2000**, *112*, 5499–5510.
114. Padilla, P.; Toxvaerd, S. J. Chem. Phys. **1991**, *95*, 509–519.
115. Toxvaerd, S. J. Chem. Phys. **1997**, *107*, 5197–5204.
116. Mackie, A.D.; Tavitian, B.; Boutin, A.; Fuchs, A.H. Mol. Simulation **1997**, *19*, 1–15.
117. Potoff, J.J.; Errington, J.R.; Panagiotopoulos, A.Z. Molec. Phys. **1999**, *97*, 1073–1083.
118. Nath, S.K.; de Pablo, J.J. J. Phys. Chem. B **1999**, *103*, 3539–3544.

10

Gibbs Ensemble Molecular Dynamics

MICHAEL KOTELYANSKII   Rudolph Technologies, Inc., Flanders, New Jersey, U.S.A.

REINHARD HENTSCHKE   Bergische Universität, Wuppertal, Germany

I. THE METHOD

The Gibbs Ensemble was introduced by A. Panagiotopoulos [1] to conveniently model the coexistence of two phases in equilibrium—having the same pressure and chemical potentials. It mimics the coexistence of two phases, a liquid and its vapor in a closed vessel. In this case the total volume occupied by the two phases and the total number of particles are constant, but the particles are allowed to transfer between the phases.

The computer model consists of two boxes, which exchange volume and particles in such a way that the total volume is constant, i.e., the increment in volume of one box is accompanied by the same decrement of the other box's volume. Particles are transferred between the simulation boxes. It is shown in [1] that the Monte Carlo algorithm can be constructed with the proper acceptance criteria for volume and particle exchange moves to simulate the two coexisting phases in equilibrium with each box containing one of the phases. The scheme is very convenient for studying phase diagrams, as chemical potential and pressure do not have to be specified. The system automatically adjusts them to the coexistence values at a given temperature by exchanging particles and volumes.

As we mentioned before in the Background chapter, any ensemble can be implemented with both Monte Carlo and Molecular Dynamics algorithms. Which of the two should be chosen depends on the problem at hand. Depending on the particular application, either Molecular Dynamics or Monte Carlo can be more complex to program, especially when complex segmental moves are involved in MC, but MD can provide more information about system dynamics. Gibbs Ensemble Monte Carlo is discussed in the previous Chapter 9 by A. Panagiotopoulos.

Here we describe an implementation of the Gibbs Ensemble Molecular Dynamics (GEMD) technique, that implements the ideas underlying Gibbs Ensemble Monte Carlo using molecular dynamics.

In molecular dynamics (MD) one numerically solves the equations of motion for a system of N particles contained in a box of volume V having the total potential energy U. Here we consider a molecular system consisting of N atoms in M molecules, and we write the total potential energy U as a sum over inter- and intramolecular interactions, i.e.,

$$U = U_{inter} + U_{intra} \tag{1}$$

The first term is the sum over atom–atom pair potentials $U_{inter} = \sum_{\alpha \in i, \beta \in j, i > j, j = 1, M} \Phi(\mathbf{r}_{\alpha\beta})$ depending on the vectors $\mathbf{r}_{\alpha\beta}$ connecting atoms α and β in the two molecules i and j. U_{intra} describes the interactions within the molecule.

In order to simulate a variable number of molecules i in each of the two boxes we introduce an extra degree of freedom ξ_i for every molecule. ξ_i can vary between 1 and 0, where $\xi_i = 1$ means that molecule i is in box one, whereas $\xi_i = 0$ means that it is in box two. For $1 > \xi_i > 0$ it is in a "transition state," where it is "felt" in both boxes. Thus, we rewrite the intermolecular potential energy of the system as a function of the coordinates and the ξ_i as

$$
\begin{aligned}
U_{inter} = &\sum_{\alpha \in i, \beta \in j, i > j, j = 1, M} \Phi(\mathbf{r}_{\alpha\beta}, V_1) * \xi_i * \xi_j \\
&+ \sum_{\alpha \in i, \beta \in j, i > j, j = 1, M} \Phi(\mathbf{r}_{\alpha\beta}, V_2) * (1 - \xi_i) * (1 - \xi_j)
\end{aligned}
\tag{2}
$$

where V_1 and V_2 are the volumes of the two boxes. The two terms represent the intermolecular potential energies of the first and the second box, respectively. Consider, for instance, two particles with $\xi_i = 1$ and $\xi_j = 1$, i.e., both particles belong to box one. In this case the product $(1 - \xi_i)(1 - \xi_j)$ vanishes, and only the first term in Eq. (2) will contribute to the nonbonded potential energy. Notice also that as soon as we apply periodic boundary conditions, and interparticle interactions are calculated involving the particles' closest images, the distance between them, and therefore the intermolecular potential energy, is a function of the box dimensions (or of the volume if the shape of the box is kept fixed).

The transfer of particles (or molecules) is controlled by the difference between their potential energies in the two boxes. The number of unphysical (but necessary) transition state molecules can be made small in comparison to the total number of molecules by introducing an additional barrier described by the transfer potential function $g(\xi_i)$.

$$U_{total} = U_{intra} + U_{inter} + \sum_i g(\xi_i) \tag{3}$$

A possible choice is $g(\xi_i) = \omega[\tanh(u\xi_i) + \tanh(u(1 - \xi_i)) - 1]$ if $0 \leq \xi_i \leq 1$ and $g(\xi_i) = \infty$ otherwise, where u is the steepness, and ω is the height of the barrier.

The rate of particle transfer between the boxes determines the rate with which the equilibrium between the boxes is reached. The higher the rate the faster the two phases equilibrate. On the other hand, particles that are in the "transition" state contribute nonphysically to both boxes at the same time, and one would try to keep their number close to zero during the production run. This tuning is achieved by properly choosing the barrier function parameters. The higher the barrier, the harder is it for the particles to cross from one box to another.

In the case of phase coexistence in a one-component system the temperature, the pressure, and the chemical potential, even though the latter two are not explicitly specified, must be equal in the two phases and thus in the two boxes. This is achieved if every change of the volume of one of the boxes is accompanied by an opposite but equal change of the volume of the other box. The total volume of the two boxes is therefore constant, while the individual volumes are variable. The GEMD equations of motion can be written as follows:

$$\mathbf{p}_\alpha = m_\alpha \dot{\mathbf{r}}_\alpha \tag{4}$$

$$\dot{\mathbf{p}}_\alpha = -\frac{\partial U}{\partial \mathbf{r}_\alpha} - \eta \mathbf{p}_\alpha$$

$$\dot{\eta} = \frac{1}{Q_T}\left[\sum_\alpha \frac{\mathbf{p}_\alpha^2}{m_\alpha} - X k_B T\right]$$

$$p_{\xi_i} = m_{\xi_i}\dot{\xi}_i$$

$$\dot{p}_\xi = -\frac{\partial U}{\partial \xi_i} = -\frac{\partial U_{inter}}{\partial \xi_i} - \frac{\partial g(\xi_i)}{\partial \xi_i}$$

$$p_{V_1} = Q_P \dot{V}_1$$

$$\dot{p}_{V_1} = -\frac{\partial U}{\partial V_1}$$

$$= -\frac{\partial}{\partial V_1}\left[\sum_{\alpha \in i,\, \beta \in j,\, i>j,\, j=1,\, M} \Phi(\mathbf{r}_{\alpha\beta}, V_1) * \xi_i * \xi_j\right.$$

$$\left. + \sum_{\alpha \in i,\, \beta \in j,\, i>j,\, j=1,\, M} \Phi(\mathbf{r}_{\alpha\beta}, V_2) * (1-\xi_i) * (1-\xi_j)\right]$$

$$= P_1^e - P_2^e$$

Here $\mathbf{p}_\alpha$ and p_{ξ_i} are the momenta conjugate to the Cartesian coordinates $\mathbf{r}_\alpha$ and transfer variable ξ_i of the particle i. The first three equations describe the evolution of a system coupled to an external heat bath with the temperature T [2,3]. η is an additional degree of freedom, describing the coupling of the system to the external thermostat, which is necessary to simulate constant temperature. Q_T describes the strength of the coupling, and X equals the number of degrees of freedom, coupled to the thermostat.

The next two equations govern the evolution of the ξ_i, and thus the transfer of the molecules between the boxes. Note that the intramolecular part of the potential energy U_{intra} is independent of the ξ_i, and only intermolecular interactions and the additional potential $g(\xi_i)$ appear on the right hand side.

The last two equations are the equations of motion of the box volume V_1, where p_{V_1} is a momentum variable conjugate to the volume of the first box V_1, and Q_P is a parameter governing the volume relaxation. Again only the intermolecular interactions U_{inter} depend on the box size, because periodic boundary conditions are not applied to the intramolecular interactions. Distances describing the intramolecular interactions U_{intra} are smaller than the box sizes. They are of the order of a few chemical bond lengths, while the box sizes are usually many times larger. As the sum of the volumes of the two boxes is constant, $p_{V_1} = -p_{V_2}$. Note that, similar to any constant-pressure MD or MC algorithm, volume changes are controlled by the difference between the instantaneous values of the "external" pressures [4,5].

II. STATISTICAL MECHANICAL FOUNDATION

As we discussed in the Background chapter, to prove that integrating this set of differential equations does indeed simulate a model system with chemical potentials and pressures equal in both boxes, we have to show that the trajectory averages calculated by integrating Eqs. (4) are the corresponding thermodynamic averages.

We start with writing the Liouville equation [6] for this model, which describes the evolution of the phase space density distribution $\rho(\Gamma)$ with time. In our case the phase space Γ includes Cartesian coordinates and momenta, thermostat coupling variable η, volume V_1, the transfer variables ξ_i and their conjugate momenta p_{ξ_i} and p_{V_1}.* The stationary (time-independent) solution of this equation, describes the distribution function for the ensemble reproduced.

$$
\frac{\partial \rho}{\partial t} + \sum_{\alpha \in i,\, i=1,\, M} \left[\dot{\mathbf{r}}_\alpha \frac{\partial \rho}{\partial \mathbf{r}_\alpha} + \dot{\mathbf{p}}_\alpha \frac{\partial \rho}{\partial \mathbf{p}_\alpha} \right] + \sum_{i=1,\, M} \left[\dot{\xi}_i \frac{\partial \rho}{\partial \xi_i} + \dot{p}_{\xi_i} \frac{\partial \rho}{\partial \mathbf{p}_{\xi_i}} \right]
$$

$$
+ \dot{V}_1 \frac{\partial \rho}{\partial V_1} + \dot{p}_{V_1} \frac{\partial \rho}{\partial p_{V_1}} + \dot{\eta} \frac{\partial \rho}{\partial \eta}
$$

$$
+ \rho \left[\sum_{\alpha \in i,\, i=1,\, M} \left(\frac{\partial \dot{\mathbf{r}}_\alpha}{\partial \mathbf{r}_\alpha} + \frac{\partial \dot{\mathbf{p}}_\alpha}{\partial \mathbf{p}_\alpha} \right) + \sum_{i=1,\, M} \frac{\partial \dot{\xi}_i}{\partial \xi_i} + \frac{\partial \dot{V}_1}{\partial V_1} + \frac{\partial \dot{p}_{V_1}}{\partial p_{V_1}} + \frac{\partial \dot{\eta}}{\partial \eta} \right] = 0 \qquad (5)
$$

By direct substitution and using Eqs. (4) to evaluate partial derivatives, one can see that the following density function does indeed satisfy the Liouville equation with $\partial \rho / \partial t = 0$. The solution is given by the following expression

$$
\rho_{GEMD}\left(\{\mathbf{r}_\alpha\}, \{\mathbf{p}_\alpha\}, \{\xi_i\}, \{p_{\xi_i}\}, V_1, p_{V_1}, \eta \right)
$$

$$
\propto \exp\left[-\frac{1}{k_B T} \left(U_{total}(\{\mathbf{r}_\alpha\}, \{\xi_i\}, V_1) \right. \right.
$$

$$
\left. \left. + \sum_{\alpha \in i,\, i=1,\, M} \frac{\mathbf{p}_\alpha^2}{2m_\alpha} + \sum_{i=1,\, M} \frac{p_\xi^2}{2m_{\xi_i}} + \frac{p_{V_1}^2}{2Q_p} + \frac{Q_T \eta^2}{2} \right) \right] \qquad (6)
$$

Here, m_{ξ_i} is a mass assigned to the particle for motion along the virtual transfer direction ξ_i. Provided the system is ergodic, the averages over trajectories obtained by integrating Eqs. (4) are equivalent to the averages calculated with the distribution function ρ_{GEMD} (6). Below we show that, provided the number of particles in the transition state is small, averages for particles in each box do in fact correspond to the ensemble with constant temperature, pressure, and chemical potential, and that in fact the pressures and chemical potentials are equal in both boxes.

The GEMD trajectory average of a property A, which depends on the coordinates and momenta of the atoms $1, \ldots, n$, constituting the

---

*See corresponding section in Chapter 1.

molecules $1, \ldots, m$ residing in the first box, out of the total N, given by

$$
\langle A \rangle_{GEMD} = \frac{1}{Q'_{GEMD}} \int d\xi^M d\mathbf{r}^N d\mathbf{p}^N dV_1 A(\{\mathbf{r}_1, \ldots, \mathbf{r}_n\}, \{\mathbf{p}_1, \ldots, \mathbf{p}_n\})
$$

$$
\times \exp\left[-\frac{1}{k_B T} \left(U_{total}(\{\mathbf{r}_\alpha\}, \{\xi_i\}, V_1) + \sum_{\alpha \in i, i=1, M} \frac{\mathbf{p}_\alpha^2}{2m_\alpha} \right) \right] \quad (7)
$$

$$
Q'_{GEMD} = \int d\xi^M d\mathbf{r}^N d\mathbf{p}^N dV_1
$$

$$
\times \exp\left[-\frac{1}{k_B T} \left(U_{total}(\{\mathbf{r}_\alpha\}, \{\xi_i\}, V_1) + \sum_{\alpha \in i, i=1, M} \frac{\mathbf{p}_\alpha^2}{2m_\alpha} \right) \right]
$$

Here we already integrated over η, p_{ξ_i}, p_{V_1} and we have canceled the respective factors in the average and in the partition function, as indicated by the primed Q'_{GEMD}. By choosing a proper $g(\xi_i)$ one can make the number of particles in the transition state negligibly small. In this case we can assume that ξ_i takes only two values 0 or 1, and therefore we can replace the integration over ξ_i by the summation over all possible combinations of the ξ_i values (0 or 1). The equation for the average becomes:

$$
\langle A \rangle_{GEMD} = \frac{1}{Q'_{GEMD}} \sum_{m=0}^{M} \frac{M!}{m!(M-m)!} \int d\mathbf{r}^n d\mathbf{p}^n dV_1 A(\{\mathbf{r}_1, \ldots, \mathbf{r}_n\}, \{\mathbf{p}_1, \ldots, \mathbf{p}_n\})
$$

$$
\times \exp\left[-\frac{1}{k_B T} \left(U_{total}(\{\mathbf{r}_\alpha\}, V_1) + \sum_{\alpha \in i, i=1}^{n} \frac{\mathbf{p}_\alpha^2}{2m_\alpha} \right) \right]
$$

$$
\times \int d\mathbf{r}^{(N-n)} d\mathbf{p}^{(N-n)} \exp\left[-\frac{1}{k_B T} \left(U_{total}(\{\mathbf{r}_\alpha\}, V_2) + \sum_{\alpha \in i, i=m+1}^{M} \frac{\mathbf{p}_\alpha^2}{2m_\alpha} \right) \right]
$$

$$
(8)
$$

Q'_{GEMD} transforms analogously. $M!/(m!(M-m)!)$ accounts for particle distinguishability. Here we also write the potential energies in the two boxes as $U_{total}(\{\mathbf{r}_\alpha\}, V_1)$ and $U_{total}(\{\mathbf{r}_\alpha\}, V_2)$, independent of the ξ_i, because we neglect particles in the "transfer" state with $0 < \xi_i < 1$, assuming ξ_i can only be equal to 0 or 1. In this case the total potential energy is a sum of potential energies due to the interactions inside each of the boxes, as seen from Eq. (2).

Now we are going to demonstrate, that the chemical potentials and the pressures are indeed the same in the two simulation boxes. The second integration over the coordinates and momenta of the $N-n$ atoms in the second box equals $Q_{M-m, V_2, T} h^{3(M-m)}(M-m)!$, where $Q_{M-m, V_2, T}$ is the partition function of the isochoric–isothermal $(M - m, V_2, T)$ ensemble. It can be expressed in terms of the partition function for the $(M - m, P_2, T)$ ensemble with constant pressure and temperature for the second box using the saddle point method (see [5,7] for details)

$$Q_{M-m, V_2, T} = Q_{M-m, P_2, T} \exp\left(\frac{P_2 V_2}{k_B T}\right) \frac{C}{(N - n)^{1/2}} \left(1 + O\left(\frac{1}{M - m}\right)\right) \quad (9)$$

Here C is a constant independent of m and M. Now we rewrite $Q_{M-m, P_2, T}$ as [6]

$$\exp\left(\frac{\mu_2 n}{k_B T}\right) = \frac{Q_{M-m, P_2, T}}{Q_{M, P_2, T}} \quad (10)$$

where μ_2 is the chemical potential in the second box. Substituting all these in the average $\langle A \rangle_{GEMD}$ (8), and canceling terms independent of n between numerator and denominator, we obtain in the thermodynamic limit:

$$\langle A \rangle_{GEMD} = \frac{1}{Q''_{GEMD}} \sum_{m=0}^{M} \frac{1}{m! h^{3M}} \exp\left(\frac{\mu_2 m}{k_B T}\right)$$

$$\times \int d\mathbf{r}^n d\mathbf{p}^n dV_1 A(\{\mathbf{r}_1, \ldots, \mathbf{r}_n\}, \{\mathbf{p}_1, \ldots, \mathbf{p}_n\})$$

$$\times \exp\left(-\frac{P_2 V_1}{k_B T}\right) \exp\left[-\frac{1}{k_B T} \left(U_{total}(\{\mathbf{r}_i\}, V_1) + \sum_{\alpha \in i, i=1}^{m} \frac{\mathbf{p}_\alpha^2}{2m_\alpha}\right)\right] \quad (11)$$

and the corresponding expression for Q''_{GEMD}, where the double prime is a reminder that the common factors are canceled and that we keep only the leading contribution in the limit of a large number of particles. The above Eq. (11) coincides with the average over the constant- (μ, P, T) ensemble [8]. Notice that the average for the first box is calculated with the pressure and chemical potential values of the second box. This proves that these values are indeed the same in both boxes.

III. IMPLEMENTATION

Two boxes in the GEMD are implemented in the following way: Cartesian coordinates of the particles do not change when the particle is transferred from one box to another. Its presence in either box is determined by the value of its "ghost" coordinate ξ_i. Thus the transfer occurs as a particle gradually entering the box from the additional space dimension, pretty much as teleportation in a science fiction novel. The other particles in the entrance box "feel" the newcomer as its ξ_i value becomes close to 0, if it enters box one, or as ξ_i approaches 1 if the "ghost" appears in box two.

Molecules are transferred as a whole, with all atoms in the molecules having the same value of ξ. Only nonbonded interactions depend on the box sizes, due to the periodic boundary conditions. Thus the nonbonded interactions depend on the "fraction" of the molecule present in box one and its fraction in box two, as determined by the ξ_i.

Given a constant temperature and pressure MD code, it is not very difficult to modify it for GEMD. Experience also shows that the same results are obtained both with the Nosé–Hoover constant temperature MD algorithm, which exactly reproduces the (NVT) ensemble [2,3], and with the approximate but numerically more stable weak coupling approach [9].

Additional variables to store energies, pressure, and other values for two different boxes are to be added. Additional arrays storing values of variables ξ_i, velocities, forces, and "masses" are to be added.

```
double precision x(N),y(N),z(N),xi(N)
double precision vx(N),vy(N),vz(N),vxi(N)
double precision fx(N),fy(N),fz(N),fxi(N)
double precision mass(N),mass_xi(N)
double precision boxx_1,boxy_1,boxz_1,invbx_1,invby_1,
   invbz_1
double precision boxx_2,boxy_2,boxz_2,invbx_2,invby_2,
   invbz_2
double precision volume_1, volume_2
double precision virial_1, virial_2
double precision pot_energy_1, pot_energy_2,
   kin_energy_1, kin_energy_2
double precision kin_energy_xi
.....................

volume_1=boxx_1*boxy_1*boxz_1
volume_2=boxx_2*boxy_2*boxz_2
invbx_1=1.d0/boxx_1
invby_1=1.d0/boxy_1
invbz_1=1.d0/boxz_1
invbx_2=1.d0/boxx_2
invby_2=1.d0/boxy_2
invbz_2=1.d0/boxz_2
..........................................
```

The nonbonded potential energy, force, and virial calculations are to be amended, incorporating Eq. (2). The code example given here is probably far from being the fastest possible way to calculate nonbonded interactions. Even though we assume that the potential is pairwise and spherically symmetric, we use function calls to calculate energy, force, and virial for the purpose of clarity. The actual implementation should be in-lined and optimized for the interatomic potential used. See for instance [10] and other chapters in this book. Calculation of the forces for the ξ_i coordinates should also be added. Here, again we use a function call to add the contribution due to the $g(\xi_i)$, which would be more efficient to be implemented in-line.

```
      do i=1,N-1
        do j=i+1,N
  c for box 1
        xi_ij = xi(i)* xi(j)
        dx = x(i)-x(j)
        dxx = dx - boxx_1*anint(dx*invbx_1)
        dy = y(i)-y(j)
        dyy = dy - boxy_1*anint(dy*invby_1)
        dz = z(i)-z(j)
        dzz = dz - boxz_1*anint(dz*invbz_1)
        r = dxx*dxx + dyy*dyy + dzz*dzz
        pot_ij_1 = potential(r)* xi_ij
        pot_energy_1 = pot_energy_1 + pot_ij_1
        force_abs = potential_deriv(r)*xi_ij
        fx = force_abs*dxx/r
        fy = force_abs*dyy/r
        fz = force_abs*dzz/r

        fx(i) = fx(i) + fx
        fy(i) = fy(i) + fy
        fz(i) = fz(i) + fz

        fx(j) = fx(i) - fx
        fy(j) = fy(i) - fy
        fz(j) = fz(i) - fz

        virial_1 = virial_1 + fx*dxx+fy*dyy+fz*dzz
  c again for box 2
        xi_ij = (1.d0-xi(i)) * (1.d0-xi(j))
        dxx = dx - boxx_2*anint(dx*invbx_2)
        dyy = dy - boxy_2*anint(dy*invby_2)
        dzz = dz - boxz_2*anint(dz*invbz_2)
        r = dxx*dxx + dyy*dyy + dzz*dzz
        pot_ij_2 = potential(r)*xi_ij
        pot_energy_2 = pot_energy_2 + pot_ij_2
```

```
      force_abs = potential_deriv(r)*xi_ij
      fx = force_abs*dxx/r
      fy = force_abs*dyy/r
      fz = force_abs*dzz/r

      fx(i) = fx(i) + fx
      fy(i) = fy(i) + fy
      fz(i) = fz(i) + fz

      fx(j) = fx(i) - fx
      fy(j) = fy(i) - fy
      fz(j) = fz(i) - fz

      virial_2 = virial_2 + fx*dxx+fy*dyy+fz*dzz
c forces along the "ghost" direction xi
      fxi(i) = fxi(i) - ( pot_energy_1 * xi(i) - pot_energy_2 *
               (1.d0-xi(j)) )
      fxi(j) = fxi(j) - ( pot_energy_1 * xi(j) - pot_energy_2 *
               (1.d0-xi(i)) )
    enddo
    enddo
c additional transfer potential for fxi
        do i=1,N
          fxi(i) = fxi(i) + deriv_g(xi(i))
        enddo
      virial_1 = virial_1/3.
      virial_2 = virial_2/3.
....................
```

The equations describing the evolution of ξ_i are to be added to the routine, calculating next time step coordinates and velocities. The same numerical integration algorithms used to integrate Cartesian coordinates in principle can be used for ξ_i. However, some care is required at the interval boundaries, where the transfer potential $g(\xi_i)$ is discontinuous [5]. Also M additional degrees of freedom are to be accounted for, and the corresponding kinetic energy is to be included, when the temperature is calculated.

The above example of the GEMD code is suitable for a system of atoms, not molecules. In the molecular system, all atoms of the same molecule have the same ξ value. Thus even though the forces driving molecule transfer are calculated for each individual atom, one has to make sure that the molecules are transferred as whole objects. This can be achieved as follows, given a working *NPT*-ensemble code for the molecular system:

1.   The same changes as described above for the atomic system are to be made.

2. All atoms in the same molecule are assigned the same "transfer mass" m_{ξ_i}.
3. After the forces fxi along the ξ_i direction are calculated for all atoms, they are redistributed evenly between the atoms of the same molecule, so that the total force along the ξ_i direction acting on the molecule remains the same. Here is how it looks in a pseudo-code:

```
for all molecules
  ftotal=0
  for all atoms in the molecule
    ftotal = ftotal+fxi(atom)
  end
  ftotal = ftotal/num_of_atoms
  for all atoms in the molecule
    fxi(atom)=ftotal
  end
end
```

4. ξ coordinate updates for the next time step, similar to those for Cartesian coordinates are to be added.

And of course statistic accumulation parts of the code have to be changed accordingly. Energy, pair radial distribution function (PRDF), or diffusion coefficient should be calculated separately for molecules in each of the boxes.

When the phase diagrams are calculated, it is important to correctly account for the long-range nature of the interaction potential. Even the nonelectrostatic interactions, usually believed to be relatively short-ranged, have to be corrected. The corrections should be applied to the pressure and to the energy. They take into account the interactions of the atoms that are farther apart than the cutoff distance. Usually at these distances the center of mass pair correlation function equals one, and the corrections are easily calculated using equations for the energy and the pressure in terms of the interaction potential alone (see equations in Section I.C.4 in Chapter 1). For the spherically symmetric pairwise interaction described by $\Phi(r)$ the long-range corrections can be calculated as:

$$E = U_{total} + \frac{1}{2} N\rho \int_{r_{cut}}^{\infty} \Phi(r) 4\pi r^2 dr \tag{12}$$

$$PV = P_{cut} V - \frac{1}{6} N\rho \int_{r_{cut}}^{\infty} r \frac{d\Phi(r)}{dr} 4\pi r^2 dr$$

Both corrections depend on the density, and they may be quite different for the gas and liquid phases. Therefore, they strongly affect the position of the critical point. This is illustrated in Figs. 1 and 2.

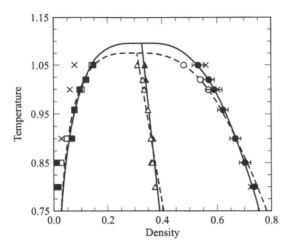

FIG. 1 Liquid–gas coexistence densities for the Lennard-Jones (LJ) fluid. Temperature and density are in reduced LJ units. Filled symbols: GEMD, open symbols: GEMC [11]. Reproduced from [5]. Note that the LJ potential in this case was cut at $2.5\,\sigma$ and shifted to zero. Crosses represent points obtained with longer LJ potential cutoff and applying the long-range correction. The critical point of the normal LJ system is close to $T_C^* = 1.35$ and $\rho_C^* = 0.307$.

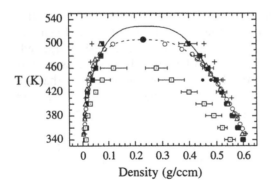

FIG. 2 Liquid–gas coexistence curve for hexane. Open circles: experimental data of reference [12]; open triangles: Gibbs Ensemble Monte Carlo result with 13.8 Å cutoff with the long-range corrections from [13]. The large solid circle corresponds to the experimental critical point. Open squares: GEMD using a cut and shifted LJ potential, where the cutoff is at 10 Å; solid squares: using a 15 Å cutoff instead; small solid circles at $T = 440$K: liquid densities for a 12 Å and 14 Å cutoff, respectively; pluses: result obtained for a 10 Å cutoff including long-range corrections. As in Fig. 1, the error bars indicate standard deviations. This is figure 3 taken from reference [5].

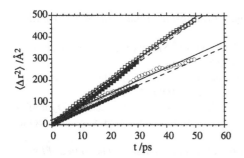

FIG. 3 Mean squared displacements $\langle \Delta r^2 \rangle$ of the hexane carbon atoms vs. time t in n-hexane calculated at coexistence in the liquid via GEMD (open symbols) and via NVT-MD (solid symbols) under corresponding conditions for $T = 400K$ (circles) and $T = 460K$ (squares). The lines (solid: GEMD; dashed: NVT-MD) are fits to the linear range of $\langle \Delta r^2 \rangle$. This is figure 6 taken from Reference [5].

IV. EXAMPLES

GEMD has been used to calculate phase diagrams for atomic as well as molecular systems, such as Lennard-Jones spheres or n-hexane, as shown in Figs. 1 and 2.

Being a molecular dynamics technique, GEMD provides dynamical information such as diffusion coefficients in the coexisting phases [5]. Shown in Fig. 3 are the mean squared displacements of the carbon atoms in n-hexane, calculated in coexisting phases with GEMD, compared to the results of the constant temperature molecular dynamics at the same conditions. This shows that diffusion coefficients and other dynamical information can be extracted from GEMD simulations, together with the thermodynamic properties.

The Gibbs Ensemble Monte Carlo technique transfers particles between the boxes in one step—as a whole. To be accepted, this insertion requires a reasonable probability to find an empty space. In the dense phases this is a problem. GEMD transfers particles continuously. Transferred atoms gradually "appear" and "disappear" from the simulation box. This may be a great advantage when modeling liquid–liquid or liquid–solid phase equilibria. More examples of GEMD applications and technical details can be found in the original papers describing applications of GEMD to the study of n-alkanes [5] and sorption in zeolites [14].

REFERENCES

1. Panagiotopoulos, A.Z. Direct Determination of Fluid Phase Equilibria by Simulating the Gibbs Ensemble: A Review. Mol. Sim. **1992**, *9*, 1–23.

2. Hoover, W.G. Canonical Dynamics: Equilibrium Phase Space Distribution. Phys. Rev. **1985**, *A31*, 1695–1697.

3. Nose, S. A Molecular Dynamics Method for Simulations in the Canonical Ensemble. Mol. Phys. **1984**, *52*, 255–268.

4. Winkler, R.G.; Moravitz, H.; Yoon, D.Y. Novel Molecular Dynamics Simulations at Constant Pressure. Mol. Phys. **1992**, *75*, 669–688.

5. Kotelyanskii, M.J.; Hentschke, R. Gibbs-Ensemble Molecular Dynamics: Liquid-Gas Equilibria for Lennard-Jones Spheres and *n*-Hexane. Molecular Simulations **1996**, *17*, 95–112.

6. Landau, L.D.; Lifshitz, E.M. Statistical Physics. *Course of Theoretical Physics*, 3rd Ed.; Pergamon Press: Oxford, Vol. 5, 1980.

7. Kotelyanskii, M.J.; Hentschke, R. Gibbs-Ensemble Molecular Dynamics: Liquid-Gas Equilibrium in a Lennard-Jones System. Phys. Rev. E **1995**, *51*, 5116–5119.

8. Hill, T.L. *Statistical Mechanics*; Dover: New York, 1987.

9. Berendsen, H.J.C.; Postma, J.P.M.; van Gunsteren, W.F.; Di Nola, A.; Haak, J.R. Molecular Dynamics with Coupling to an External Bath. J. Chem. Phys. **1984**, *81*, 3684–3690.

10. Allen, M.P.; Tildesley, D.J. *Computer Simulation of Liquids*; Clarendon Press: Oxford, 1993.

11. Smit, B. Phase Diagrams of Lennard-Jones Fluids. J. Chem. Phys. **1992**, *96*, 8639–8640.

12. Smith, B.D.; Srivastava, R. *Thermodynamic Data for Pure Compounds: Hydrocarbons and Ketones*; Elsevier: Amsterdam, 1986.

13. Siepmann, S.; Karaborni, J.I.; Smit, B. Simulating the Critical Behavior of Complex Fluids. Nature **1993**, *365*, 330–332.

14. Hentschke, R.; Bast, T.; Aydt, E.; Kotelyanskii, M. Gibbs-Ensemble Molecular Dynamics: A New Method for Simulations Involving Particle Exchange. J. Mol. Mod. **1996**, *2*, 319–326.

11

Modeling Polymer Crystals

GREGORY C. RUTLEDGE   Massachusetts Institute of Technology, Cambridge, Massachusetts, U.S.A.

I. INTRODUCTION

Crystallizable polymers constitute the majority of plastics in current use. Nevertheless, the study of polymer crystals by molecular simulations requires concepts and techniques somewhat different from those invoked by conventional views of polymer melts and glasses. Amorphous polymers are liquid-like in their structure; they exhibit only short range order. Amorphous melts are liquid-like in their dynamics as well. The methods employed to study amorphous polymers thus rely heavily on the Monte Carlo and molecular dynamics methods developed in the 1950s and 1960s to study simple liquids, and on refinements which permit efficient sampling of the vast conformation space available to a polymer "coil" in the amorphous state. Polymer crystals, on the other hand, exhibit long range order similar to that found in small molecule solids; they are characterized by high density, limited conformation space, and pronounced anisotropy. The study of crystalline polymers benefits considerably from the solid-state physics developed in the first half of the twentieth century. However, polymer crystals differ in several important respects from simple atomic crystals, and the modern simulation of polymer crystals draws upon ideas from both solid-state and fluid-state physics.

II. STRUCTURE OF POLYMER CRYSTALS

As with any material, the molecular simulation of a crystalline solid requires the specification of the nuclear coordinates of all the atoms in the solid, or a reasonable approximation thereof, and a force field to quantify the interaction between all the atoms in the solid. The simplification of defining a material structure in terms of the coordinates of the nuclei of the constituent

359

atoms follows from the Born–Oppenheimer approximation, which separates the full quantum mechanical problem, based on relative mass and velocity of the subatomic particles, into an electronic part and a nuclear part. The electronic problem consists of describing the electronic state of the solid in the field of stationary nuclei. The nuclear problem consists of describing the state of the nuclei in the potential field created by the rapidly equilibrated electrons. In atomistic simulation, attention is usually on the nuclear problem, which is used to specify the structure of the material, while the electronic energy is represented approximately by an analytical force field. In amorphous materials, it often suffices to specify structure in terms of pair correlations over short distances. Crystalline solids, on the other hand, possess both short range and long range order; in the ideal crystal, the long range order is limited only by the extent of the crystal itself. Long range order may persist in a material in one or two dimensions, as in liquid crystals and some rotator phases, or in three dimensions ("true" crystals). Just as in atomic crystals, polymer crystals exhibit a periodic structure, the "lattice," which describes how unit cells repeat in space. The literature on point and space group symmetries in crystals and the role of the unit cell is extensive and dates back to the early development of X-ray diffraction methods for solids. A complete review is beyond the scope of this chapter, but the interested reader is referred to several excellent introductory texts [1–3].

The atomic structure of the crystalline solid may be described completely by the extent or size of the crystal, the geometry of the unit cell, and the coordinates of the atoms in one unit cell (see Fig. 1). The unit cell may be of arbitrary shape and size; the periodicity of the lattice is defined by the 3×3 matrix $\mathbf{h}$ whose columns are the three vectors $\mathbf{a}$, $\mathbf{b}$, and $\mathbf{c}$ defining the edges of the unit cell. The entire solid is composed of such unit cells, and the extent of the crystal along the directions specified by $\mathbf{a}$, $\mathbf{b}$, and $\mathbf{c}$ is given by the indices h, k, and l, which take integer values between $-\infty$ and $+\infty$. Within each unit cell, the coordinates of atom j are often expressed as fractions of $\mathbf{a}$, $\mathbf{b}$, and $\mathbf{c}$, respectively, which are collected into the vector $\mathbf{s}_j$, the fractional coordinate of atom j. The coordinate of any atom i in the crystal, which we designate generally as $\mathbf{q}_i$, may be expressed as the image coordinate of the jth atom in the Lth unit cell:

$$\mathbf{q}_i = \mathbf{q}_{j,L} = \mathbf{h}(\mathbf{s}_j + \mathbf{L}) \tag{1}$$

where

$$\mathbf{h} = [\mathbf{a}, \mathbf{b}, \mathbf{c}] \quad \text{and} \quad \mathbf{L}^t = [h, k, l]$$

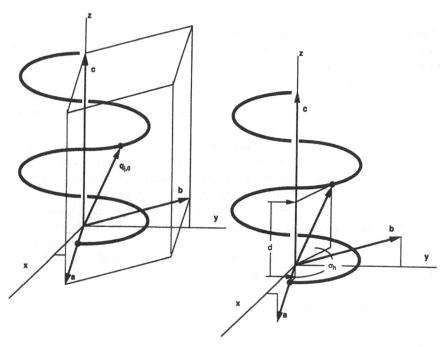

FIG. 1 Definition of crystallographic parameters. The points designate equivalent locations in successive helical repeat units. **a**, **b**, and **c** are the unit cell vectors; $q_{j,0}$ is the coordinate of the jth atom in the $\mathbf{L} = \mathbf{0}$ unit cell; σ and d are the helix parameters.

The superscript t designates the vector transpose. The perfect periodicity of the crystal lattice exists only in an average sense. Dynamically, each atom oscillates about its position in the lattice so that at any instant the periodicity of the crystal may be broken. However, the persistence of long range order implies that the forces on the atoms tend to restore each atom to its lattice site. The relevant phase space (i.e., the set of positions and momenta accessed by the atoms) of crystals, therefore, is considerably smaller in comparison to fluid phases. In the crystal, most or all of the atoms are localized about well-defined points in configuration space. Long range motion of atoms in crystals may be quite significant, but usually consists of discrete, activated jumps between what are essentially equivalent states of the crystal. In some cases, atomic motion is more continuous, and description of the solid as a disordered crystal may be more appropriate (see below).

Polymer crystals exhibit several characteristics not typically encountered elsewhere in simulations. First, the unit cells of polymers tend to consist of tens of atoms, while the molecules themselves comprise 10^4 to 10^6 atoms.

Rather than having a situation where a unit cell consists of one or more molecules, as in small molecule crystals, the situation is reversed: a single molecule participates in many unit cells. This has significant ramifications for the correlations, both structural and dynamical, between unit cells. Second, the periodicity of the crystal lattice usually implies a periodicity for the conformation of the polymer chain itself. This condition places a severe restriction on the conformation space available to the chain in a crystalline solid. Polymer chains in crystals are more appropriately described as "helices" rather than "coils." A helix conformation consists of a regular repetition of the torsion angles characteristic of a small subsection of the chain, the helix repeat unit [4,5]. Each helix has an axis and a handedness of rotation (the direction in which the chain backbone winds around the helix axis). The helix itself is characterized by the translational displacement parallel to the helix axis, d, and the angle of rotation about the helix axis, σ, executed by a single helix repeat unit. The coordinates of Eq. (1) may be expressed in terms of these helical parameters as follows:

$$\mathbf{q}_{h, L_h, n} = \mathbf{R}_{h, n}\mathbf{hs}_h + \mathbf{hL}_h + \mathbf{T}_n \tag{2}$$

where

$$\mathbf{R}_{h, n} = \begin{bmatrix} \cos(n\sigma + \omega_h) & -\sin(n\sigma + \omega_h) & 0 \\ \sin(n\sigma + \omega_h) & \cos(n\sigma + \omega_h) & 0 \\ 0 & 0 & 1 \end{bmatrix}$$
$$\mathbf{L}_h = [h', k', 0]^t \quad \text{and} \quad \mathbf{T}_n = [0, 0, nd]^t$$

Here, the subscript h indexes different helices in the unit cell. In this convention, the fractional coordinates for each unique helix in the unit cell are specified with the helix axis directed along the z-coordinate and passing through the origin. Each helix may have a distinct setting angle, ω_h, and h' and k' may take fractional values to indicate helices at nonprimitive lattice positions. The lattice index l is replaced by the helix repeat index, n.

Third, nearly all known polymer crystal polymorphs exhibit chain helices packed into the solid with axes parallel.* As a result, there exists a unique chain continuation direction, which gives rise to a highly anisotropic

---

*The γ phase of isotactic polypropylene presents a notable exception. In this crystal form, the polymer chains all have the same conformation, but are arranged in layers with an angle of about 80° between the axes of chains in successive layers [6].

material on the molecular scale. In the chain direction the crystal is held together by chemical bonds, while in the other two directions it is held together by weaker, chain-packing forces, such as van der Waals forces or hydrogen bonds. This combination of conformational periodicity, through-crystal chemical bonding and directionality of strong and weak interactions makes polymer crystals unique.

To specify a starting configuration for any polymer simulation, a reasonably good estimate of the unit cell geometry, and in many cases restrictions on the atomic coordinates s, may be obtained by wide angle X-ray diffraction. Estimates of the crystal size can be obtained by wide or small angle X-ray scattering, electron or optical microscopy. One of the unique features of crystalline polymers is the chain folded lamellar morphology, common among flexible polymers. These plate-like crystallites are the result of kinetic factors that dominate during crystallization itself. The thicknesses of such lamellae are also considerably less than the contour length of the chain molecules. The ramification of this length mismatch is that the surfaces of such lamellae involve reentry of the same molecule into the crystal, in the form of loops or folds. The thinness of the lamellae and the molecular connectivity of the loop or fold surfaces may have a significant effect on the dynamical behavior of real polymer crystals. Detailed atomistic studies of entire fold surfaces in polymer crystals are relatively few [7,8]. More recently, atomistic simulations of the crystal/amorphous interface of a semicrystalline polymer matrix have been presented that provide insight into the preferred topology for chain reentry along several relevant crystallographic facets [9–11]. The presence of disordered surfaces or interfaces may affect the relative stability of finite-sized crystallites. Nevertheless, the lowest energy forms of large polymer crystals are expected to be those with fully extended chains (i.e., no fold or loop surfaces), and it is these that have generally been the object of simulation.

Finally, a few comments about the uniqueness of polymer crystal structures and phase space localization are warranted. Almost all crystallizable polymers exhibit polymorphism, the ability to form different crystal structures as a result of changes in thermodynamic conditions (e.g., temperature or pressure) or process history (e.g., crystallization conditions) [12]. Two or more polymorphs of a given polymer result when their crystal structures are nearly iso-energetic, such that small changes in thermodynamic conditions or kinetic factors cause one or another, or both, to form. Polymorphism may arise as a result of competitive conformations of the chain, as in the case of syndiotactic polystyrene, or as a result of competitive packing modes of molecules with similar conformations, as in the case of isotactic polypropylene. In some instances, the conformational change may be quite subtle; isotactic polybutene, for example, exhibits three crystal polymorphs in which the number of constitutional repeat units per turn of the helix varies only from 3 to 4. Typically, the thermodynamic transition between polymorphs is discontinuous; each polymorph represents a distinctly different local region in the total phase space of the polymer. However, polymer crystals may also exhibit internal forms of disorder within a particular unit cell geometry. Common types of disorder include (i) the statistical, iso-energetic packing of helices of opposite handedness or axis direction within the unit cell, (ii) conformationally disordered chains which are not true helices, but nevertheless satisfy approximately the lattice periodicity along the chain direction, (iii) constitutional disordered chains composed of copolymers with irregular monomer sequences, and (iv) thermal disorder due to rapid interconversion between iso-energetic conformational or packing states. Examples of the last include free rotation of the pendant methyl group in isotactic polypropylene or the rotation of entire chains in polyethylene (i.e., rotator phases) at elevated temperatures. Algorithms must be evaluated carefully when dealing with different types of disorder.

III. COMPUTATIONAL METHODS

We distinguish between two classes of calculation or simulation appropriate for the study of polymer crystals: *optimization methods* and *sampling methods*. Optimization methods take explicit advantage of the localization of crystals in phase space; numerical optimization techniques are used to identify and characterize a specific region of phase space. Molecular mechanics and lattice dynamics fall in this group. Sampling methods generally do not take advantage of, and thus are not restricted by, assumptions of localization. Monte Carlo and molecular dynamics fall in this group. Each of these computational approaches is considered below. Since optimization methods, particularly lattice dynamics, are used almost exclusively for crystalline solids, these will be presented in some detail. Sampling methods are discussed to the extent that their use for simulation of polymer crystals requires some special considerations.

A. Optimization Methods

Consider an ideal case where the crystal exhibits none of the forms of disorder mentioned in the previous section. A given polymorph is then characterized by localization about an equilibrium point in phase space where the potential energy surface is concave. The relevant part of the energy surface can be approximated to arbitrary accuracy using a Taylor series expansion in small displacements about the local minimum energy structure, $\mathbf{q}_0$:

$$U(\{q_i\}) = U(\{q_{i0}\}) + \sum_{i=1}^{3N} \frac{\partial U}{\partial q_i}\bigg|_0 (q_i - q_{i0}) + \frac{1}{2}\sum_{i=1}^{3N}\sum_{j=1}^{3N} \frac{\partial^2 U}{\partial q_i \partial q_j}\bigg|_0 (q_i - q_{i0})(q_j - q_{j0})$$

$$+ \frac{1}{6}\sum_{i=1}^{3N}\sum_{j=1}^{3N}\sum_{k=1}^{3N} \frac{\partial^3 U}{\partial q_i \partial q_j \partial q_k}\bigg|_0 (q_i - q_{i0})(q_j - q_{j0})(q_k - q_{k0}) + \cdots$$

(3a)

or in tensor notation:

$$U(\mathbf{q}) = U(\mathbf{q}_0) + \Delta\mathbf{q}^t \cdot \nabla U_0 + \frac{1}{2}\Delta\mathbf{q}^t \cdot \nabla^2 U_0 \cdot \Delta\mathbf{q} + \cdots$$

$$= \sum_{i=0}^{\infty} \frac{(\Delta\mathbf{q}^t \cdot \nabla)^i}{i!} U_0$$

(3b)

Here, $\mathbf{q}$ represents the $3N$ vector of coordinates for a crystal composed of N atoms and $\Delta\mathbf{q} = \mathbf{q} - \mathbf{q}_0$ is the displacement from the local minimum. There are $3N-6$ molecular degrees of freedom in $\mathbf{s}$ and 6 lattice degrees of freedom in $\mathbf{h}$. In classical mechanics, in the limit that temperature T approaches 0, $\Delta\mathbf{q}$ also approaches 0. In this limit of small displacements, the energy and geometry of the crystal may be approximated well by $U(\mathbf{q}_0)$ and $\mathbf{q}_0$, respectively. The problem of computing the properties of a crystal phase simplifies in this case to identification of a single point in configuration space where the structure is in mechanical equilibrium, characterized by all the forces on the atoms, ∇U_0, equal to zero and all the second derivatives $\nabla\nabla U_0$ positive, indicative of a stable point on the potential energy surface. This kind of numerical optimization problem goes by the name of *molecular mechanics*. Standard techniques [13] such as steepest descent, conjugate gradient, or modified Newton methods may be used to find the minimum in $U(\mathbf{q})$ as a function of the degrees of freedom $\mathbf{q} = \{\mathbf{h}, \mathbf{L}, \mathbf{s}\}$. For large crystals, since every unit cell in the crystal is identical, it suffices to minimize $U(\mathbf{q})$ with respect to $\mathbf{q} = \{\mathbf{h}, \mathbf{s}\}$ with $\mathbf{L} = \mathbf{0}$ (i.e., the degrees of freedom of a single unit cell), provided that care is taken to evaluate $U(\mathbf{q})$ over a sufficiently large number of unit cells, using techniques such as Ewald summation [14]. It is important that the conditions of both zero gradient and positive second derivatives be met, in order to ensure that the structure is indeed a stable one, and not a saddle point (cf. Section IV). Taking the limit of zero displacement rules out any explicit thermal (e.g., entropy) or kinetic contributions to the structure and properties of the crystal. However, this does not imply that the calculation is necessarily representative of a crystal at 0K. Thermal contributions may creep into the calculation implicitly, through the parameterization of the force field or the choice of lattice parameters **h**.

The relatively simple molecular mechanics calculation has proven very useful in the study of polymer crystals. It is generally the method of first resort for any new crystal study. The primary result of such a calculation is structural information on a very detailed level, for purposes of gaining insight into the balance of forces or obtaining structure factors to compare with experiments. For very large problems, the calculations may be speeded up through the use of one or more common simplifications invoked to reduce the length of the vector **q**. First, the problem of determining the helical conformation of the chain may be decoupled from that of the intermolecular packing structure. The chain conformation having the lowest intramolecular energy is determined first using isolated chain calculations, appropriately constrained to ensure helicity. (This calculation can usually be accomplished simply by setting the lattice vectors **a** and **b** to be very large.) Subsequently, the energy of the crystal is minimized with respect to the intermolecular degrees of freedom and with the chains themselves treated as rigid molecules having only three translational and one rotational degree of freedom per chain [15–17]. However, such rigid chain calculations do not reflect the effects of strong intermolecular packing interactions on conformation. Packing interactions have been shown in some cases to distort the conformation of the chain [18,19] or even to stabilize a crystal structure containing a completely different and less stable conformation [20]. Second, the simultaneous optimization of intramolecular and intermolecular degrees of freedom may be retained at minimal cost by taking advantage of known symmetries within the unit cell, usually deduced from X-ray results [21,22]. In crystallography, the smallest group of atoms that contains no internal symmetry is called the asymmetric unit and, in many instances, comprises a fraction of the total number of atoms in the unit cell. In this way, the number of degrees of freedom is reduced to the coordinates of the asymmetric unit, the lattice vectors of the unit cell, and the symmetry operations defining the coordinates of the remaining atoms within the same unit cell.

Thermal contributions to the properties of the crystal are introduced by taking higher order terms in the Taylor series expansion [Eq. (3)] for the energy surface in the vicinity of the equilibrium structure. The classical theory of lattice dynamics was originally formulated for crystals by Born and von Kármán [23,24] and later presented in the comprehensive text of Born and Huang [25]. There are several excellent texts that discuss lattice dynamics in great detail [25–28]; here, we will only summarize the main points.

In the harmonic approximation, all contributions to the energy beyond second order in atomic displacements are neglected. Since the gradient ∇U_0 evaluated at equilibrium is zero, this yields:

$$\Delta U(\mathbf{q}) = U(\mathbf{q}) - U(\mathbf{q}_0) \cong \frac{1}{2} \Delta \mathbf{q}^t \cdot \nabla\nabla U_0 \cdot \Delta \mathbf{q}$$

$$= \frac{1}{2} \sum_{i=1}^{3N} \sum_{j=1}^{3N} \frac{\partial^2 U}{\partial q \partial q}\bigg|_0 \Delta q_i \Delta q_j \tag{4}$$

Differentiating both sides of Eq. (4) with respect to Δq_i, one obtains an equation for the force $\partial \Delta U / \partial \Delta q_i$ acting on coordinate q_i, and hence the equations of motion describing oscillatory displacement of the atoms about their equilibrium positions:

$$m_i \frac{d^2(\Delta q_i(t))}{dt^2} = -\sum_{j=1}^{3N} \frac{\partial^2 U}{\partial q_i \partial q_j}\bigg|_0 \Delta q_j(t) \tag{5}$$

The periodicity of the crystal permits description of the displacement of atoms in the form of traveling waves with wavevector $\mathbf{k}$. Solution for the oscillatory displacement of coordinate i takes the form:

$$\Delta q_i(t) = u_i \exp\left[i(\mathbf{k} \cdot \mathbf{q}_j - \omega(\mathbf{k})t)\right] \tag{6}$$

where u_i is the amplitude of displacement of coordinate i and $\omega(\mathbf{k})$ is the frequency of oscillation. There are $3N$ such solutions. The harmonic motion of each atom is, in general, coupled with the motions of the other atoms in the crystal. However, there exists a particular set of linear combinations of these solutions, called normal modes, which are independent of each other. Substitution of Eq. (6) into Eq. (5) yields:

$$m_i u_i \omega^2 = \sum_{j=1}^{3N} \frac{\partial^2 U}{\partial q_i \partial q_j}\bigg|_0 \exp\left[-i\mathbf{k} \cdot (\mathbf{q}_j - \mathbf{q}_i)\right] u_j \tag{7}$$

This equation can now be expressed in terms of the displacements of the N_u atoms of the unit cell, rather than the total number of atoms of the entire crystal. In matrix form, this becomes:

$$(\mathbf{D} - \omega^2\mathbf{I})\mathbf{u}' = 0 \tag{8}$$

where the vector $\mathbf{u}'$ consists of the $3N_u$ mass-weighted elements of one unit cell, $u'_i = u_i\sqrt{m_i}$, and $\mathbf{D(k)}$ is the $3N_u \times 3N_u$ mass-weighted dynamical matrix, whose elements are defined as:

$$D_{ij}(\mathbf{k}) = \frac{\sum_\mathbf{L} \left.\frac{\partial^2 U}{\partial q_i \partial q_j}\right|_0 \exp[i\mathbf{k} \cdot \mathbf{h}(\mathbf{s}_j - \mathbf{s}_i + \mathbf{L})]}{\sqrt{m_i m_j}} \tag{9}$$

The sum over $\mathbf{L}$ corresponds to a triple summation over all values of h, k, and l in the crystal. The second derivatives of potential energy with respect to the components of $\mathbf{q}$ are obtained from the converged Hessian matrix at the potential energy minimum. The roots of Eq. (8) yield the frequencies $\omega_i(\mathbf{k})$ for the $3N_u$ independent 1-dimensional oscillators, or normal modes of vibration. Associated with each frequency is a polarization vector $\boldsymbol{\xi}_i(\mathbf{k})$ which is a linear combination of the original mass-weighted atomic displacements $\mathbf{u}'$. The polarization vectors are orthonormal, such that $\xi'_i\xi^*_j = \delta_{ij}$ (here, the superscript * denotes the complex conjugate). These are found by diagonalizing the matrix $\mathbf{D(k)}$:

$$\boldsymbol{\xi}^{-1}\mathbf{D}\boldsymbol{\xi} = \boldsymbol{\Lambda} \tag{10}$$

$\boldsymbol{\Lambda}(\mathbf{k})$ is the diagonal matrix whose elements are the $3N_u$ values $\omega_i^2(\mathbf{k})$. $\mathbf{D(k)}$ is Hermitian, i.e., $\mathbf{D} = (\mathbf{D}^*)'$, so values of ω_i^2 are always real and positive for solution about a minimum in energy. It is apparent that the set of frequencies $\omega_i(\mathbf{k})$ is a function of the wavevector, $\mathbf{k}$. However, the periodicity of the lattice reduces the set of unique wavevectors to those located in the first Brillouin Zone (BZ), the region of reciprocal space whose boundaries lie halfway between reciprocal lattice points, eg., $-\pi/a < k \leq \pi/a$ along the $\mathbf{a}^*$ dimension, and similarly for $\mathbf{b}^*$ and $\mathbf{c}^*$; $\mathbf{a}^* = (\mathbf{b} \times \mathbf{c})/V$, $\mathbf{b}^* = (\mathbf{c} \times \mathbf{a})/V$ and $\mathbf{c}^* = (\mathbf{a} \times \mathbf{b})/V$ are the reciprocal lattice vectors, while $V = (\mathbf{a} \cdot \mathbf{b} \times \mathbf{c})$ is the unit cell volume. The k-dependence of the $3N_u$ values $\omega_i(\mathbf{k})$ is called a dispersion curve, of which there are $3N_u$ "branches" in the BZ. Of these $3N_u$ branches, three are "acoustic" modes, which have the property that ω_i approaches zero as the magnitude of $\mathbf{k}$ approaches zero; the other $3N_u - 3$ branches are called "optic" modes. Furthermore, most polymer crystal structures have more than one chain participating in the unit cell. Each chain has four molecular degrees of freedom (three translational, and one rotational about the chain axis), giving rise to $4N_{chain} - 3$ "molecular" optic modes, which tend to be lower in frequency than the remaining $3N_u - 4N_{chain}$ "atomic" optic modes. For the "finite but unbounded" crystal having h_{max} unit cells along **a**, k_{max} unit cells along **b**, and l_{max} unit cells along **c**, and periodic boundary conditions, the wavevectors along each branch are restricted to discrete values: $\mathbf{k} = 2\pi((h\mathbf{a}^*/h_{max}) + (k\mathbf{b}^*/k_{max}) + (l\mathbf{c}^*/l_{max}))$, where h takes integer values running from $-h_{max}/2$ to $+h_{max}/2$, and similarly for k and l, for a total of $3N$ vibration frequencies in the finite crystal. In the limit that the crystal becomes infinitely large, the interval between allowed wavevectors goes to zero, and both $\omega_i(\mathbf{k})$ and $\xi_i(\mathbf{k})$ become continuous functions of $\mathbf{k}$.

Each of the normal modes in the harmonic approximation is a traveling wave in the crystal and represents a packet of energy; by analogy with the wave/photon duality of electromagnetic radiation, the wave packets in crystals are called phonons. The energy of each phonon can be quantified either classically or quantum mechanically. Since the normal modes are independent, the vibrational partition function for the crystal is simply the product of the individual partition functions for each phonon. These are known analytically for harmonic oscillators. The corresponding classical and quantum mechanical partition functions for the crystal are:

$$\text{classical}: \quad \ln Q_{cm}^{vib} = -\int_{BZ} d\mathbf{k} \sum_{i=1}^{3N_u} \ln\left(\frac{\hbar\omega_i(\mathbf{k})}{k_B T}\right) \tag{11a}$$

$$\text{quantum}: \quad \ln Q_{qm}^{vib} = -\int_{BZ} d\mathbf{k} \left[\sum_{i=1}^{3N_u} \frac{\hbar\omega_i(\mathbf{k})}{2k_B T} + \ln\left(1 - \exp\left(\frac{-\hbar\omega_i(\mathbf{k})}{k_B T}\right)\right)\right] \tag{11b}$$

The free energy is thus computed explicitly (N_k is the normalization factor $\int_{BZ} d\mathbf{k}$):

$$A(\mathbf{h}, T) = U(\mathbf{q}_0) + A_{vib}(\mathbf{q}_0, T) = U(\mathbf{q}_0) - \frac{k_B T}{N_k} \ln Q_{vib} \tag{12}$$

The size of crystal simulated in this fashion is determined by both the range of interactions included in the computation of the potential $U(\mathbf{q})$, and the resolution of wavevectors in the BZ. As mentioned earlier, for periodic systems like crystals, Ewald summation permits the efficient evaluation of interactions for a material of infinite extent. Evaluation of thermodynamic properties for an infinite crystal requires accurate evaluation of integrals involving the dispersion relations, $\omega_i(\mathbf{k})$. Standard quadrature methods such as Gauss–Legendre are used to get the highest accuracy with the fewest number of integration points. Generally, different densities of integration (mesh) points should be tried, to ensure convergence. Some quantities, such as the atomic displacement tensor discussed below, may be dominated by particular subregions of the BZ, and require special attention to precise evaluation in these regions.

In the study of polymer crystals, lattice dynamical calculations have been implemented in different ways. Straightforward minimization of the potential energy with respect to both fractional coordinates $\mathbf{s}$ and lattice parameters $\mathbf{h}$ simultaneously yields a single, temperature-independent structure. The resulting normal modes are thus also temperature-independent, giving rise to a strictly harmonic crystal [29–31]. In the quasiharmonic approximation (QHA), vibrational frequencies for any lattice $\mathbf{h}$ are equated with those of the harmonic crystal at the state of strain defined by $\mathbf{h}$. QHA thus allows for anharmonicities that arise due to changes in size or shape of the unit cell. Early implementations of lattice dynamics with QHA relied on experimental X-ray diffraction measurements for appropriate values of $\mathbf{h}$ at finite temperatures; the potential energy was then minimized with respect to internal degrees of freedom $\mathbf{s}$ at fixed external strain, $\varepsilon(\mathbf{h}, \mathbf{h}_0)$, referred to as the zero static internal stress approximation (ZSISA) [32]. Reviews of these types of calculations for numerous polymers are available [21,22]. A more predictive and self-consistent implementation of QHA lattice dynamics, within the ZSISA, is possible at little additional effort by minimizing the potential energy with respect to $\mathbf{s}$ at fixed $\mathbf{h}$; the vibrational contribution to free energy is then computed and the total free energy $A(\mathbf{h}, T)$ is minimized with respect to the external variables in $\mathbf{h}$, under the condition that potential energy is minimal with respect to $\mathbf{s}$ for each choice of $\mathbf{h}$. The relative contribution of A_{vib} to A varies with temperature, resulting in a shift towards lower frequencies, and hence larger $\mathbf{h}$ in most cases, at higher temperature; the contraction observed along the chain axis in polymer crystals which possess fully extended chains is an interesting exception. The shift in lattice frequencies for polyethylene between 0K and 300K is shown in Fig. 2. This shift gives rise to thermal expansion behavior. This self-consistent approach permits estimation of material properties based on the force field alone, without additional input from experimental data. It thus provides a powerful means for checking the validity of a force field. This self-consistent version of lattice dynamics has also been applied to numerous polymers [19,33–36]. Finally, it is possible within the QHA to circumvent the ZSISA, by formulating analytical derivatives of the normal mode frequencies with respect to both $\mathbf{s}$ and $\mathbf{h}$, and then minimizing the total free energy with respect to all degrees of freedom, both internal and external,

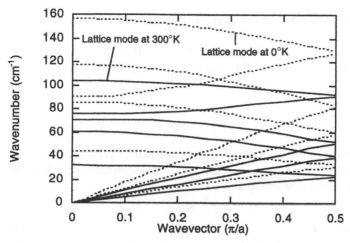

FIG. 2 $4N_{chain}$ lattice (acoustic and external optic) normal mode branches for polyethylene along the [100] lattice direction. Each branch is symmetric about the Γ ($\mathbf{k} = \mathbf{0}$) point. 0K (dashed lines) and 300K (solid lines).

simultaneously [37]. This can be important for quantities that depend sensitively on internal strains. To date, only properties for polyethylene-like crystals described by relatively simple potentials have been obtained by this method and evaluating the analytical derivatives at the static geometry [38].

Although we have not discussed the specific form of the potential energy function used to describe $U(\mathbf{q})$, other than to suggest that it is an effective potential described parametrically by the positions of atomic nuclei, it is common to treat electrostatic forces in atomistic calculations using partial atomic charges assigned to the various nuclei. However, this description for electrostatics does not account for distortion of the electron distribution (the electronic part of the Born–Oppenheimer separation) in the environment of the crystal. The presence of dipoles in the anisotropic crystal can lead to significant mutual induction, or polarization, effects. The lattice dynamical description presented up to this point can be easily extended to include mutual induction through use of the shell model for atomic polarizability [39]. This model introduces an additional $3N$ coordinates to $\mathbf{q}$, corresponding to the center of action of the electron distribution about each atom, which may polarize the atom by situating off-center from the nucleus itself. This results in an additional $3N$ normal modes, called plasmon modes, which are considerably higher in frequency than phonon modes. The shell model has been used to study poly(vinylidene fluoride) using the harmonic [40] and self-consistent quasiharmonic approximations [35,41].

The free energy of the crystal corresponding to other ensembles may be computed through inclusion of additional work terms in the equation for free energy, for example:

$$G(\boldsymbol{\sigma}, T) = U(\mathbf{h}) + A_{vib}(\mathbf{h}, T) - V_0 \boldsymbol{\sigma}^t \cdot \boldsymbol{\varepsilon} \tag{13}$$

For mechanical work, $\boldsymbol{\sigma}$ is the applied stress and $\boldsymbol{\varepsilon}$ is the deformation strain, defined with respect to a reference cell $\mathbf{h_0}$. For a cell deformed from $\mathbf{h_0}$ to $\mathbf{h}$, the (external) strain may be computed as:

$$\boldsymbol{\varepsilon} = \frac{1}{2} \left[\mathbf{h}_0^{t,-1} \mathbf{G} \mathbf{h}_0^{-1} - \mathbf{I} \right] \tag{14}$$

$\mathbf{G}$ is the metric tensor $\mathbf{h^t h}$ and V_0 is the cell volume. A wide range of thermal and mechanical properties may be estimated through optimization of $A(\mathbf{h}, T)$ or $G(\boldsymbol{\sigma}, T)$ at different temperatures and states of deformation. Among these, the constant volume heat capacity may be computed using:

$$C_V = \frac{k_B}{N_k} \int_{BZ} d\mathbf{k} \sum_{i=1}^{3N_a} x^2 e^x [e^x - 1]^{-2} \quad \text{where} \quad x = \frac{\hbar \omega_i(\mathbf{k})}{k_B T} \tag{15}$$

Thermal expansion coefficients may be computed as thermally induced strains:

$$\boldsymbol{\alpha} = \frac{1}{2} \left[\mathbf{h}_0^{t-1} \frac{d\mathbf{G}}{d\mathbf{T}} \mathbf{h}_0^{-1} \right] \tag{16}$$

Similarly, elastic stiffness and compliance constants can be computed using the relations

$$\mathbf{C} = \frac{\partial \boldsymbol{\sigma}}{\partial \boldsymbol{\varepsilon}} \quad \text{i.e.,} \quad C_{ij} = \frac{\partial \sigma_i}{\partial \varepsilon_j} \tag{17a}$$

and

$$\mathbf{S} = \mathbf{C}^{-1} = \frac{\partial \boldsymbol{\varepsilon}}{\partial \boldsymbol{\sigma}} \tag{17b}$$

or

$$\mathbf{C} = \frac{1}{V_0}\frac{\partial^2 A(\mathbf{h}, T)}{\partial \boldsymbol{\varepsilon} \partial \boldsymbol{\varepsilon}} \quad \text{and} \quad C_{ij} = \frac{1}{V_0}\frac{\partial^2 A(\mathbf{h}, T)}{\partial \varepsilon_i \partial \varepsilon_j} \tag{17c}$$

Here, $\boldsymbol{\sigma}$ and $\boldsymbol{\varepsilon}$ are the 6×1 vectors for stress and strain, respectively, in Voigt notation (i.e., $\varepsilon_i = \varepsilon_{ii}$ for $i = 1,2,3$; $\varepsilon_4 = 2\varepsilon_{23}$, $\varepsilon_5 = 2\varepsilon_{13}$, $\varepsilon_6 = 2\varepsilon_{12}$). Lattice Grüneisen coefficients offer a measure of the thermal stress on a crystal:

$$\gamma = \frac{V}{C_V}\mathbf{C}\boldsymbol{\alpha} \tag{18}$$

The anisotropy typical of the polymer crystal is exemplified by the material properties of several polymer crystals studied to date. These are shown in Table 1.

The elastic modulus of the crystal in the chain direction is one of the most important properties of a semicrystalline polymer and has received considerable theoretical attention. It also serves to illustrate some of the differences between various methods of calculation. Figure 3 illustrates the results for C_{33} for polyethylene from several different calculations: potential energy minimization, quasiharmonic lattice dynamics, Monte Carlo and molecular dynamics. The open symbols in Fig. 3 are all indicative of optimization-based calculations, while the filled symbols are derived from sampling-based methods (cf. Section III.B). For clarity, only the results from lattice dynamics using classical vibrations are shown; calculations on polyethylene using quantum mechanical vibrations typically yield lower values of C_{33} at low temperatures, but both classical and quantum mechanical estimates of elastic constants converge around 300K [33,45,46]. However, other properties, notably thermal expansion coefficients and heat capacity, have been shown to exhibit quantum mechanical effects well past 300K [33]; vibration modes with frequencies greater than $200\,\text{cm}^{-1}$, predominantly intramolecular vibrations, are highly nonclassical at room temperature. The anisotropic nature of the crystal precludes the kind of orientational averaging, present in amorphous simulations, which helps to mitigate the error associated with treating these high frequency vibrations classically. Thermal vibration accounts for about 12%, a 40 GPa drop, in the estimates of modulus at 300K. Above 300K, the lattice dynamics estimates fall off more rapidly. Inspection of Eq. (17c) reveals that the drop in modulus may be traced to three interrelated factors: (i) an increase in reference cell volume with temperature; (ii) a change in the potential energy surface with increasing volume

TABLE 1 Properties of Several Polymer Crystals Computed at 300K Using a Self-Consistent QHA Lattice Dynamics Calculation, with Quantum Mechanical Treatment of Vibrational Free Energy

	PE[a]	iPP[b]	PVDF[c]	PPTA[d]	PET[e]
C_V (kcal mol^{-1} K^{-1})[f]	0.0196	0.00703	0.0306	0.1145	0.0470
$\alpha_1 (\times 10^{-5}$ K$^{-1})$	28.3	8.2	6.56	7.9	11.4
$\alpha_2 (\times 10^{-5}$ K$^{-1})$	10.7	8.7	5.13	2.9	4.12
$\alpha_3 (\times 10^{-5}$ K$^{-1})$	−2.4	−0.84	−0.38	−0.57	−1.07
γ_1	1.2	0.68	0.79	0.76	1.13
γ_2	0.97	0.79	0.66	1.07	1.02
γ_3	−2.3	0.39	−0.040	−0.24	−0.69
C_{11} (GPa)	6.0	9.9	24.5	11.9	14.4
C_{22} (GPa)	7.3	10.8	26.1	32.2	17.3
C_{33} (GPa)	280	55.1	276	284	178
C_{44} (GPa)	2.7	4.0	na	14.3	6.6
C_{55} (GPa)	1.7	4.9	na	0.5	1.4
C_{66} (GPa)	2.9	1.7	na	6.1	1.2
C_{12} (GPa)	3.4	3.4	1.2	11.1	6.4
C_{13} (GPa)	8.7	7.1	7.6	10.5	3.4
C_{23} (GPa)	9.5	4.4	9.1	14.2	9.5

[a]PE (polyethylene), orthorhombic. *Source*: Ref. 33.
[b]iPP (isotactic polypropylene), α-phase, monoclinic (not all properties shown). *Source*: Ref. 34.
[c]PVDF [poly(vinylidene fluoride)], β-phase, orthorhombic. *Source*: Ref. 41.
[d]PPTA [poly(p-phenylene terephthalamide)], orthorhombic. *Source*: Ref. 19.
[e]PET [poly(ethylene terephthalate)], triclinic (not all properties shown). *Source*: Ref. 36.
[f]Per mole of unit cells.
na: not available.

to a surface of lower curvature; (iii) the reduction in vibrational frequencies and their contribution to elastic stiffness. Comparison between lattice dynamics and Monte Carlo results for a comparable force field (open and filled circles, triangles in Fig. 3) indicates reasonable agreement at low temperature, followed by a breakdown of the quasiharmonic approximation at high temperatures. Results for different force fields, on the other hand, may differ by as much as 50–100 GPa (e.g., diamonds vs. circles or triangles in Fig. 3).

Thermoelectrical and electromechanical properties may be computed in similar fashion by including an electrical work term, $-V_0\mathbf{E}\cdot\mathbf{d}$, in Eq. (13). $\mathbf{E}$ is the applied electric field and $\mathbf{d}$ is the resultant electric displacement,

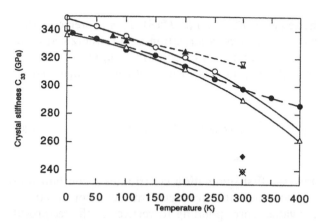

FIG. 3 C_{33} axial modulus of polyethylene obtained using different force fields and computational methods. All calculations are classical. Open symbols are from optimizations (circles, triangles, and diamond are self-consistent QHA lattice dynamics, all others are plotted at temperature of experimental lattice parameters; potential energy minimizations are plotted at 0K). Filled symbols are Monte Carlo (circles) or molecular dynamics (diamond, triangles). Force field of Ref. 55: open and filled circles (from Ref. 46); force field of Ref. 30: open (from Ref. 33) and filled (from Ref. 68) triangles; *pcff* force field of Biosym/MSI: open and filled diamonds (from Ref. 66); × (from Ref. 42); + (from Ref. 43); open inverted triangle (from Ref. 44); open square (from Ref. 29). Curves are drawn as guides to the eye.

related to the polarization of the crystal. For poly(vinylidene fluoride), piezoelectric and pyroelectric properties were computed without resort to an electric field term by computing the change in polarization with temperature and strain [40,41,47].

Wide angle X-ray diffraction provides an experimental probe of atomic scale structure through the structure factor $F(\mathbf{S})$, which is defined as:

$$F(\mathbf{S}) = \sum_i f_i(\mathbf{S}) T_i(\mathbf{S}) \exp(-2\pi i \mathbf{S}' \mathbf{q}_i) \qquad (19)$$

Here, $\mathbf{S} = (\mathbf{S}' - \mathbf{S}_0')/\lambda$ and $|\mathbf{S}| = 2\sin\theta/\tilde{\lambda}$ $\mathbf{S}$ refers to the scattering vector of X-ray analysis, with $\mathbf{S}_0'$, and $\mathbf{S}'$ being the direction of incident and scattered radiation, respectively. Here, λ is the radiation wavelength and θ is the scattering angle. $f_i(\mathbf{S})$, $T_i(\mathbf{S})$, and $\mathbf{q}_i$ are the atomic structure factor, the Debye–Waller temperature factor, and the position, respectively, of atom i. The intensity observed in an X-ray experiment is related to the power spectrum of $F(\mathbf{S})$ by the relation: $I(\mathbf{S}) = I_0 F(\mathbf{S}) F^*(\mathbf{S})$. I_0 is a prefactor that accounts for such effects as absoprtion, polarization, and Lorentz scattering [1–3]. Taking advantage of the separation of coordinate $\mathbf{q}_i$ into a unit cell translation and placement of the atom within its unit cell, as suggested by Eq. (1), one obtains:

$$F(\mathbf{S}) = \sum_{\mathbf{L}} \exp(-2\pi i \mathbf{S}'\mathbf{hL}) \sum_j f_j(\mathbf{S}) T_j(\mathbf{S}) \exp(-2\pi i \mathbf{S}'\mathbf{hs}_j) \tag{20}$$

For large, perfect crystals, the first summation contributes a constant pre-factor $F_c = (h_{max} k_{max} l_{max})^2$ and the scattering vector interferes constructively (i.e., is nonzero) only for values corresponding to vertices of the reciprocal lattice: $\mathbf{S} = h\mathbf{a}^* + k\mathbf{b}^* + l\mathbf{c}^* = \mathbf{h}^*\mathbf{L}$. The static structure factor then becomes simply:

$$F(\mathbf{S}) = F_c \sum_j f_j(\mathbf{S}) T_j(\mathbf{S}) \exp(-2\pi i \mathbf{L}'\mathbf{s}_j) \tag{21}$$

The signature of thermal motion is observed experimentally in X-ray diffraction as a change (usually a reduction) in $F(\mathbf{S})$, and hence in the observed intensity. The structure factor for each atom is modified by the factor $T_i(\mathbf{S}) = \exp(-2\pi^2 \mathbf{S}'\mathbf{B}_i\mathbf{S})$, which depends on both the magnitude and orientation of $\mathbf{S}$. The crystal dynamics are embodied in the direction and amplitude of normal mode vibrations. The amplitude of vibration depends on the energy content of that phonon, $E_{vib,j} = k_B T^2(\partial \ln Q_j / \partial T)$. The orientation of displacement is described by the polarization vector ξ_j for that mode. From this, one can compute the probability density function for an atom vibrating under the influence of a single normal mode. Since the phases of the modes are independent, the average displacement for atom i due to the action of all normal modes can be written as:

$$\mathbf{B}_i = \langle \mathbf{u}_i \mathbf{u}_i^t \rangle = \frac{1}{N_k N_u m_i} \int_{BZ} d\mathbf{k} \sum_{j=1}^{3N_u} \frac{E_{vib,j}(\mathbf{k})}{\omega_j^2(\mathbf{k})} \xi_{i,j}(\mathbf{k}) \left(\xi_{i,j}^*(\mathbf{k}) \right)^t \tag{22}$$

A plot of the orthorhombic unit cell for polyethylene is illustrated in Fig. 4, with the 50% probability density surfaces shown for each atom of the unit cell. The anisotropic dynamics of the polyethylene crystal are immediately apparent. Significantly, the largest contributions to $\mathbf{B}_i$ come from the long wavelength acoustic modes, due to the inverse dependence on ω^2. In experimental studies on polymers, the temperature factor is often

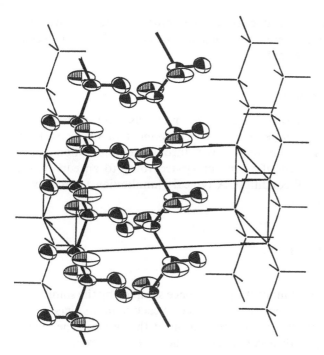

FIG. 4 Orthorhombic unit cell of polyethylene, with 50% probability thermal vibration ellipsoids for each atom, computed using the force field of Ref. 55 with 8000 normal mode vibrations sampled in the Brillouin Zone. Thermal vibration ellipsoids were plotted using Ortep-III [48].

treated as isotropic and equivalent for all atoms; in this case the corresponding B-factor is simply $8\pi^2(\mathrm{tr}(B_i))$, averaged over all the atoms of the unit cell.

The principal limitations of lattice dynamics, of course, are the shape of the potential energy surface implied by the harmonic approximation and the restriction to localized phase spaces. When polymorphism is an issue, each polymorph may be examined independently using lattice dynamics. However, crystals with internal disorder at elevated temperatures include significant contributions from configurations that are not accurately represented by the quadratic form of the Taylor series expansion for potential energy. Studies of atomic crystals, as well as a recent study of polyethylene [34,49], suggest that QHA is reasonably accurate up to about 2/3 of the melting temperature, barring the presence of solid–solid transformations at lower temperatures. Higher order anharmonic contributions may be treated methodically through the use of perturbation theory [50–53]. However, the improvement usually comes at considerable computational expense. For these situations, sampling methods such as Monte Carlo or molecular dynamics are usually preferred.

B. Sampling Methods

Both Monte Carlo and molecular dynamics methods sample directly the phase space of a small but representative component of the crystal, the former by performing stochastic moves through configuration space, the latter by following a specified trajectory according to an equation of motion and chosen initial condition. A typical Hamiltonian for molecular dynamics simulation is [14]:

$$H(\mathbf{p}, \mathbf{q}) = \sum_{i=1}^{3N} \frac{p_i^2}{2m_i} + U(\mathbf{q}) \tag{23}$$

The Hamiltonian is a constant of the phase space sampled by the simulation. A "mechanical" property is one that is defined at each point in phase space, i.e., it can be computed for each configuration of the system. The macroscopic observable F_{obs} is then computed as the ensemble average (Monte Carlo) or time average (molecular dynamics) of the mechanical property F_i; according to the ergodic hypothesis, these should be equivalent.

$$F_{obs} \approx \langle F \rangle = \lim_{N \to \infty} \frac{\sum_{i=1}^{N} F_i e^{-\beta E_i}}{\sum_{i=1}^{N} e^{-\beta E_i}} = \lim_{t \to \infty} \frac{1}{t} \int_0^t dt' F(t') \tag{24}$$

Combinatorial quantities such as entropy and free energy, which depend on the entire distribution of states sampled, require further effort to extract. Methods based on thermodynamic relations which express these quantities as integrals over ensemble averages of mechanical quantities, e.g., $\partial H(\lambda)/\partial \lambda$, where λ defines the state at which the Hamiltonian is evaluated, are most often used to extract thermal properties [14]. λ may be temperature, volume, or even a change in the force field itself.

$$A(\lambda_2) - A(\lambda_1) = \int_1^2 d\lambda' \left\langle \frac{\partial H(\lambda')}{\partial \lambda'} \right\rangle \tag{25}$$

Since much of this book is dedicated to the description of powerful Monte Carlo and molecular dynamics algorithms for the study of chain molecules, we will not go into these methods in as much detail as with lattice dynamics. Nevertheless, a few comments are in order.

By sampling phase space directly, the methods of Monte Carlo and molecular dynamics can avoid the restrictive approximations of lattice dynamics. In principle, at least, polymorphism, anharmonicity, static and dynamic disorder may all be rigorously captured by direct simulation. However, other limitations arise which differ from those encountered in a lattice dynamical approach. Comparison of both approaches can provide insight into the significance of these limitations [46].

First, both Monte Carlo and molecular dynamics simulations must approximate the macroscopic crystal using a very small number of atoms, typically on the order of 10^3–10^4 with current computers. With periodic boundary conditions, atoms outside of the simulation box are restricted to remain dynamically in phase with those inside the simulation box. For a simulation cell consisting of $h_{max} \times k_{max} \times l_{max}$ unit cells, only those wave-vectors whose magnitudes are integer multiples of $2\pi\mathbf{a}^*/h_{max}$, $2\pi\mathbf{b}^*/k_{max}$, and $2\pi\mathbf{c}^*/l_{max}$ are sampled in the simulation. Thus, unless h_{max}, k_{max}, and l_{max} are chosen suitably large, the dynamical resolution is rather poor. This finite size effect may affect both lattice properties, such as thermal expansion, and atomic displacements, which as noted previously are most affected by long wavelength acoustic modes. For polyethylene, finite size effects are noticeable in small simulations containing $2 \times 3 \times 6$ unit cells (432 atoms) at temperatures around 100K, but become progressively worse at higher temperatures [54,55]. Furthermore, finite size effects in polyethylene were found to be quite anisotropic, being largest along the b direction of the lattice, but also strongly coupled to the lattice size in the chain direction. Free or rigid boundary conditions [56] may be used instead of periodic boundaries, but these incur surface as well as finite size effects, the implications of which are less thoroughly quantified.

The anisotropy of polymer crystals has other implications for simulation, as well. Along the chain direction, intramolecular degrees of freedom may dominate the dynamical behavior, while dynamics lateral to the chain are predominantly intermolecular in nature. As evidenced by the large separation between internal and external mode frequencies in lattice dynamics, these two groups are largely decoupled. In molecular dynamics, this separation of dynamical time scales manifests itself as an inefficient transport of energy between high frequency and low frequency motions. This may be observed as poor equilibration of kinetic energy between intramolecular and intermolecular degrees of freedom. To overcome this problem, Ryckaert and Klein used massive stochastic collisions [57] in their molecular dynamics simulation [58]. The collisions serve to redistribute the kinetic energy among all the modes of motion. In Monte Carlo simulations, this effect is observed as inefficient sampling of local coordinate **s** and cell **h** degrees of freedom simultaneously. Since the problem is one of sampling internal and external degrees of freedom, most polymer algorithms geared towards efficient simulation of conformation space are of little use in this respect. Instead, one introduces a combination of atomic and molecular moves. Atomic moves consist of small displacements of individual atoms, as was done in early simulations of atomic liquids and solids, or of small groups of bonded atoms. This permits rapid sampling of high frequency, intramolecular degrees of freedom. Molecular moves consist of translational displacements of the entire chain as a rigid body, as well as rotations of chains about their axes during a simulation. This permits rapid sampling of lower frequency lattice degrees of freedom [54,59–62]. Displacements on the order of 0.05 Å and 0.2 Å are typical for atomic and molecular moves, respectively, and 5–10 degrees for chain rotation. For sampling over large energy barriers in crystals with some disorder, molecular moves which are some fraction of the lattice spacing, or which involve a symmetry operation in the crystal have been used [61,62].

A third, less obvious limitation of sampling methods is that, due to the heavy computational burden involved, simpler interatomic potential models are more prevalent in Monte Carlo and molecular dynamics simulations. For example, polarizability may be an important factor in some polymer crystals. Nevertheless, a model such as the shell model is difficult and time-consuming to implement in Monte Carlo or molecular dynamics simulations and is rarely used. United atom models are quite popular in simulations of amorphous phases due to the reduction in computational requirements for a simulation of a given size. However, united atom models must be used with caution in crystal phase simulations, as the neglect of structural detail in the model may be sufficient to alter completely the symmetry of the crystal phase itself. United atom polyethylene, for example, exhibits a hexagonal unit cell over all temperatures, rather than the experimentally observed orthorhombic unit cell [58,63]; such a change of structure could be reflected in the dynamical properties as well.

The properties of polymer crystals may be computed using the same relations as presented earlier for lattice dynamics. However, it is generally more expedient to take advantage of the fluctuation-dissipation theorem [64] to relate response functions to the magnitude of fluctuations at equilibrium. For example, the fluctuation formula for the constant volume heat capacity computed in the canonical (NVT) ensemble is

$$C_V = \frac{\langle E^2 \rangle - \langle E \rangle^2}{k_B T} \qquad (26)$$

where it is understood that the energy E includes both kinetic and potential contributions. A similar useful relation exists for the elastic stiffness moduli (in the $N\sigma T$ ensemble) [65]:

$$\mathbf{C} = \frac{k_B T}{\langle V \rangle}\langle \varepsilon\varepsilon' \rangle^{-1} \quad \text{or} \quad C_{ij} = \frac{k_B T}{\langle V \rangle}\langle \varepsilon_i\varepsilon_j \rangle^{-1} \tag{27}$$

As indicated earlier, accurate calculation of the elastic moduli by this formula requires a simulation cell large enough to sample adequately the lattice modes of vibration. For polyethylene and polypropylene, simulations consisting of $4 \times 6 \times 12$ unit cells and $3 \times 1 \times 3$ unit cells, respectively (corresponding to simulation cells approximately 20 Å on a side in each case) have been used with reasonable success [66]. Whereas simulations on the order of 10^4 to 10^5 Monte Carlo or molecular dynamics steps are typically required to obtain good averages for energy, density, and many other properties, fluctuation averages such as those used in Eqs. (27) typically require simulations one to two orders of magnitude longer in order to achieve accuracies around 5% [67,68]. This has been attributed to slow convergence of high frequency motions [68], as well as the difficulty in sampling external variables mentioned above. A formula that takes advantage of the correlated fluctuations of stress and strain has been proposed to improve the convergence characteristics [67].

IV. CRYSTAL IMPERFECTIONS AND RELATED PROCESSES

Up to this point, we have focused on the restrictive nature of the phase space of polymer crystals. The harmonic approximation of lattice dynamics assumes that atoms execute only small displacements from their equilibrium positions, such that the effects of anharmonicity are low. Monte Carlo and molecular dynamics, of course, make no such approximations. Many processes in polymer crystals involve significant rearrangements of atoms and passage through a (series of) metastable configuration(s). These processes include relaxations, mass transport, and plasticity, as well as lamellar thickening in flexible chain polymer crystallites. Molecular simulations can be used to study the mechanisms by which these processes occur. The mechanism may be characterized by the (one or more) trajectories through phase space which carry the crystal structure from a "reactant" state, belonging to the region of space occupied before the process, to a "product" state, belonging to the region of space occupied after the process. Short of complete melting and recrystallization of the polymer, the process must occur through the action of a more or less localized region of disorder in the crystal, which constitutes a crystallographic defect. The type of simulation used to analyze the process depends significantly on the nature of the trajectory.

For trajectories involving configurations with energies on the order of the thermal energy, k_BT, a molecular dynamics simulation can efficiently sample configuration space both close to an equilibrium point as well as along the (classical) trajectories between equilibrium points. The great power of this approach is that little or no prior knowledge of the trajectory is required, so long as the energy requirements are low and passage along the trajectory is sufficiently frequent to ensure observation of one or more passages during the typical 1 ns simulation. A rule of thumb based on transition state theory (see below) is that the potential energy along the trajectory does not exceed about $10\,k_BT$ at any point.

Wunderlich and co-workers have reported extensively on the dynamics of a united atom model of polyethylene using molecular dynamics [63]. In this model, the crystal structure resembles the high temperature hexagonal phase observed in polyethylene. Simply by analyzing the structure and energy of configurations sampled in a conventional molecular dynamics simulation, one can identify crystallographic defects, conformational relaxations, and diffusion processes. With the united atom model, these authors observed that gauche conformational defects formed in the otherwise all-trans chain with a frequency on the order of 10 GHz per bond at 350K. The estimated activation energy is on the order of 5–$8\,k_BT$. These gauche states were observed to be short-lived and uncorrelated. Kink defects (i.e., having conformation $g^{\pm}tg^{\mp}$) were observed to diffuse rapidly and annihilate at the free surface of the simulation. Sequential deviation in the same direction by a string of bonds along the chain gives rise to a "twist," by which the chain may execute a 180 degree flip about its axis, with a passage time on the order of 2 ps at 300K. Longitudinal diffusion of the chain was observed to couple to a lattice acoustic mode. All-atom simulations revealed the onset of translational diffusion in polyethylene above 250K, and evolution to four-fold rotational disorder above 375K, on the time scale of picoseconds [58].

For processes whose frequency of occurrence is less than 1 GHz (i.e., once per nanosecond), the events are sufficiently rare to render molecular dynamics of little use as a means for direct simulation, given current computer limitations. Rare events are more often modeled using some form of transition state theory (TST) [69]. The basic assumptions of TST are that there exists a dividing surface which separates phase space into a reactant region and a product region, that the region of space comprising the reactant state and the dividing surface are in equilibrium, and that trajectories passing through the dividing surface are captured in the product state. In its conventional form, TST places this dividing surface at a first order saddle point in the potential energy surface, defined as a stationary point with one and only one negative eigenvalue of the dynamical matrix. The problem then centers on the characterization of the reactant and saddle point (i.e., transition state) configurations. For simplicity, the present discussion adheres to this conventional interpretation.

In conventional TST, the one-way flux, or reaction rate, from reactant to product may be written as:

$$k^{TST} = \frac{k_B T \kappa}{h} \frac{Q_{vib}^{\ddagger}}{Q_{vib,0}} \exp\left(-\frac{\Delta E}{k_B T}\right) \tag{28}$$

where h is Planck's constant and κ is the transmission coefficient, usually taken to be on the order of unity. Characterizing the kinetic rates of processes dominated by rare events reduces to estimating the activation energy $\Delta E = (U^{\ddagger} - U_0)$ and the ratio, $Q_{vib}^{\ddagger}/Q_{vib,0}$, of the vibrational partition functions for reactant (0) and transition ($\ddagger$) states. If the reactant and transition states are reasonably well known, the methods of lattice dynamics may be used to compute the four quantities required, $U^{\ddagger}$, $Q_{vib}^{\ddagger}$, U_0, and $Q_{vib,0}$. It is worth noting that the specification of the trajectory and the associated one-way flux in TST constitutes a classical treatment of the one degree of freedom associated with motion along the trajectory; $Q_{vib}^{\ddagger}$ may be evaluated either classically or quantum mechanically using Eqs. (11), but possesses one fewer degree of freedom than $Q_{vib,0}$ due to the restriction of motion to a specifed trajectory. At a saddle point, one imaginary frequency is obtained in the dynamical matrix and is discarded.

The identification of the transition state configuration(s) for rare events in polymer crystals can be a complicated task. Conformational defects of the types discussed above in relation to molecular dynamics simulations have received the most attention in polymer crystals, due to their association with experimentally observed mechanical and dielectric relaxations, crystal thickening, etc. These are also illustrative of the application of TST to polymer crystal studies.

It is generally believed that relaxations in the crystal phase occur through the formation and propagation of local regions of disorder, called defects. The gauche bonds, kinks, and twists observed in the molecular dynamics simulations cited earlier are all examples of conformational defects. Conformational defects in polymer chains are classified by the net translational and rotational lattice mismatch incurred by a defect involving a particular length of the chain (the "extent" of the defect). Such translational mismatches have been called "dislocations," by analogy to dislocations in the solid-state physics literature. However, studies of relaxations in polymer crystals have typically focused on point defects, and should not be confused with the line defects of the latter; similarly, rotational mismatches are termed "disclinations." In polymers, the helical symmetry of the chain permits a combined translational/rotational mismatch that has been termed a "dispiration" [70]. The identification of the structure of defects which satisfy the constraints of a dislocation, disclination, or dispiration typically requires considerable insight, inspiration, or brute force searching of conformation space [71–73]. For highly localized conformational defects, the process of defect identification can be simplified by solving the conformational problem in torsion space alone, for defects up to six bonds long [74,75]. The method involves solving a set of nonlinear equations, and also serves as the basis of the Concerted Rotation Monte Carlo algorithm [76]. Once the conformational problem is solved, one can construct a cell containing a single reference chain containing the conformational defect, and thereby compute the energy and distortion of the defect due to packing in the lattice. Unlike perfect crystal simulations, simulations to date involving defects have been performed almost exclusively using finite cells with rigid or free boundary conditions so as to avoid defect–defect interactions.

Having once identified a stable defect in the crystal lattice and computed its energy of formation, determining the kinetics of a process based on this defect in TST requires the further identification of a saddle point along a trajectory which displaces the defect by one lattice period; the periodicity of the lattice permits a scenario where subsequent propagation of the defect occurs through a sequence of similar "hops" in the lattice. Identification of a suitable trajectory for defect propagation in polyethylene was performed by introducing the defect into the lattice, as described above, and then driving one or more coordinates (e.g., a torsion angle in the chain or setting angle in the crystal) so that the defect is forced to move to the next unit cell in the lattice. This type of calculation is essentially quasi-static; a constraint on one or more rotation angles in the chain is incremented, and the potential energy of the defect reminimized [73,77]. The process is repeated stepwise until a potential energy barrier is crossed (i.e., the defect "falls" into the product state upon minimization). However, this approach does not generally locate a first order saddle point on the potential energy surface, but rather a maximum along the trajectory chosen by the coordinate-incrementing protocol employed. The resulting trajectories are thus sensitive to how the defect is driven along the chain and provide upper bounds on the true saddle point energy [77]. Alternative numerical approaches can be used to refine the estimate of a true first order saddle point based on an adiabatic reaction path connecting reactant to product. One particularly useful algorithm for this purpose is the conjugate peak refinement algorithm of Fischer and Karplus [78], which requires knowledge only of the equilibrium conformation of the defect in order to identify a lowest energy path from

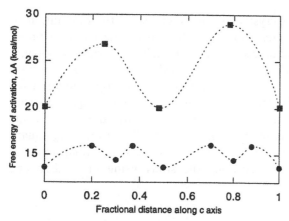

FIG. 5 Free energy of formation and transition state barriers ($\Delta A = (Q^{\ddagger}_{\text{vib}}/Q_{\text{vib},0})\exp(-\Delta E/k_B T)$) to transport for two types of defects capable of producing the α-relaxation in poly(vinylidene fluoride) (from Ref. 79). tgtg* $\rightarrow$ g*tgt (circles); g*tgt $\rightarrow$ tgtg* (squares). Dashed curves are drawn as guides to the eye.

reactant to product. This saddle point search algorithm was used in a study of poly(vinylidene fluoride) [79] and, more recently, polyethylene [80]. For cases where a good initial guess of the saddle point configuration is available, more efficient gradient algorithms can be used to refine the estimate of the saddle point configuration and energy [81]. Once the saddle point configuration is identified, $U^{\ddagger}$ and $Q^{\ddagger}$ are readily computed using the methods of lattice dynamics, and TST applied. Figure 5 illustrates two trajectories computed for propagation of defects in poly(vinylidene fluoride) to account for the α-relaxation in the α-crystalline form of that polymer. It is interesting to note that, whereas heats of formation for defects may be fairly large, the methods described above can be used to characterize the trajectory of the defect through the lattice, regardless of whether it involves a hopping mechanism between adjacent lattice sites, separated by large energy barriers, or a relatively facile free-streaming motion involving only low energy barriers to propagation. Both hopping [70] and free-streaming [73] mechanisms have also been proposed for motion of defects in polyethylene.

V. SUMMARY

There is currently a variety of methods for studying polymer crystals by molecular simulation. As with other materials, the usual caveats regarding force field accuracy and proper execution of the simulation apply.

The anisotropy of polymer crystals is characteristic of the orientation of intra- and intermolecular interactions, which is also reflected in a broad separation on the dynamical time scale. It is especially important in polymer crystals to be cognizant of the limitations imposed by either the assumptions on which a method is based (e.g., the quasiharmonic approximation for lattice dynamics) or the robustness of the simulation method (e.g., ergodicity of the simulation in Monte Carlo or molecular dynamics). In some instances, valuable checks on these assumptions and limitations (e.g., quantum mechanical vs. classical dynamics, finite size effects and anharmonicity) can and should be made by repeating the study using more than one method.

REFERENCES

1. Cullity, B.D. *Elements of X-Ray Diffraction*, 2nd Ed.; Addison-Wesley: Reading, 1978.
2. Alexander, L.E. *X-Ray Diffraction Methods in Polymer Science*; Wiley Interscience: New York, 1969.
3. Guinier, A. *X-Ray Diffraction in Crystals, Imperfect Crystals, and Amorphous Bodies*; W.H. Freeman and Co.: San Francisco, 1963.
4. Miyazawa, T. J. Polym. Sci. **1961**, *55*, 215.
5. Sugeta, H.; Miyazawa, T. Biopolymers **1965**, *5*, 673.
6. Meille, S.V.; Brückner, S.; Porzio, W. Macromolecules **1990**, *23*, 4114–4121.
7. Davé, R.S.; Farmer, B.L. Polymer **1988**, *29*, 1544–1554.
8. Fisher, H.P.; Eby, R.K.; Cammarata, R.C. Polymer **1994**, *35*, 1923–1930.
9. Balijepalli, S.; Rutledge, G.C. J. Chem. Phys. **1998**, *109*, 6523–6526.
10. Balijepalli, S.; Rutledge, G.C. Comput. Theor. Polym. Sci. **2000**, *10* (1,2), 103–113.
11. Gautam, S.; Balijepalli, S.; Rutledge, G.C. Macromolecules **2000**, *33* (24), 9136–9145.
12. Corradini, P.; Guerra, G. Advances in Polymer Science **1991**, *100*, 185.
13. Press, W.H.; Teukolsky, S.A.; Vetterling, W.T.; Flannery, B.P. *Numerical Recipes in C: The Art of Scientific Computing*, 2nd Ed.; Cambridge University Press: Cambridge, 1992.
14. Allen, M.P.; Tildesley, D.J. *Computer Simulation of Liquids*; Clarendon Press: Oxford, 1987.
15. McCullough, R.L. J. Macromol. Sci. Phys. **1974**, *9*, 97.
16. Tripathy, S.K.; Hopfinger, A.J.; Taylor, P.L. J. Phys. Chem. **1981**, *85*, 1371–1380.
17. Bacon, D.J.; Geary, N.A. J. Mater. Sci. **1983**, *18*, 853–863.
18. Rutledge, G.C.; Suter, U.W. Macromolecules **1991**, *24*, 1921–1933.
19. Lacks, D.J.; Rutledge, G.C. Macromolecules **1994**, *27*, 7197–7204.
20. Zhang, R.; Taylor, P.L. J. Chem. Phys. **1991**, *94*, 3207–3212.

21. Tadokoro, H. *Structure of Crystalline Polymers*; Wiley and Sons: New York, 1979.
22. Tashiro, K. Prog. Polym. Sci. **1993**, *18*, 377–435 and references therein.
23. Born, M.; von Kármán, T. Phys. Zeit. **1912**, *13*, 297.
24. Born, M.; von Kármán, T. Phys. Zeit. **1913**, *14*, 15.
25. Born, M.; Huang, K. *Dynamical Theory of Crystal Lattices*; Oxford University Press: Oxford, 1954.
26. Willis, B.T.M.; Pryor, A.W. *Thermal Vibrations in Crystallography*; Cambridge University Press: Cambridge, 1975.
27. Horton, G.K.; Maradudin, A.A. Eds. *Dynamical Properties of Solids*; North-Holland: Amsterdam, 1974–1990; Vol. 1–7.
28. Dove, M.T. *Introduction to Lattice Dynamics*; Cambridge University Press: Cambridge, 1993.
29. Sorensen, R.A.; Liau, W.B.; Kesner, L.; Boyd, R.H. Macromolecules **1988**, *21*, 200–208.
30. Karasawa, N.; Dasgupta, S.; Goddard, W.A. J. Phys. Chem. **1991**, *95*, 2260–2272.
31. Boyd, R.H. In *Advances in Polymer Science*, Monnerie, L., Suter, U.W., Eds.; Springer Verlag: Berlin, 1994; Vol. 116, 1–25.
32. Allan, N.L.; Barron, T.H.K.; Bruno, J.A.O. J. Chem. Phys. **1996**, *105*, 8300–8303.
33. Lacks, D.J.; Rutledge, G.C. J. Phys. Chem. **1994**, *98*, 1222–1231.
34. Lacks, D.J.; Rutledge, G.C. Macromolecules **1995**, *28*, 1115–1120.
35. Carbeck, J.D.; Lacks, D.J.; Rutledge, G.C. J. Chem. Phys. **1995**, *103*, 10347–10355.
36. Rutledge, G.C. Macromolecules **1997**, *30*, 2785–2791.
37. Taylor, M.B.; Barrera, G.D.; Allan, N.L.; Barron, T.H.K. Phys. Rev. B **1997**, *56* (22), 14380–14390.
38. Bruno, J.A.O.; Allan, N.L.; Barron, T.H.K.; Turner, A.D. Phys. Rev. B **1998**, *58* (13), 8416–8427.
39. Dick, B.; Overhauser, A. Phys. Rev. **1958**, *112*, 90.
40. Karasawa, N.; Goddard, W.A. Macromolecules **1992**, *25*, 7268–7281.
41. Carbeck, J.D.; Rutledge, G.C. Polymer **1996**, *37*, 5089–5097.
42. Anand, J.N. J. Macromol. Sci.-Phys. B1 **1967**, (3), 445.
43. Wobser, V.G.; Blasenbrey, S. Kolloid. Z. Z. Polym. **1970**, *241*. 985.
44. Tashiro, K.; Kobayashi, M.; Tadokoro, H. Macromolecules **1978**, *11* (5), 914.
45. Martoňák, R.; Paul, W.; Binder, K. Phys. Rev. E **1998**, *57*, 2425–2437.
46. Rutledge, G.C.; Lacks, D.J.; Martoňák, R.; Binder, K. J. Chem. Phys. **1998**, *108*, 10274–10280.
47. Tashiro, K.; Kobayashi, M.; Tadokoro, H.; Fukada, E. Macromolecules **1980**, *13*, 691.
48. Burnett, M.N.; Johnson, C.K. Ortep-III: Oak Ridge Thermal Ellipsoid Plot Program for Cyrstal Stucture Illustration. Oak Ridge National Laboratory Report ORNL-6895, 1996.
49. Lacks, D.J.; Rutledge, G.C. J. Chem. Phys. **1994**, *101*, 9961–9965.

50. Shukla, R.C.; Cowley, E.R. Phys. Rev. B **1985**, *31*, 372–378.
51. Shukla, R.C.; Shanes, F. Phys. Rev. B **1985**, *32*, 2513–2521.
52. Shukla, R.C.; Bose, S.K.; Delogu, R.F. Phys. Rev. B **1992**, *45*, 12812–12820.
53. Shukla, R.C.; Lacks, D.J. J. Chem. Phys. **1997**, *107*, 7409–7417.
54. Martonák, R.; Paul, W.; Binder, K. Comput. Phys. Commun. **1996**, *99*, 2–8.
55. Martonák, R.; Paul, W.; Binder, K. J. Chem. Phys. **1997**, *106*, 8918–8930.
56. Liang, G.L.; Noid, D.W.; Sumpter, B.G.; Wunderlich, B. Comput. Polym. Sci. **1993**, *3*, 101.
57. Andersen, H.C. J. Chem. Phys. **1980**, *72*, 2384.
58. Ryckaert, J.-P.; Klein, M.L. J. Chem. Phys. **1986**, *85*, 1613–1620.
59. Yamamoto, T. J. Chem. Phys. **1985**, *82*, 3790–3794.
60. Yamamoto, T. J. Chem. Phys. **1988**, *89*, 2356–2365.
61. Foulger, S.H.; Rutledge, G.C. Macromolecules **1995**, *28*, 7075–7084.
62. Foulger, S.H.; Rutledge, G.C. J. Polym. Sci.-Phys. **1998**, *36*, 727–741.
63. Sumpter, B.G.; Noid, D.W.; Liang, G.L.; Wunderlich, B. In *Advances in Polymer Science*; Monnerie, L., Suter, U.W., Eds.; Springer Verlag: Berlin, 1994; Vol. 116, 27–72 and references therein.
64. Van Kampen, N.G. *Stochastic Processes in Physics and Chemistry*; North-Holland: Amsterdam, 1992.
65. Parrinello, M.; Rahman, A. J. Chem. Phys. **1982**, *76* (5), 2662–2666.
66. Zehnder, M.M.; Gusev, A.A.; Suter, U.W. Rev. Inst. Français Petrole **1996**, *51* (1), 131–137.
67. Gusev, A.; Zehnder, M.M.; Suter, U.W. Phys. Rev. B. **1996**, *54*, 1–4.
68. Cagin, T.; Karasawa, N.; Dasgupta, S.; Goddard, W.A. Mat. Res. Soc. Symp. Proc. **1992**, *278*, 61.
69. Glasstone, S.; Laidler, K.; Eyring, H. *The Theory of Rate Processes*; McGraw-Hill: New York, 1941.
70. Reneker, D.H. J. Mazur. Polymer **1988**, *29*, 3–13.
71. Reneker, D.H. J. Polym. Sci. **1962**, *59* (168), S39.
72. Boyd, R.H. J. Polym. Sci.-Phys. **1975**, *13*, 2345.
73. Mansfield, M.; Boyd, R.H. J. Polym. Sci.-Phys. **1978**, *16*, 1227.
74. McMahon, P.E.; McCullough, R.L.; Schlegel, A.A. J. Appl. Phys. **1967**, *38* (11), 4123.
75. Go, N.; Scheraga, H.A. Macromolecules **1970**, *3*, 178.
76. Dodd, L.R.; Boone, T.D.; Theodorou, D.N. Mol. Phys. **1993**, *78*, 961–996.
77. Reneker, D.H.; Fanconi, B.M.; Mazur, J. J. Appl. Phys. **1977**, *48* (10), 4032.
78. Fischer, S.; Karplus, M. Chem. Phys. Lett. **1992**, *194*, 252–261.
79. Carbeck, J.D.; Rutledge, G.C. Macromolecules **1996**, *29* (15), 5190–5199.
80. Mowry, S.W.; Rutledge, G.C. Macromolecules **2002**, *35* (11), 4539–4549.
81. Schlegel, H.B. In *Ab Initio Methods in Quantum Chemistry*; Lawley, K.P, Ed.; Wiley and Sons: New York, 1987; 249 pp.

12

Plastic Deformation of Bisphenol-A-Polycarbonate: Applying an Atomistic-Continuum Model

JAE SHICK YANG and WON HO JO Seoul National University, Seoul, Korea

SERGE SANTOS and ULRICH W. SUTER Eidgenössische Technische Hochschule (ETH), Zurich, Switzerland

I. INTRODUCTION

Atomistic modeling techniques have been employed for the investigation of mechanical properties of amorphous polymeric solids for years [1–5]. Theodorou and Suter [1] successfully calculated the elastic constants of glassy atactic polypropylene (aPP) by molecular mechanics techniques, surmising that the entropic contributions to the elastic response are negligible for such glasses. Mott et al. [2] strained polypropylene glasses beyond the macroscopic elastic range and found evidence for the occurrence of diffuse plastic unit events and estimated, by comparison with experimental results, that the size of the plastically transforming domains in elementary shear-transformation events extends over roughly 100 Å (well beyond the system size that can conveniently be simulated). Plastic relaxation occurs through cooperative rearrangements with rather small local transformation strains [$\mathcal{O}(2\%)$]. These findings were assumed to be true for all polymeric glasses, in contrast to the long-held belief that the elementary process of plastic transformation is a well-localized conformational transition. Hutnik et al. [3] used the same technique on glassy bisphenol-A-polycarbonate (BPA-PC) and confirmed the results on aPP. All these calculations suffered from small system size and the fact that the atomistic box shape was used as a simulation variable, prescribed and controlled during the athermal simulations. While this is perfectly appropriate for the elucidation of elastic properties, it is not optimal when the "natural" system path is unknown, as in plastic deformation. It would be preferable to be able to simulate the behavior of atomistic subsystems embedded in a medium (of identical properties) and free to follow the driving forces that the gradient of (free) energy suggests. The overall system would have to be considerably larger, however, than the ones employed for these atomistic simulations.

The demand for larger systems in the modeling of mechanical properties of polymeric materials requires the consideration of multiple length scales. For very large length scales it suffices to think of the material as an elastic medium and an atomistic description is unnecessary. However, for the plasticity to be captured at the molecular level, atomistically detailed modeling is required and several efforts to combine the atomistic and continuum levels for the modeling of viscoelastic properties of materials have been published [6–10]. The fundamental difficulty in this endeavor lies in the way in which the length scales are coupled. We have presented [11] an approach at atomistic-continuum modeling, where the system consists of a continuum matrix and an atomistic inclusion, and the inclusion-boundary behavior and the atomistic cell behavior are connected via the common strain transformation. This new model has proven to be consistent for modeling the elastic deformation of a nearest-neighbor fcc crystal of argon and glassy BPA-PC. It has the advantage of being free of artificial constraints and assumptions beyond those intrinsic to the descriptions on the two levels being connected.

Here, the atomistic-continuum model is applied to the study of plastic deformation of BPA-PC. First, the elastic constants of BPA-PC are calculated by atomistic simulation. These values are used as the elastic constants for the matrix throughout the simulated deformation. Then, the system, i.e., the continuum matrix with its atomistic inclusion, is deformed stepwise up to a strain of about 0.2. The overall system is constrained to exactly follow a predescribed deformation sequence, but the atomistic inclusion is free to follow any strain path consistent with the misfit stresses acting between it and the matrix. The result is a new look at the behavior of a glassy polymeric inclusion deformed plastically in a surrounding elastic medium.

II. MODEL

We briefly recapitulate the model, detailed previously [11], which is based on the variable-metric total-energy approach by Gusev [12] and consists of a system comprising an inclusion embedded in a continuous medium. The inclusion behavior is described in atomistic detail whereas the continuum is modeled by a displacement-based finite-element method. Since the atomistic model provides configuration-dependent material properties inaccessible to continuum models, the inclusion in the atomistic-continuum model acts as a magnifying glass, which allows observation of the molecular properties of the material.

A. Continuum Model

The variable-metric total-energy approach [12] is adopted here for the matrix. As illustrated in Fig. 1, the periodic system is described by the scaling matrix [13,14] $\mathbf{H} = [\mathbf{ABC}]$, where $\mathbf{A}$, $\mathbf{B}$, and $\mathbf{C}$ are the overall system's continuation vectors. Two kinds of nodal points are specified in the matrix: One ($\mathbf{x}^b$) on the inclusion boundary and the other ($\mathbf{x}^c$) in the continuum (throughout this text, vectors are written as column matrices). For convenience, the scaled coordinates [13,14] ($\mathbf{s}^b$ and $\mathbf{s}^c$) are chosen as degrees of freedom via $\mathbf{x}^b = \mathbf{H}\mathbf{s}^b$ and $\mathbf{x}^c = \mathbf{H}\mathbf{s}^c$. These nodal points are used as vertices of a periodic network of Delaunay tetrahedra. The Delaunay network uniquely tessellates space without overlaps or fissures; the circumsphere of the four nodal points of any tetrahedron does not contain another nodal point of the system [15,16]. For each tetrahedron β, the local scaling matrix $\mathbf{h}^\beta = [\mathbf{a}^\beta \mathbf{b}^\beta \mathbf{c}^\beta]$ is defined in the same manner as $\mathbf{H}$ for the overall system. The local (Lagrange) strain $\boldsymbol{\varepsilon}^\beta$ is assumed to be constant inside each tetrahedron and defined as [13,14]

$$\boldsymbol{\varepsilon}^\beta = \frac{1}{2}\left[\left(\mathbf{G}^\beta\right)^T \mathbf{G}^\beta - \mathbf{I}\right], \qquad \mathbf{G}^\beta = \mathbf{h}^\beta \left(\mathbf{h}_0^\beta\right)^{-1} \tag{1}$$

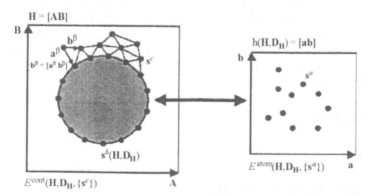

FIG. 1 Sketch of atomistic-continuum model, for illustrative purposes drawn for a two-dimensional version only. All degrees of freedom in the system are of one of four types: $\mathbf{H}$, $\mathbf{D}_H$, $\mathbf{s}^c$, and $\mathbf{s}^a$.

where the subscript 0 represents the reference state before deformation and $\mathbf{I}$ is the unit matrix of order 3. The system strain ε^{sys} is calculated from Eq. (1) by substituting $\mathbf{H}$ and $\mathbf{H}_0$ for $\mathbf{h}^\beta$ and h_0^β, respectively. The strain energy in the continuum E^{cont} can be evaluated from the sum over all tetrahedra:

$$E^{cont}\left(\mathbf{H},\{\mathbf{s}^b\},\{\mathbf{s}^c\}\right) = \frac{1}{2}\sum_\beta V_\beta \left(\varepsilon^\beta\right)^T \mathbf{C}^\beta \varepsilon^\beta \tag{2}$$

where V_β is the volume of the reference state and $\mathbf{C}^\beta$ the matrix of elastic constants of tetrahedron β.

B. Atomistic Model

A spatially periodic simulation box represents the contents of the atomistic inclusion. Again, the shape of this box is expressed by a scaling matrix $\mathbf{h} = [\mathbf{abc}]$ (cf. Fig. 1). The scaled coordinates of atom $\mathbf{s}^a$ are used as degrees of freedom via $\mathbf{x}^a = \mathbf{h}\mathbf{s}^a$. The Lagrange strain tensor of the atomistic box, ε^{atom}, can be calculated from Eq. (1) by replacing $\mathbf{h}^\beta$ and h_0^β with $\mathbf{h}$ and $\mathbf{h}_0$, respectively. The energy in the atomistic box can be expressed as E^{atom} $(\mathbf{h},\{\mathbf{s}^a\})$ and can be obtained via any energy model for atomistic simulation, e.g., a force-field approach.

C. Atomistic-Continuum Model

To connect the two models with different length scales, the displacement of the nodal points on the inclusion boundary is associated with the deformation of the atomistic box by $\mathbf{x}^b = \mathbf{h}\mathbf{h}_0^{-1}\mathbf{x}_0^b$. This means that the inclusion boundary is constrained to follow the homogeneous deformation of the atomistic cell vectors. The degree of freedom $\mathbf{D}_\mathbf{H}$ is introduced to relate $\mathbf{h}$ with $\mathbf{H}$ through

$$\mathbf{h} = \mathbf{H}(\mathbf{I} + \mathbf{D}_\mathbf{H})\mathbf{H}_0^{-1}\mathbf{h}_0 \tag{3}$$

Thus, $\mathbf{s}^b$ can be expressed by

$$\mathbf{s}^b = \mathbf{H}^{-1}\mathbf{x}^b = \mathbf{H}^{-1}\mathbf{h}\mathbf{h}_0^{-1}\mathbf{H}_0\mathbf{s}_0^b = (\mathbf{I} + \mathbf{D}_\mathbf{H})\mathbf{s}_0^b \tag{4}$$

There are four types of degrees of freedom in the system: $\mathbf{H}$ and $\mathbf{s}^c$ for the continuum model; $\mathbf{h}$ and $\mathbf{s}^a$ for the atomistic model.

The total potential energy in the system E^{sys} is consequently expressed as

$$E^{sys}(\mathbf{H},\mathbf{D_H},\{\mathbf{s}^c\},\{\mathbf{s}^a\}) = E^{cont}(\mathbf{H},\mathbf{D_H},\{\mathbf{s}^c\}) + \alpha_V E^{atom}(\mathbf{H},\mathbf{D_H},\{\mathbf{s}^a\}),$$

$$\alpha_V = \frac{V^{inc}}{V^{atom}} \tag{5}$$

where V^{inc} and V^{atom} are the volumes of the inclusion and atomistic box, respectively (evidently, the inclusion must have the same energy density as the atomistic box). Note that α_V is kept constant during deformation because $\mathbf{x}^b$ deforms according to $\mathbf{h}$.

The length scales in the model are given by the inclusion volume fraction f_I and α_V, e.g., in a cubic system box of edge length L with a spherical inclusion of diameter d, defined by an atomistic box of edge length a, $d = (6\alpha_V/\pi)^{1/3}a$ and $L = (\alpha_V/f_I)^{1/3}a$.

While this two-scale model allows for a "natural" connection between the two scales (i.e., the inclusion-boundary behavior and the atomistic cell behavior are connected via the common strain transformation), it also is limited by the required uniformity of strain in the atomistic inclusion (through the periodic continuation conditions on the atomistic cell) and in each of the tetrahedra of the Delaunay tessellation.

III. SIMULATION METHOD

A. Model System

The system was a cube with an edge length $L = 100\,\text{Å}$ and one spherical inclusion (volume fraction $f_I = 0.18$). We specified 1000 nodal points on the inclusion boundary ($\mathbf{x}^b$) and 6259 in the continuum ($\mathbf{x}^c$). The atomistic box consists of one BPA-PC chain with degree of polymerization $X = 74$, resulting in 2454 atoms. The molecular structure of BPA-PC is sketched in Fig. 2; the dihedral angles are numbered as in previous publications [3], even though a flexible-chain, Cartesian-coordinate approach is used here. Three independent atomistic structures were generated and equilibrated at the experimental density of $1.20\,\text{g cm}^{-3}$ (300 K) [3], following the procedure of [11], whereupon the energies of the atomistic structures were minimized with respect to all degrees of freedom, including $\mathbf{h}^{atom}$, to remove "residual stress" inside the atomistic boxes [17] through the Polak-Ribiere conjugate gradient minimization method [18]. The *pcff* force field [19–21] was used to calculate the atomistic energy, which is composed of 14 components:

$$\begin{aligned} E^{atom} = {} & E_b + E_\theta + E_\phi + E_\chi + E_{elec} + E_{vdW} \\ & + E_{bb'} + E_{bb''} + E_{b\theta} + E_{\theta\theta'} + E_{b\phi} + E_{b'\phi} + E_{\theta\phi} + E_{\phi\theta\theta'} \end{aligned} \tag{6}$$

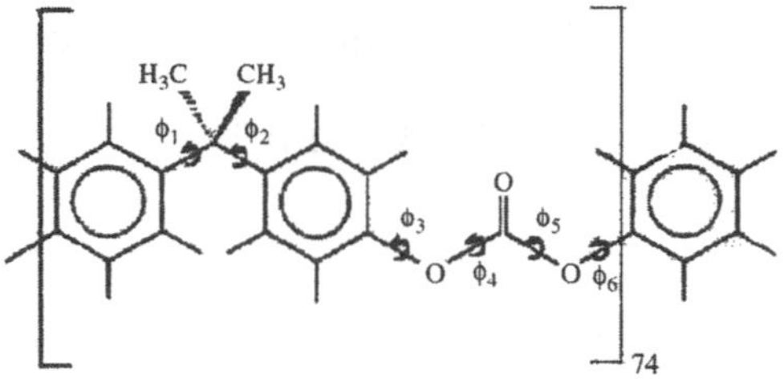

FIG. 2 The constitutional repeat unit of BPA-PC.

where the first four terms are "valence" terms (bond length, bond angle, dihedral angle, and out-of-plane angle), the next two terms nonbonded interaction terms (electrostatic and van der Waals), and the last eight terms "coupling valence" terms. For efficiency, an in-house code was constructed that yields the same energy and partial derivatives with respect to coordinates and components of the scaling matrix as the commercial program *Discover* [19]. The nonbonded energy cutoff and the spline width were 8 Å and 2 Å, respectively. The charged-group based method [19] was adopted for the calculation of nonbonded energies to keep the total charge inside the cutoff to zero.

B. Elastic Deformation of the Atomistic Model

The constant strain method was used to determine the elastic constants $\mathbf{C}^{atom}$ of three independent atomistic boxes. The average values of these elastic constants were assigned to the matrix [i.e., to all $\mathbf{C}^{\beta}$ in Eq. (2)]. A set of small strains ($\varepsilon \approx 10^{-4}$) were applied to the atomistic box in its minimum-energy configuration and the energy of the atomistic box was minimized with respect to all degrees of freedom, except $\mathbf{h}$. The elastic constants of the atomistic box $\mathbf{C}^{atom}$ were evaluated by parabolic fits to

$$E^{atom} - E_0^{atom} = \frac{1}{2} V^{atom} \left(\varepsilon^{atom} \right)^T \mathbf{C}^{atom} \varepsilon^{atom} \tag{7}$$

In the Voigt notation, elastic constants can be expressed as 6×6 matrices [22]. For an isotropic material, $\mathbf{C}$ takes the form

$$\mathbf{C} = \begin{pmatrix} 2\mu + \lambda & \lambda & \lambda & 0 & 0 & 0 \\ \lambda & 2\mu + \lambda & \lambda & 0 & 0 & 0 \\ \lambda & \lambda & 2\mu + \lambda & 0 & 0 & 0 \\ 0 & 0 & 0 & \mu & 0 & 0 \\ 0 & 0 & 0 & 0 & \mu & 0 \\ 0 & 0 & 0 & 0 & 0 & \mu \end{pmatrix} \tag{8}$$

where λ and μ are the so-called Lamé constants. The Young's modulus E, the shear modulus G, the bulk modulus B, and the Poisson's ratio ν are obtained by

$$E = \mu\frac{3\lambda + 2\mu}{\lambda + \mu}, \qquad G = \mu$$

$$B = \lambda + \frac{2}{3}\mu, \qquad \nu = \frac{\lambda}{2(\lambda + \mu)} \tag{9}$$

C. Plastic Deformation of the Atomistic-Continuum Model

The deformation method developed by Mott et al. [2] was used here. The system was subjected to deformation in small steps up to a system strain of 0.2, the overall energy being minimized with respect to all degrees of freedom (except **H**) after each step. The strain increment of the system, $\triangle\varepsilon^{sys}$, was

$$\triangle\varepsilon^{sys} = \begin{pmatrix} 2 & 0 & 0 \\ 0 & -1+q & 0 \\ 0 & 0 & -1+q \end{pmatrix} \times 10^{-3} \text{ (for uniaxial extension)} \tag{10}$$

$$\triangle\varepsilon^{sys} = \begin{pmatrix} 2 & 0 & 0 \\ 0 & -2+r & 0 \\ 0 & 0 & 0 \end{pmatrix} \times 10^{-3} \text{ (for pure shear)} \tag{11}$$

where q and r are small values adjusted to keep the system volume constant. Thus, for both, uniaxial extension and pure shear, the system volume V^{sys} was held constant while the atomistic box was free to change its volume V^{atom} [through the coupling degree of freedom $\mathbf{D_H}$, Eq. (3)].

The work-equivalent tensile strains in the system ε_{eq}^{sys} and in the atomistic box ε_{eq}^{atom} are given by [2]

$$\varepsilon_{eq}^{sys} = \left[\frac{2}{3}\text{Tr}(\hat{\varepsilon}^{sys})^2\right]^{1/2} \quad \text{where} \quad \hat{\varepsilon}^{sys} = \varepsilon^{sys} - \frac{1}{3}\text{Tr}(\varepsilon^{sys})\mathbf{I} \tag{12}$$

$$\varepsilon_{eq}^{atom} = \left[\frac{2}{3}\text{Tr}(\hat{\varepsilon}^{atom})^2\right]^{1/2} \quad \text{where} \quad \hat{\varepsilon}^{atom} = \varepsilon^{atom} - \frac{1}{3}\text{Tr}(\varepsilon^{atom})\mathbf{I} \tag{13}$$

The internal stress tensors in the system $\boldsymbol{\sigma}^{sys}$ and in the atomistic box $\boldsymbol{\sigma}^{atom}$ are calculated by [13,14]

$$\sigma^{sys} = -\frac{1}{V^{sys}}\mathbf{H}\left(\frac{\partial E^{sys}}{\partial \mathbf{H}}\right)^T, \qquad \sigma^{atom} = -\frac{1}{V^{atom}}\mathbf{h}\left(\frac{\partial E^{atom}}{\partial \mathbf{h}}\right)^T \qquad (14)$$

The von Mises-equivalent tensile stresses of the system σ_{eq}^{sys} and of the atomistic box σ_{eq}^{atom} are given by [2]

$$\sigma_{eq}^{sys} = \left[\frac{3}{2}\mathrm{Tr}(\hat{\sigma}^{sys})^2\right]^{1/2} \quad \text{where} \quad \hat{\sigma}^{sys} = \sigma^{sys} - \frac{1}{3}\mathrm{Tr}(\sigma^{sys})\mathbf{I} \qquad (15)$$

$$\sigma_{eq}^{atom} = \left[\frac{3}{2}\mathrm{Tr}(\hat{\sigma}^{atom})^2\right]^{1/2} \quad \text{where} \quad \hat{\sigma}^{atom} = \sigma^{atom} - \frac{1}{3}\mathrm{Tr}(\sigma^{atom})\mathbf{I} \qquad (16)$$

IV. RESULTS AND DISCUSSION

A. Elastic Deformation

In the Voigt notation, the average elastic constants for three atomistic boxes are (accurate to approximately $\pm 0.1\,\mathrm{GPa}$)

$$\mathbf{C}^{atom} = \begin{pmatrix} 7.46 & 4.63 & 4.52 & 0.00 & 0.11 & -0.16 \\ 4.63 & 7.08 & 4.38 & -0.02 & 0.15 & 0.01 \\ 4.52 & 4.38 & 7.03 & -0.06 & 0.10 & -0.03 \\ 0.00 & -0.02 & -0.06 & 1.47 & -0.01 & 0.03 \\ 0.11 & 0.15 & 0.10 & -0.01 & 1.51 & -0.06 \\ -0.16 & 0.01 & -0.03 & 0.03 & -0.06 & 1.55 \end{pmatrix} (\mathrm{GPa}) \qquad (17)$$

Comparing Eq. (17) with Eq. (8) leads us to accept the BPA-PC models as isotropic glasses with Lamé constants and moduli as listed in Table 1. Comparison with experimental values [23] indicates that the agreement is satisfactory.

B. Plastic Deformation

Figure 3 shows stress–strain curves from the simulation runs. Because more than 80% of the system volume is assumed to behave perfectly elastically, the system stress increases almost linearly with the system strain [cf. Eq. (5)].

TABLE 1 Comparison of Calculated Elastic Constants with Experimental Values

Property	Calculated	Experimental [23]
Lamé constant λ (GPa)	4.35 ± 0.10	4.27–5.55
Lamé constant μ (GPa)	1.43 ± 0.06	0.8–1.1
Young's modulus E (GPa)	3.93 ± 0.14	2.3–2.5
Shear modulus G (GPa)	1.43 ± 0.06	0.8–1.1
Bulk modulus B (GPa)	5.30 ± 0.10	5.0–6.1
Poisson's ratio ν	0.38 ± 0.01	0.42–0.43

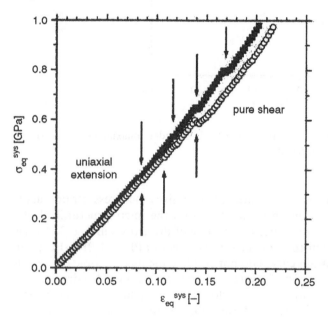

FIG. 3 Stress–strain curves of the system under uniaxial extension (filled squares) and pure shear (open circles).

However, the lines are not smooth but contain numerous small discontinuities where the system stress decreases precipitously after a small strain increment owing to a plastic response of the inclusion; arrows mark the particularly conspicuous "drops." The two curves for uniaxial and pure-shear deformations would superpose perfectly if the system behaved totally elastically; deviations are evident after a sufficient number of plastic unit events, which are irreversible and dissipative and probably different for

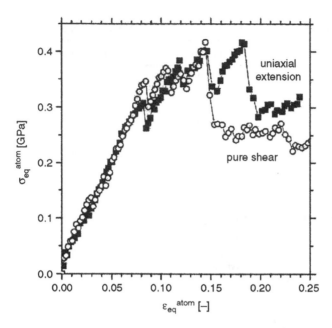

FIG. 4 Stress–strain curves of the atomistic box under uniaxial extension (filled squares) and pure shear (open circles).

different deformation modes. Figure 4, which gives the stress–strain curves for the atomistic inclusion only, clearly shows the microstructural plastic behavior, particularly pronounced at the loci of the arrows in Fig. 3. As has been observed before in the simpler simulations of aPP and BPA-PC [2,3], the atomistic stress–strain curves do not emulate the macroscopic elastic-to-plastic transition, but consist of several straight lines with sudden drops in the stress at some points. Note that the slopes of those lines between plastic events remain nearly constant during deformation; the glassy phases elastically strained between plastic transformations all have the same moduli.

In Fig. 5, the stress of the system is compared with that in the atomistic box. It is apparent that this ratio assumes almost constant values between plastic events with sudden increases at these events. Because the elastic constants of the matrix are those calculated from the atomistic box at the outset of the simulations, the system is homogeneous at small strains. As the system strain increases further, the atomistic box becomes softer than the perfectly elastic matrix ("strain softening").

Interestingly (cf. Fig. 6), the work-equivalent tensile strain of the atomistic box, ε_{eq}^{atom}, is equal to that of the system, ε_{eq}^{sys}, until far beyond the first plastic unit events (at about 8% strain) and only starts to deviate at

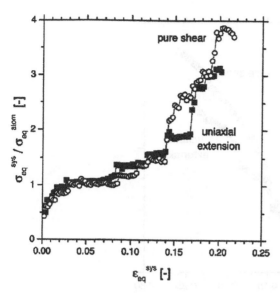

FIG. 5 The ratio of the von Mises-equivalent tensile stress of the system, σ_{eq}^{sys}, to that in the atomistic box, σ_{eq}^{atom}, against the system strain under uniaxial extension (filled squares) and pure shear (open circles).

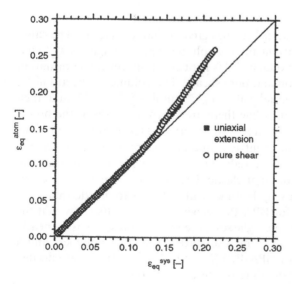

FIG. 6 The work-equivalent tensile strain of the atomistic inclusion, ε_{eq}^{atom}, as a function of the system strain, ε_{eq}^{sys}, under uniaxial extension (filled squares) and pure shear (open circles).

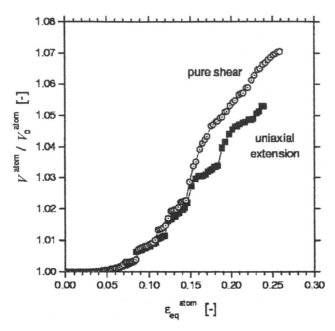

FIG. 7 The volume of the atomistic box, V^{atom}/V_0^{atom}, as a function of the atomistic strain, ε_{eq}^{atom}, under uniaxial extension (filled squares) and pure shear (open circles).

about 15% strain (and thereafter becomes greater than the latter). The cause of this is evident when the atomistic box volume is plotted against ε_{eq}^{atom} (cf. Fig. 7, note that the system volume V^{sys} is kept constant during deformation under both uniaxial extension and pure shear). The volume of the atomistic box V^{atom} remains at its initial value V_0^{atom} until $\varepsilon_{eq}^{atom} \simeq 0.07$ and then gradually increases with strain while there are sharp increases at the points where the sudden drops in the stress are observed. Hence, our simulations indicate that the plastic unit event is a dilatational process, in which the transforming domain assumes a reduced density while the surrounding "matrix" takes up the concomitant elastic deformation. This is in apparent contrast with the deductions by Hutnik et al. [3] who found densification during plastic deformation for BPA-PC (experimental results show that the plastic deformation of polymer glasses at constant pressure is a nearly constant-volume process, with a small dilatation for vinyl polymers and a small volume contraction for BPA-PC) [24–28]. Further studies are needed to gain a satisfactory understanding of these observations.

Because the matrix is modeled as a perfectly elastic medium, the matrix energy is quadratic with respect to the system strain and, consequently, the energy of the continuum portion of the model against strain must exhibit a

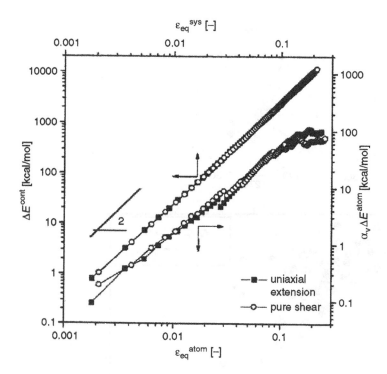

FIG. 8 Energy change in the continuum matrix, $\triangle E^{cont}$, and in the atomistic inclusion, $\alpha_V \triangle E^{atom}$, as a function of the corresponding work-equivalent tensile strains, under uniaxial extension (filled squares) and pure shear (open circles).

slope of 2, plotted doubly logarithmically (cf. Fig. 8). The inclusion energy ($\alpha_V E^{atom}$) shows the same tendency at low strain, but deviates as plastic events begin to soften the material.

The backbone dihedral angles in the constitutional repeat unit can be categorized into three groups [29]: (i) ϕ_1 and ϕ_2, which determine the conformation of the phenylene rings with respect to the isopropylidene group; (ii) ϕ_3 and ϕ_6, which specify the conformation of the phenylene rings with respect to the carbonate group; (iii) ϕ_4 and ϕ_5, which give the conformation of the carbonate group. We now analyze the distribution of the angles in these groups and their changes during deformation.

ϕ_1 and ϕ_2 are strongly interdependent and prefer "propeller-like" conformations. Figure 9 contains a plot of all pairs of values of ϕ_1 and ϕ_2 angles concurrently occurring at one isopropylidene moiety. The intramolecular ground states are located at (45°, 45°) and (135°, −45°) [29]. Upon deformation to $\varepsilon_{eq}^{atom} = 0.256$, these distributions do not change perceptibly

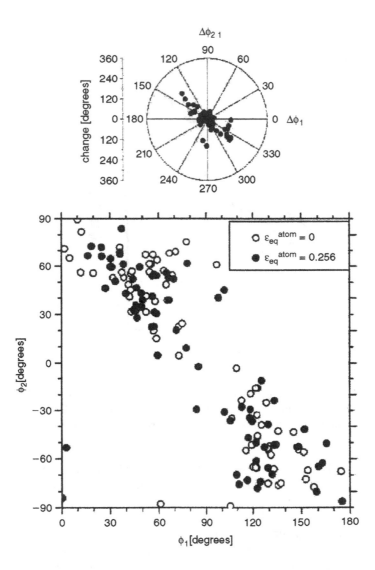

FIG. 9 Distribution of the dihedral angles ϕ_1 and ϕ_2. The large plot contains the values for the pairs (ϕ_1, ϕ_2) before deformation ($\varepsilon_{eq}^{atom} = 0.0$, open circles) and after deformation under pure shear ($\varepsilon_{eq}^{atom} = 0.256$, filled circles). The polar plot indicates the changes in ϕ_1 and ϕ_2, where the angle on the perimeter denotes the direction of the change and the distance from the center denotes the magnitude of the change.

(the pairs of angles are marked with open circles for the unstrained and with filled circles for the deformed state). An analysis of the changes in ϕ_1 and ϕ_2 during this massive deformation experiment reveals that the conformations change mostly from one propeller-like state to a symmetry-related one and that the magnitude of this change often exceeds 180°; the polar plot included in Fig. 9 indicates the direction of the change in the (ϕ_1, ϕ_2) plane on the circumference and its magnitude as the distance from the center. The orientation of this plot is such that the direction of change conforms to the large plot in Fig. 9.

Angles ϕ_3 and ϕ_6 assume values of ±45° at conformations of minimal energy, while ϕ_4 and ϕ_5 are preferably at 0° (trans) and 180° (cis). Figure 10

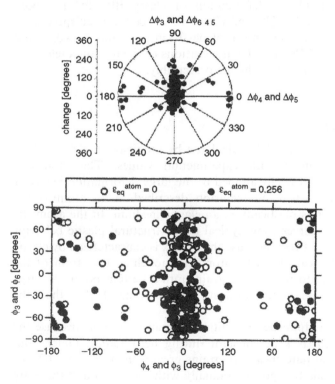

FIG. 10 Distribution of the dihedral angles ϕ_3 (or ϕ_6) and ϕ_4 (or ϕ_5). The large plot contains the values for the pairs [ϕ_3 (or ϕ_6), ϕ_4 (or ϕ_5)] before deformation ($\varepsilon_{eq}^{atom} = 0.0$, open circles) and after deformation under pure shear ($\varepsilon_{eq}^{atom} = 0.256$, filled circles). The polar plot indicates the changes in ϕ_3 (or ϕ_6) and ϕ_4 (or ϕ_5), where the angle on the perimeter denotes the direction of the change and the distance from the center denotes the magnitude of the change.

is constructed similarly to Fig. 9: The large plot exhibits pairs of dihedral angles joined at the same oxygen atom (i.e., ϕ_3 and ϕ_4 or ϕ_6 and ϕ_5), again with open circles for the undeformed state and with filled circles for $\varepsilon_{eq}^{atom} = 0.256$. The polar plot shows the changes in these angle pairs upon strong deformation, again in an orientation that conforms directly to the large plot. It is obvious that a large number of significant conformational transitions occur during the deformation runs, but most involve state changes of the angles adjoining the phenylene units only; changes of the carbonate torsion angle from trans to cis (or vice versa) are rare. And again, the torsion angle distributions do not change perceptibly upon deformation to $\varepsilon_{eq}^{atom} = 0.256$.

The absence of significant change in the distribution of the backbone dihedral angles upon plastic deformation of glassy BPA-PC has been experimentally observed. Utz [30] and Utz et al. [31] compressed specifically ^{13}C labeled samples uniaxially and in plane strain to strains of -0.68 and found, by solid-state NMR, that the distributions of the torsion angle pairs ϕ_3 and ϕ_4 (or ϕ_6 and ϕ_5) do not change perceptibly.

V. CONCLUSIONS

The elastic constants of BPA-PC, calculated by atomistic modeling, show relatively good agreement with experimental values. The atomistic-continuum model is successfully applied to the large deformation behavior of a BPA-PC system where an inclusion described in an atomistically detailed manner deforms plastically in an elastic medium. In the atomistic box, the stress–strain curves show typical microstructural plastic behavior with sharp drops in the stress. Because the matrix is characterized by elastic constants of BPA-PC calculated by atomistic modeling, the strain of the atomistic box is equal to the system strain until a certain point. At large deformations, the strain in the atomistic box becomes higher than the system strain.

The volume of the atomistic box initially remains constant because the volume of the system is kept constant during deformation under both uniaxial extension and pure shear. At strains greater than $\varepsilon_{eq}^{atom} \approx 0.07$, the volume of the atomistic box grows gradually with the strain and there are sharp increases at several points where sudden drops in the stress in the atomistic box are observed. No considerable change in the distribution of the backbone dihedral angles is observed during deformation up to the system strain 0.2.

The emerging picture of plastic unit events in the polymeric glass is that of a sudden dilatation followed by melt-like flow.

REFERENCES

1. Theodorou, D.N.; Suter, U.W. Macromolecules **1986**, *19*, 139.
2. Mott, P.H.; Argon, A.S.; Suter, A.S. Phil. Mag. A **1993**, *67*, 931.
3. Hutnik, M.; Argon, A.S.; Suter, U.W. Macromolecules **1993**, *26*, 1097.
4. Fan, C.F. Macromolecules **1995**, *28*, 5215.
5. Brown, D.; Clarke, J.H.R. Macromolecules **1991**, *24*, 2075.
6. Tadmor, E.B.; Phillips, R.; Ortiz, M. Langmuir **1996**, *12*, 4529.
7. Miller, R.; Phillips, R. Phil. Mag. A **1996**, *73*, 803.
8. Tadmor, E.B.; Ortiz, M.; Phillips, R. Phil. Mag. A **1996**, *73*, 1529.
9. Shenoy, V.B.; Miller, R.; Tadmor, E.B.; Phillips, R.; Ortiz, M. Phys. Rev. Lett. **1998**, *80*, 742.
10. Rafii-Tabar, H.; Hua, L.; Cross, M. J. Phys.: Condens. Matter **1998**, *10*, 2375.
11. Santos, S.; Yang, J.S.; Gusev, A.A.; Jo, W.H.; Suter, U.W. submitted for publication.
12. Gusev, A.A. J. Mech. Phys. Solids **1997**, *45*, 1449.
13. Parrinello, M.; Rahman, A. J. Appl. Phys. **1981**, *52*, 7182.
14. Parrinello, M.; Rahman, A. J. Chem. Phys. **1982**, *76*, 2662.
15. Tanemura, M.; Ogawa, T.; Ogita, N. J. Comp. Phys. **1983**, *51*, 191.
16. Mott, P.H.; Argon, A.S.; Suter, A.S. J. Comp. Phys. **1992**, *101*, 140.
17. Hutnik, M.; Argon, A.S.; Suter, U.W. Macromolecules **1991**, *24*, 5956.
18. Press, W.H.; Teukolsky, S.A.; Vetterling, W.T.; Flannery, B.P. *Numerical Recipes*, 2nd Ed.; Cambridge University Press: Cambridge, 1992.
19. *Discover 2.9.8/96.0/4.0.0 User Guide*; Molecular Simulations Inc.: San Diego, CA, 1996.
20. Maple, J.R.; Hwang, M.-J.; Stockfisch, T.P.; Dinur, U.; Waldman, M.; Ewig, C.S.; Hagler, A.T. J. Comp. Chem. **1994**, *15*, 162.
21. Sun, H.; Mumby, S.J.; Hagler, A.T. J. Am. Chem. Soc. **1994**, *116*, 2978.
22. Nye, J.F. *Physical Properties of Crystals*; Oxford University Press: Oxford, 1985.
23. Van Krevelen, D.W. *Properties of Polymers*, 3rd Ed.; Elsevier: New York, 1990.
24. Christ, B. In *The Physics of Glassy Polymers*, 2nd Ed.; Haward, R.N., Young, R.J., Eds.; Chapman & Hall: London, 1997; Chapter 4.
25. Spitzig, W.A.; Richmond, O. Polym. Eng. Sci. **1979**, *19*, 1129.
26. Ruan, M.Y.; Moaddel, H.; Jamieson, A.M. Macromolecules **1992**, *25*, 2407.
27. Xie, L.; Gidley, D.W.; Hristov, H.A.; Yee, A.F. J. Polym. Sci., Polym. Phys. Ed. **1995**, *33*, 77.
28. Hasan, O.A.; Boyce, M.C.; Li, X.S.; Berko, S. J. Polym. Sci., Polym. Phys. Ed. **1993**, *31*, 185.
29. Hutnik, M.; Gentile, F.T.; Ludovice, P.J.; Suter, U.W.; Argon, A.S. Macromolecules **1991**, *24*, 5962.
30. Utz, M. Ph.D. thesis, Eidgenössische Technische Hochschule, Zürich, 1998.
31. Utz, M.; Atallah, A.S.; Robyr, P.; Widmann, A.H.; Ernst, R.R.; Suter, U.W. Macromolecules **1999**, *32*, 6191.

13

Polymer Melt Dynamics

WOLFGANG PAUL Johannes-Gutenberg-Universität Mainz, Mainz, Germany

MARK D. EDIGER University of Wisconsin–Madison, Madison, Wisconsin, U.S.A.

GRANT D. SMITH University of Utah, Salt Lake City, Utah, U.S.A.

DO Y. YOON Seoul National University, Seoul, Korea

I. INTRODUCTION

Polymer melts and glassy amorphous polymers constitute many of the "plastic" materials we are used to in everyday life. Besides this technological importance, there is also fundamental interest in the structural and dynamical properties of complex molecules with their connectivity constraints. We will focus in this chapter on the use of chemically realistic polymer models for the study of the static and dynamic behavior of polymer melts with an excursion into the glass transition dynamics of polymer melts as revealed by simulations of a coarse-grained bead-spring model. The basic concepts and algorithms we will present are valid for both types of models. With the chemically realistic models we can aim for a quantitative comparison with experiment and a detailed mechanistic understanding of segmental motions in real polymers. The bead-spring type models allow for studies of universal polymer properties and in general for simulations in a broader parameter range (lower temperatures, larger chains, longer simulation times).

II. MODELS AND DATA STRUCTURES

The types of models we will be using have already been described in Chapter 1. The potential energy can be written as

$$\mathcal{V} = \mathcal{V}_B + \mathcal{V}_{NB} \tag{1}$$

407

with the bonded energy terms

$$V_B = \sum_b V_b + \sum_\theta V_\theta + \sum_\phi V_\phi$$

given as

$$V_b = \text{constraint} \quad \text{or} \quad V_b = \frac{1}{2}k_b(b - b_0)^2 \tag{2}$$

$$V_\theta = \frac{1}{2}k_\theta(\theta - \theta_0)^2 \quad \text{or} \quad V_\theta = \frac{1}{2}k'_\theta(\cos(\theta) - \cos(\theta_0))^2 \tag{3}$$

$$V_\phi = \sum_{n=0}^{6} A_n \cos(n\phi) \tag{4}$$

The force constants in the bond length potential of chemically realistic models generally are the largest of all the energy parameters and therefore lead to high frequency oscillations. These very fast motions are often constrained in the simulation for performance purposes. For small oscillations around the equilibrium angle the two forms for the bond angle potential V_θ can be transformed into each other. Typically one considers at most six terms in the expansion of the dihedral angle potential V_ϕ (see Fig. 1) and some of the coefficients may be zero. The parameters in the bonded interaction potentials can be mostly obtained from spectroscopic information, with the exception of the height of the barriers in the dihedral potential. A quantitatively accurate determination of these is only possible from quantum chemical *ab initio* calculations [1,2]. Structural relaxation of the chain conformation as well as of the configuration of the melt requires transitions between the isomeric states of the dihedral potential. The crossing of the barriers is a thermally activated process and therefore exponentially sensitive to the height of these barriers.

For the nonbonded interactions in Eq. (1) we will consider the following contributions. A repulsion/dispersive interaction either of the Lennard-Jones form

$$V_{LJ}(r_{ij}) = 4\epsilon \left(\left(\frac{\sigma}{r_{ij}}\right)^{12} - \left(\frac{\sigma}{r_{ij}}\right)^6 \right) \tag{5}$$

or of the exponential-6 form

$$V_{E6}(r_{ij}) = \epsilon \left[\exp\{-r_{ij}/\xi\} - \left(\frac{r_0}{r_{ij}}\right)^6 \right] \tag{6}$$

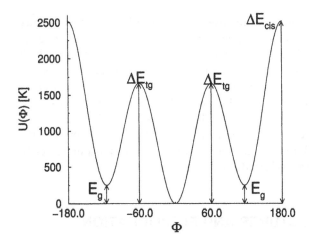

FIG. 1 Dihedral potential for polyethylene (PE). E_g is the trans–gauche energy difference and ΔE_{tg} and ΔE_{cis} are the barriers for torsional transitions.

where ξ and r_0 are the length scales of the repulsive and dispersive parts, respectively, and a Coulomb term

$$\mathcal{V}_q(r_{ij}) = -\frac{1}{4\pi\epsilon_0} \frac{q_i q_j}{r_{ij}} \tag{7}$$

The partial charges q_i as well as repulsive interactions can be parameterized from quantum chemical calculations. The parameters of the dispersion forces are not true two-body interaction terms but rather are parameters for an effective two-body potential appropriate for a condensed phase environment: they are parameterized by comparison with thermodynamic properties such as pVT data, heats of vaporization, or phase behavior. In general one would also need to consider the polarizability of the atoms to account for time dependent partial charges, but we can neglect these effects for the systems we will discuss in the following.

We will consider three different classes of models: all or explicit atom models, united atom models, and bead-spring models. All fit into the above description and can be simulated using the same MD code such as described for instance in [3]. The nonbonded interactions are calculated from the atom positions and atom types of the interacting pairs. The long-range Coulombic forces need a special treatment for an efficient simulation, and this problem is discussed in Chapter 8 by Andrey V. Dobrynin. The standard way to treat the bonded interactions is to use topology files for the 2-body (bond lengths), 3-body (bond angles), and 4-body (dihedral angles) interactions. These files describe the chain structure in specifying which atom is bonded to which others, which three atoms participate in a given bond angle degree of freedom, and which four atoms participate in a given dihedral angle degree of freedom. The bonded interactions can then be calculated by working through these lists. In this way a bead-spring model and an explicit atom model just differ in the topology files specifying the dynamic degrees of freedom, and the potential form and parameter set to use with one specific degree of freedom.

III. STARTING STRUCTURES AND EQUILIBRATION

When the model is specified, we have to generate an equilibrated starting configuration, from which a time series of configurations will be calculated for later measurement of static and dynamic properties. The thermodynamic state point for the simulation can be specified by fixing the number of particles in the simulation box, N, and any of the pairs energy/volume (NVE), temperature/volume (NVT), or temperature/pressure (NpT) and, of course, the chain length, K. Experiments are done at constant pressure and temperature. For the simulations one usually fixes the temperature by using a thermostat like the Nosé–Hoover [4] one and proceeds in the following way. The simulation is performed at a preselected pressure until the time averaged density reaches a stationary value and then one switches to a constant density simulation at the thus determined density. For this simulation the pressure will fluctuate around the selected equilibrium pressure, for instance ambient pressure, and the static and dynamic properties in the simulation will be representative of that pressure. This argument assumes that the system size is reasonably large, so that the fluctuations are small compared to the average value. When one uses a chemically realistic model with a carefully validated force field, one will find that the equilibrium density obtained in that way will only differ by about 1% from the experimental density of that polymer at the selected temperature and pressure [2]. If the force field is not quantitatively correct, comparison with experiment will turn out better when performed at the same pressure and not at the same density.

To perform the equilibration discussed above, we first have to generate a dense configuration of nonoverlapping polymers. There are two standard ways to do this. The first is to start with a big simulation volume in a dilute situation and to stepwise decrease the volume in a first part of the simulation until the desired pressure is approximately reached. The second way is to estimate the target density and grow the chains into a box of the corresponding volume without considering the nonbonded interactions. In the first part of the simulation the repulsive part of the nonbonded interaction is then regularized to a finite value at distance zero, which is in turn increased in a stepwise fashion until all overlaps have disappeared [5].

When we have generated the system at the desired temperature and pressure/density we still have to ensure that it is equilibrated with respect to the other observables of the system. In general, equilibration of a system can be judged by looking at the quantity with the longest relaxation time, which in an amorphous polymer melt is the end-to-end distance of the chains. When we are looking at chains which are long enough to obey random walk statistics for the end-to-end distance, this equilibration can be quantified by looking at the Gaussianity of the end-to-end vector distribution $P(R)$ [6] or at the decay of the end-to-end vector autocorrelation function $\langle \mathbf{R}(t) \cdot \mathbf{R}(0) \rangle$. Typically, properties which are sensitive to shorter distance behavior equilibrate faster than those living on larger scales. When there is no coupling between the length scales, it may be sufficient to equilibrate the scale one wants to study. For instance, when looking at the structural relaxation in a polymer melt as a function of temperature, it may be sufficient to ensure equilibration on the scale of one (or a few) statistical segments of the chain, but this can only be decided *a posteriori*, if one has independent measurements to compare with. If there is a coupling to larger scale behavior one has a quenched structure and static as well as dynamic quantities will depend on the quenching rate.

IV. STATIC PROPERTIES

The first static property characterizing the system is the pVT behavior, which was already discussed in the last section. Besides this thermodynamic information, an amorphous polymer melt is mainly characterized through two types of structural information: the single chain structure factor and the overall structure factor of the melt.

The single chain structure factor in Fig. 2 [7] is calculated using the united atom positions with an unspecified value for the coherent neutron scattering length and therefore normalized such that the $q = 0$ value is unity. This is valid if we are only interested in the low q behavior of the scattering of a homopolymer (one type of scatterer), i.e., the first curved part, and the straight line in the double logarithmic plot indicating a self-similar structure on these length scales. This part of the structure factor can be fitted with the

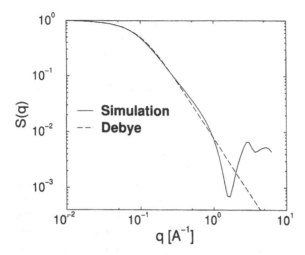

FIG. 2 Single chain structure factor for a united atom model of polyethylene calculated for neutron scattering [7].

Debye function, which is the analytical expression for the structure factor of a random walk,

$$S_{Deb}(qR_g) = \frac{2}{q^4 R_g^4}\left(\exp\{-q^2 R_g^2\} - 1 + q^2 R_g^2\right) \tag{8}$$

which is the best experimental way to determine the radius of gyration of the chains. In the simulation it can be obtained in the same way and also by direct calculation

$$R_g^2 = \frac{1}{K}\sum_{i=1}^{K}(\mathbf{r}_i - \mathbf{R}_{cm})^2 \tag{9}$$

where K is the number of atoms per chain and $\mathbf{R}_{cm}$ is the center of mass of the chain. For larger q the structure factor will be sensitive to the actual positions of the true scattering centers in the melt, i.e., in this case the positions of the carbons and deuteriums and their respective coherent scattering lengths. This is also true for the global structure factor of the melt, which therefore has to be calculated as a superposition of the partial structure factors of all the atom types

$$S_{coh}(q) = 1 + \frac{\rho}{\sigma^{coh}}\sum_{\nu=1}^{M}\sum_{\mu=1}^{M}x_\nu x_\mu \sqrt{\sigma_\nu^{coh}\sigma_\mu^{coh}}\,\hat{h}_{\nu\mu}(q) \tag{10}$$

Here ν and μ label the M different atom types, the x_ν are their number fractions, the σ_ν^{coh} are their coherent scattering cross sections, ρ is the density of scattering centers, and the average coherent scattering cross section is defined as

$$\overline{\sigma^{coh}} = \sum_{\nu=1}^{M} x_\nu \sigma_\nu^{coh} \tag{11}$$

The functions $\hat{h}_{\nu\mu}(q) = \mathcal{F}[g_{\nu\mu}(r) - 1]$ are the Fourier transforms of the pair correlation functions

$$g_{\nu\mu}(r) = \frac{1}{N\rho x_\nu x_\mu} \left\langle \sum_{i=1}^{N_\nu} \sum_{j=1}^{N_\mu} \delta\big(|\mathbf{r}_i - \mathbf{r}_j| - r\big) \right\rangle \tag{12}$$

For the calculation of the melt structure factor the indices i and j in the above equation run over all atoms in the simulation volume, whereas for the calculation of the single chain structure factor they only run over the atoms of that chain. Equation (10) is written in a notation suggestive of neutron scattering, when we only have nuclear scattering of the neutrons and the scattering amplitude can be expressed as a scattering length. When we want to compare a simulated structure with X-ray scattering experiments on that structure, we have to take into account the wavelength dependence of the atomic form factor and replace $\sqrt{\sigma_\nu^{coh}}$ by $f_\nu(q)$.

The above discussed information is, of course, contained in an explicit atom simulation, and to a certain degree also contained quantitatively in a united atom simulation. This information is only qualitatively contained in a bead-spring type simulation. For the united atom model we can go beyond calculating the scattering of the united atom centers—which would be similar to the bead-spring system—and reinsert the hydrogen atoms into the simulated configurations [8]. For this, the hydrogen atoms are placed into their $T=0$ mechanical equilibrium positions (C-H bond length, C-C-H bond angle, and H-C-H bond angle are fixed to their mechanical equilibrium values) so that all thermal fluctuations of these degrees of freedom are neglected. Nevertheless, the distribution of scattering lengths in the simulated volume resembles closely that of an explicit atom simulation. Figure 3 shows the X-ray structure factor of the alkane $C_{44}H_{90}$ from a united atom [8] and an explicit atom simulation [9]. The intensities are not normalized as in Eq. (10). For both simulations the position and width of the amorphous halo (first peak) correspond well with experimental results. The amorphous halo is the most prominent feature of the structure factor

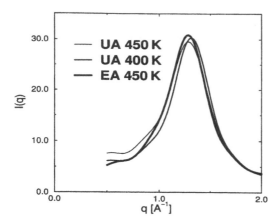

FIG. 3 X-ray structure factor of $C_{44}H_{90}$ compared between a united atom and an explicit atom simulation [8].

and representative of the amorphous state (melt as well as glass). Its properties qualitatively agree between chemically realistic and coarse-grained models. Upon constant pressure cooling of a polymer melt, the position of the amorphous halo moves to larger q and its width decreases, but there is no additional structure appearing upon glassy solidification.

V. DYNAMIC PROPERTIES

In the discussion of the dynamic behavior of polymer melts we will start with the local reorientational motion of chain segments. Experimentally this motion can for example be observed in Nuclear Magnetic Resonance (NMR) experiments measuring the ^{13}C spin lattice relaxation time. Owing to the different chemical shift for different environments of these nuclei, this measurement can be very specific for selected positions in the polymer chain [8–10]. It turns out that this reorientation dynamics is mainly determined by torsional transitions of dihedral angles adjacent to the relaxing CH vector. We discussed above that these transitions are exponentially sensitive to the barrier heights in the torsion potentials. These in turn can only be determined with an uncertainty of about 100K even through high-level quantum chemical calculations. Furthermore, we introduce some error in the torsional dynamics when we use a united atom representation of the chain. Although both errors may be small, they get exponentially amplified in the torsional transition rate and therefore a comparison with experimental spin lattice relaxation time measurements (or their equivalent) is a crucial test to validate the torsional force fields and to perform some final small adjustments of barrier heights, when necessary.

For the dipolar relaxation mechanism, the spin lattice relaxation time is sensitive to the reorientation dynamics of the CH bond vectors. Different orientations of the CH bond result in slightly different magnetic fields at the carbon nucleus and the modulation of this field allows the spin flips to occur. When we define $\hat{e}_{CH}$ as the unit vector along a CH bond, the second Legendre polynomial of its autocorrelation function is given by

$$P_2(t) = \frac{1}{2}\left(3\langle(\hat{e}_{CH}(t) \cdot \hat{e}_{CH}(0))^2\rangle - 1\right) \tag{13}$$

For an isotropic system the spin lattice relaxation time T_1 and a related quantity, the Nuclear Overhauser Enhancement (NOE), can be determined from the power spectrum of this relaxation function.

$$J(\omega) = \frac{1}{2}\int_{-\infty}^{\infty} P_2(t)e^{i\omega t}dt \tag{14}$$

$$\frac{1}{nT_1} = K\left[J(\omega_H - \omega_C) + 3J(\omega_C) + 6J(\omega_H + \omega_C)\right] \tag{15}$$

$$\text{NOE} = 1 + \frac{\gamma_H}{\gamma_C}\frac{6J(\omega_H + \omega_C) - J(\omega_H - \omega_C)}{J(\omega_H - \omega_C) + 3J(\omega_C) + 6J(\omega_H + \omega_C)} \tag{16}$$

Here ω_C and ω_H are the Larmor frequencies of carbon and hydrogen atoms respectively, n is the number of bound hydrogens per carbon, $K = 2.29 \times 10^9\,\text{s}^{-2}$ for sp^3 hybridization, $K = 2.42 \times 10^9\,\text{s}^{-2}$ for sp^2 hybridization, and γ_H and γ_C are the gyromagnetic ratios of hydrogen and carbon ($\gamma_H/\gamma_C = 3.98$).

The CH vector autocorrelation function can be readily measured in a simulation and resolved for specific positions on the chain. The angle brackets in Eq. (13) indicate an averaging procedure over chemically identical carbons in the system and over multiple time origins along the generated trajectory of the model system. To calculate its Fourier transform, this relaxation function is then best fit either with some model predictions, for instance based on local conformational dynamics plus Rouse mode contributions [11], or with some phenomenological fit function. In our experience, $P_2(t)$ is generally very well reproduced by a superposition of an

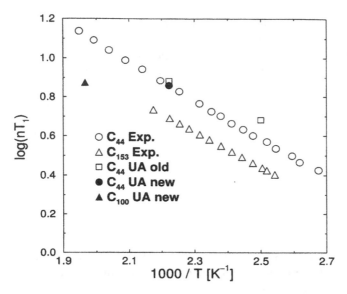

FIG. 4 Temperature dependence of the spin lattice relaxation time for two different alkanes as seen in experiment (performed at 75 MHz carbon resonance frequency) and MD simulation [11].

exponential part and a stretched exponential Kohlrausch–Williams–Watts (KWW) part.

$$P_2(t) = A \exp\left\{-\left(\frac{t}{\tau_{KWW}}\right)^{\beta}\right\} + (1 - A)\exp\left\{-\frac{t}{\tau_{exp}}\right\} \tag{17}$$

For short PE chains the relaxation behavior of the four terminal carbons at the chain end can be resolved in the NMR experiment and the inner part of the chain gives rise to a collective resonance. In Fig. 4 we show the behavior of the spin lattice relaxation time for these inner carbons as a function of temperature for two alkanes of different chain length, n-$C_{44}H_{90}$ and n-$C_{100}H_{202}$. This seemingly local quantity depends strongly on the chain length even for these center carbons [11]. Inspecting directly the P_2 function obtained in the simulation we can see that this is due to the fact that a few percent of the correlations measured by this function decay not through local motion but through a coupling to the much slower overall conformational relaxation of the chains. This effect is also nicely captured in the simulations. For the united atom simulations of C_{44} we can furthermore see two sets of data points. The one agreeing perfectly with the experimental results stems from an improved force field, where the torsional barriers in

Fig. 1 were slightly raised to improve the agreement with the experimental data. Other experimental observables such as dielectric relaxation [12] and the coherent dynamic structure factor for neutron spin echo [13] have also been calculated from simulations and used to judge the realism of the local relaxation processes in the simulations.

Simulations of local relaxation in polymers have provided several important qualitative insights about the mechanisms of molecular motion. When a conformational transition occurs, nearby atoms along the chain backbone must adjust their positions to accommodate the transition. This cooperativity length scale is not accessible experimentally and identifies the fundamental kinetic unit (which may be the relevant motion for dynamics near the glass transition temperature). Simulations on polyisoprene [14], polyethylene [8,15], and other flexible polymers indicate that 4–6 carbon atoms along the chain backbone are involved in localizing a conformational transition. Simulations have also shown significant spatial heterogeneity in conformational transition rates [15] and the importance of motions other than conformational transitions in local relaxation processes [16,17].

For an analysis of the overall conformational relaxation of a polymer chain we focus here on the chain dynamics on length scales which show the self-similar random walk behavior statically. Imagine looking at a given polymer chain in a stroboscopic fashion and from a certain distance. The center of mass of the chain will move (at long times it has to diffuse) and the conformation of the chain will change from one random walk realization to a different one. The simplest model for polymer melt dynamics which captures these effects is the Rouse model [18], in which the polymers are modeled by a chain of phantom particles connected by entropic springs (this generates the Gaussian random walk conformations) moving under the impact of stochastic forces in an unstructured background characterized by a friction coefficient.

$$\xi d\mathbf{r}_i(t) = -\frac{3k_B T}{b_K^2}(\mathbf{r}_{i+1}(t) - 2\mathbf{r}_i(t) + \mathbf{r}_{i-1}(t))dt + \sqrt{2\xi k_B T}\, d\mathbf{W}_i(t) \tag{18}$$

The statistical segment length or Kuhn length b_K is the step length of the random walk with the same large scale structure as the chain under study and the segmental friction ξ and the strength of the stochastic force are connected through the fluctuation-dissipation theorem

$$\langle dW_{i\alpha}(t)dW_{j\beta}(t')\rangle = 2\xi k_B T dt \delta_{ij}\delta_{\alpha\beta}\delta(t-t') \tag{19}$$

Equation (18) can be solved analytically through a transformation to the normal modes of a harmonic chain and many exact predictions can be derived. Since there is only a random force acting on the center of mass of a chain moving according to the Rouse model, the center of mass motion is purely diffusive with a diffusion coefficient $D = k_B T/\xi K$. Of interest to us is an analytical prediction for the single chain coherent intermediate dynamic structure factor. This is the dynamic counterpart to the static single chain structure factor we discussed earlier on and this quantity is measured in Neutron Spin Echo (NSE) experiments. The Rouse model yields [18]

$$
\begin{aligned}
S(q, t) = \frac{1}{K} \exp\{-q^2 D t\} \\
\times \sum_{n,m=1}^{K} \exp\left\{ -\frac{q^2 b_K^2}{6} |n - m| - \frac{2 K q^2 b_K^2}{3\pi^2} \right. \\
\left. \times \sum_{p=1}^{K} \cos\left(\frac{p\pi n}{K}\right) \cos\left(\frac{p\pi m}{K}\right)\left(1 - e^{-p^2 t/\tau_R}\right) \right\}
\end{aligned}
\tag{20}
$$

where $\tau_R = \xi K^2 b_K^2/3\pi^2 k_B T$ is the so-called Rouse time. For small momentum transfers q, one is only sensitive to the motion of the polymer chain as a whole and the above equation simplifies to the scattering law for diffusive motion $S(q, t) = S(q, 0) \exp(-Dq^2 t)$. In this way one can determine the center of mass self-diffusion coefficient from the NSE data and of course also from the computer simulation. The computer simulation furthermore also has access to this property by measuring directly the average mean square displacement of the center of mass of a chain and using Einstein's relation

$$
D = \lim_{t \to \infty} \frac{\langle \Delta R_{cm}^2(t) \rangle}{6t}
\tag{21}
$$

Through the diffusion coefficient the segmental friction ξ is determined and from the single chain static structure factor Eq. (10) one can obtain the radius of gyration and from that the statistical segment length b_K entering the Rouse model. Again, in the simulation this quantity can be independently determined by direct calculation of the mean squared end-to-end distance of the chains. In this way both parameters (b_K, ξ) entering the analytical predictions of the Rouse model can be determined and the model critically tested. This can be done on the coarse-grained bead-spring model level as well as with a fully atomistic model and it has been found that the Rouse model gives a good quantitative description only on its largest intrinsic length and time scales. On small length scales (large q) the local structure and stiffness of the chains cannot be captured by entropic springs and on small time scales it turns out that the interactions between the chains in the melt cannot be captured by a homogeneous background friction ξ but lead to a subdiffusive motion of the centers of mass of the chains [19–22]. It also leads to a non-Gaussian distribution for the mutual displacement $|\mathbf{r}_i(t)-\mathbf{r}_j(0)|$ of two atoms on the same chain [22].

This, in principle, invalidates all theoretical predictions like the Rouse result for the intermediate dynamic structure factor, because they all rely on the dynamic Gaussian assumption

$$\langle\exp[i\mathbf{q}\cdot(\mathbf{r}_i(t)-\mathbf{r}_j(0))]\rangle = \exp\left[-\frac{q^2}{6}\langle(\mathbf{r}_i(t)-\mathbf{r}_j(0))^2\rangle\right] \tag{22}$$

In the simulation one can prove that the above assumption is not fulfilled by calculating the scattering function correctly averaging over the phase factors [left hand side of Eq. (22)] or by employing the dynamic Gaussian assumption [right hand side of Eq. (22)]. Figure 5 shows this comparison for C_{100} at 509K. One can clearly see, that the dynamic Gaussian assumption leads to an overestimation of the decay of correlations and to a wrong prediction of the scattering.

A quantitative comparison with experiments, however, can only be performed for a chemically realistic simulation. For this the positions of the scatterers in the time dependent generalization of Eqs. (10) and (12) run over

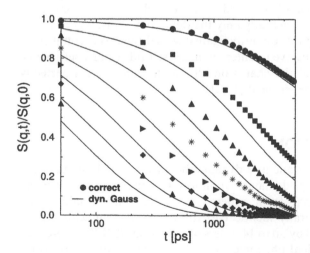

FIG. 5 Comparison of single chain intermediate dynamic structure factor of polyethylene at $T=509\mathrm{K}$ for the correct calculation and one assuming Gaussian distributed relative atom displacements.

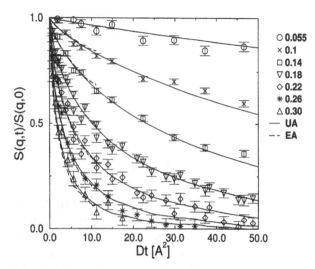

FIG. 6 Comparison of single chain intermediate dynamic structure factor of polyethylene at $T = 509K$ between NSE experiments and chemically realistic MD simulations [21].

the carbon and deuterium atoms of a single chain and are evaluated with a time displacement t. Figure 6 shows a comparison of the experimental result for the single chain intermediate dynamic structure factor of polyethylene at $T = 509K$ with simulations of a united atom as well as an explicit atom model. For this comparison actually only the deuterium scattering has been evaluated. When plotting this quantity against scaled time to account for a 20% difference in the observed center of mass self-diffusion coefficients, the relaxation behavior of the chains on all length scales agrees perfectly between simulation and experiment [10,21].

VI. GLASS TRANSITION

So far we have discussed static and dynamic behavior of high temperature polymer melts of unentangled chains. Upon lowering the temperature many polymeric systems do not form ordered structures but remain in an amorphous state, structurally identical to the melt state but with typical relaxation times increased by up to 14 orders of magnitude [23]. This so-called glass transition is a gradual phenomenon with an onset that is observable in the slightly undercooled melt around a temperature where the typical time scales have increased by about 3 orders of magnitude from their high temperature values. This is a range of time scales that at the moment is accessible only for simulations of coarse-grained models when one wants to perform equilibrium cooling. If one performs simulations at lower temperatures, one typically is in a quench situation where the results in general depend on the quenching rate one uses. For MD simulations this quench rate is at least 9 orders of magnitude larger than experimental rates. There may be properties of the supercooled melts which decouple from the slowest degrees of freedom in the melt but it is *a priori* unclear which these are and the assignment may even differ from one polymer to another. We have already seen that even a seemingly local quantity like the CH vector reorientation as measured by spin lattice relaxation time experiments couples to the overall conformational relaxation of the polymer chain, which defines the slowest mode in the system. It has been found phenomenologically in some polymers [24], that MD simulations employing high quench rates were actually able to identify the glass transition temperature of these polymers through the jump in the thermal expansion coefficient rather accurately, but this is not a general finding.

We will focus now on the temperature region where the onset of the glassy freezing first starts to be felt. This temperature region is the regime of applicability of mode coupling theory (MCT) [25], which predicts a slowing down of the relaxation processes in the system because particles get trapped in cages formed by their neighbors. This caging process leads to the development of a two-stage relaxation behavior, which is, for instance, observable in intermediate scattering functions (experiment and simulation) or directly in the mean square displacement of the atoms (simulation). The theory is formulated for simple liquids like hard-sphere or soft-sphere systems and we will therefore ask ourselves in the following why and to what degree it can describe polymer behavior.

For this we will use a type of scattering function we have not discussed so far, which is the incoherent intermediate scattering function. Due to its large incoherent scattering amplitude it is mainly the self-correlation of the hydrogen motion that is seen in incoherent neutron scattering experiments.

$$S_{inc}(\mathbf{q}, t) = \frac{1}{M} \sum_{j=1}^{M} \exp\{i\mathbf{q} \cdot (\mathbf{r}_j(t) - \mathbf{r}_j(0))\} \tag{23}$$

where M is the number of hydrogens in the system and where we dropped the scattering amplitude which only adjusts the absolute scale. For an isotropic amorphous melt one can perform a spherical average of this equation to get

$$S_{inc}(q, t) = \frac{1}{M} \sum_{j=1}^{M} \frac{\sin(q|\mathbf{r}_j(t) - \mathbf{r}_j(0)|)}{q|\mathbf{r}_j(t) - \mathbf{r}_j(0)|} \tag{24}$$

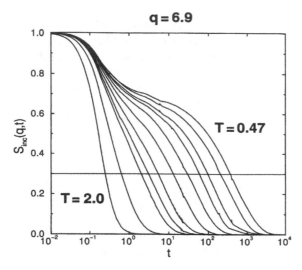

FIG. 7 Intermediate incoherent scattering function of a bead-spring polymer of length 10 beads, approaching the mode coupling temperature which is $T = 0.45$ in Lennard-Jones units [26].

We will use this equation to look at the incoherent intermediate structure factor of a bead-spring model without any bond angle or dihedral potential as we approach this MCT temperature region [26,27]. Here the index j runs over all beads in the system. We will do this for the q-value of the position of the amorphous halo in the liquid structure factor of the model. As we can see in Fig. 7 a two-step relaxation process develops upon cooling which can also—for this simplified polymer model—be reasonably well analyzed in terms of the predictions of MCT [26]. The long time behavior of these curves can be fitted by a KWW form and the time scale entering this function is the so-called structural or α-relaxation time scale that shows the dramatic increase by more than 14 orders of magnitude between the melt and the glass. This time can also be defined by looking at the decay to the 30% level indicated by the horizontal line.

When we look directly at the mean square displacements of the beads in the simulation which is shown in Fig. 8, we can see the caging by the neighbors resulting in a plateau regime. The height of the plateau is only a fraction of the bond length in the model ($l = 1$), so that effectively the beads have actually "not yet noticed" that they are bound to a chain and are not free particles, which is why the theory which was developed for simple liquids could be reasonably applied in this case [26,27]. The same behavior can be observed in the center of mass motion of the chains. The long time

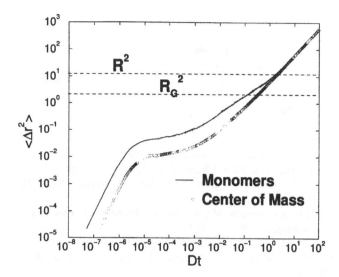

FIG. 8 Master curve for the mean square displacements of the beads and of the centers of mass of the chains obtained by plotting the common part of the mean square displacement curves for all temperatures as a function of time scaled by the respective diffusion coefficient of that temperature [27].

behavior of the mean square displacements is still the same as would be predicted by the Rouse model we discussed above. The monomer friction coefficient, however, that sets the typical time scales for the Rouse modes, has been strongly increased due to this caging effect that is responsible for the glass transition.

VII. OUTLOOK

In this section we have discussed simulations of polymer melt dynamics using chemically realistic as well as coarse-grained polymer models. We have seen that both have their complementary uses and that these studies can help to test theories as well as help in the interpretation of experimental data. With the development of ever faster computers and the invention of highly sophisticated equilibration techniques it will become possible to extend the range of temperatures and chain lengths that can be studied in simulations significantly and this will further enhance the contribution that computer simulations of polymer melts and glasses can make to our understanding especially of the relaxation processes occurring in these systems.

REFERENCES

1. Smith, G.D.; Jaffe, R.L. J. Phys. Chem. **1996**, *100*, 18718.
2. Smith, G.D.; Paul, W. J. Phys. Chem. A **1998**, *102*, 1200.
3. Lucretius MD code and sample data files for PE: http://www.che.utah.edu/~gdsmith/mdcode/main.html.
4. Nosé, S. J. Chem. Phys. **1984**, *81*, 511.
5. Kremer, K.; Grest, G.S. J. Chem. Phys. **1990**, *92*, 5057.
6. Yoon, Do, Y. unpublished.
7. Paul, W.; Smith, G.D.; Yoon, Do.Y. Macromolecules **1997**, *30*, 7772.
8. Paul, W.; Yoon, Do.Y.; Smith, G.D. J. Chem. Phys. **1995**, *103*, 1702.
9. Smith, G.D.; Yoon, Do.Y.; Zhu, W.; Ediger, M.D. Macromolecules **1994**, *27*, 5563.
10. Smith, G.D.; Paul, W.; Monkenbusch, M.; Willner, L.; Richter, D.; Qiu, X.H.; Ediger, M.D. Macromolecules **1999**, *32*, 8857.
11. Qiu, X.H.; Ediger, M.D. Macromolecules **2000**, *33*, 490.
12. Smith, G.D.; Yoon, Do.Y.; Wade, C.G.; O'Leary, D.; Chen, A.; Jaffe, R.L. J. Chem. Phys. **1997**, *106*, 3798.
13. Moe, N.E.; Ediger, M.D. Phys. Rev. E **1999**, *59*, 623.
14. Moe, N.E.; Ediger, M.D. Polymer **1996**, *37*, 1787.
15. Boyd, R.H.; Gee, R.H.; Han, J.; Hin, Y. J. Chem. Phys. **1994**, *101*, 788.
16. Moe, N.E.; Ediger, M.D. Macromolecules **1995**, *28*, 2329.
17. Moe, N.E.; Ediger, M.D. Macromolecules **1996**, *29*, 5484.
18. Doi, M.; Edwards, S. *The Theory of Polymer Dynamics*; Clarendon Press: Oxford, 1986.
19. Paul, W.; Heermann, D.W.; Kremer, K.; Binder, K. J. Chem. Phys. **1991**, *95*, 7726.
20. Kopf, A.; Dünweg, B.; Paul, W. J. Chem. Phys. **1997**, *107*, 6945.
21. Paul, W.; Smith, G.D.; Yoon, Do.Y.; Farago, B.; Rathgeber, S.; Zirkel, A.; Willner, L.; Richter, D. Phys. Rev. Lett. **1998**, *80*, 2346.
22. Smith, G.D.; Paul, W.; Monkenbusch, M.; Richter, D. J. Chem. Phys., **2001**, *114*, 4285.
23. Proceedings of the Third International Discussion Meeting on Relaxation in Complex Systems, J. Non-Cryst Solids, 1998, 235–237.
24. Han, J.; Ge, R.H.; Boyd, R.H. Macromolecules **1994**, *27*, 7781.
25. Götze, W.; Sjögren, L. In *Transport Theory and Statistical Physics*; Yip, S., Nelson, P., Eds.; Marcel Dekker: New York, 1995; 801–853.
26. Bennemann, C.; Baschnagel, J.; Paul, W. Eur. Phys. J. B **1999**, *10*, 323.
27. Bennemann, C.; Baschnagel, J.; Paul, W.; Binder, K. Comp. Theor. Polym. Sci. **1999**, *9*, 217.

14

Sorption and Diffusion of Small Molecules Using Transition-State Theory

MICHAEL L. GREENFIELD University of Rhode Island, Kingston, Rhode Island, U.S.A.

I. INTRODUCTION

In numerous applications, it would be useful if molecular simulation could be used to predict how the solubility or diffusivity of a small molecule in a polymer film varied with respect to changes in the molecular architecture of the film, or with different film processing. For example, this ability would enable prescreening among the many possible chain structures that could be devised, so chemical synthesis efforts could be targeted towards the most promising candidates. Polymer materials for sorption–diffusion membranes are one of many fields in which such large synthetic efforts have been conducted [1–4]. However, two primary barriers prevent accurate predictions from becoming routine: (1) the absence of reliable potential energy functions for penetrant–polymer systems of interest, and (2) the need for methods that access the long time scales inherent in small-molecule diffusion. The development of potential energy functions is the subject of ongoing research. Efforts focus on general force fields applicable to wide classes of molecules [5–14] or to particular types of polymer chains [15–21]. (These are only some of many possible examples.) This chapter discusses small-molecule sorption and diffusion, with a focus on simulating penetrant diffusion through polymers over long time scales.

Different simulation methods are appropriate for different penetrant–polymer systems, depending on the dynamical state of the polymer chains. Within a rubbery polymer or an amorphous polymer above its glass temperature, penetrants diffuse quickly enough ($D \sim 10^{-6} \, \text{cm}^2 \, \text{s}^{-1}$) to use molecular dynamics simulations [22] (see also Chapter 6 of this book).

425

The diffusive regime is reached after tens to hundreds of picoseconds [23–25], with longer simulation times required in order to accumulate good statistics. Molecular dynamics has been applied to small-molecule penetrants in numerous polymers. Early work focused on polyolefins (PE, PP, PBD, PIB) [17,23–37], and work on those systems has continued [38–45]. Other systems studied include polydimethylsiloxane [36,46–48], polystyrene [49,50], polyesters [51–53], polyimides and polyamide-imides [54–58], polysulfones [59], poly(aromatics) [60], polybenzoxazine [61], polyphosphazenes [62], poly(ether-ether-ketone) [63], poly(trimethyl-silylpropyne) [64], and more generic representations [65,66]. Recently, molecular dynamics simulations of charged species in polymers (for fuel cell applications) have been conducted, [17,67–76]. Molecular dynamics is also appropriate for moderate to high penetrant concentrations in glassy polymers, at which point the polymer chains become plasticized and the penetrant diffusion rate increases to a liquid-like value. Equivalently, the penetrant concentration suppresses the glass temperature to below the temperature simulated. Nonequilibrium molecular dynamics simulations [77,78] have been generally unsuccessful: forces beyond the linear response regime were required to impact the diffusion rate, relatively simple polymer models have been required, and computational times have not provided speedups over equilibrium MD.

In a glassy polymer (amorphous and below its concentration-dependent glass temperature), diffusion is much slower—$10^{-9}\,cm^2s^{-1}$ is typical—and molecular dynamics is incapable of displacing penetrant molecules sufficiently over accessible simulation times. As an order-of-magnitude estimate, after 10 ns an average penetrant would be displaced by

$$r \sim \sqrt{6Dt} \sim \left(6 \times 10^{-9} cm^2 s^{-1} \times 10^{-8} s\right)^{1/2} \sim 0.75\,\text{Å}$$

This distance is 30–100 times smaller than the typical simulation cell sizes used. Furthermore, since diffusion through a glassy polymer is thought to occur by a hopping mechanism [79], such an average displacement more likely would result only from statistics: one molecule with a 7.5 Å jump and nine molecules with net displacements of zero, for example. Direct molecular dynamics simulations are thus prone to large errors in predicted diffusivity in a glassy polymer, due to (1) mismatches between the simulation time and the expected waiting time for a jump to occur and (2) poor sampling of the many jump rates that would be found in a macroscopic sample.

One method for predicting diffusion rates in glassy polymers has been proposed by Boyd and coworkers [49,52]. As in glassy polymers, a hopping mechanism has been found in melt polymers not far above their glass temperatures [23,33,80,81].* Molecular dynamics simulations can be conducted at several temperatures above T_g, and an Arrhenius dependence can be used to extrapolate (by ~50–100K [49,52]) into the glassy region. However, this method requires the diffusion mechanism to remain unchanged throughout this large extrapolation.

As an alternative, this chapter describes methods for predicting small-molecule diffusivity that are based on transition-state theory (TST) [82–84] and kinetic Monte Carlo (KMC) [85,86]. These methods capitalize on the proposed penetrant jump mechanism. TST was described in Chapter 1 and is typically used to estimate the rates of chemical reactions from first principles; here we use TST to calculate the rate of characteristic jumps for each penetrant in a host polymer matrix. The collection of jump rates can be combined with the penetrant jump topology and KMC to obtain the penetrant diffusion coefficient. Other results obtainable from these simulations are physical aspects related to the jump mechanism: the sizes and shapes of voids accessible to penetrant molecules [87], enthalpic and entropic contributions to the penetrant jump rate [88,89], the extent and characteristics of chain motions that accompany each jump [90], and the shape and structure of the jump network itself [91].

Three approaches have been applied for describing the coupling between the jumping of a penetrant and the motions of nearby polymer chains. Each approach leads to a different approximation for the TST rate constant. The *frozen polymer method* [92–94] is analogous to methods used successfully to simulate penetrant diffusion through porous crystalline materials, such as zeolites (see [95,96] for reviews). All polymer chains are considered fixed in place, and transition-state theory-based rate constants are calculated from the energy barriers found for a penetrant to pass from one local potential energy minimum to another. The jump path is intrinsically three-dimensional: (x, y, z) position of the penetrant. This method is the most straightforward; however, in polymers it yields rate constants that lead to diffusion coefficients much lower (by factors of 10^3–10^6) than experimental values, because neglecting chain fluctuation contributions is physically unrealistic in a polymeric material. However, we discuss it here in order to introduce several concepts and techniques used in other TST methods. Next, a modification that we refer to here as the *average fluctuating polymer method* was proposed [93,97]. Harmonic fluctuations of the polymer matrix about its zero-force position were allowed, with the magnitude of the fluctuations controlled by a parameter similar to the Debye–Waller factor

---

*Explicit descriptions of the hopping mechanism can also be found in most of the recent MD papers. References [80,81] focus specifically on the jump mechanism.

in X-ray scattering. The jump path was again confined to the three degrees of freedom of the penetrant, but now within the free energy field characteristic of the vibrating polymer. This method has led to reasonable agreement with experiment in particular cases [24,63,97–100] and has been implemented in a popular molecular simulation software package [101].* However, the predictions are sensitive to the value of the Debye–Waller parameter, and hence some judgment is required when using this technique. In the most recent method [88,91,102], which we refer to as the *explicit polymer method*, the dimensionality of the jump path was increased to include explicitly both penetrant displacements and polymer chain motions. Each resulting rate constant then captures the details of the particular chain motions that accompany the jump, rather than mean-field average motions, but at the expense of a much higher computational cost. This method has been applied to a few penetrants in polypropylene [88,89,91], with reasonable agreement compared to experiment and correlation.

In this chapter we will begin by formulating the transition-state theory method within the context of small-molecule diffusion in a polymer matrix, and then show how the transition-state theory equations are applied to a single jump in each of these three approaches. We then compare different methods for establishing the network of possible penetrant jumps and explain how diffusion can be simulated on that network. We close with a summary and a discussion of the outlook for other systems. Several appendices describe aspects of particular approaches in more detail.

II. FORMULATION OF TST METHOD

Transition-state theory is a classical methodology for treating infrequent event processes that involve crossing a barrier significantly higher than the thermal energy $k_B T$ [82]. The complicated dynamical paths followed by an N-body system are reduced to a statistical description in which sampling over local fluctuations is performed in order to answer the question, "How much time goes by (on average) before the system resides in a definably different portion of configuration space?" Such a question may be answered phenomenologically with a set of first-order rate constants, each describing the rate of evolution along a different available pathway [103]. Transition-state theory provides a means to calculate such rate constants by comparing the probability of the system lying on a dividing surface, which separates two states, to the probability of occupying an initial state. This makes TST

---

*The average fluctuating polymer method has been incorporated into several products of Accelrys, Inc., such as Cerius^2.

a bridge between kinetics and thermodynamics, since the qualities of a rate process are determined from equilibrium information.*

A general formulation of transition-state theory was derived in Chapter 1 and will not be repeated here. We focus instead on applying a TST approach to diffusion in polymers. Several steps are required: stable regions in configuration space must be located (sorption states for a penetrant molecule), transition states and dividing surfaces between stable regions must be identified, pathways along which the system passes between sorption states must be followed, and the rate of each process must be calculated. Calculating the small-molecule diffusion rate through a polymer then requires an appropriate averaging technique for converting the jump rate constants $k^{TST}[=]s^{-1}$ and jump lengths $\ell[=]\text{Å}$ to a diffusion coefficient $D[=]\text{cm}^2\text{s}^{-1}$.

A. Transition State

For transition-state theory to be applicable, the physical system must include some high-energy region through which the system must pass in order to convert from "reactant" to "product." Numerous such regions exist for a small molecule diffusing through a glassy polymer, and the so-called transition states lie within them. Mathematically, a transition state is usually taken as a first-order saddle point on the $3N$-dimensional potential energy hypersurface, meaning that all elements of the potential energy gradient equal zero and all eigenvalues of the Hessian matrix of second derivatives but one are positive. In Cartesian coordinates,

$$\frac{\partial \mathcal{V}}{\partial X^i} = 0 \quad \text{for all positions } X^i \tag{1}$$

The notation X^i indicates the ith of the $3N$ Cartesian coordinates. Writing the index as a superscript follows the conventions of Riemannian geometry [104]. Each Eq. (1) can be written in mass-weighted Cartesian coordinates, $x^i \equiv m_i^{1/2} X^i$; these are useful in regard to following the diffusion path. An equivalent expression can then be formulated in generalized coordinates (bond angles, torsion angles, etc.) by writing [88]

$$\frac{\partial \mathcal{V}}{\partial x^i} = \sum_{j=1}^{3N} \frac{\partial q^j}{\partial x^i} \frac{\partial \mathcal{V}}{\partial q^j} \tag{2}$$

---

*While the dividing surface technically describes a hypothetical equilibrium, it can be sampled using equilibrium statistical mechanics.

for each mass-weighted Cartesian coordinate. In matrix notation,

$$\mathcal{J}^T \nabla_{\mathbf{q}} \mathcal{V} = \mathbf{0} \tag{3}$$

where $\mathcal{J}$ equals the Jacobian matrix of partial derivatives for transforming from mass-weighted Cartesian coordinates to generalized coordinates,

$$\mathcal{J} = \begin{pmatrix} \dfrac{\partial q^1}{\partial x^1} & \cdots & \dfrac{\partial q^1}{\partial x^{3N}} \\ \vdots & & \vdots \\ \dfrac{\partial q^{3N}}{\partial x^1} & \cdots & \dfrac{\partial q^{3N}}{\partial x^{3N}} \end{pmatrix} \tag{4}$$

The scalar "Jacobian" factor that more typically arises in a coordinate transformation equals det $\mathcal{J}$. For a nondegenerate coordinate system, det $\mathcal{J} \neq 0$; thus the potential energy gradient with respect to the generalized coordinates equals the zero vector at the transition state.

Generalized coordinates are useful when simulating penetrant–polymer systems because many points in configuration space are not physically realistic: they correspond to undesired values for the bond lengths, for example. Such generalized coordinates may be kept at constant values using the concept of infinitely stiff degrees of freedom [105]. These degrees of freedom are considered flexible (i.e., assuming any value according to the Boltzmann-weighted probability density) during the formulation of the methodology. However, they are effectively constrained to the desired positions using springs with force constants that approach infinity. Compared to an approach using rigid constraints throughout, this "infinitely stiff" model has been shown to lead to classical statistical mechanics results that more closely resemble the true underlying quantum-mechanical statistical mechanics [105]. In practice, these generalized coordinates are kept constant throughout the penetrant jump process; their flexibility only arises in formulating the method.

Finding a saddle point is similar to finding a local minimum or maximum: an initial guess is refined using steps based on local function information. However, standard methods for finding stationary points (where the gradient equals zero) typically are designed for finding maxima or minima, so special saddle-point-focused algorithms are required. Most common in a chemical context are those based on work by Cerjan and Miller [106–108]. Rather than taking the step defined by a quasi-Newton approach, these methods use a shift in eigenvalues to take a step such as [108]

$$\delta \mathbf{h} = \sum_i -(\delta \mathbf{q}_i \cdot \nabla_{\mathbf{q}} \mathcal{V}) \frac{\delta \mathbf{q}_i}{(\lambda_i - \lambda^s)} \tag{5}$$

that maximizes in the direction of one approximate eigenvector δq_1 and minimizes along all other eigenvectors. The shift factor λ^s for the first eigenvector (with a negative eigenvalue) differs from that for all other eigenvectors [107,108]. An initial estimate of the Hessian matrix (and its eigenvalues and eigenvectors) is required. "Updated" Hessian estimates [109] can be used in later steps, but an exact Hessian (calculated analytically or estimated numerically) leads to the best results, especially near the transition state. Baker's description [108] of this procedure is particularly targeted towards implementation issues. Quick convergence to a saddle point is aided by a good initial guess; several techniques for doing so are discussed in Section IV. Methods for obtaining the transition state in conjunction with the entire path [110,111] are discussed below as well.

B. Jump Pathway—the Intrinsic Reaction Coordinate (IRC)

The transition state defines an unstable position in configuration space. A system displaced slightly from the transition state will generally experience a force that directs it even further away, and the net displacement will increase until the system thermalizes in the region surrounding a local minimum.* Depending on the direction of the initial displacement, the system will find itself near one of two local minima (assuming the transition state was a first-order saddle point). These two minima are associated with this transition state in that they share the common dividing surface that passes through it.

In applying transition-state theory, it is traditional to calculate the Intrinsic Reaction Coordinate (IRC) [112], which is the lowest-energy path that connects the transition state to its adjacent local minima. The term "reaction" refers to the common application of TST to chemical reaction dynamics. Here, the IRC refers to the changes in penetrant and polymer degrees of freedom that describe the mechanism of a particular penetrant jumping motion. This exact pathway would be followed in the limit of infinitely small displacements with continuous application of a frictional force equal and opposite to the forces arising from the potential energy gradient. A true system undergoing molecular dynamics would not follow this path exactly; rather, it would fluctuate about it, much like a single melt polymer chain fluctuates about the chain contour in the reptation theory of long chain diffusion [113,114].

---

*It is theoretically possible to displace the system such that forces redirect it to the transition state. In practice, however, numerical imprecision in following such paths will likely lead to forces directing the system away from the transition state.

It is convenient to begin calculating the IRC at the transition state. Near the transition state, fluctuations in the mass-weighted atomic positions can be combined (by a linear transformation with a Jacobian of unity) into fluctuations along the set of $3N$ eigenvectors (or normal modes) $\delta\mathbf{x}$ of the Hessian matrix evaluated at the transition state, which are found by solving

$$\mathbf{H}\delta\mathbf{x} = \lambda\delta\mathbf{x} \tag{6}$$

From the transition state, the IRC is directed along the eigenvector corresponding to the negative eigenvalue [115,116]. Motions along all other eigenvectors locally define a multidimensional dividing surface that delineates the reactant and product regions. The IRC is orthogonal to this dividing surface. In other words, the IRC step through the transition state ("over" the pass in a mountain crossing analogy) is directed *along* the single direction of decreasing local curvature and *orthogonal* to all directions of increasing local curvature.

Away from the transition state, the IRC is formally defined by a steepest descent trajectory* in mass-weighted Cartesian coordinates [112],

$$\delta\mathbf{x} = \nabla_{\mathbf{x}}\mathcal{V}\,\delta\tau \tag{7}$$

and corresponds physically to a sequence of infinitesimal displacements along the local force vector. (Higher-order terms that require the Hessian matrix are discussed below.) The factor $\delta\tau$ (of units time squared) scales the size of each step taken [112]. Since the velocity is set to zero in each step, the IRC equation has also been called a path with continuous dissipation of kinetic energy [117,118].

Extensions of the IRC to generalized coordinates have been discussed in the literature [88,102,117]. Physically, the IRC pathway and the vibrational frequencies and normal modes at the transition state should be independent of the coordinate system chosen, so Eqs. (6) and (7) must be reformulated in terms of generalized coordinates. Detailed steps are shown in Appendix A. The resulting expression at the transition state is

$$\mathbf{H}_{qq}\,\delta\mathbf{q} = \lambda\,\mathbf{a}\,\delta\mathbf{q} \tag{8}$$

---

*This usage of "steepest descent" refers to a path constructed by taking a sequence of small steps, each directed opposite to the local gradient. This usage differs from that in the optimization literature; there [109], the magnitude of each step is chosen such that the objective function is minimized with respect to motion along the step vector.

in which $\mathbf{H}_{qq} = \partial^2 \mathcal{V} / \partial q^i \partial q^j$. The covariant metric tensor $\mathbf{a}$ [119] is calculated most easily from the Jacobian matrix for transforming from generalized to mass-weighted Cartesian coordinates,

$$
\mathcal{J}' = \begin{pmatrix} \dfrac{\partial x^1}{\partial q^1} & \cdots & \dfrac{\partial x^1}{\partial q^{3N}} \\ \vdots & & \vdots \\ \dfrac{\partial x^{3N}}{\partial q^1} & \cdots & \dfrac{\partial x^{3N}}{\partial q^{3N}} \end{pmatrix}
\tag{9}
$$

$\mathcal{J}'$ is the matrix inverse of $\mathcal{J}$, introduced in Eq. (4),

$$
\mathcal{J} \mathcal{J}' = \mathcal{J}' \mathcal{J} = \mathcal{I}
\tag{10}
$$

Starting from Eq. (9), $\mathbf{a}$ is calculated by

$$
\mathbf{a} = \mathcal{J}'^T \mathcal{J}'
$$

or

$$
a_{ij} = \sum_k \frac{\partial x^k}{\partial q^i} \frac{\partial x^k}{\partial q^j}
\tag{11}
$$

Equation (8) is an example of a generalized eigenvalue problem, whose numerical solution provides the eigenvectors $\delta \mathbf{q}$ and eigenvalues λ of the Hessian in the space of all generalized coordinates. The units on each side match exactly,

$$
\left(\mathbf{H}_{qq} \delta \mathbf{q} \right)_i [=] \sum_j \frac{\text{energy}}{q_i \, q_j} q_j = \frac{\text{kg m}^2}{\text{s}^2 \, q_i}
$$

$$
(\mathbf{a} \, \delta \mathbf{q})_i [=] \sum_j \frac{(\text{kg}^{1/2} \text{ m})^2}{q_i \, q_j} q_j = \frac{\text{kg m}^2}{q_i}
$$

and λ has units of s^{-2}, or frequency squared, as expected.

"Canned routines" that solve the generalized eigenvalue problem are available in standard numerical libraries.* Because $\mathbf{a} = \mathcal{J}'^T \mathcal{J}'$, the covariant metric tensor $\mathbf{a}$ must be symmetric and positive definite, which facilitates finding the eigenvalues and eigenvectors, compared to a more general case in which the matrix characteristics of $\mathbf{a}$ are unknown.

---

*The routines REDUC, TRED2, and TQL2 of EISPACK, available at [120], are examples of canned routines. Matrix routines from the LAPACK [121] and LINPACK [122] packages are also available.

Away from the transition state, the IRC equation is written as [88,102]

$$\mathbf{a} \, \delta\mathbf{q} = \nabla_{\mathbf{q}} \mathcal{V} \, \delta\tau \tag{12}$$

Taking a step thus requires calculating the gradient and covariant metric tensor, and then solving a set of linear equations.

C. Narrowing the Diffusion Path to a Localized Region

Since a penetrant jump occurs over a relatively small region within a simulation cell, it is useful to restrict the atoms involved in the TST approach, leaving only the penetrant and some of the polymer chain segments. This can significantly reduce the number of variables and thus the size of the matrix calculations (which scale as N^3). Constraining the system requires a careful manipulation of the mass-weighted Cartesian coordinates and the generalized coordinates. The origins of the nontrivial nature lie in the covariant metric tensor $\mathbf{a}$. Since this matrix is not diagonal in general, it couples the step along one coordinate dq^i to the gradient with respect to many $q^j (j \neq i)$. The coordinates q^i and q^j must correspond to part of the same chain for a_{ij} to be nonzero.

Choosing which generalized coordinates to include is facilitated by dividing them into three sets. One set, the bond lengths, are considered infinitely stiff and should remain unchanged between the transition state and the local minima. In a sufficiently large penetrant–polymer system, another set is comprised of the many generalized coordinates defined by atoms far from the penetrant. If a local fluctuation among polymer chains allows a penetrant to jump, the values of far-off coordinates should not be appreciably different at the origin state, transition state, or destination state. A third set, those nonconstrained generalized coordinates defined by atoms near a jumping penetrant, are "flexible" when describing a transition state and IRC. Both other sets of degrees of freedom are held constrained to their initial values (in the energy-minimized neat polymer structure) by springs with force constants whose strengths approach infinity.

The details of narrowing the diffusion path are explained in Appendix B. Here we quote the resulting TST equations and algorithms.

At the transition state, Eq. (3) is replaced by the similar expression

$$\nabla_{\mathbf{q}_f} \mathcal{V} = \mathbf{0} \tag{13}$$

meaning that the potential energy gradient with respect to each *flexible* coordinate equals zero.

The eigenvalues and eigenvectors at the transition state, with respect to the flexible coordinates, are calculated from

$$\mathbf{H}^0_{qq} \, \delta\mathbf{q}_f = \lambda \, \mathbf{a}^0 \, \delta\mathbf{q}_f \tag{14}$$

The step to take away from the transition state is the eigenvector $\delta\mathbf{q}_f$ corresponding to the negative eigenvalue of the generalized eigenvalue problem Eq. (14). The first step along the IRC (from the transition state) is found by the following steps:

1. calculate the Hessian matrix of second derivatives with respect to the flexible generalized coordinates, $\mathbf{H}^0_{qq}$,
2. calculate the covariant metric tensor in the flexible generalized coordinates, $\mathbf{a}^0$,
3. solve for the eigenvector that corresponds to the one negative eigenvalue.

A small step along this eigenvector $\delta\mathbf{q}_f$ defines the first step along the IRC. Along the step, changes occur in only the flexible coordinates; infinitely stiff coordinates remain unchanged.

Away from the transition state, the remaining steps along the diffusion path are calculated from

$$\mathbf{a}^0 \, \delta\mathbf{q}_f = \nabla_{\mathbf{q}_f} \mathcal{V} \, \delta\tau \tag{15}$$

Tracking all but the first step along the IRC thus requires these steps:

1. calculate the potential energy gradient in the flexible subset of generalized coordinates, $\nabla_{\mathbf{q}_f} \mathcal{V}$,
2. calculate the covariant metric tensor $\mathbf{a}^0$ in the subset of flexible coordinates,
3. solve a series of f linear equations for the step vector $\delta\mathbf{q}_f$,
4. scale each term by the same small constant $\delta\tau$.

The elements of the Jacobian matrices (and thus the elements of $\mathbf{a}^0$) vary with chain conformation and must be recalculated after each step. The IRC ends when it reaches a local minimum (the gradient $\nabla_{\mathbf{q}_f} \mathcal{V}$ approaches zero).

D. Final State(s)

The two IRCs that begin at a transition state each terminate at a local potential energy minimum, where the gradient equals zero. Near the minimum, the gradient becomes sufficiently small (a step size of order 10^{-10}, for example [88]) that each IRC step [Eq. (15)] is excessively short. It would then take a large number of steps to travel the small remaining distance to the local minimum configuration. Instead, a quasi-Newton scheme [109] can be used at this point to reach the local minimum. Relatively few steps should be required, and the resulting potential energy decreases should be negligible. If large decreases in energy do occur, that indicates larger-scale structural relaxation of the polymer matrix, and the system configuration prior to quasi-Newton steps should be restored [i.e., the last configuration found via Eq. (15)]. In such a case, additional regular IRC steps should be taken despite the low gradient, and the step size should soon increase again. This approach maintains the validity of the calculated IRC. Such behavior could occur along a near-flat plateau on the potential energy landscape that is surrounded by more steep regions of similar orientation. After continued energy decreases, a small gradient should recur, and quasi-Newton steps can be attempted again.

E. Rate Constant

The pair of rate constants for each jump path (IRC) can be calculated in one of several ways. A formal, microscopic formulation [83] was discussed in Chapter 1 for the rate constant,

$$k_{A \to B}(t) = \frac{\langle \dot{q}(0)\delta(q^{\ddagger} - q(0))\, \theta(q(t) - q^{\ddagger})\rangle}{\langle \theta(q^{\ddagger} - q)\rangle} \tag{1.101}$$

The denominator equals the contribution Q_A of the origin sorption state A to the entire canonical partition function Q. Evaluating Q_A requires some definition of the bounds on the state volume in configuration space. The numerator equals the average velocity along the reaction coordinate for state points within the dividing surface region at $t = 0$ that terminate in the product state. Transition-state theory assumes that *all* state points in this region with $\dot{q} > 0$ terminate in the product state, i.e., $(q(t) - q^{\ddagger}) > 0$ for all t [83]. If this velocity along the IRC (in some combination of mass-weighted Cartesian coordinates) is position-independent and separable from all other velocities, it may be integrated over $0 \leq \dot{q} < \infty$, leading to a prefactor $k_B T/h$, where h is Planck's constant. The remaining numerator terms, in combination with a definition of the dividing surface, equal the dividing surface region contribution $Q^{\ddagger}$ to the canonical partition function.

Equation (1.101) can thus be rewritten as

$$k^{TST} = \frac{k_B T}{h} \frac{Q^{\ddagger}}{Q} \exp\left(-\frac{\mathcal{V}^{\ddagger} - \mathcal{V}_0}{k_B T}\right) \tag{16}$$

or as the canonical ensemble equivalent of Eq. (1.117),

$$k^{TST} = \frac{k_B T}{h} \exp\left(-\frac{A^{\ddagger} - A_0}{k_B T}\right)$$

(from Chapter 1). The exponential in Eq. (16) arises from a difference in the zero of energy between $Q^{\ddagger}$ and Q.

Several approximations to Eq. (16) can then be made. The simplest approximation is to assume a constant prefactor and to use the potential energy difference between the transition state and the reactant sorption state,

$$k^{TST} = k_0 \exp\left(-\frac{\mathcal{V}^{\ddagger} - \mathcal{V}_0}{k_B T}\right) \tag{17}$$

This approach can be interpreted via Eq. (1.117) as assuming that the free energy difference equals the potential energy difference (entropy change equals zero) and that the temperature dependence in the exponential dominates that in the prefactor.

A less drastic approximation is to assume that (1) potential energy increases relative to the origin or transition state energies are adequately described locally by a harmonic approximation, and (2) potential energies outside these harmonic regions are sufficiently large such that the Boltzmann factor approaches zero. With this combination of assumptions, the origin state and dividing surface boundaries can be extended to $\pm\infty$, and the partition functions can be calculated from the vibrational frequencies at the transition state and the reactant state, with respect to the flexible degrees of freedom. The frequencies are obtained from the eigenvalues of Eq. (14) by

$$\nu_i = \frac{\sqrt{\lambda_i}}{2\pi} \qquad i = 2, \ldots, f \text{ (transition state)}$$
$$i = 1, \ldots, f \text{ (reactant state)} \tag{18}$$

Note that no frequency calculation is required for the negative eigenvalue at the transition state. Two alternate forms of the rate constant can then be obtained. Within a classical statistical-mechanics model, infinitely stiff coordinates [105] contribute a factor related to the product of their normal mode frequencies, each of which approaches infinity [88]. The resulting rate constant is

$$
k^{TST} = \left(\frac{\prod_{\alpha=1}^{f} \nu_\alpha}{\prod_{\alpha=2}^{f} \nu_\alpha^\ddagger} \right) \frac{\det \mathbf{a}^{-1''}}{\det \left(\mathbf{a}^{-1''} \right)^\ddagger} \exp\left(-\frac{\mathcal{V}^\ddagger - \mathcal{V}_0}{k_B T} \right) \tag{19}
$$

The matrix $\mathbf{a}^{-1''}$, defined in Appendix B, is a contravariant metric tensor in the infinitely stiff coordinates. Each $\det(\mathbf{a}^{-1''})$ term can be evaluated directly using a recursion relation [102,123] or via the Fixman relation, Eq. (82). In a quantum mechanics-based statistical mechanics model, the effect of infinite stiffness on each partition function is not*

$$
\lim_{\nu \to \infty} \frac{k_B T}{h \nu_i} = 0
$$

but instead

$$
\lim_{\nu \to \infty} \left[1 - \exp\left(-\frac{h \nu_i}{k_B T} \right) \right] = 1
$$

Consequently the contributions from infinitely stiff modes become unity and the rate constant can be written as

$$
k^{TST} = \frac{k_B T}{h} \frac{\prod_{\alpha=1}^{f} \left[1 - \exp\left(-\frac{h \nu_\alpha}{k_B T} \right) \right]}{\prod_{\alpha=2}^{f} \left[1 - \exp\left(-\frac{h \nu_\alpha^\ddagger}{k_B T} \right) \right]} \exp\left(-\frac{\mathcal{V}^\ddagger - \mathcal{V}_0}{k_B T} \right) \tag{20}
$$

The jump rate constants reported in our previous work [88] were calculated using Eq. (20). Within the harmonic approximation, the difference in

---

*Despite this zero limit, the ratio of the products of these frequencies, calculated via $(\det \mathbf{a}^{-1''\ddagger} / \det \mathbf{a}^{-1''}_0)$, is well-defined with a value near unity. See [88,123] for details.

entropy between the reactant and transition states can be approximated by [88,124]

$$
\frac{\Delta S}{k_B} = \ln\left(\frac{h}{k_B T}\frac{\left(\prod_{\alpha=1}^{f}\nu_\alpha\right)\det\left(\mathbf{a}^{-1''}\right)^{\ddagger}}{\left(\prod_{\alpha=2}^{f}\nu_\alpha^{\ddagger}\right)\det\left(\mathbf{a}^{-1''}\right)^{\min}}\right) - 1
\tag{21}
$$

(classical mechanics-based) or

$$
\frac{\Delta S}{k_B} = \ln\left(\frac{\prod_{\alpha=2}^{f}\left[1 - \exp\left(-\dfrac{h\nu_\alpha^{\ddagger}}{k_B T}\right)\right]}{\prod_{\alpha=1}^{f}\left[1 - \exp\left(-\dfrac{h\nu_\alpha}{k_B T}\right)\right]}\right) - 1
\tag{22}
$$

(quantum mechanics-based). The physical interpretation is that a higher frequency corresponds to a more narrow potential well, which allows fewer conformations. An advantage of the harmonic approximation is its speed: only frequencies are required after the transition state and IRC are found. One disadvantage is the harmonic approximation for the energy, which breaks down within the origin state or dividing surface. A second disadvantage is that it neglects any contributions along the IRC itself.

A more exact approach is to use sampling techniques to calculate the free energy difference between reactant and transition states, with the rate constant calculated by Eq. (1.117). One possibility is to calculate free energy differences between hyperplanes orthogonal to the IRC [125,126] (see also Chapters 9 and 10 in this book). Another is to use "chain-of-states" methods to sample fluctuations among different states along the IRC simultaneously [111,118,127,128]. A difficulty with the latter for explicit inclusion of polymer chain fluctuations is the high dimensionality of each state point.

Finally, an improvement to the transition-state theory-based rate constant is to use the formalism in Chapter 1 [Eqs. (1.103) and (1.104)] to calculate the dynamical correction factor κ [83,84]. Several simulations beginning from the dividing surface and directed towards the final state are conducted, and κ equals the fraction of jumps that thermalize in the final state. Each such simulation is short (10–20 ps) because it follows the system to a point of lower energy and higher entropy. A separation of time scales is required so subsequent jumps out of the product state are unlikely at short times.

III. STARTING POINT: POLYMER MOLECULAR STRUCTURES

For using any of the diffusion simulation methods described in this chapter, it is necessary to begin with a set of chain configurations that represent the polymer of interest on the molecular level. For a bulk polymer these are likely to be at the bulk density, within a simulation box of edge length 20–50 Å and with periodic boundary conditions in two or three dimensions. Larger edge lengths are more desirable in that they allow larger natural length scales to be probed within the simulation. They are less desirable in that they require larger simulation times, since calculation times scale with $N^m \sim L^{3m}$ and m is typically between 1 and 2.

Constructing and equilibrating molecular-level polymer configurations are described in detail in the literature [129–135]. Such techniques are important because much longer times than can be achieved in practice are required in order to equilibrate the entire chain, starting from a random configuration in a room-temperature simulation. The objective of those methods is to ensure that representative polymer chain configurations are used for the sorption and diffusion calculations.

For the purpose of studying diffusion, it is necessary to have a penetrant molecule amidst the polymer chains in the molecular simulation cell. It is possible to add the penetrant during the diffusion calculations. However, it is also advantageous to equilibrate the polymer sample to some extent in the presence of the penetrant molecule. This will ensure that fast ($\tau < 0.1$ ns), penetrant-induced polymer relaxations occur before beginning the diffusion calculations. Sometimes, and especially in molecular dynamics simulations, several penetrant molecules are added in order to improve sampling. (Of order 10 sets of penetrant mean-squared displacements are sometimes obtained from one simulation.) One risk of this approach is that the locally high penetrant concentration could plasticize the polymer, artificially increasing the penetrant diffusivity. Plasticization is desirable if this is the concentration range of interest, but it should be avoided if the objective is calculating diffusivity in the infinite dilution limit.

For accumulating good statistics, several different sets of polymer chain configurations should be used. Each resulting simulation cell will likely have a different resistance to diffusion. This is physically reasonable, because molecular-level imaging (AFM, STM, etc.) reveals heterogeneity over nm to μm length scales [136]. The question then arises of how the results from different calculations can be averaged. For polymer samples at overall thermal equilibrium, the Boltzmann factor for the total energy is an appropriate weighting factor, and lower energy structures will occur a higher proportion of the time. For glasses, physics-based studies have shown that an arithmetic average over different structures is most appropriate [137]. We have applied the latter when calculating the distribution of jump rate constants for methane in glassy atactic polypropylene [88]. In a real glassy polymer, the overall density (and thus the volume available to a penetrant) depends on the sample preparation, such as the quench rate from above T_g, the strain history, and the amount of physical aging [1]. Consequently the real distribution of jump rate constants could be expected to depend on all of these variables. Detailed accounting for these effects is beyond the current scope of molecular simulation.

IV. FROZEN POLYMER METHOD

The most straightforward approach for simulating diffusion through a glassy polymer is the frozen polymer method. The essential idea of the frozen polymer method is that polymer chains remain fixed in place and provide a static external field, through which a small-molecule penetrant can diffuse.

This method is the least accurate for simulating small-molecule diffusion in a polymer, because neglecting contributions from polymer chain fluctuations is unrealistic. These limitations were learned from its initial applications: argon in polyethylene [92]; helium, oxygen, and nitrogen in polypropylene [93]; helium, oxygen, nitrogen, hydrogen, and argon in polycarbonate [93,94]; helium, oxygen, nitrogen, hydrogen, and argon in polyisobutylene [94]. Materials such as zeolites are more rigid than polymers, making neglecting fluctuations in the surrounding material more physically realistic. Consequently an analogous approach has been used successfully for alkanes [85] or aromatics [138] in zeolites; see [95,96] for reviews. Despite these limitations, the frozen polymer method provides a useful framework for introducing steps used in more sophisticated TST approaches. A frozen polymer has also aided with interpreting different contributors to the diffusion mechanism, within recent molecular dynamics simulations of helium in polypropylene [44].

Removing the polymer chains from explicit consideration directly affects the list of flexible coordinates in the TST formulation. For a monatomic penetrant atom or a pseudo-monatomic united atom, such as methane, the only flexible coordinates are the mass-weighted position (x, y, z) of the penetrant. The fixed polymer coordinates can be considered part of the force field defining the penetrant potential energy. The covariant metric tensor $\mathbf{a}^0$ equals a three-by-three identity matrix.

For a linear penetrant molecule such as oxygen or nitrogen, its position can be represented using the mass-weighted position of the first atom (x_1, y_1, z_1) and the orientation of the bond vector in spherical coordinates,

$$x_2 = x_1 + \ell \cos\phi \sin\theta \qquad (23)$$
$$y_2 = y_1 + \ell \sin\phi \sin\theta \qquad (24)$$
$$z_2 = z_1 + \ell \cos\theta \qquad (25)$$

ϕ and θ are the projection of the bond on the x-y plane $(0 \le \phi < 2\pi)$ and its azimuthal angle relative to the z-axis $(0 \le \theta \le \pi)$. The mass-weighted bond length is

$$\ell = m^{1/2}[(X_2 - X_1)^2 + (Y_2 - Y_1)^2 + (Z_2 - Z_1)^2]^{1/2} \qquad (26)$$

The Jacobian $\mathcal{J}'$ is found [using Eq. (9)] by partial differentiation of the mass-weighted Cartesian coordinates $(x_1, y_1, z_1, x_2, y_2, z_2)$ (one per row) with respect to the generalized coordinates $(x_1, y_1, z_1, \phi, \theta, \ell)$ (one per column), leading to

$$\mathcal{J}' = \begin{pmatrix} 1 & 0 & 0 & 0 & 0 & 0 \\ 0 & 1 & 0 & 0 & 0 & 0 \\ 0 & 0 & 1 & 0 & 0 & 0 \\ 1 & 0 & 0 & -\ell\sin\phi\sin\theta & \ell\cos\phi\cos\theta & \cos\phi\sin\theta \\ 0 & 1 & 0 & \ell\cos\phi\sin\theta & \ell\sin\phi\cos\theta & \sin\phi\sin\theta \\ 0 & 0 & 1 & 0 & -\ell\sin\theta & \cos\theta \end{pmatrix} \qquad (27)$$

The covariant metric tensor is found by matrix multiplication [Eq. (11)],

$$\mathbf{a} = \begin{pmatrix} 2 & 0 & 0 & -\ell\sin\phi\sin\theta & \ell\cos\phi\cos\theta & \cos\phi\sin\theta \\ 0 & 2 & 0 & \ell\cos\phi\sin\theta & \ell\sin\phi\cos\theta & \sin\phi\sin\theta \\ 0 & 0 & 2 & 0 & -\ell\sin\theta & \cos\theta \\ -\ell\sin\phi\sin\theta & \ell\cos\phi\sin\theta & 0 & \ell^2\sin^2\theta & 0 & 0 \\ \ell\cos\phi\cos\theta & \ell\sin\phi\cos\theta & -\ell\sin\theta & 0 & \ell^2 & 0 \\ \cos\phi\sin\theta & \sin\phi\sin\theta & \cos\theta & 0 & 0 & 1 \end{pmatrix} \qquad (28)$$

The bond length could also be considered infinitely stiff, in which case $\mathcal{J}'_f$ would lack the last column of $\mathcal{J}'$, and $\mathbf{a}^0$ (a 5×5 matrix) would lack the last row and column of $\mathbf{a}$.

For a nonlinear rigid molecule such as aromatic compounds (benzene, toluene, xylenes), coordinates corresponding to the center of mass and the molecular orientation can be used [138]. The required Eulerian angles ϕ, ψ, θ denote rotations about the z axis, the x axis in this first intermediate coordinate system, and the z axis in this second intermediate coordinate system [139]. The resulting covariant metric tensor is*

$$
\mathbf{a} = \begin{pmatrix}
1 & 0 & 0 & 0 & 0 & 0 \\
0 & 1 & 0 & 0 & 0 & 0 \\
0 & 0 & 1 & 0 & 0 & 0 \\
0 & 0 & 0 & \begin{bmatrix} I_1 \sin^2\theta \sin^2\psi \\ +I_2 \sin^2\theta \cos^2\psi \\ +I_3 \cos^2\theta \end{bmatrix} & I_3 \cos\theta & (I_1 - I_2)\sin\theta \sin\psi \cos\psi \\
0 & 0 & 0 & I_3 \cos\theta & I_3 & 0 \\
0 & 0 & 0 & (I_1 - I_2)\sin\theta \sin\psi \cos\psi & 0 & I_1 \cos^2\psi + I_2 \sin^2\psi
\end{pmatrix}
\tag{29}
$$

This metric tensor can then be used in Eqs. (14) and (15). Its block-diagonal character separates translation and rotation steps, an advantage for deciphering diffusion mechanisms. One complication, however, is that the Eulerian angles must be carefully monitored and redefined in case any pass through zero [138].

In principle, the transition-state theory equations developed above [Eqs. (13), (14), (15), (16), (1.117), (20)] can be used for each jump in the frozen polymer method. While the individual components of the saddle-point search have a reputation of being slow (calculating the Hessian matrix, evaluating all eigenvalues and eigenvectors), in practice these steps are fast because the dimensionality is small. Standard matrix routines [120–122,140] are recommended for the matrix calculations. Typically following the IRC will require significantly more computational effort than finding the transition state.

A good estimate for the transition state location will speed up the search process and help to ensure a successful search. Geometric interpretations of the simulation cell are often useful for this purpose. The voxel method (useful throughout the frozen polymer method) is discussed below. Another approach is to subdivide (or *tessellate*) the periodic simulation cell into tetrahedra [87,93,141–145] or cubes [146,147], with a hard-sphere representation of the atoms replacing an energy calculation. Sorption states can be assigned to the center of tetrahedra or cubes not fully occupied by atoms of the polymer chain, with transition states assigned to the center of surface triangles [93]. Alternatively, tetrahedra or cubes can be lumped into sorption states, and initial guesses for transition state locations can be taken at the center of tetrahedra/cubes that form the intersection of sorption states

---

*Equation 12 in [138] [the equivalent of Eq. (29) substituted into Eq. (15)] contains a typographical error. This form of $\mathbf{a}^0$ is correct.

[88,102]. The important point is using such criteria to recognize the three-dimensional regions in a polymer structure that most closely resemble the necks between (presumably) larger sorption state regions.

An alternative approach typically used within the frozen polymer method and the average fluctuating polymer method is the so-called "voxel" approach [85,94]. A finely spaced cubic grid (lattice spacing of 0.1 Å, for example) is overlaid onto the simulation cell, and the energy required to insert a penetrant at the center of each cube is calculated. For each voxel (i.e., cube) i, the lowest-energy neighboring cube j is identified, and the connectivity $i \rightarrow j$ is stored. Alternatively [85], a steepest descent calculation can be conducted from each voxel. After looping over all voxels, the net result is a connectivity map for the particular polymer structure chosen. Voxels with no neighbors of lower energy correspond to local minima. Each voxel can then be labeled as leading (or "belonging") to the state defined by the local minimum in which a connectivity path terminates. (If $i \rightarrow j \rightarrow k \rightarrow \ell$, then voxels i, j, k, and ℓ belong to state ℓ.) The dividing surface between states is defined by the common faces of voxels in different states, and the center of the face between the lowest-energy voxels on the dividing surface approximates the position of the transition state. This position is not an exact solution to Eq. (13), but it is close to one. Similarly, the connectivity path defined by adjoining, lowest-in-energy voxels approximates the IRC defined by Eqs. (14) and (15).

For a spherical penetrant, the energy at each voxel can be used to calculate each rate constant through numerical integration of Eq. (16). The kinetic energy contributions to each partition function are integrated analytically in mass-weighted Cartesian coordinates,

$$Q^{\ddagger} = \left(\frac{2\pi k_B T}{h^2}\right) Z_{A \rightarrow B}^{\ddagger} \tag{30}$$

$$Q = \left(\frac{2\pi k_B T}{h^2}\right)^{3/2} Z_A \tag{31}$$

The remaining configurational integral Z_A for each initial state is a sum of the contributions from each voxel, multiplied by the mass-weighted voxel volume

$$Z_A = \sum_{\substack{\text{voxel } j \text{ in} \\ \text{state } A}} Z_j(\mathbf{r}) V_j m^{3/2} \tag{32}$$

$$= \sum_{\substack{\text{voxel } j \text{ in} \\ \text{state } A}} \exp\left(-\frac{\mathcal{V}_j}{k_B T}\right) V_j m^{3/2} \tag{33}$$

The energy $\mathcal{V}_j$ corresponds to a ghost penetrant molecule placed at the center of voxel j, i.e., the voxel energy calculated previously. The configurational integral for each dividing surface is similarly evaluated, using contributions from the voxels that define the interstate boundary,

$$Z^{\ddagger}_{A \to B} = \sum_{\substack{\text{voxel } j \text{ on} \\ A \to B \\ \text{boundary}}} Z^{\ddagger}_j A_j m = \sum_{\substack{\text{voxel } j \text{ on} \\ A \to B \\ \text{boundary}}} \exp\left(-\frac{\mathcal{V}_j}{k_B T}\right) A_j m \tag{34}$$

The voxel energy $\mathcal{V}_j$ is assumed constant within the volume V_j and across the surface area A_j.

The dividing surface is a curved region, which voxels approximate as a "zig-zag" of cubic surfaces. To calculate each A_j (the area of the dividing plane that cuts through voxel j), Gusev and Suter [97] proposed

$$A_j = KL_v^2 \tag{35}$$

where L_v is the voxel edge length and K (see Table 1) depends on the number of neighboring voxels assigned to state B (the destination state). The fewer neighboring voxels in the destination state, the better this approximation. Gusev et al. [94] estimated area errors δA of under 7% based on a voxel edge of $L_v = 0.2$ Å. From Eqs. (16), (30), and (34), these errors propagate directly into the rate constant. Smaller voxels would lead to smaller area errors, according to

$$|\delta A| \leq L_v^2 |\delta K| \tag{36}$$

where δK is the difference between the value listed in Table 1 and the true ratio of dividing surface size to voxel face area. Consequently decreasing

TABLE 1 Coefficients Used when Estimating the Contribution of Each Voxel to the Dividing Surface Area (Equation 35 from [97]). A Voxel can Have a Maximum of 6 Nearest Neighbors

Number of neighboring voxels in state B	K
1	1
2	$\sqrt{2}$
3	1.41
4	2.0
5	0
6	0

voxel size can significantly reduce dividing surface area errors. The final voxel-based expression for the rate constant is thus

$$k^{TST} = \left(\frac{k_B T}{2\pi m}\right)^{1/2} \frac{\sum_{\substack{\text{voxel } j \text{ on} \\ \text{boundary}}} \exp\left(-\frac{V_j}{k_B T}\right) K_j L_v^2}{\sum_{\substack{\text{voxel } j \\ \text{in state}}} \exp\left(-\frac{V_j}{k_B T}\right) L_v^3} \tag{37}$$

for all voxels having the same edge length L_v.

The relationship between individual sorption states and a fully connected jump network will be discussed in Section IX. Methods for calculating a diffusion coefficient from the structure and rate constants are discussed in Section X.

V. AVERAGE FLUCTUATING POLYMER METHOD

Compared to the rigid polymer method, the average fluctuating polymer method improves the treatment of how polymer chains move during the penetrant diffusion process. Rather than remaining fixed in place, polymer chains execute harmonic vibrations about their equilibrium positions. Penetrant jumps are then coupled to elastic fluctuations of the polymer matrix and are independent of structural relaxation of the polymer chains [24,97]. After a penetrant jump completes, chains near the final sorption state will likely show slight elastic deviations as they swell to accommodate the penetrant molecule. Since no chain conformation relaxations are allowed, other polymer chains will essentially retain their initial conformation. The penetrant jump rate then depends only on the local, quasiharmonic fluctuations in the sorption state and the transition state [24,97].

The magnitude of elastic vibrations is controlled by a parameter $\langle \Delta^2 \rangle^{1/2}$, of units Å. While the original formulation of this method allowed for different parameter values for each atom type (or even every atom), traditionally a single size is used for all atoms in the polymer matrix. The fluctuations impose an additional equilibrium probability density distribution

$$P(\mathbf{x} - \langle \mathbf{x} \rangle) d(\mathbf{x} - \langle \mathbf{x} \rangle)$$

$$= \prod_i \left(2\pi \langle \Delta^2 \rangle\right)^{-3/2} \exp\left(-\frac{(x_i - \langle x_i \rangle)^2 + (y_i - \langle y_i \rangle)^2 + (z_i - \langle z_i \rangle)^2}{2\langle \Delta^2 \rangle}\right) d(\mathbf{x}_i - \langle \mathbf{x}_i \rangle)$$

$$\tag{38}$$

where $\langle \mathbf{x} \rangle$ is the vector of all polymer atom positions in the penetrant-free equilibrated (i.e., preaveraged) polymer sample. The three-dimensional position of a single atom is written $\mathbf{x}_i$. The mean-squared fluctuation along each average position (x, y, or z) equals $\langle \Delta^2 \rangle$; hence the parameter notation as an RMS value. The unnormalized probability density for all penetrant positions $\mathbf{r}_p$ and polymer atom positions $\mathbf{x}$ is then written as

$$P(\mathbf{r}_p, \mathbf{x})d(\mathbf{x} - \langle \mathbf{x} \rangle)$$

$$= \prod_i (2\pi\langle \Delta^2 \rangle)^{-3/2} \exp\left(-\frac{\mathcal{V}_{ip}(\mathbf{r}_p, \mathbf{x}_i)}{k_B T}\right) \exp\left(-\frac{|\mathbf{x}_i - \langle \mathbf{x}_i \rangle|^2}{2\langle \Delta^2 \rangle}\right) d(\mathbf{x}_i - \langle \mathbf{x}_i \rangle)$$

$$(39)$$

The parameter $\langle \Delta^2 \rangle^{1/2}$ can be interpreted [24] as being proportional to $T^{1/2}$, its overall effect being an additive elastic contribution to the total energy.

The modified position distribution function is incorporated directly into the rate constant calculation based on the voxel method, with the penetrant positioned at the voxel center. Each single-voxel configuration integral [Z_j in Eq. (32)], referenced to the energy of the local minimum origin state, contains an integral over polymer conformation fluctuations sampled via the $\langle \Delta^2 \rangle^{1/2}$ parameter,

$$Z_j(\mathbf{r}_p) = \int \exp\left(\frac{-\mathcal{V}(\mathbf{r}_p, \mathbf{x})}{k_B T}\right) P(\mathbf{x} - \langle \mathbf{x} \rangle) d(\mathbf{x} - \langle \mathbf{x} \rangle) \qquad (40)$$

The Boltzmann factor depends on the positions of all atoms that interact with the penetrant. For a pairwise additive penetrant–polymer force field, such as the Lennard-Jones expression

$$\mathcal{V}(\mathbf{r}_p) = \sum_i 4\epsilon_{ip}\left[\left(\frac{\sigma_{ip}}{|\mathbf{r}_p - \mathbf{x}_i|}\right)^{12} - \left(\frac{\sigma_{ip}}{|\mathbf{r}_p - \mathbf{x}_i|}\right)^6\right] \qquad (41)$$

the configurational integral can be written as a product of separable terms,

$$Z_j(\mathbf{r}_p) = \prod_i Z_{ji}(\mathbf{r}_p) = \prod_i (2\pi\langle \Delta^2 \rangle)^{-3/2} \int \exp\left(-\frac{\mathcal{V}_{ip}(\mathbf{r}_p, \mathbf{x}_i)}{k_B T}\right)$$

$$\times \exp\left(-\frac{|\mathbf{x}_i - \langle \mathbf{x}_i \rangle|^2}{2\langle \Delta^2 \rangle}\right) d(\mathbf{x}_i - \langle \mathbf{x}_i \rangle) \qquad (42)$$

While Eq. (42) cannot be solved analytically, it is amenable to an accurate numerical solution, discussed in Appendix D. Applying the average fluctuating polymer method thus involves

1. evaluating the single voxel contributions for each voxel in each sorption state A [see Eq. (105)],
2. multiplying all contributions to evaluate Z_j for each state and $Z^{\ddagger}_{A \to B}$ for each dividing surface [Eq. (42)],
3. calculating the partition functions for the state and dividing surface [Eqs. (32) and (34)],
4. calculating the rate constant for each $A \to B$ transition by Eq. (16),
5. using the resulting jump network to calculate the diffusivity.

The last point is discussed below in Section X.

The effect of different $\langle \Delta^2 \rangle$ values on the normalized probability density of different separations between a fixed penetrant atom and a single fluctuating atom of the polymer is shown in Fig. 1. Only one dimension is shown to simplify the plot, and parameters are chosen such that

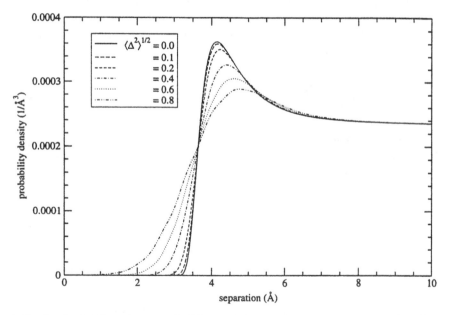

FIG. 1 Probability density of different interatomic separations occurring for a combination of the Lennard-Jones potential and the pseudo-elastic energy provided by the Debye–Waller factor. Normalization is such that the integral over a spherical region from 0 to $10\,\text{Å}$ equals unity. Different lines indicate different $\langle \Delta^2 \rangle^{1/2}$ values (in Ångstroms).

$\epsilon/k_BT=0.437$, $\sigma=3.69$ Å (methane–methyl interaction in atactic polypropylene at 233K [88]). The larger the value of $\langle\Delta^2\rangle^{1/2}$, the "softer" the effective potential, since smaller separations can be achieved with nonzero probability. In other words, larger polymer chain fluctuations facilitate a penetrant squeezing through a tight channel.

The choice of $\langle\Delta^2\rangle^{1/2}$ is the most difficult part of applying the fluctuating polymer method. As seen in Fig. 1, changes in $\langle\Delta^2\rangle^{1/2}$ of order 0.2 Å cause qualitative differences in the distribution of atom positions. Too small a value of $\langle\Delta^2\rangle^{1/2}$ leads to energy barriers that are too high and diffusion coefficients that are too low; the limit $\langle\Delta^2\rangle^{1/2} \to 0$ corresponds to the frozen polymer method. Too high a value of $\langle\Delta^2\rangle^{1/2}$ corresponds to channel openings being too facile, with diffusion coefficients that are artificially large. $\langle\Delta^2\rangle^{1/2}$ values are also penetrant-dependent [97].

One approach that would come to mind for choosing $\langle\Delta^2\rangle^{1/2}$ would be to run a molecular dynamics simulation on the pure polymer, setting $\langle\Delta^2\rangle^{1/2}$ equal to an asymptotic value. However, this approach fails in an amorphous polymer. Over long times, mean-square atom displacements increase linearly with time due to diffusive chain motions, leading to

$$\langle\Delta^2\rangle^{1/2} = \sqrt{6Dt} \tag{43}$$

The value of $\langle\Delta^2\rangle^{1/2}$ at 1 ns will be of order 10 times larger than the value at 10 ps. Even in the reptation regime [113,114], in which $\langle r^2\rangle \sim t^{1/2}$, $\langle\Delta^2\rangle^{1/2}$ would appear to increase by about $3\times$ over this time scale, which based on Fig. 1 would have a big effect on diffusion. Clearly the value of $\langle\Delta^2\rangle^{1/2}$ depends on the time scale of interest.

Several methods have been proposed for choosing the parameter $\langle\Delta^2\rangle^{1/2}$. Its value is crucial, because the magnitude and characteristics of mean-squared displacement (normal vs. anomalous diffusion) are sensitive functions of $\langle\Delta^2\rangle^{1/2}$. Regrettably, several practitioners of this method omit $\langle\Delta^2\rangle^{1/2}$ values from their publications. Arizzi [93] noted that an analytic form is available from the X-ray scattering literature [148],

$$\langle\Delta^2\rangle = \frac{3h^2T}{4\pi^2mk_B\theta_D^2}\left[\frac{T}{\theta_D}\int_0^{\theta_D/T}\frac{\xi d\xi}{e^\xi - 1} + \frac{\theta_D}{4T}\right] \tag{44}$$

where θ_D equals the Debye temperature. This approach yields a different value for each atom (due to their differing masses) and $\langle\Delta^2\rangle^{1/2}$ values slightly larger than used in later simulation work. Gusev and Suter [97] initially proposed that $\langle\Delta^2\rangle^{1/2}$ should be chosen to correspond to the elapsed time at which a penetrant is most likely to jump into another state.

They reasoned that since the jump significantly changes the list of neighboring polymer chain atoms (and especially the distances from the atoms that remain neighbors), it is appropriate for the elastic effects to be considered afresh. This process [97] then entails self-consistency:

1. guess $\langle \Delta^2 \rangle^{1/2}$,
2. calculate the escape time distribution,
3. rechoose $\langle \Delta^2 \rangle^{1/2}$ based on the most likely escape time,
4. recalculate the escape time distribution,
5. iterate to convergence.

This method implicitly requires the peak and breadth of the escape time distribution to be equally sensitive to $\langle \Delta^2 \rangle^{1/2}$, since the average and breadth of a waiting time distribution are governed by only one parameter [103]. However, the sensitivities are not necessarily the same, and Gusev et al. [24] noted that the $\langle \Delta^2 \rangle^{1/2}$ value so-calculated is only approximate. They next suggested [24] conducting both MD and TST simulations, using the moderate time behavior ($t < 10$ ns) to provide a target for choosing $\langle \Delta^2 \rangle^{1/2}$:

1. obtain $\langle r^2(t) \rangle$ from MD,
2. guess $\langle \Delta^2 \rangle^{1/2}$,
3. calculate $\langle r^2(t) \rangle$ from TST and kinetic Monte Carlo (a technique described below),
4. modify $\langle \Delta^2 \rangle^{1/2}$, recalculate, and iterate to obtain better agreement.

Such a combination of MD and TST has promise for systems in which the assumption of uncoupled elastic fluctuations is valid.

VI. EXPLICIT POLYMER METHOD

The explicit polymer method provides more molecular-level detail about each penetrant jump, at the expense of a higher computational cost compared to the frozen polymer and average fluctuating polymer methods. The polymer chains participate explicitly in defining the path of a penetrant jump. In contrast to the average fluctuating polymer method, however, the *particular* conformation fluctuations that open the interstate channel are tabulated as part of the jump path. In other words, the steps $\delta \mathbf{q}_f$ along the IRC include penetrant degrees of freedom and *some* of the many polymer degrees of freedom. The IRC thus traces out a multidimensional pathway that starts in a local potential energy minimum (described as a penetrant residing in a sorption state), passes through a multidimensional transition state as polymer chain fluctuations open a channel and the penetrant moves through it, and terminates in another local potential energy minimum.

The general approach for applying the explicit polymer method involves

1. identifying multidimensional transition states,
2. following the IRC from each transition state to its affiliated local minima,
3. calculating the rate constant for each jump,
4. using a kinetic Monte Carlo approach to calculate the diffusion coefficient.

A molecular scale network spanning the original simulation cell, with its jump paths and rate constants, can then be used when calculating the diffusivity. Alternatively, a larger network can be constructed [91] (see Section IX) that incorporates molecular-level correlations.

The task of searching for a penetrant jump transition state was discussed in Section II. Each multidimensional transition state is a solution to Eq. (13), which is solved most efficiently in stages [88]. First a low-dimensional transition state is found for the penetrant by keeping the polymer frozen (as above). Next, polymer degrees of freedom can be introduced into the transition state search, using the transition state found in the lower dimensionality as an initial guess. The appropriate degrees of freedom to include are those defined by chain segments near the penetrant. All of the selected degrees of freedom (including the penetrant) are allowed to evolve during the transition state search in the higher dimension. The resulting transition state has two elements of unstable equilibrium. Forces on the penetrant are balanced such that it senses neither a force towards the origin state nor one towards the destination state. Simultaneously, neighboring polymer chains have withdrawn in response to penetrant–polymer forces until they are balanced by the force required to compress the polymer matrix locally. The result is a more open channel for penetrant diffusion.

To some extent, as the number of coordinates involved in calculating the transition state increases, the transition state energy decreases. As an example, Fig. 2 shows the decrease in transition state energy for a methane penetrant in atactic polypropylene at three different transition states, as a function of transition state dimensionality. The greater the number of degrees of freedom, the larger the chain region able to take up this compressive force and the smaller the transition state energy. However, the decrease in energy differs from transition state to transition state; no simple, quantitative correlation for the magnitude of the energy decrease exists. The overall energy decrease relative to that in the initial state (about an order of magnitude smaller than for the frozen polymer case) corresponds to a rate increase by a factor of about 10^6. A penetrant jump assisted by polymer motions is thus much more likely than an

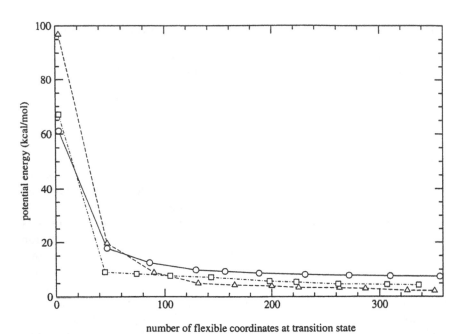

FIG. 2 Penetrant diffusion transition state energy relative to the local minimum energy of pure polymer, as a function of the number of coordinates considered flexible during the transition state search. Different symbols correspond to different transition states. The transition state energy at three flexible coordinates would be used in the frozen polymer method.

unassisted jump, as found for the fluctuating vs. frozen polymer methods. As seen in the figure, the transition state energy ultimately reaches a plateau value, after which it is unaffected by including additional degrees of freedom.

In order to reach such a plateau, the initial polymer structure (no penetrant present) must be in a local potential energy minimum. Otherwise, each additional polymer coordinate included will contribute a new opportunity for chain relaxation, leading to a transition state lower in energy than the original structure but higher than a local potential energy minimum.

For conducting numerous penetrant diffusion simulations, it isn't reasonable to inspect Fig. 2 by eye to choose when to stop adding degrees of freedom; an automated procedure is required instead. One approach [88] is to continue until all degrees of freedom defined by atoms within the penetrant neighbor list have been added. The remaining atoms are affected only indirectly: the penetrant interacts with atom i, which then interacts with atom j. This approach led to about 350 degrees of freedom (penetrant position, bond angles, torsion angles, sometimes chain start positions and orientations) for methane in glassy atactic polypropylene. A second approach is to identify all atoms that the penetrant is likely to encounter during its jump between two sorption states, and then to include all of them in the transition state search. Such a list could be created by following the IRC in only the penetrant dimensionality, in the spirit of the frozen polymer method. An advantage of this approach would be that all degrees of freedom that would participate in the IRC would be identified in advance and would move towards a zero-force conformation at the transition state. A disadvantage is the large number of degrees of freedom involved (of order 500 for methane in atactic polypropylene). In either case, we found [88] it was worthwhile to increase the included number of degrees of freedom slowly (5, 15, . . . , etc.), creating a sequence of higher dimensional initial guesses, rather than to attempt the complete search initially. While this would seem to require more calculation, in practice the sequential initial guesses were good enough such that in most cases no single search required a long time.

Next, the multidimensional IRC must be determined for this jump. The Jacobian and the covariant metric tensor include contributions from all f flexible coordinates, and the IRC is found by following a single step away from the transition state [Eq. (14)] and then multiple steps along the IRC [Eq. (15)]. Qualitative descriptions of polymer contributions to the diffusion mechanism can be taken directly from how each flexible coordinate changes along the IRC. For example, Fig. 3 shows different chain conformations along one diffusion jump. Dashed and solid triangles indicate original and instantaneous conformations for selected chain segments. Due to coupled changes in polymer bond and torsion angles, chains move back from the penetrant path as the jump occurs, creating a wider channel.

Next the rate constant is calculated for each jump. In prior work [88] we used a harmonic approximation in the flexible degrees of freedom, with the quantum mechanics-based partition function [Eq. (20)]. Such a rate constant calculation is fast in comparison to tracking the IRC. Using a free energy method (Eq. (1.117) or sampling as in Ref. [111]) would yield more accurate rate constants at a cost of a larger, more time consuming calculation. Dynamical corrections [Eqs. (1.103) and (1.104)] would increase accuracy but add additional computational requirements. If the software and run time are available, these more advanced methods are recommended.

Finally, the rate constants can be used to calculate the diffusivity. Methods based on jump network structure and kinetic Monte Carlo simulation are discussed below.

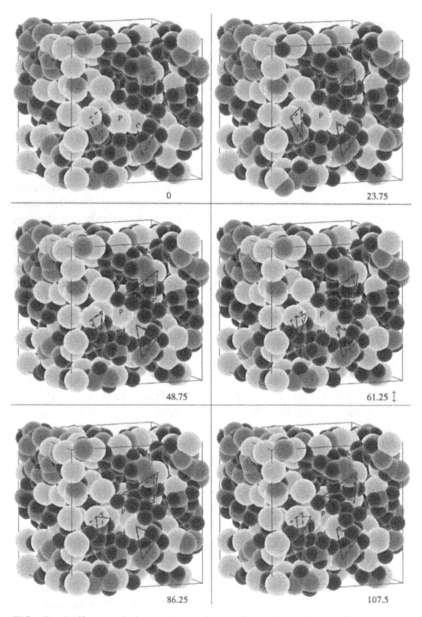

FIG. 3 Different chain conformations of atactic polypropylene as a methane penetrant (sphere labeled p) jumps between two sorption states. Text labels indicate the extent along the reaction coordinate. Differences between the positions of dashed and solid triangles indicate the motion of chain segments to create a more open channel for diffusion. Conformations are adapted from Movie 1 in [90].

VII. OTHER IRC METHODS

Numerical artifacts can result from curvature along the IRC. For example, too large a step along the gradient can lead to a departure from the true diffusion path. The gradient at the end of the step will direct the system back towards the IRC, but due to components orthogonal to the IRC it is likely that so-called "overshoot" will occur again. The net result is a sequence of steps that repeatedly cross the IRC [126]. This leads to two disadvantages:

1. No individual calculated point is actually on the IRC itself.
2. Several more steps are required, leading to additional calculations.

A cartoon of this effect is shown in Fig. 4.

One proposed solution to this problem is methods that incorporate both the local gradient and the Hessian into steps along the IRC [116,149,150]. These allow for direct incorporation of curvature within each step. Each step requires significantly more calculation, since the Hessian is required. However, IRC curvature is likely, and since repeated crossings can be avoided, fewer larger steps may be necessary. Which approach is more efficient (based on both the gradient and Hessian or on only the gradient) depends on whether an analytic Hessian calculation is available and if so then how long it takes.

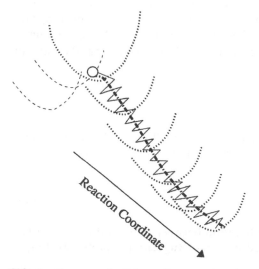

FIG. 4 Cartoon of a "zig-zag" approximation of the IRC (bold dot–dash line). The thin solid path that repeatedly crosses the IRC (dot–dash path) is likely to result when no Hessian information is used, such as with Eqs. (14) and (15). Dashed and dotted lines indicate contours of constant potential energy.

Path-based methods [110,111,118,127,128,151–153] differ from the approaches discussed above in that they attempt to find an entire reaction/ diffusion path (the IRC) between local minima, without first performing a transition state search. Instead, the IRC is represented by a sequence of order $N^{st} = 10$–50 state points in configuration space. Each state point includes all f flexible mass-weighted Cartesian or generalized coordinates. The different methods correspond to different algorithms, typically [110,118,152] some sort of minimization, for solving for the resulting $N^{st}f$ coordinates. On a multidimensional potential energy surface, the solution corresponds to a discretized approximation to a low-energy pathway traversing as directly as possible between local minima. The $f = 2$ analog inspires descriptive names for these approaches: "chain of states" [118], "nudged elastic band" [110,152], "throwing ropes over mountain passes in the dark" [111]. In model problems, the local minima have not even had to neighbor one another; the path obtained could pass through multiple transition states and intermediate local minima. However, the high dimensionality has prevented path-based methods from being applied to diffusion in polymers with a flexible matrix. For example, 20 state points along an IRC in [88] (about 350 flexible coordinates) would require a minimization with respect to 7000 variables. Evaluating the Hessian matrix is required in some methods.

It is also possible that via thermal motions a penetrant–polymer system would sample different *sets* of pathways, each with its own transition state structure, for a penetrant jump between two associated local minima. The methods in [111] are particularly suited to calculating the rate in that case.

VIII. SORPTION

Penetrant sorption has three impacts on small-molecule permeability. A direct impact is the proportionality between permeability P and solubility in solution-diffusion transport [1,154]

$$P = DS \qquad (45)$$

The sorption coefficient S equals the ratio between penetrant concentration in the polymer phase and imposed penetrant fugacity (or pressure) [154]

$$S \equiv C/f \qquad (46)$$

The higher the sorption coefficient, the larger the number of penetrant molecules dissolved in the polymer sample and able to diffuse across it.

Another direct effect is that high penetrant concentrations can increase polymer chain mobility (plasticization), thereby increasing the diffusion coefficient [1]. An indirect effect concerns relative sorption among different sorption states. From microscopic reversibility [22], the ratio of equilibrium solubilities within different sorption states equals the ratio of jump rate constants between them,

$$\frac{S_A}{S_B} = \frac{k_{B \to A}}{k_{A \to B}} \qquad (47)$$

Consequently a jump network with a distribution of rate constants must also have a distribution of solubility coefficients. The more broad the sorption coefficient distribution, the wider the imbalance in the forward and reverse jump rates along a single path. Conversely, assigning the same sorption coefficient to each sorption state would require all jumps to have the same rate constant. In addition, most sorption states will have more than one neighboring sorption state (if each state had only one neighbor in a random orientation, it would be geometrically unlikely to form a percolating network), and thus most solubility coefficients are governed by multiple equivalents of Eq. (47): one per neighboring state. These latter effects of the sorption coefficients are indirect because they are independent of the total amount of sorption; the rate constants are unaffected if all sorption coefficients are scaled by the same value [see Eq. (47)].

Sorption predictions rely on calculating the chemical potential of the penetrant molecule in each sorption state. Details are described in Chapters 9 and 10 of this book. In the low-concentration limit, the required solubility coefficients for use in Eq. (47) and in diffusion calculations can be calculated by [154,155]

$$S_A = \lim_{p \to 0} \left(\frac{\langle N_{pen} \rangle_A}{\langle V_{polym} \rangle} \right) \Big/ P = \frac{1}{k_B T} \left\langle \exp\left(-\frac{\mathcal{V}_{pen}}{k_B T} \right) \right\rangle = \frac{1}{k_B T} \exp\left(-\frac{\mu_A^{ex}}{k_B T} \right) \qquad (48)$$

The ratio $\langle N_{pen} \rangle_A / \langle V_{polym} \rangle$ equals the number of penetrant molecules dissolved in a particular sorption state A per total volume of polymer, and P equals the total pressure. The ensemble average brackets $\langle \cdots \rangle$ imply averaging over all penetrant positions within a sorption state A while the polymer conformation fluctuates within the canonical ensemble according to $\mathcal{V}^{polym}$, the total potential energy in the absence of a penetrant molecule.

μ_A^{ex} equals the excess chemical potential for a penetrant in sorption state A. While each sorption coefficient S_A depends inversely on the system volume, the total sorption coefficient $S = \sum_A S_A$ is an intensive quantity, since the number of sorption states varies linearly with the total polymer volume. Equation (48) applies to the low-concentration limit because it neglects penetrant–penetrant interactions.

Several methods are available for calculating the excess chemical potential of a penetrant molecule using molecular simulation. Widom particle insertion [156,157] is the most straightforward. The potential energy change that *would* occur *if* a penetrant were inserted at a randomly chosen position is calculated, and the excess chemical potential is proportional to the average of the resulting Boltzmann factor,

$$\mu^{ex} = -k_B T \ln\left\langle \exp\left(-\frac{\mathcal{V}_{pen}}{k_B T}\right)\right\rangle \tag{49}$$

This approach has been applied several times to small-molecule sorption in polymers [46,77,158,159]. However, a difficulty arises for moderate-sized or large penetrants: there would be insufficient space for a penetrant to be inserted; hence the increase in potential energy is large and the predicted sorption is low. Several alternative approaches exist, and here we list just a few that have been applied to penetrant sorption. In mixed canonical/grand canonical simulations, the penetrant chemical potential is specified and its concentration fluctuates, but the amount of polymer remains constant [155,160–162]. In Gibbs ensemble Monte Carlo [163,164], molecule exchanges between two simulation cells are used to achieve phase equilibrium. Thermodynamic integration [22,165] is based on using a coupling parameter to relate a more easily simulated system to the system of interest, and several recent penetrant–polymer implementations show promise [166–168]. Extended ensembles treat these coupling parameters as variables in the simulation [168–171]. For chemically similar small and large molecules, chain scission/regrowth or connectivity reassignment can be used to estimate the chemical potential [172,173]. Combinations, such as using an extended ensemble within a Gibbs ensemble framework [45,174], are also useful. Other approaches [161,171,175] lead to segment-level incremental chemical potentials, which are useful for calculating relative solubility. Finally, it was recently proposed that a polymer membrane in contact with a penetrant phase can be simulated directly [176].

Techniques that can improve statistical sampling are often crucial in these approaches. An excluded-volume map can improve efficiency by focusing particle insertions into free volume voids [159,177–179]. Specially designed Monte Carlo moves allow polymer chain fluctuations within a densely packed system [180–182] (see also Chapter 7). Recent reviews [183,184] and Chapters 9 and 10 provide in-depth descriptions of methods for calculating chemical potentials in polymer systems, and their applications.

Another approach to calculating penetrant sorption [185] is to calculate the penetrant energy as in the Widom insertion method, but at points along a regularly spaced lattice, using the results to calculate the Henry's Law constant. A related, popular technique is Gusev and Suter's theory [186], which couples particle insertion concepts and the voxel method described above for finding transition states in three dimensions. Their approach is particularly convenient when applying the frozen polymer method or the fluctuating polymer method, in part because the voxel calculations are performed anyway and in part because it has been implemented in a popular software package [101]. It differs from the prior approach by applying a local partition function to each sorption state. The resulting functional form is similar to the popular dual-mode sorption model [1], with a different Langmuir parameter b_i for each state.

One aspect of a typical sorption calculation may require some modification in order to use the results in a diffusion simulation. As suggested by Eq. (47) and illustrated below, the sorption coefficient *distribution* is required for formulating the jump network and thereby obtaining the penetrant diffusion coefficient. Thus the sorption calculation must be formulated such that individual results are obtained for each sorption state, rather than a single sorption result for an entire simulation cell. If the network generation method described below will be used (rather than the molecular-scale jump network that emerges directly from simulated molecular structures), it is convenient to accumulate sorption correlations during molecular-level simulations. For example, mutual distributions of sorption magnitude and sorption state connectivity (number of neighbors accessible via low-energy paths from each sorption state) can be later incorporated into a large-scale network.

IX. NETWORK STRUCTURE

For conducting a kinetic Monte Carlo simulation, the most straightforward choice for a network of sorption states and rate constants is that of the original molecular structure. Its key advantage is its one-to-one correspondence with the detailed polymer configuration. However, the small size of a typical simulation box is a disadvantage. For example, in [97] it was observed that anomalous diffusion continued until the root-mean-squared displacement equaled the box size. From this match in length scales, it is not clear if the anomalous region truly ends or is artificially terminated by periodic boundary conditions.

The effect of periodic boundary conditions on the network structure of an amorphous polymer is shown in Fig. 5. The nine images depict a three-dimensional, 23 Å edge-length simulation cell of atactic polypropylene, with period replication shown explicitly in two dimensions. Different colors indicate different sorption states of accessible volume [87]. Within a single cell, no positional order is apparent. Within the set of periodically replicated cells, however, an imposed crystallinity is apparent: there is one disordered box per unit cell, and one's eye quickly recognizes the periodic array of

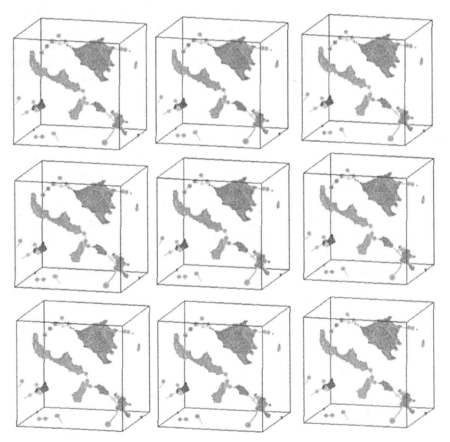

FIG. 5 Nine periodic replications of geometric sorption states in a particular configuration of atactic polypropylene. Penetrant diffusion occurs via jumps among these states. The states within an individual simulation cell appear disordered, but the crystalline nature of a periodically replicated set is more visually noticeable.

sorption states. A dynamical realization of this effect is shown in Figure 2 of [99]: a hydrogen penetrant sequentially visits periodic images of the same sorption state in a polyimide matrix.

In [91] we proposed an alternative approach for creating a penetrant jump network. The original, detailed molecular simulation provides averages and distributions for the number of sorption states per unit volume, the number of connections (via low-energy diffusion paths) to neighboring sorption states, the center-to-center distances along each connection, and the angles between connections that share a common sorption state endpoint. These molecular-level correlations can then be used to build up a larger network using reverse Monte Carlo [91,187]. Sorption state positions and connectivities are first chosen at random, and Monte Carlo-type moves are used to improve agreement between the instantaneous characteristics of the jump network and the overall distributions obtained from detailed molecular simulations. Such moves entail translating a sorption state position by ± 0.5 Å, adding a connection between two previously unconnected sorption states, or removing an existing connection. The objective function (analogous to the potential energy) quantifies the similarity between the instantaneous and target distributions in a least-squares sense. A schedule of decreasing pseudo-temperatures can then be used to ease the network into the desired configuration. In previous work [91] we generated networks with edge lengths of 141 Å, with each network connection corresponding to a methane pathway within glassy atactic polypropylene. Larger networks are possible as well; the limiting requirement is sufficient memory to store the array of angles between jump paths. The computer time required to construct a network is much smaller than that required for the molecular simulations; typically an overnight run is sufficient.

Next it is necessary to assign relative sorption probabilities and jump rate constants to the jump network. The distributions of solubility coefficients [calculated by Eq. (48)] and rate constants are required. First, a cumulative solubility coefficient distribution is calculated: $P^{S,tot}(S)$ equals the probability that a sorption state has a solubility coefficient of S or smaller. Next, a uniformly distributed random number $0 \leq \xi < 1$ is selected in turn for each sorption state. That sorption state is assigned the value S_A for which

$$P^{S,tot}(S_A) = \xi \tag{50}$$

If enough simulation data are available, separate distributions $P^{S,tot}(S)$ could be tabulated for each amount of connectivity. For example, high interstate connectivity can be correlated with higher relative sorption capacity [91]. By design, the resulting network will have a distribution of sorption coefficients that matches the desired target values.

Rate constants must then be found that simultaneously satisfy their desired distribution and Eq. (47). All rate constants in one direction (from lower to higher sorption state number, for example) can start at a fixed value, say $1\mu s^{-1}$, and reverse rate constants can then be calculated from Eq. (47). Another reverse Monte Carlo simulation may then be conducted, with the attempted move being a simultaneous increase or decrease in the magnitude of a randomly selected pair of forward and reverse rate constants. Again, repeating many times and using a schedule of decreasing pseudo-temperatures can lead to reasonable agreement with the desired rate constant distribution.

The resulting jump networks are then ready for use in a diffusion calculation. Using different molecular-level polymer configurations to obtain the target structure, sorption, and rate distributions increases the likelihood that the resulting large jump network will properly reflect the characteristics of a polymer sample.

X. KINETIC MC TO DIFFUSION COEFFICIENT

The final step in obtaining a diffusion coefficient is to simulate the dynamics of a penetrant molecule on the network of sorption states and rate constants. Analogous to the "frozen" positions of voids and channels in a glassy polymer, the relative sorption probabilities and jump rate constants typically remain constant throughout the diffusion simulation. For uniform rate constants on an ordered lattice, it is possible to solve the dynamics analytically. For the disordered network found for voids through a polymer matrix, a numerical solution is required.

Because a penetrant is likely to explore a sorption state for a long time before jumping, it is reasonable at low concentrations to assume that individual penetrant jumps are uncoupled from one another and that the sequential visiting of states is a Markov process [103]. Since each jump occurs independently, the probability density $p^{W}(t)$ that a time t elapses before the next jump occurs (the waiting time) is distributed according to a continuous-time Poisson process [103],

$$p^{W}(t)\,dt = R\exp(-Rt)\,dt \qquad (51)$$

$R[=]s^{-1}$ equals the overall jumping rate out of all states,

$$R = \sum_{i}\sum_{j} R_{i \to j}(t) \qquad (52)$$

and

$$R_{i \to j} = N_i(t) k_{i \to j} \tag{53}$$

equals the flux from state i to state j. The average waiting time until the next jump will occur anywhere on the network equals

$$\int_0^\infty t p^W(t) \, dt = \int_0^\infty Rt \exp(-Rt) \, dt$$

$$= \left(-te^{-Rt} - \frac{e^{-Rt}}{R} \right) \bigg|_0^\infty = (-0 - 0) - \left(0 - \frac{1}{R} \right) = \frac{1}{R} \tag{54}$$

The amount of time that elapses between penetrant jumps thus depends explicitly on the instantaneous sorption state population and on the distribution of jump rate constants out of occupied states.

The sorption state populations evolve according to a master equation. For a system in which each sorption state can be represented by a single local minimum,

$$\frac{dp_i}{dt} = -\sum_j k_{i \to j} p_i + \sum_j k_{j \to i} p_j \tag{55}$$

Networks generated using methods of Section IX fall in this category. For some systems, there is a separation of time scales and jumps can be separated into two categories: fast jumps among different local minima in the same sorption state and slow jumps between connected sorption states. For this case, the master equation can be written

$$\frac{dp_A}{dt} = -\sum_B \left(\sum_{j=1}^{N_B} \sum_{i=1}^{N_A} k_{i \to j} \frac{p_i}{p_A} \right) p_A + \sum_B \left(\sum_{j=1}^{N_B} \sum_{i=1}^{N_A} k_{j \to i} \frac{p_j}{p_B} \right) p_B \tag{56}$$

A and B represent different overall sorption states and i and j are individual regions of states A and B, respectively. The ratio p_i/p_A equals the fraction of sorption in state A that occurs within region i. After some time, the sorption state probabilities stabilize at their equilibrium values p_i^{eq} or p_A^{eq} that satisfy microscopic reversibility.

The master equations can be solved numerically using kinetic Monte Carlo simulation [85,86,155]. To begin, a number of independent, noninteracting "ghost" penetrants are placed in the sorption states of each network, according to the equilibrium distribution $p_A^{eq} = S_A / \sum S_A$.

These initial positions are stored in a vector $\mathbf{r}_0$. The distribution can be sampled using the same method as described above when assigning the sorption coefficients to each sorption state:

- Create a cumulative distribution,

$$P_n = \sum_{i=1}^{n} S_i \Big/ \sum_{i=1}^{N_s} S_i \tag{57}$$

- Choose a uniform random number $0 \leq \xi < 1$ for each ghost particle and assign it to the state n for which $P_{n-1} \leq \xi < P_n$.

The number of penetrants N_p should be much larger than the number of sorption states N_s in order to sample the probabilities p_A^{eq} correctly. The number of penetrants in a state i at a time t is denoted below by $N_i(t)$.

To begin a kinetic Monte Carlo simulation step, the flux along each jump and their ratio, the probabilities

$$q_{i \to j}(t) = \frac{R_{i \to j}(t)}{R(t)} \tag{58}$$

are calculated. Next, the expected waiting time until the next event, τ, is found as follows. The instantaneous waiting time distribution [Eq. (51)] leads to a cumulative waiting time distribution by

$$P^W(t) = \int_0^t p^W(t) \, dt = 1 - \exp(-Rt) \tag{59}$$

Choosing a probability $0 \leq \xi < 1$ then implies a particular time τ at which this cumulative probability $\xi = P^W(\tau)$ is reached, and solving using Eq. (59) leads to

$$\tau = -\frac{1}{R} \ln(1 - \xi) \tag{60}$$

The jump that occurred in this event is chosen from the probabilities $q_{i \to j}(t)$ [from Eqs. (53) and (58) a larger rate constant implies a larger probability], and the particular ghost penetrant leaving state i (if more than one was present) is chosen uniformly. Selecting from the $q_{i \to j}(t)$ distribution is performed in the same manner as above, using the cumulative probability density distribution. The state label for the jumping penetrant is updated, the occupancies $N_i(t)$ and $N_j(t)$ are reset to account for the jump, and the process is then repeated many times. Each jump changes the average mean-squared displacement (it can increase or decrease) and increases the elapsed time by τ. In practice, only fluxes out of the sorption states i and j that involve the most recent jump have to be recalculated according to Eq. (53); this decreases the number of computations significantly for a large system.

The kinetic Monte Carlo simulation should begin by conducting many steps in order to equilibrate the penetrant sorption state positions, after which r_0 is reset to zero. A production run can then begin. In the production run, the net result of many steps (10^3–10^4 per particle) is a list of penetrant sorption state occupancies as a function of time. The mean-squared displacement can be calculated (during or after the simulation) at each time as

$$\left\langle |\mathbf{r}(t) - \mathbf{r}_0|^2 \right\rangle = \frac{1}{N} \sum_{i=1}^{N} [(x(t)_i - x_0)^2 + (y(t)_i - y_0)^2 + (z(t)_i - z_0)^2] \qquad (61)$$

The time-dependent mean-squared displacement will fluctuate from step to step and will increase over the long term.

Two characteristics of the relationship between the mean-squared displacement and time are usually of interest. Over long times, mean-squared displacement will increase linearly with time, and the diffusion coefficient D can be extracted from the results as

$$D = \frac{1}{2d} \lim_{t \to \infty} \frac{\partial \langle r^2 \rangle}{\partial t} = \frac{1}{2d} m \qquad (62)$$

where m is the long-time slope on a linear plot of $\langle r^2 \rangle$ vs. t, and d is the dimensionality.* Over shorter times, diffusion may be anomalous, meaning

$$\langle r^2 \rangle \propto t^n \qquad (63)$$

with $n < 1$. (Note that this is opposite to so-called case II diffusion [188], in which penetrant mean-squared displacement increases with time squared. Here diffusion is slower than in the Fickian case.) Features of interest are the value of the exponent n and the times and distances over which diffusion is anomalous. Anomalous diffusion can occur due to a distribution of rate constants [189–191] or network connectivity [24,91,192]; it can also depend on the penetrant size [42]. The essential feature is a limitation of the diffusion paths such that they cannot span the full dimensionality over the simulation time scales available.

An example of the change in mean-squared displacement with time is shown in Fig. 6 for methane in atactic polypropylene [91]. These calculations were performed using a network generated via the methods described in Section IX. At times of less than 1 ns, little information is

---

*Even for a network whose connectivity can be described using a fractal dimension, d is the dimensionality of the space in which the network is represented; i.e., $d = 2$ for surface diffusion and $d = 3$ for bulk diffusion.

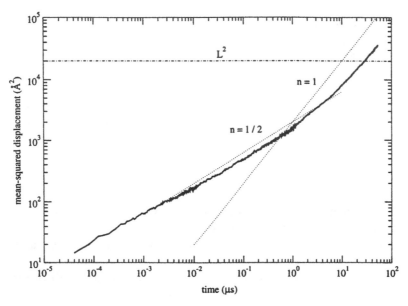

FIG. 6 Mean-squared displacement for methane in glassy atactic polypropylene, from calculations in [91]. The dot–dash line indicates the box size squared. Dotted lines indicate diffusion exponents of $n = 1/2$ and $n = 1$. A turnover from anomalous to Fickian diffusion is predicted over times of $1–5\,\mu s$.

available. These time scales accessible to MD have been "coarse grained" away in calculating the rate constants. For nanosecond-scale dynamics, MD is preferable to TST. For times up to $1\,\mu s$ or so, an anomalous regime is predicted in which mean-square displacement increases with time to the 1/2 power ($n \approx 1/2$). This value is consistent with Monte Carlo simulations of anomalous diffusion [193]. After about $5\,\mu s$, Fickian diffusion is predicted, with $D \approx 10^{-8}\,cm^2\,s^{-1}$ at $T{=}233K$. The mean-squared displacement reaches the box size squared (shown by the dot–dash line) after about $30\,\mu s$, well after the anomalous regime has ended. While data for this particular system are not available, the prediction is consistent with extrapolated data for an ethylene/propylene copolymer [91,194].

XI. SUMMARY AND OUTLOOK FOR OTHER SYSTEMS

Several different methods for simulating small-molecule diffusion in a glassy polymer have been discussed. The first involved MD simulations at and above T_g, using an Arrhenius relationship to extrapolate to temperatures well below the glass temperature. This approach is straightforward with typical simulation tools, but it involves a large extrapolation.

Four methods based on transition-state theory were then discussed. The frozen polymer method is inappropriate for penetrant diffusion in glassy polymers, because chain fluctuations are intricately coupled to the diffusion mechanism. Path-based methods are very sound, but the resulting high dimensionality could pose computational problems.

The average fluctuating polymer method and the explicit polymer method both account to some degree for polymer chain contributions to each rate constant. However, there are several differences between the two approaches. In the average fluctuating polymer method, chains experience harmonic position fluctuations, with nonlinear resulting energy changes used in the rate constant calculation. In the explicit polymer method, anharmonic position fluctuations occur, with harmonic energy fluctuations about the transition state and local minimum configurations (as the method has been applied previously). In the average fluctuating polymer method, all elastic position fluctuations contribute to the rate constant. However, *which* fluctuation is coupled to diffusion is not clear. In the explicit polymer method, a complete set of orthogonal fluctuations (normal modes) about one particular configuration is included, but no other configurations are sampled (in the harmonic approximation). However, this one configuration sampled corresponds to the lowest-energy mechanism for the jump itself. In this sense, the choice of configuration is analogous to the maximum term approximation in statistical mechanics: while only one set is included, that set is (at least one of) the most important. The average fluctuating polymer method depends on a time scale over which the magnitude of fluctuations is appropriate, and results vary strongly with a single, hard-to-predict parameter. The only time scale in applying the explicit polymer method to a jump is the inverse of the jump rate constant itself, and it is not sensitive to such a parameter.

The proper choice among these methods depends on the problem of interest and the software available. While the average fluctuating polymer method has been applied to single united-atom penetrants, how well it would work for larger, flexible penetrants is not clear. To date, the explicit polymer method has been applied to united atom and dimer penetrants. Since the method already handles hundreds of polymer chain degrees of freedom, it provides a framework for treating larger penetrants. Indeed, there are no intrinsic scientific barriers to introducing additional penetrant coordinates, such as bond or torsion angles. The effects of local, anharmonic changes in polymer chain configuration should be even more important in those cases than they are for monatomic and diatomic penetrant molecules, which will require transition-state theory treatments beyond the harmonic approximation. Instead, the barriers are methodological: energy and coordinate derivatives must be calculated for complicated force fields and polymer structures, and software that facilitates incorporating such modifications needs to be developed. Reformulating in other coordinate systems could also simplify some stages while complicating others. Working in mass-weighted Cartesian coordinates would simplify the derivations but increase the matrix computation requirements by about 3.5 times (3 instead of 2 degrees of freedom per flexible atom, in computations that scale as N^3).

Regardless of the method, applications of transition-state theory for simulating penetrant diffusion are ultimately limited by the physical diffusion mechanism. For sufficiently large penetrant molecules, the separation of time scales that accompanies the jump mechanism could diminish, making it more difficult to distinguish between events that do or don't contribute to diffusive motion. How large a penetrant corresponds to this limit is still a research question.

APPENDIX A: IRC DERIVATION IN GENERALIZED COORDINATES

To utilize generalized coordinates, it is necessary to reformulate Eqs. (6) and (7). First, the Hessian with respect to the generalized coordinates must be related to the Hessian with respect to mass-weighted Cartesian coordinates. Expanding $\partial/\partial x^i$ as $\sum_k (\partial q^k/\partial x^i)(\partial/\partial q^k)$ (as done for the gradient at the transition state) leads to

$$H_{ij} = \sum_{k=1}^{3N} \sum_{l=1}^{3N} \frac{\partial q^k}{\partial x^i} \frac{\partial q^l}{\partial x^j} \frac{\partial^2 V}{\partial q^k \partial q^l} + \sum_{l=1}^{3N} \frac{\partial^2 q_l}{\partial x^i \partial x^j} \frac{\partial V}{\partial q^l}$$

or

$$
H_{ij} = \left(\frac{\partial q^1}{\partial x^i}, \frac{\partial q^2}{\partial x^i}, \dots, \frac{\partial q^{3N}}{\partial x^i} \right)
\begin{pmatrix}
\frac{\partial^2 V}{\partial q^1 \partial q^1} & \cdots & \frac{\partial^2 V}{\partial q^1 \partial q^{3N}} \\
\vdots & & \vdots \\
\frac{\partial^2 V}{\partial q^{3N} \partial q^1} & \cdots & \frac{\partial^2 V}{\partial q^{3N} \partial q^{3N}}
\end{pmatrix}
\begin{pmatrix}
\frac{\partial q^1}{\partial x^j} \\
\vdots \\
\frac{\partial q^{3N}}{\partial x^j}
\end{pmatrix}
$$

$$
+ \left(\frac{\partial^2 q^1}{\partial x^i \partial x^j}, \frac{\partial^2 q^2}{\partial x^i \partial x^j}, \dots, \frac{\partial^2 q^{3N}}{\partial x^i \partial x^j} \right)
\begin{pmatrix}
\frac{\partial V}{\partial q^1} \\
\vdots \\
\frac{\partial V}{\partial q^{3N}}
\end{pmatrix}
\tag{64}
$$

for each term H_{ij} of the Hessian. Different rows and columns of $\mathbf{H}$ result from taking all $i = 1, \ldots, 3N$ and $j = 1, \ldots, 3N$ components, respectively. The resulting matrix equation is

$$\mathbf{H} = \mathcal{J}^T \mathbf{H}_{qq} \mathcal{J} + \left(\frac{\partial^2 \mathbf{q}^T}{\partial x^i \partial x^j} \right) \nabla_{\mathbf{q}} \mathcal{V} \tag{65}$$

$\mathbf{H}_{qq}$ is the Hessian of second derivatives with respect to generalized coordinates and $\partial^2 \mathbf{q}^T / \partial x^i \partial x^j$, a third-rank tensor, may be thought of as a $3N \times 3N$ matrix, in which each component is itself a vector; the ijth vector contains the coordinate transformation second derivatives of $\mathbf{q}$ with respect to x^i and x^j. The resulting rightmost term is a matrix for which the ijth element is the dot product of the ijth vector with $\nabla_{\mathbf{q}} \mathcal{V}$. This tensor complicates the use of an analytic Hessian matrix in generalized coordinates, since additional derivatives are required compared to typical simulation needs.

Next, the proper equation for determining eigenvalues and eigenvectors must be reformulated. The Hessian $\mathbf{H}$ can be replaced using Eq. (65), and $\mathbf{x}$ (renormalized to a small size $\delta\mathbf{x}$ rather than unity) is replaced using

$$\delta x^i = \sum_{j=1}^{3N} \frac{\partial x^i}{\partial q^j} \delta q^j$$

or

$$\delta\mathbf{x} = \mathcal{J}' \delta\mathbf{q} \tag{66}$$

$\mathcal{J}'$ (introduced in the main text) is the Jacobian matrix of partial derivatives for transforming from generalized coordinates to mass-weighted coordinates. These substitutions result in

$$\mathcal{J}^T \mathbf{H}_{qq} \mathcal{J} \; \mathcal{J}' \delta\mathbf{q} + \left(\frac{\partial^2 \mathbf{q}^T}{\partial x^i \partial x^j} \right) \nabla_{\mathbf{q}} \mathcal{V} \; \mathcal{J}' \delta\mathbf{q} = \lambda \; \mathcal{J}' \delta\mathbf{q} \tag{67}$$

This expression can be simplified after premultiplying by $\mathcal{J}'^T$ and using the identity $\mathcal{J} \mathcal{J}' = \mathcal{I}$, leading to

$$\mathbf{H}_{qq} \delta\mathbf{q} + \mathcal{J}'^T \left(\frac{\partial^2 \mathbf{q}^T}{\partial x^i \partial x^j} \right) \nabla_{\mathbf{q}} \mathcal{V} \; \mathcal{J}' \delta\mathbf{q} = \lambda \mathcal{J}'^T \mathcal{J}' \; \delta\mathbf{q} \tag{68}$$

The product of Jacobian matrices can then be replaced with the covariant metric tensor $\mathbf{a}$ [Eq. (11)]. If we restrict attention to the transition state, where $\nabla_{\mathbf{q}} \mathcal{V} = 0$, we obtain Eq. (8).

Reformulating the IRC step away from the transition state [Eq. (7)] is similar. The step $\delta\mathbf{x}$ is replaced using Eq. (66) and the gradient is replaced using Eq. (2), leading to

$$\mathcal{J}' \delta \mathbf{q} = \mathcal{J}^T \nabla_{\mathbf{q}} \mathcal{V} \, \delta \tau \qquad (69)$$

Premultiplying by $\mathcal{J}'^{T}$, replacing the $\mathcal{J}'^{T} \mathcal{J}^{T}$ product with the identity matrix, and substituting for the covariant metric tensor using Eq. (11) leads to Eq. (12).

APPENDIX B: IRC IN A SUBSET OF COORDINATES

The key to reducing the dimensionality of a TST problem is to reformulate the transition state criterion and the IRC equation such that flexible coordinates evolve as necessary, while stiff coordinates remain implicitly at their original values. This suggests formulating incremental position changes in terms of generalized coordinate changes $\delta \mathbf{q}$ and derivatives in terms of $\partial/\partial \mathbf{q}$. The latter is particularly convenient, since a partial derivative $\partial/\partial q^i$ is taken with all other $q^j (j \neq i)$, both flexible and stiff, held constant.

Narrowing the total step involves dividing it into flexible ($\delta \mathbf{q}_f$) and stiff ($\delta \mathbf{q}_s$) components,

$$\delta \mathbf{q} = \begin{pmatrix} \delta \mathbf{q}_f \\ \delta \mathbf{q}_s \end{pmatrix}$$

Changes can be made along the flexible coordinates without affecting the stiff coordinates. The gradient vector is divided as

$$\nabla_{\mathbf{q}} \mathcal{V} = \begin{pmatrix} \nabla_{\mathbf{q}_f} \mathcal{V} \\ \nabla_{\mathbf{q}_s} \mathcal{V} \end{pmatrix}$$

Following Weiner [119, § 6.31], we use Greek indices to denote flexible coordinates, uppercase indices to denote stiff coordinates, and lowercase indices to denote the entire set.

The Jacobian matrix for transforming from mass-weighted Cartesian to generalized coordinates can be narrowed in a similar manner. A derivative $\partial x^k/\partial q^\alpha$ is taken with respect to flexible generalized coordinate q^α with all other generalized coordinates held constant, so no stiff coordinates change along the direction $\partial \mathbf{x}/\partial q^\alpha$. Thus the Jacobian matrix $\mathcal{J}'$ may be divided as

$$\mathcal{J}' = \begin{pmatrix} \dfrac{\partial x^1}{\partial q^1} & \cdots & \dfrac{\partial x^1}{\partial q^f} & \dfrac{\partial x^1}{\partial q^{f+1}} & \cdots & \dfrac{\partial x^1}{\partial q^{3N}} \\ \vdots & \vdots & \vdots & \vdots & \vdots & \vdots \\ \dfrac{\partial x^{3N}}{\partial q^1} & \cdots & \dfrac{\partial x^{3N}}{\partial q^f} & \dfrac{\partial x^{3N}}{\partial q^{f+1}} & \cdots & \dfrac{\partial x^{3N}}{\partial q^{3N}} \end{pmatrix} = (\mathcal{J}'_f \ \mathcal{J}'_s) \qquad (70)$$

where $\mathcal{J}_f'$ and $\mathcal{J}_s'$ are rectangular $3N \times f$ and $3N \times (3N - f)$ matrices, respectively.

The covariant metric tensor, $\mathbf{a}$, can be narrowed by subdividing it based on the flexible and stiff coordinates [119, p. 217]

$$\mathbf{a} = \begin{pmatrix} \mathbf{a}^0 & \vdots & \mathbf{a}' \\ -- & -- & -- \\ \mathbf{a}'^T & \vdots & \mathbf{a}'' \end{pmatrix} \tag{71}$$

The $f \times f$ submatrix $\mathbf{a}^0$ in the upper left is the covariant metric tensor in only the flexible coordinates. Each element

$$a_{\alpha\beta} = \sum_k \frac{\partial x^k}{\partial q^\alpha} \frac{\partial x^k}{\partial q^\beta}$$

is unaffected when q^α and q^β remain flexible, since constrained generalized coordinates (such as q^A) were already held constant implicitly when taking the partial derivatives. The elements of $\mathbf{a}^0$ may also be found from the Jacobian matrix in the reduced dimensionality as

$$\mathbf{a}^0 = \mathcal{J}_f'^T \mathcal{J}_f' \tag{72}$$

using the Jacobian subsets introduced above. Each element of the submatrix $\mathbf{a}'$ and its transpose $\mathbf{a}'^T$ involve one flexible coordinate and one stiff coordinate. All derivatives in $\mathbf{a}''$ are with respect to stiff coordinates.

The contravariant metric tensor $\mathbf{a}^{-1}$ in the full space can be subdivided in a similar manner. However, a subset of $\mathbf{a}^{-1}$ in the reduced space is not simply a single submatrix, since the partial derivatives that define each term

$$a_{\alpha\beta}^{-1} = \sum_k \frac{\partial q^\alpha}{\partial x^k} \frac{\partial q^\beta}{\partial x^k}$$

are taken with all other *mass-weighted positions* held constant, not all other generalized coordinates. Instead, the reduced-space contravariant metric tensor equals the matrix inverse in the reduced dimensionality,

$$\overline{\mathbf{a}^{-1}}^0 = (\mathbf{a}^0)^{-1} \tag{73}$$

The matrix $\overline{\mathbf{a}^{-1}}^0$ is sometimes denoted $\mathbf{G}$ [105].

Several relationships among the submatrices of the covariant and contravariant metric tensors can also be determined [105]. Since the matrix product of the covariant and contravariant metric tensors is the identity matrix [using Eq. (10), $\mathbf{aa}^{-1} = \boldsymbol{J}'^{T}\boldsymbol{J}'$ $\boldsymbol{J}\boldsymbol{J}^{T} = \boldsymbol{J}'^{T}$ $(\boldsymbol{I})$ $\boldsymbol{J}^{T} = \boldsymbol{I}$), direct matrix multiplication

$$\mathbf{a}\,\mathbf{a}^{-1} = \begin{pmatrix} \mathbf{a}^{0} & \mathbf{a}' \\ \mathbf{a}'^{T} & \mathbf{a}'' \end{pmatrix} \begin{pmatrix} \mathbf{a}^{-1^{0}} & \mathbf{a}^{-1'} \\ \mathbf{a}^{-1'T} & \mathbf{a}^{-1''} \end{pmatrix} = \boldsymbol{I} = \begin{pmatrix} \boldsymbol{I}_{f} & \mathbf{0} \\ \mathbf{0} & \boldsymbol{I}_{3N-f} \end{pmatrix} \tag{74}$$

($\boldsymbol{I}_{f}$ is an $f \times f$ identity matrix) leads to the simultaneous equations

$$\mathbf{a}^{0}\mathbf{a}^{-1^{0}} + \mathbf{a}'\mathbf{a}^{-1'^{T}} = \boldsymbol{I}_{f} \tag{75}$$

$$\mathbf{a}^{0}\mathbf{a}^{-1'} + \mathbf{a}'\mathbf{a}^{-1''} = 0 \tag{76}$$

Rearranging Eq. (76) and postmultiplying by $(\mathbf{a}^{-1''})^{-1}$ yields

$$\mathbf{a}' = -\mathbf{a}^{0}\mathbf{a}^{-1'}(\mathbf{a}^{-1''})^{-1}$$

and substituting this into Eq. (75) leads to

$$\mathbf{a}^{0}\left(\mathbf{a}^{-1^{0}} - \mathbf{a}^{-1'}(\mathbf{a}^{-1''})^{-1}\mathbf{a}^{-1'^{T}}\right) = \boldsymbol{I}_{f}$$

and thus

$$\overline{\mathbf{a}^{-1^{0}}} = \left(\mathbf{a}^{-1^{0}} - \mathbf{a}^{-1'}(\mathbf{a}^{-1''})^{-1}\mathbf{a}^{-1'^{T}}\right) \tag{77}$$

which, after converting the nomenclature, is equation 10 in [105]. Thus the contravariant metric tensor in the flexible degrees of freedom may be calculated directly from the contravariant metric tensor in the full dimensionality. From the similar expression

$$\begin{pmatrix} \mathbf{a}^{-1^{0}} & \mathbf{a}^{-1'} \\ \mathbf{a}^{-1'T} & \mathbf{a}^{-1''} \end{pmatrix} \begin{pmatrix} \mathbf{a}^{0} & \mathbf{a}' \\ \mathbf{a}'^{T} & \mathbf{a}'' \end{pmatrix} = \boldsymbol{I} = \begin{pmatrix} \boldsymbol{I}_{f} & \mathbf{0} \\ \mathbf{0} & \boldsymbol{I}_{3N-f} \end{pmatrix} \tag{78}$$

multiplying the second row of $\mathbf{a}^{-1}$ and the first column of $\mathbf{a}$ leads to

$$\mathbf{a}^{-1'^{T}}\mathbf{a}^{0} + \mathbf{a}^{-1''}\mathbf{a}'^{T} = 0 \tag{79}$$

Rearranging and premultiplying yields

$$\mathbf{a}'^T = -(\mathbf{a}^{-1''})^{-1}\mathbf{a}^{-1'^T}\mathbf{a}^0 \tag{80}$$

which is used below. Finally, each side of Eq. (74) can be postmultiplied by

$$\begin{pmatrix} \mathbf{a}^0 & 0 \\ \mathbf{a}'^T & \mathcal{I} \end{pmatrix}$$

Expanding the matrix products on the left hand side leads to

$$\mathbf{a}\begin{pmatrix} \mathbf{a}^{-1}{}^0\mathbf{a}^0 + \mathbf{a}^{-1'}\mathbf{a}'^T & \mathbf{a}^{-1'} \\ \mathbf{a}^{-1'^T}\mathbf{a}^0 + \mathbf{a}^{-1''}\mathbf{a}'^T & \mathbf{a}^{-1''} \end{pmatrix} = \begin{pmatrix} \mathbf{a}^0 & 0 \\ \mathbf{a}'^T & \mathcal{I} \end{pmatrix} \tag{81}$$

Substituting using the transpose of Eq. (75) and Eq. (79) leads to

$$\mathbf{a}\begin{pmatrix} \mathcal{I}_f & \mathbf{a}^{-1'} \\ 0 & \mathbf{a}^{-1''} \end{pmatrix} = \begin{pmatrix} \mathbf{a}^0 & 0 \\ \mathbf{a}'^T & \mathcal{I} \end{pmatrix} \tag{82}$$

and taking the determinant of each side yields

$$\det \mathbf{a} \, \det \mathbf{a}^{-1''} = \det \mathbf{a}^0 \tag{83}$$

This interrelationship among the magnitude of different metric tensor determinants was first derived by Fixman [195] and is useful in discussions of the rigid and flexible models of polymer conformation (see Appendix C).

Next, the saddle point and Hessian calculations can be narrowed to a subset of all coordinates. At a saddle point in the full dimensionality, the potential energy gradient equals zero with respect to all generalized coordinates. The same criterion holds for the flexible coordinates in a subset of the full dimensionality. For constrained degrees of freedom $A = f+1, \ldots, 3N$, the stiff potential maintaining the constraints is of the form

$$\mathcal{V} = \sum_A \frac{1}{2} k_{stiff\,A} (q^A - q_0^A)^2$$

with each k_{stiff_A} conceptually taking on values that approach infinity [105]. Even for a differential displacement δq^A, this energy contribution

$\frac{1}{2}k_{stiff\,A}(\delta q^A)^2$ would still approach infinity, as would the force $k_{stiff\,A}\delta q^A$. Hence the only states of finite probability with finite forces correspond to $\delta q^A = 0$: stiff coordinates retain their original values at saddle points in the reduced dimensionality. The Hessian submatrix associated with the constrained degrees of freedom also takes on a diagonal form, with eigenvalues given by the stiff spring constants. All these eigenvalues are positive; thus it is meaningful mathematically to discuss the transition state with respect to a flexible subset of generalized coordinates. To summarize, a saddle point in this system of reduced dimensionality is a point at which each element of the potential energy gradient $\partial V / \partial q^\alpha$ equals zero for all flexible generalized coordinates q^α, and each eigenvalue of the Hessian in q-space

$$H_{\alpha\beta} \equiv \frac{\partial^2 V}{\partial q^\alpha \partial q^\beta}$$

is positive except for the lowest, which must be negative. The form of the eigenvalue equation to solve will be discussed next.

As in the full dimensionality, the first step of the IRC is directed along the eigenvector corresponding to the negative eigenvalue. The eigenvalue equation to solve in the subset of flexible coordinates may be found from the submatrix partitions of $\mathbf{a}$, $\mathbf{H}$, and $\delta \mathbf{q}$. In order to distinguish between the flexible and stiff coordinates, it is convenient to premultiply Eq. (8) by $\mathbf{a}^{-1}$, leading to

$$\begin{pmatrix} \mathbf{a}^{-1\,0} & \mathbf{a}^{-1\,\prime} \\ \mathbf{a}^{-1\,\prime T} & \mathbf{a}^{-1\,\prime\prime} \end{pmatrix} \begin{pmatrix} \mathbf{H}_{qq}^0 & \mathbf{H}_{qq}^\prime \\ \mathbf{H}_{qq}^{\prime T} & \mathbf{H}_{qq}^{\prime\prime} \end{pmatrix} \begin{pmatrix} \delta\mathbf{q}_f \\ \delta\mathbf{q}_s \end{pmatrix} = \lambda \begin{pmatrix} \delta\mathbf{q}_f \\ \delta\mathbf{q}_s \end{pmatrix} \tag{84}$$

Expanding the left hand side leads to separate equations for the flexible and stiff coordinates,

$$\left(\mathbf{a}^{-1\,0}\mathbf{H}_{qq}^0 + \mathbf{a}^{-1\,\prime}\mathbf{H}_{qq}^{\prime T}\right)\delta\mathbf{q}_f + \left(\mathbf{a}^{-1\,0}\mathbf{H}_{qq}^\prime + \mathbf{a}^{-1\,\prime}\mathbf{H}_{qq}^{\prime\prime}\right)\delta\mathbf{q}_s = \lambda\,\delta\mathbf{q}_f \tag{85}$$

$$\left(\mathbf{a}^{-1\,\prime T}\mathbf{H}_{qq}^0 + \mathbf{a}^{-1\,\prime\prime}\mathbf{H}_{qq}^{\prime T}\right)\delta\mathbf{q}_f + \left(\mathbf{a}^{-1\,\prime T}\mathbf{H}_{qq}^\prime + \mathbf{a}^{-1\,\prime\prime}\mathbf{H}_{qq}^{\prime\prime}\right)\delta\mathbf{q}_s = \lambda\,\delta\mathbf{q}_s \tag{86}$$

As described above, $\delta q_s^A \to 0$ for each stiff coordinate A in the infinite stiffness limit. Even with the force constants approaching infinity, the right hand side of Eq. (86) approaches zero, as does the second term on the left hand side of the same equation. Since the step $\delta\mathbf{q}_f$ in the flexible coordinates is nonzero, for Eq. (86) to be satisfied the metric tensor submatrices and the Hessian submatrices must be related such that

$$\mathbf{a}^{-1''}\mathbf{H}_{qq}^{'T} = -\mathbf{a}^{-1'T}\mathbf{H}_{qq}^{0} \tag{87}$$

After solving for $\mathbf{H}_{qq}^{'T}$ in this limit and substituting the result into Eq. (85), rearrangement leads to

$$\left[\mathbf{a}^{-1^0}\mathbf{H}_{qq}^0 - \mathbf{a}^{-1'}\mathbf{a}^{-1''^{-1}}\mathbf{a}^{-1'T}\mathbf{H}_{qq}^0\right]\delta\mathbf{q}_f = \overline{\mathbf{a}^{-1^0}}\,\mathbf{H}_{qq}^0\,d_f = \lambda\,\delta\mathbf{q}_f \tag{88}$$

Premultiplying by $\mathbf{a}^0$ leads to Eq. (14).

Another alternative is to derive Eq. (14) from Eq. (8) by expanding directly in terms of submatrices of $\mathbf{a}$. Matrix multiplication leads to two sets of equations, each of which contains $\delta\mathbf{q}_f$ and $\delta\mathbf{q}_s$,

$$\mathbf{H}_{qq}^0\delta\mathbf{q}_f + \mathbf{H}_{qq}'\delta\mathbf{q}_s = \lambda\mathbf{a}^0\delta\mathbf{q}_f + \lambda\mathbf{a}'\delta\mathbf{q}_s \tag{89}$$

$$\mathbf{H}_{qq}^{'T}\delta\mathbf{q}_f + \mathbf{H}_{qq}''\delta\mathbf{q}_s = \lambda\mathbf{a}^{'T}\delta\mathbf{q}_f + \lambda\mathbf{a}''\delta\mathbf{q}_s \tag{90}$$

Setting $\delta\mathbf{q}_s \to \mathbf{0}$ in the first leads directly to Eq. (14). However, the same limit for the second leads to the seemingly contradictory

$$\mathbf{H}_{qq}^{'T}\delta\mathbf{q}_f = \lambda\mathbf{a}^{'T}\delta\mathbf{q}_f \tag{91}$$

an overspecified system of equations for a typical case in which there are more stiff coordinates than flexible coordinates. However, this equation is equivalent to Eq. (14). The proof first requires substituting Eq. (80) into Eq. (91) and premultiplying by $\mathbf{a}^{-1''}$, which leads to

$$\mathbf{a}^{-1''}\mathbf{H}_{qq}^{'T}\delta\mathbf{q}_f = -\lambda\mathbf{a}^{-1'T}\mathbf{a}^0\delta\mathbf{q}_f \tag{92}$$

The left hand side can then be replaced using Eq. (87), leading to

$$-\mathbf{a}^{-1'T}\mathbf{H}_{qq}^0\delta\mathbf{q}_f = -\lambda\mathbf{a}^{-1'T}\mathbf{a}^0\delta\mathbf{q}_f \tag{93}$$

Removing the common matrix $\mathbf{a}^{-1'T}$ from the leftmost side of each side of this equation yields Eq. (14).

Theodorou (D. N. Theodorou, personal communication, 1993) has obtained the same results with a derivation that remains in the Cartesian coordinate system. First, he partitioned the eigenvectors into flexible and stiff components, finding that deviations along the stiff eigenvectors must equal zero. Next, he invoked a harmonic approximation with respect to the full set of mass-weighted Cartesian coordinates, ultimately obtaining the same equations as those presented here, within a change of coordinate system.

The IRC step in a flexible subset of coordinates away from the transition state also follows from subdividing the step vector and the Jacobian matrix. Begin with the IRC equation in mass-weighted Cartesian coordinates, Eq. (7). Substitute for $\delta\mathbf{x}$ using Eq. (66), but note that only a subset of $\mathcal{J}'$ is required, since $\delta q^A = 0$ for each stiff coordinate A. The elements of $\mathcal{J}'$ that multiply a nonzero value of δq^i correspond exactly to the elements of $\mathcal{J}'_f$ defined in Eq. (70). Similarly, only the gradients with respect to flexible generalized coordinates are nonzero

$$\nabla_{\mathbf{q}_f}\mathcal{V} = \mathcal{J}'^T_f \nabla_{\mathbf{x}}\mathcal{V} \tag{94}$$

since the gradient with respect to stiff degrees of freedom equals zero in either the rigid or the infinitely stiff model. Premultiplying Eq. (7) by $\mathcal{J}'^T_f$, substituting with the flexible equivalent of Eq. (66), and using Eq. (94) results in

$$\mathcal{J}'^T_f \, \mathcal{J}'_f \, \delta\mathbf{q}_f = \nabla_{\mathbf{q}_f}\mathcal{V} \, \delta\tau \tag{95}$$

From Eq. (72) above, the product $\mathcal{J}'^T_f \mathcal{J}'_f$ equals the covariant metric tensor in the reduced coordinate system, yielding Eq. (15).

To evaluate the rate, it is necessary to combine expressions for the partition functions at the local minima and at the transition state. The latter partition function is one dimension smaller than that of the former, since integration is limited to the $(3N-1)$-dimensional dividing surface that passes through the transition state. An exact differential equation for this surface exists [112], but its evaluation is difficult; it amounts to following the $3N-1$ intrinsic reaction coordinates defined by taking the first step along an eigenvector associated with a positive eigenvalue. A common assumption is to approximate the dividing surface near the transition state with a hyperplane orthogonal to the eigenvector corresponding to the negative eigenvalue; the dividing plane is orthogonal to the step taken away from the transition state. The resulting equation can be written in a flexible subset of generalized coordinates as

$$(\mathbf{x} - \mathbf{x}^\ddagger) \cdot \delta\mathbf{x}_{-\lambda} = (\mathbf{q}_f - \mathbf{q}_f^\ddagger)^T \, \mathbf{a}_0 \, \delta\mathbf{q}_{f_{-\lambda}} = 0 \tag{96}$$

where $\delta\mathbf{x}_{-\lambda}$ and $\delta\mathbf{q}_{f_{-\lambda}}$ are the eigenvectors that correspond to the negative eigenvalue in mass-weighted Cartesian coordinates and in the flexible subset of generalized coordinates. In terms of the normal modes that correspond to the flexible generalized coordinates, this surface restricts integration to modes with real frequencies. From the orthogonality of eigenvectors, all these modes are normal to the first step along the reaction coordinate.

APPENDIX C: CHOICE OF POLYMER MODEL— FLEXIBLE, RIGID, OR INFINITELY STIFF

An important question first recognized in the 1960s and 1970s [105,195,196] concerns whether statistical mechanical models of polymers consider certain degrees of freedom (traditionally bond lengths and sometimes bond angles) to be either (1) rigidly constrained to their initial values or (2) theoretically allowed to fluctuate, but under a harmonic potential with force constants that approach infinity. The two approaches lead to different relative probabilities for each polymer chain conformation and hence to different partition functions, and Gō and Scheraga [105] showed that option 2, the "flexible model in the limit of infinite stiffness," more closely resembles the true quantum-mechanical partition function.

In Monte Carlo simulation, the choice of polymer model is governed by the choice of attempted moves. Typically kinetic energy is integrated analytically over all modes, and the partition function for the flexible model in the limit of infinite stiffness results [105]. In molecular dynamics, constraints (SHAKE, etc.) freeze kinetic energy contributions and the partition function for the rigid model results [105,197]. To achieve sampling from the desired partition function, it is necessary to add a pseudopotential based on the covariant metric tensor $\mathbf{a}^0$ [198].

For modeling penetrant diffusion in polymers using transition-state theory, it is important to show that the approaches for following the IRC correspond to the flexible model in the limit of infinite stiffness, rather than to the rigid model. At first, it would seem that reducing the dimensionality by constraining particular generalized coordinates *must* correspond to the rigid model. In the development above, the constraints on the reaction path were incorporated directly into the basis vectors of configuration space [via Eqs. (11), (14), (15)] without introducing a form of the potential. What would happen if the full dimensionality were retained while force constants for stiff degrees of freedom increased in magnitude? For an infinitely stiff chain, any slight perturbations of the bond length would increase the potential energy towards infinity and the probability of observing such a state would drop exponentially to zero. However, if the basis vectors in the full dimensionality are followed (i.e., use of $\mathbf{a}^{-1}$ or $\mathbf{a}$ rather than $\overline{\mathbf{a}^{-1}{}^{0}}$ or $\mathbf{a}^{0}$ in the IRC equations), even infinitesimal steps would induce changes in the stiff degrees of freedom, since each change depends on all elements of the gradient,

$$
dq^A = \sum_j \left(\sum_k \frac{\partial q^A}{\partial x^k} \frac{\partial q^j}{\partial x^k} \right) \frac{\partial \mathcal{V}}{\partial q^j}
\tag{97}
$$

and the coordinate transformation derivatives $\partial q^A/\partial x^k$ are independent of the potential energy and are conformation-dependent. The only way to maintain a reasonable energy after a step is to ensure that the stiff degrees of freedom do not change, and hence to step along the basis vectors found using only the flexible generalized coordinates. These basis vectors are set by the matrix $\mathbf{a}^0$ and its contravariant inverse, $\mathbf{a}^{-1}{}^{0}$, as suggested by Fixman [195]. Thus, both a "rigid" polymer and an "infinitely stiff" polymer step along the same basis vectors, and the above development of the IRC equation does not select one model over the other.

Why is this IRC formulation insensitive to the choice of polymer model? In Gō and Scheraga's treatment [105], differences between the two models originated when integrating the kinetic energy [197]. In the flexible model, the independent variables were the Cartesian coordinates $\mathbf{X}$ and their conjugate momenta $\mathbf{p}$. The kinetic energy terms are uncoupled ($\mathbf{a}$ and $\mathbf{a}^{-1}$ are diagonal) in these coordinates (equation 4 in [105]), and integration is analytic. The Jacobian determinant $\det \mathcal{J}'$ that appears when converting from the mass-weighted Cartesian representation $\mathbf{x}$ to the generalized coordinates $\mathbf{q}$ is only a function of the bond lengths and bond angles, and the partition function may be written as [105]

$$
Q = \left(\frac{2\pi k_B T}{h^2} \right)^{3N/2} \int d\mathbf{q}_f \; \det \mathcal{J}' \; \exp\left(-\frac{\mathcal{V}(\mathbf{q}_f)}{k_B T} \right)
\tag{98}
$$

where the Jacobian determinant is intentionally kept inside the integral. (Only parts of it may be removed if some bond angle force constants do not approach infinity.) The masses have been absorbed into the generalized coordinates $\mathbf{q}$.

In the rigid model, the independent variables are the generalized coordinates $\mathbf{q}$ and their conjugate momenta P_r (equation 12 in [105]). The kinetic energy

$$
T = \frac{1}{2} \dot{\mathbf{q}}^T \, \mathbf{a}^0 \, \dot{\mathbf{q}} = \frac{1}{2} P_r{}^T \, \overline{\mathbf{a}^{-1}{}^{0}} \, P_r
\tag{99}
$$

couples all generalized momenta on a given chain, since the contravariant metric tensor is nondiagonal, and integration of the kinetic energy requires yet another change of coordinates, now from the conjugate momenta P_r to the normal modes ξ of the kinetic energy. The Jacobian determinant of this transformation is $[\det \mathbf{a}^{-1^0}]^{-1/2}$, resulting in the partition function for the rigid model,

$$Q = \left(\frac{2\pi k_B T}{h^2}\right)^{3N/2} \int d\,\mathbf{q}_f \left[\frac{1}{\det \mathbf{a}^{-1^0}}\right]^{1/2} \exp\left(-\frac{\mathcal{V}(\mathbf{q}_f)}{k_B T}\right) \tag{100}$$

or

$$Q = \left(\frac{2\pi k_B T}{h^2}\right)^{3N/2} \int d\,\mathbf{q}_f \left(\det \mathbf{a}^0\right)^{1/2} \exp\left(-\frac{\mathcal{V}(\mathbf{q}_f)}{k_B T}\right) \tag{101}$$

Hence the appearance of the metric tensor inside the integral is due to the nondiagonal nature of the kinetic energy in the constrained space.

Gō and Scheraga also showed that the Jacobian determinant in the flexible model could be expressed as the Jacobian determinant used in the rigid model multiplied by a number of classical harmonic oscillators, each contributing kinetic and potential energy to the Hamiltonian. This relationship is summarized by Eq. (83). Integration over the kinetic energy part led to cancellation of $(\det \mathbf{a}^0)^{1/2}$ in the configurational integral, while integration over the potential energy led to an arbitrary constant multiplying Q. The treatments of the two polymer models are identical in terms of only the potential energy. Within the IRC formalism, dissipation of kinetic energy is instantaneous, and the kinetic energy need not be treated until rate constants (and partition functions) are evaluated. In summary, the method of following the IRC described above only specifies the Riemannian geometry of the reaction path; it does not choose a physical model of the classical statistical mechanics. Such a choice is made only when calculating the rate constants.

APPENDIX D: EVALUATING THE SINGLE VOXEL PARTITION FUNCTION

To apply the average fluctuating polymer method, it is necessary to integrate the Boltzmann factor over fluctuations in each polymer atom position [Eq. (42)]. Each single voxel partition function appears to depend on the penetrant position (i.e., the voxel location) and the locations of all neighboring atoms. Fortunately, Eq. (42) can be reformulated into two different, simpler forms, each amenable to numerical solution. Below we detail how to calculate each individual integral in Eq. (42).

The first approach is to replace each polymer atom coordinate $\mathbf{x}_i$ by the sum of its original position $\langle \mathbf{x}_i \rangle$ and a deviation vector,

$$\mathbf{x}_i = \langle \mathbf{x}_i \rangle + \left(2\langle \Delta^2 \rangle \right)^{1/2} \begin{pmatrix} t_i' \\ u_i' \\ v_i' \end{pmatrix} \tag{102}$$

Below we leave off each atom subscript i from the deviation vector in order to simplify the notation. Differential changes in $(\mathbf{x}_i - \langle \mathbf{x}_i \rangle)$ and $(t', u', v')^T$ are related by a Jacobian

$$d(x_i - \langle x_i \rangle) d(y_i - \langle y_i \rangle) d(z_i - \langle z_i \rangle) = \left(2\langle \Delta^2 \rangle \right)^{3/2} dt' \, du' \, dv'$$

The square of each penetrant–polymer atom separation can then be written as

$$D_i^2 = \left(\mathbf{r}_{p_x} - \langle x_i \rangle - \sqrt{2\langle \Delta^2 \rangle} t' \right)^2 + \left(\mathbf{r}_{p_y} - \langle y_i \rangle - \sqrt{2\langle \Delta^2 \rangle} u' \right)^2$$
$$+ \left(\mathbf{r}_{p_z} - \langle z_i \rangle - \sqrt{2\langle \Delta^2 \rangle} v' \right)^2$$

The contribution to the configurational integral can then be written as (using a Lennard-Jones potential as an example)

$$Z_{ji}(\mathbf{r}_p) = \pi^{-3/2} \int_{-\infty}^{\infty} \int_{-\infty}^{\infty} \int_{-\infty}^{\infty} \exp\left[-\frac{4\epsilon_{ip}}{k_B T} \left(\frac{\sigma_{ip}^{12}}{[D^2]^6} - \frac{\sigma_{ip}^6}{[D^2]^3} \right) \right] e^{-t'^2} e^{-u'^2} e^{-v'^2} dt' \, du' \, dv'$$

$$\tag{103}$$

and its value depends on the orientation vector $\mathbf{r}_p - \langle \mathbf{x}_i \rangle$. Three-dimensional numerical integration could then yield the value of the integral.

Fortunately a simpler integration scheme is possible. Prior to substituting as in Eq. (102), the coordinate system can be rotated, without any loss in generality, such that the vector $\mathbf{r}_p - \langle \mathbf{x}_i \rangle$ lies along the z axis in the rotated coordinate system. The penetrant–polymer atom distance simplifies to

$$D_i^2 = 2\langle \Delta^2 \rangle t'^2 + 2\langle \Delta^2 \rangle u'^2 + \left(\mathbf{r}_{p_z} - \langle z_i \rangle - \sqrt{2\langle \Delta^2 \rangle} v' \right)^2$$

and the contribution to the configurational integral depends *only* on the original distance between the penetrant and polymer atom, $|\mathbf{r}_p - \langle\mathbf{x}_i\rangle|$. Z_{ji} can then be evaluated using Gauss–Hermite integration [199], in which an integrand containing e^{-x^2} is approximated by

$$\int_{-\infty}^{\infty} f(x)\exp(-x^2)\,dx \approx \sum_{k=1}^{n} w_k f(x_k) \tag{104}$$

i.e., the function f evaluated at the roots x_k of a Gauss–Hermite polynomial and scaled by weighting factors w_k. For the contribution of one penetrant–polymer atom separation to the overall configurational integral,

$$
\begin{aligned}
Z_{ji}&\left(|\mathbf{r}_p - \langle\mathbf{x}_i\rangle|\right)\\
&= \pi^{-3/2}\int_{-\infty}^{\infty}\int_{-\infty}^{\infty}\int_{-\infty}^{\infty}\exp\left(-\frac{V_{ip}}{k_BT}\right)\exp(-t'^2)\exp(-u'^2)\exp(-v'^2)\,dt'\,du'\,dv'\\[2mm]
&= \pi^{-3/2}\sum_{k_z}w_{k_z}\left[\sum_{k_y}w_{k_y}\left(\sum_{k_x}w_{k_x}\exp\left[-\frac{4\epsilon_{ip}}{k_BT}\right.\right.\right.\\[2mm]
&\quad\times\left(\frac{\sigma_{ip}^{12}}{\left[2\langle\Delta^2\rangle t_k'^2 + 2\langle\Delta^2\rangle u_k'^2 + \left(\mathbf{r}_{p_z} - \langle z_i\rangle - \sqrt{2\langle\Delta^2\rangle}v_k'\right)^2\right]^6}\right.\\[2mm]
&\quad\left.\left.\left.\left.-\frac{\sigma_{ip}^{6}}{\left[2\langle\Delta^2\rangle t_k'^2 + 2\langle\Delta^2\rangle u_k'^2 + \left(\mathbf{r}_{p_z} - \langle z_i\rangle - \sqrt{2\langle\Delta^2\rangle}v_k'\right)^2\right]^3}\right)\right]\right)\right]
\end{aligned}
\tag{105}
$$

For an odd order n, one Gauss–Hermite root equals zero, meaning the energy is evaluated at the voxel center $(t',u',v') = (0,0,0)$. Each individual configurational integral is thus determined by the summation in Eq. (105). The rotation of the interatomic vector to along the z axis is useful because it allows the function $Z_{ji}(|\mathbf{r}_p - \langle\mathbf{x}\rangle|)$ in Eq. (105) to be pretabulated in one dimension (the separation), rather than evaluated from scratch for each penetrant–polymer atom separation in each voxel. The pretabulated function thus depends on $|\mathbf{r}_p - \langle\mathbf{x}\rangle|/\sigma$, $\langle\Delta^2\rangle^{1/2}/\sigma$, and ϵ/k_BT. The same procedure can be applied for other two-body potentials by making appropriate substitutions for $\mathcal{V}_{ip}$ in Eq. (105).

The second approach is to express the deviation vector between $\mathbf{x}_i$ and $\langle\mathbf{x}_i\rangle$ in spherical coordinates,

$$x - \langle x_i\rangle = \left(2\langle\Delta^2\rangle\right)^{1/2} r \sin\theta\cos\phi$$

$$y - \langle y_i\rangle = \left(2\langle\Delta^2\rangle\right)^{1/2} r \sin\theta\sin\phi$$

$$z - \langle z_i\rangle = \left(2\langle\Delta^2\rangle\right)^{1/2} r \cos\theta$$

with Jacobian

$$d(x_i - \langle x_i\rangle)d(y_i - \langle y_i\rangle)d(z_i - \langle z_i\rangle) = \left(2\langle\Delta^2\rangle\right)^{3/2} r^2 \sin\theta\, dr d\theta d\phi$$

The coordinate r is allowed to vary over $-\infty \le r \le \infty$, restricting the θ coordinate to $0 \le \theta \le \pi$. The penetrant–polymer atom separation can then be written

$$D_i^2 = (\mathbf{r}_{p_x} - \langle x_i\rangle)^2 + (\mathbf{r}_{p_y} - \langle y_i\rangle)^2 + (\mathbf{r}_{p_z} - \langle z_i\rangle)^2 + 2\langle\Delta^2\rangle r^2 - 2\sqrt{2}\langle\Delta^2\rangle^{1/2} r$$

$$\times \left((\mathbf{r}_{p_x} - \langle x_i\rangle)\sin\theta\cos\phi + (\mathbf{r}_{p_y} - \langle y_i\rangle)\sin\theta\sin\phi + (\mathbf{r}_{p_z} - \langle z_i\rangle)\cos\theta\right)$$

As above, the penetrant–polymer separation vector can be rotated, without loss in generality, to lie on the z axis of the rotated coordinate system. The components $(\mathbf{r}_{p_x} - \langle x_i\rangle)$ and $(\mathbf{r}_{p_y} - \langle y_i\rangle)$ then both equal zero, and the separation distance is

$$D_i^2 = (\mathbf{r}_{p_z} - \langle z_i\rangle)^2 + 2\langle\Delta^2\rangle r^2 - 2\sqrt{2}\langle\Delta^2\rangle^{1/2} r(\mathbf{r}_{p_z} - \langle z_i\rangle)\cos\theta \tag{106}$$

The ϕ coordinate has no impact on the distance in this rotated frame, so integration is analytic. The remaining integral for each polymer atom contribution to the single voxel partition function is

$$Z_{ji}(|\mathbf{r}_p - \langle\mathbf{x}_i\rangle|) = \frac{2}{\sqrt{\pi}}\int_{-\infty}^{\infty} dr\, r^2 \exp(-r^2)\int_0^1 d(\cos\theta)\exp\left[-\frac{4\epsilon_{ip}}{k_B T}\left(\frac{\sigma_{ip}^{12}}{(D_i^2)^6} - \frac{\sigma_{ip}^{6}}{(D_i^2)^3}\right)\right]$$

$$\tag{107}$$

A numerical scheme such as the trapezoidal rule or Simpson's rule [109] can be used for the integral over $\cos\theta$, with Gauss–Hermite integration used to integrate over r.

In comparing the two methods at a fixed separation of $5\,\text{Å}$, we found essentially identical results so long as enough terms were used. For the first approach, 7–9 terms within each sum were sufficient ($7^3 = 343$ terms in all). For the second approach, 7–9 Gauss–Hermite terms were sufficient as well, but in combination with of order 10^4 trapezoidal terms. The first method (wholly in Cartesian coordinates) is thus recommended for pretabulating the function $Z_{ji}(D_i)$.

Even within the approach based on Cartesian coordinates, more Gauss–Hermite terms are required for each coordinate at shorter separations. Figure 7 compares the modified potential energy $-k_BT\ln Z_{ji}(r)$ as calculated using 3–16 terms. Within the repulsive regime, using too few terms leads to oscillations about the potential of mean force calculated with more terms. The smaller the separation included, the more terms required.

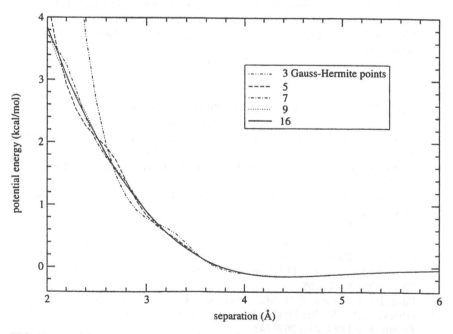

FIG. 7 Modified potential energy between a penetrant and a single polymer atom, within the average fluctuating polymer method. Different lines correspond to the numerical results based on different numbers of Gauss–Hermite terms. The same number of terms was used for each of the x, y, and z coordinates. Increasing the number of terms decreases oscillations at small separations.

In practice, this additional requirement is not a large burden if a pretabulated integral Z_{ji} is constructed.

REFERENCES

1. Vieth, W.R. *Diffusion in and Through Polymers: Principles and Applications*; Hansen: New York, 1991.
2. Paul, D.R.; Yampol'skii, Y.P. *Polymeric Gas Separation Membranes*; CRC Press: Boca Raton, FL, 1994; Chapter 1.
3. Koros, W.J.; Fleming, G.K. J. Membr. Sci. **1993**, *83*, 1–80.
4. Nagai, K.; Masuda, T.; Nakagawa, T.; Freeman, B.D.; Pinnau, I. Prog. Polym. Sci. **2001**, *26*, 721–798.
5. Jorgensen, W.L.; Maxwell, D.S.; Tirado-Rives, J. J. Am. Chem. Soc. **1996**, *118*, 11225–11236.
6. Rigby, D.; Sun, H.; Eichinger, B.E. Polym. Internat. **1997**, *44*, 311–330.
7. Sun, H. J. Phys. Chem. B **1998**, *102*, 7338–7364.
8. Martin, M.G.; Siepmann, J.I. J. Phys. Chem. B **1998**, *102*, 2569–2577.
9. Martin, M.G.; Siepmann, J.I. J. Phys. Chem. B **1999**, *103*, 4508–4517.
10. Chen, B.; Siepmann, J.I. J. Phys. Chem. B **1999**, *103*, 5370–5379.
11. Wick, C.D.; Martin, M.G.; Siepmann, J.I. J. Phys. Chem. B **2000**, *104*, 8008–8016.
12. Chen, B.; Potoff, J.J.; Siepmann, J.I. J. Phys. Chem. B **2001**, *105*, 3093–3104.
13. Nath, S.K.; Escobedo, F.A.; de Pablo, J.J. J. Chem. Phys. **1998**, *108*, 9905–9911.
14. Nath, S.K.; de Pablo, J.J. Mol. Phys. **2000**, *98*, 231–238.
15. Smith, G.D.; Jaffe, R.L.; Yoon, D.Y. Macromolecules **1993**, *26*, 298–304.
16. Smith, G.D.; Jaffe, R.L.; Yoon, D.Y. J. Phys. Chem. **1993**, *97*, 12752–12759.
17. Müller-Plathe, F. Acta Polymerica **1994**, *45*, 259–293.
18. Smith, G.D.; Borodin, O.; Bedrov, D. J. Phys. Chem. A **1998**, *102*, 10318–10323.
19. Hedenqvist, M.S.; Bharadwaj, R.; Boyd, R.H.; Macromolecules **1998**, *31*, 1556–1564.
20. Holt, D.B.; Farmer, B.L. Polymer **1999**, *40*, 4667–4772.
21. Neyertz, S.; Pizzi, A.; Merlin, A.; Maigret, B.; Brown, D.; Deglise, X. J. Appl. Polym. Sci. **2000**, *78*, 1939–1946.
22. Allen, M.P.; Tildesley, D.J. *Computer Simulation of Liquids*; Oxford University Press: New York, 1987.
23. Pant, P.V.K.; Boyd, R.H. Macromolecules **1993**, *26*, 679–686.
24. Gusev, A.A.; Müller-Plathe, F.; van Gunsteren, W.F.; Suter, U.W. Adv. Polym. Sci. **1994**, *116*, 207–247.
25. Gee, R.H.; Boyd, R.H. Polymer **1995**, *36*, 1435–1440.
26. Takeuchi, H. J. Chem. Phys. **1990**, *93*, 4490–4491.
27. Takeuchi, H.; Okazaki, K. J. Chem. Phys. **1990**, *92*, 5643–5652.
28. Takeuchi, H.; Roe, R.-J.; Mark, J.E. J. Chem. Phys. **1990**, *93*, 9042–9048.

29. Boyd, R.H.; Pant, P.V.K. Macromolecules **1991**, *24*, 6325–6331.
30. Pant, P.V.K.; Boyd, R.H. Macromolecules **1992**, *25*, 494–495.
31. Müller-Plathe, F. J. Chem. Phys. **1991**, *94*, 3192–3199.
32. Müller-Plathe, F. Chem. Phys. Lett. **1991**, *177*, 527–535.
33. Müller-Plathe, F. J. Chem. Phys. **1992**, *96*, 3200–3205.
34. Müller-Plathe, F.; Rogers, S.C.; van Gunsteren, W.F. Macromolecules **1992**, *25*, 6722–6724.
35. Trohalaki, S.; Kloczkowski, A.; Mark, J.E.; Roe, R.J.; Rigby, D. Comput. Theor. Polym. Sci. **1992**, *2*, 147.
36. Tamai, Y.; Tanaka, H.; Nakanishi, K. Macromolecules **1994**, *27*, 4498–4508.
37. Han, J.; Boyd, R.H. Macromolecules **1994**, *27*, 5365–5370.
38. Chassapis, C.S.; Petrou, J.K.; Petropoulos, J.H.; Theodorou, D.N. Macromolecules **1996**, *29*, 3615–3624.
39. Takeuchi, H.; Okazaki, K. Mol. Simul. **1996**, *16*, 59–74.
40. Fukuda, M.; Kuwajima, S. J. Chem. Phys. **1997**, *107*, 2149–2158.
41. Fukuda, M.; Kuwajima, S. J. Chem. Phys. **1998**, *108*, 3001–3009.
42. Cuthbert, T.R.; Wagner, N.J.; Paulaitis, M.E.; Murgia, G.; D'Aguanno, B. Macromolecules **1999**, *32*, 5017–5028.
43. van der Vegt, N.F.A. Macromolecules **2000**, *33*, 3153–3160.
44. Boshoff, J.H.D.; Lobo, R.F.; Wagner, N.J. Macromolecules **2001**, *34*, 6107–6116.
45. Nath, S.K.; Banaszak, B.J.; de Pablo, J.J. Macromolecules **2001**, *34*, 7841–7848.
46. Sok, R.M.; Berendsen, H.J.C.; van Gunsteren, W.F. J. Chem. Phys. **1992**, *96*, 4699–4704.
47. Fritz, L.; Hofmann, D. Polymer **1997**, *38*, 1035–1045.
48. Charati, S.G.; Stern, S.A. Macromolecules **1998**, *31*, 5529–5535.
49. Han, J.; Boyd, R.H. Polymer **1996**, *37*, 1797–1804.
50. Müller-Plathe, F. Macromolecules **1996**, *29*, 4782–4791.
51. Li, T.; Kildsig, D.O.; Park, K. J. Control. Release **1997**, *48*, 57–66.
52. Bharadwaj, R.K.; Boyd, R.H. Polymer **1999**, *40*, 4229–4236.
53. Shanks, R.; Pavel, D. Mol. Simul. **2002**, *28*, 939–969.
54. Smit, E.; Mulder, M.H.V.; Smolders, C.A.; Karrenbeld, H.; van Eerden, J.; Feil, D. J. Membr. Sci. **1992**, *73*, 247–257.
55. Zhang, R.; Mattice, W.L. J. Membr. Sci. **1995**, *108*, 15–23.
56. Hofmann, D.; Ulbrich, J.; Fritz, L.; Paul, D. Polymer **1996**, *37*, 4773–4785.
57. Hofmann, D.; Fritz, L.; Paul, D. J. Membr. Sci. **1998**, *144*, 145–159.
58. Hofmann, D.; Fritz, L.; Ulbrich, J.; Schepers, C.; Böhning, M. Macromol. Theory Simul. **2000**, *9*, 293–327.
59. Niemelä, S.; Leppänen, J.; Sundholm, F. Polymer **1996**, *37*, 4155–4165.
60. Fried, J.R.; Sadat-Akhavi, M.; Mark, J.E. J. Membr. Sci. **1998**, *149*, 115–126.
61. Kim, W.-K.; Mattice, W.L. Macromolecules **1998**, *31*, 9337–9344.
62. Fried, J.R.; Ren, P. Comput. Theor. Polym. Sci. **2000**, *10*, 447–463.
63. Tocci, E.; Hofmann, D.; Paul, D.; Russo, N.; Drioli, E. Polymer **2001**, *42*, 521–533.
64. Fried, J.R.; Goyal, D.K. J. Polym. Sci. Polym. Phys. **1998**, *36*, 519–536.

65. Sonnenburg, J.; Gao, J.; Weiner, J.H. Macromolecules **1990**, *23*, 4653–4657.
66. Rottach, D.R.; Tillman, P.A.; McCoy, J.D.; Plimpton, S.J.; Curro, J.G. J. Chem. Phys. **1999**, *111*, 9822–9831.
67. Müller-Plathe, F. J. Chem. Phys. **1995**, *103*, 4745–4756.
68. Catlow, C.R.A.; Mills, G.E. Electrochim. Acta **1995**, *40*, 2057–2062.
69. Payne, V.A.; Lonergan, M.C.; Forsyth, M.; Ratner, M.A.; Shriver, D.F.; de Leeuw, S.W.; Perram, J.W. Solid State Ionics **1995**, *81*, 171–181.
70. Neyertz, S.; Brown, D. J. Chem. Phys. **1996**, *104*, 3797–3809.
71. Ennari, J.; Elomaa, M.; Sundholm, F. Polymer **1999**, *40*, 5035–5041.
72. Borodin, O.; Smith, G.D. Macromolecules **2000**, *33*, 2273–2283.
73. Snyder, J.F.; Ratner, M.A.; Shriver, D.F. J. Electrochem. Soc. **2001**, *148*, A858–A863.
74. Aabloo, A.; Thomas, J. Solid State Ionics **2001**, *143*, 83–87.
75. Vishnyakov, A.; Neimark, A.V. J. Phys. Chem. B **2001**, *105*, 9586–9594.
76. Ferreira, B.A.; Müller-Plathe, F.; Bernardes, A.T.; De Almeida, W.B. Solid State Ionics **2002**, *147*, 361–366.
77. Müller-Plathe, F.; Rogers, S.C.; van Gunsteren, W.F. J. Chem. Phys. **1993**, *98*, 9895–9904.
78. Sunderrajan, S.; Hall, C.K.; Freeman, B.D. J. Chem. Phys. **1996**, *105*, 1621–1632.
79. Takeuchi, H. J. Chem. Phys. **1990**, *93*, 2062–2067.
80. Müller-Plathe, F.; Laaksonen, L.; van Gunsteren, W.F. J. Mol. Graphics **1993**, *11*, 118–120.
81. Müller-Plathe, F.; Laaksonen, L.; van Gunsteren, W.F. J. Mol. Graphics **1993**, *11*, 125–126.
82. Glasstone, S.; Laidler, K.J.; Eyring, H. *The Theory of Rate Processes*; McGraw-Hill: New York, 1941.
83. Chandler, D. J. Chem. Phys. **1978**, *68*, 2959.
84. Voter, A.F.; Doll, J.D. J. Chem. Phys. **1985**, *82*, 80–92.
85. June, R.L.; Bell, A.T.; Theodorou, D.N. J. Phys. Chem. **1991**, *95*, 8866–8878.
86. Fichthorn, K.A.; Weinberg, W.H. J. Chem. Phys. **1991**, *95*, 1090–1096.
87. Greenfield, M.L.; Theodorou, D.N. Macromolecules **1993**, *26*, 5461–5472.
88. Greenfield, M.L.; Theodorou, D.N. Macromolecules **1998**, *31*, 7068–7090.
89. Rallabandi, P.S.; Thompson, A.P.; Ford, D.M. Macromolecules **2000**, *33*, 3142–3152.
90. Greenfield, M.L.; Theodorou, D.N. Mol. Simul. **1997**, *19*, 329–361.
91. Greenfield, M.L.; Theodorou, D.N. Macromolecules **2001**, *34*, 8541–8553.
92. Jagodic, F.; Borstnik, B.; Azman, A. Makromol. Chem. **1973**, *173*, 221–231.
93. Arizzi, S. Diffusion of Small Molecules in Polymeric Glasses: A Modelling Approach. Ph.D. thesis, Massachusetts Institute of Technology, 1990.
94. Gusev, A.A.; Arizzi, S.; Suter, U.W.; Moll, D.J. J. Chem. Phys. **1993**, *99*, 2221–2227.
95. Bell, A.T.; Maginn, E.J.; Theodorou, D.N. In *Handbook of Heterogeneous Catalysis*, Ertl, G., Knozinger, H., Weitkamp, J., Eds.; VCH: Weinheim, 1997.
96. Auerbach, S.M. Int. Rev. Phys. Chem. **2000**, *19*, 155–198.

97. Gusev, A.A.; Suter, U.W. J. Chem. Phys. **1993**, *99*, 2228–2234.
98. Gusev, A.A.; Suter, U.W.; Moll, D.J. Macromolecules **1995**, *28*, 2582–2584.
99. Hofmann, D.; Fritz, L.; Ulbrich, J.; Paul, D. Polymer **1997**, *38*, 6145–6155.
100. Hofmann, D.; Fritz, L.; Ulbrich, J.; Paul, D. Comput. Theor. Polym. Sci. **2000**, *10*, 419–436.
101. Products from Accelrys, Inc., http://www.accelrys.com/chemicals/polymers/diffusion.html.
102. Greenfield, M.L. Molecular Modeling of Dilute Penetrant Gas Diffusion in a Glassy Polymer using Multidimensional Transition-State Theory. Ph.D. thesis, University of California, Berkeley, 1996.
103. van Kampen, N.G. *Stochastic Processes in Physics and Chemistry*, Revised Ed.; North Holland: Amsterdam, 1992.
104. Synge, J.L.; Schild, A. *Tensor Calculus*; University of Toronto Press: Toronto, 1949.
105. Gō, N.; Scheraga, H.A. Macromolecules **1976**, *9*, 535–542.
106. Cerjan, C.J.; Miller, W.H. J. Chem. Phys. **1981**, *75*, 2800–2806.
107. Banerjee, A.; Adams, N.; Simons, J.; Shepard, R. J. Phys. Chem. **1985**, *89*, 52–57.
108. Baker, J. J. Comput. Chem. **1986**, *7*, 385–395.
109. Press, W.H.; Flannery, B.P.; Teukolsky, S.A.; Vetterling, W.T. *Numerical Recipes. The Art of Scientific Computing*; Cambridge University Press: Cambridge, 1989.
110. Henkelman, G.; Jónsson, G.; Jónsson, H. In *Theoretical Methods in Condensed Phase Chemistry* (*Progress in Theoretical Chemistry and Physics*), Schwartz, S.D., Ed.; Kluwer Academic Publishers: New York, 2000; Vol. 5, 269–300.
111. Bolhuis, P.G.; Chandler, D.; Dellago, C.; Geissler, P.L. Ann. Rev. Phys. Chem. **2002**, *53*, 291–318.
112. Fukui, K. Acc. Chem. Res. **1981**, *14*, 363–368.
113. deGennes, P.-G. *Scaling Concepts in Polymer Physics*; Cornell University Press: Ithaca, 1979.
114. Doi, M.; Edwards, S.F. *The Theory of Polymer Dynamics*; Oxford University Press: New York, 1986; Chapter 6.
115. Pechukas, P. J. Chem. Phys. **1976**, *64*, 1516–1521.
116. Page, M.; McIver, J.W., Jr. J. Chem. Phys. **1988**, *88*, 922–935.
117. Banerjee, A.; Adams, N.P. Int. J. Quantum Chem. **1992**, *43*, 855–871.
118. Sevick, E.M.; Bell, A.T.; Theodorou, D.N. J. Chem. Phys. **1993**, *98*, 3196–3212.
119. Weiner, J.H. *Statistical Mechanics of Elasticity*; Wiley-Interscience: New York, 1983.
120. http://www.netlib.org/eispack
121. http://www.netlib.org/lapack
122. http://www.netlib.org/linpack
123. Michael L. Greenfield, Doros N. Theodorou, J. Chem. Phys. (*manuscript in preparation*).

124. Starkweather, H.W. Polymer **1991**, *32*, 2443–2448.
125. Elber, R. J. Chem. Phys. **1990**, *93*, 4312–4321.
126. Lazaridis, T.; Tobias, D.J.; Brooks, C.L.; Paulaitis, M.E. J. Chem. Phys. **1991**, *95*, 7612–7625.
127. Pratt, L.R. J. Chem. Phys. **1986**, *85*, 5045–5048.
128. Gillilan, R.; Wilson, K.R. J. Chem. Phys. **1992**, *97*, 1757–1772.
129. Theodorou, D.N.; Suter, U.W. Macromolecules **1985**, *18*, 1467–1478.
130. Rapold, R.F.; Mattice, W.L. J. Chem. Soc. Faraday Trans. **1995**, *91*, 2435–2441.
131. Rapold, R.F.; Mattice, W.L. Macromolecules **1996**, *29*, 2457–2466.
132. Kotelyanskii, M.; Wagner, N.J.; Paulaitis, M.E. Macromolecules **1996**, *29*, 8497–8506.
133. Kotelyanskii, M. Trends Polym. Sci. **1997**, *5*, 192–198.
134. Santos, S.; Suter, U.W.; Müller, M.; Nievergelt, J. J. Chem. Phys. **2001**, *114*, 9772–9779.
135. Müller, M.; Nievergelt, J.; Santos, S.; Suter, U.W. J. Chem. Phys. **2001**, *114*, 9764–9771.
136. Ratner, B.D.; Tsukruk, V.V. *Scanning Probe Microscopy of Polymers*; American Chemical Society: Washington, DC, 1998; Symposium Series, No. 694.
137. Fischer, K.H.; Hertz, J.A. *Spin Glasses*; Cambridge University Press: Cambridge, 1991.
138. Snurr, R.Q.; Bell, A.T.; Theodorou, D.N. J. Phys. Chem. **1994**, *98*, 11948–11961.
139. Goldstein, H. *Classical Mechanics*, 2nd Ed.; Addison-Wesley: Reading, MA, 1980.
140. Anderson, E.; Bai, Z.; Bischof, C.; Blackford, S.; Demmel, J.; Dongarra, J.; Croz, J.D.; Greenbaum, A.; Hammarling, S.; McKenney, A.; Sorensen, D. *LAPACK Users' Guide*; Society for Industrial and Applied Mathematics: Philadelphia, 1999; available on-line at http://www.netlib.org/lapack/lug/
141. Voronoï, G. Z. Reine Angew. Math. **1908**, *134*, 198–211.
142. Tanemura, M.; Ogawa, T.; Ogita, N. J. Comput. Phys. **1983**, *51*, 191–207.
143. Arizzi, S.; Mott, P.H.; Suter, U.W. J. Polym. Sci. Polym. Phys. Ed. **1992**, *30*, 415–426.
144. Gentile, F.T.; Arizzi, S.; Suter, U.W.; Ludovice, P.J. Ind. Eng. Chem. Res. **1995**, *34*, 4193–4201.
145. Putta, S.; Nemat-Nasser, S. Materials Sci. Eng. A **2001**, *317*, 70–76.
146. Takeuchi, H.; Okazaki, K. Makromol. Chem. Macromol. Symp. **1993**, *65*, 81–88.
147. Misra, S.; Mattice, W.L. Macromolecules **1993**, *26*, 7274–7281.
148. Cullity, B.D. *Elements of X-ray Diffraction*, 2nd Ed.; Addison-Wesley: Reading, MA, 1978.
149. Gonzalez, C.; Schlegel, H.B. J. Phys. Chem. **1990**, *94*, 5523–5527.
150. Fischer, S.; Karplus, M. Chem. Phys. Lett. **1992**, *194*, 252–261.
151. Czerminski, R.; Elber, R. Int. J. Quant. Chem. **1990**, *24*, 167–186.

152. Mills, G.; Jónsson, H.; Schenter, G.K. Surf. Sci. **1995**, *324*, 305–337.

153. Zimmer, M.F. Phys. Rev. Lett. **1995**, *75*, 1431–1434.

154. Petropoulos, J.H. J. Membr. Sci. **1990**, *53*, 229–255.

155. Theodorou, D.N. In *Diffusion in Polymers*; Neogi, P., Ed.; Marcel Dekker: New York, 1996; 67–142.

156. Widom, B. J. Chem. Phys. **1963**, *39*, 2808–2812.

157. Widom, B. J. Phys. Chem. **1982**, *86*, 869–872.

158. Müller-Plathe, F. Macromolecules **1991**, *24*, 6475–6479.

159. Tamai, Y.; Tanaka, H.; Nakanishi, K. Macromolecules **1995**, *28*, 2544–2554.

160. Boone, T.D. Prediction of Glass-Melt Behavior and Penetrant Sorption Thermodynamics in Vinyl Polymers via Molecular Simulations. Ph.D. thesis, University of California, Berkeley, 1995.

161. Spyriouni, T.; Economou, I.G.; Theodorou, D.N. Phys. Rev. Lett. **1998**, *80*, 4466–4469.

162. Kofke, D.A. Adv. Chem. Phys. **1999**, *105*, 405–441.

163. Panagiotopoulos, A.Z. Mol. Phys. **1987**, *61*, 813–826.

164. Panagiotopoulos, A.Z. Mol. Simul. **1992**, *9*, 1–24.

165. Shing, K.S.; Gubbins, K.E. Mol. Phys. **1982**, *46*, 1109–1128.

166. Knopp, B.; Suter, U.W.; Gusev, A.A. Macromolecules **1997**, *30*, 6107–6113.

167. Knopp, B.; Suter, U.W. Macromolecules **1997**, *30*, 6114–6119.

168. van der Vegt, N.F.A.; Briels, W.J. J. Chem. Phys. **1998**, *109*, 7578–7582.

169. Escobedo, F.A.; de Pablo, J.J. J. Chem. Phys. **1995**, *103*, 2703–2710.

170. Escobedo, F.A.; de Pablo, J.J. J. Chem. Phys. **1996**, *105*, 4391–4394.

171. Nath, S.K.; de Pablo, J.J.; DeBellis, A.D. J. Am. Chem. Soc. **1999**, *121*, 4252–4261.

172. de Pablo, J.J.; Laso, M.; Suter, U.W.; Cochran, H.D. Fluid Phase Equilibria **1993**, *83*, 323–331.

173. Zervopoulou, E.; Mavrantzas, V.G.; Theodorou, D.N. J. Chem. Phys. **2001**, *115*, 2860–2875.

174. Nath, S.K.; de Pablo, J.J. J. Phys. Chem. B **1999**, *103*, 3539–3544.

175. Kumar, S.K.; Szleifer, I.; Panagiotopoulos, A.Z. Phys. Rev. Lett. **1991**, *66*, 2935–2938.

176. Kikuchi, H.; Kuwajima, S.; Fukuda, M. J. Chem. Phys. **2001**, *115*, 6258–6265.

177. Deitrick, G.L.; Scriven, L.E.; Davis, H.T. J. Chem. Phys. **1989**, *90*, 2370–2385.

178. Stapleton, M.R.; Panagiotopoulos, A.Z. J. Chem. Phys. **1990**, *92*, 1285–1293.

179. Fukuda, M. J. Chem. Phys. **2000**, *112*, 478–486.

180. de Pablo, J.J.; Laso, M.; Suter, U.W. J. Chem. Phys. **1992**, *96*, 6157–6162.

181. Siepmann, J.I.; Frenkel, D. Mol. Phys. **1992**, *75*, 59–70.

182. Dodd, L.R.; Boone, T.D.; Theodorou, D.N. Mol. Phys. **1993**, *78*, 961–996.

183. de Pablo, J.J.; Yan, Q.; Escobedo, F.A. Annu. Rev. Phys. Chem. **1999**, *50*, 377–411.

184. Ferguson, D.M.; Siepmann, J.I.; Truhlar, D.G., Eds. *Monte Carlo Methods in Chemical Physics* (*Adv. Chem. Phys.*); John Wiley & Sons: New York, 1999; Vol. 105.

185. Bezus, A.G.; Kiselev, A.V.; Lopatkin, A.A.; Du, P.Q. J. Chem. Soc. Faraday Trans. II **1978**, *74*, 367–379.

186. Gusev, A.A.; Suter, U.W. Phys. Rev. A **1991**, *43*, 6488–6494.

187. McGreevy, R.L.; Howe, M.A. Annu. Rev. Mater. Sci. **1992**, *22*, 217–242.

188. Alfrey, T., Jr.; Gurnee, E.F.; Lloyd, W.G. J. Polym. Sci. C **1966**, *12*, 249–261.

189. Havlin, S.; Ben-Avraham, D. Adv. Phys. **1987**, *36*, 695–798.

190. Haus, J.W.; Kehr, K.W. Phys. Rep. **1987**, *150*, 263–406.

191. Stauffer, D.; Aharony, A. *Introduction to Percolation Theory*, 2nd Ed.; Taylor & Francis: London, 1992.

192. Müller-Plathe, F.; Rogers, S.C.; van Gunsteren, W.F. Chem. Phys. Lett. **1992**, *199*, 237–243.

193. Weber, H.; Paul, W. Phys. Rev. E **1996**, *54*, 3999–4007.

194. Brandrup, J.; Immergut, E.H.; Grulke, E.A. *Polymer Handbook*, 4th Ed.; Wiley–Interscience: New York, 1999.

195. Fixman, M. Proc. Nat. Acad. Sci. USA **1974**, *71*, 3050–3053.

196. Gō, N.; Scheraga, H.A. J. Chem. Phys. **1969**, *51*, 4751–4767.

197. Helfand, E. J. Chem. Phys. **1979**, *71*, 5000–5007.

198. Fixman, M. J. Chem. Phys. **1978**, *69*, 1527–1537.

199. Abramowitz, M.; Stegun, I.A. *Handbook of Mathematical Functions*; National Bureau of Standards: Washington, DC, 1964; Reprint published by Dover Publications: Mineola, New York, 1965.

15

Coarse-Graining Techniques

K. BINDER and WOLFGANG PAUL   Johannes-Gutenberg-Universität Mainz, Mainz, Germany

SERGE SANTOS and ULRICH W. SUTER   Eidgenössische Technische Hochschule (ETH), Zurich, Switzerland

I. INTRODUCTION AND OVERVIEW

For many materials, it is difficult to span the wide range of scales of length and time from the atomistic description to that of macroscopic phenomena. For polymers, this "multiscale problem" is particularly severe, since a single flexible macromolecule already exhibits structure on many length scales (Fig. 1), and is a slowly relaxing object [1,2]. In addition, one often wishes to consider systems with multiple constituents (concentrated polymer solutions or blends, mixtures of homopolymers and block copolymers or polymers of different architecture) which exist in several phases, and then the scales of length and time that are of physical interest truly extend from the subatomic scale (relevant for the electronic structure of polymers, their chemical reactions, excited states, etc.) to the macroscopic scale (e.g., phase separated domains of the scale of micrometers or larger, glassy relaxation on the scale of seconds or hours, etc.) [3].

   Since there is no hope of treating all these scales equally well in a single type of computer simulation, the idea of "coarse-graining" has emerged [4,5], where a mapping was attempted from an atomistically realistic description of polymers such as polyethylene (PE) or bisphenol-A-polycarbonate (BPA-PC) towards the bond fluctuation model [6,7]; see Fig. 2. Now coarse-grained lattice models such as the self-avoiding walk model or the bond fluctuation model in its standard form have been used for a long time to model polymers qualitatively, but this has the obvious drawback that the information on the chemical structure of the considered polymer is lost completely, and thus nothing can be said on structure–property relationships. Though these standard coarse-grained models are

491

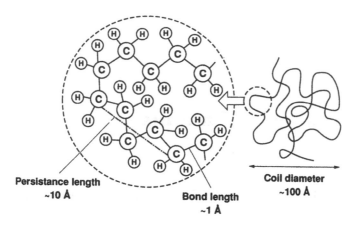

FIG. 1 Schematic illustration of the various length scales of a macromolecule. Polyethylene is used as an example. From Binder [1].

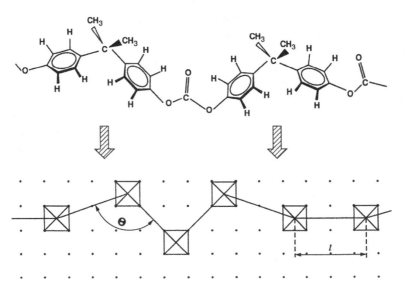

FIG. 2 A chemically realistic description of a polymer chain (bisphenol-A-polycarbonate [BPA-PC] in the present example) is mapped approximately onto the bond fluctuation model by using suitable potentials for the length ℓ of the effective bonds and the angles between them. In this example (3:1 mapping) one chemical repeat unit of BPA-PC containing $n = 12$ covalent bonds along the backbone of the chain is translated into three effective bonds. From Paul et al. [5].

nevertheless useful for some purposes (see Chapter 4), the idea to be described here is the attempt to keep some information about the chemical structure indirectly also on the coarse-grained level. Thus, although the lattice spacing of the model in Fig. 2 corresponds to about 2 Å, and hence clearly the detailed packing of the various atoms or atomic groups in the polymer melt or glassy state cannot be directly represented, some information on such properties is kept indirectly, via the potentials $U(\ell)$, $V(\vartheta)$ for the lengths, ℓ, of the effective bonds and the angles, ϑ, between them. Chemically different polymers are then described by different potentials.

Obviously, constructing the effective potentials explicitly will not be very straightforward [8,9], and also the description of physical properties by means of such a coarse-grained lattice model cannot be expected to be perfect. Thus, the idea has arisen to use the coarse-grained model only as an intermediate step for a simulation that equilibrates the long wavelength degrees of freedom of the polymer and then to carry out an "inverse mapping" where in Fig. 2 the arrow is simply read backwards! However, this step is technically very difficult, and although it has been under discussion for the bond fluctuation model for more than a decade now, it has not yet actually been performed successfully (only for another type of lattice model on the face-centered cubic lattice, obtained from chains on the diamond lattice where every second monomer is eliminated, see [10,11], has a reverse mapping been possible). The explicit construction of potentials for coarse-grained degrees of freedom [such as $U(\ell)$, $V(\vartheta)$] and the inverse mapping has successfully been carried out for a coarse-grained continuum model rather than a lattice model, e.g., a bead-spring model [12,13]. The basic idea is the same in all these methods (Fig. 3), but there is considerable freedom in defining the details of this procedure. For example, in the case of bisphenol-A-polycarbonate a 2:1 mapping to a bead-spring chain was performed [12,13], i.e., one chemical monomer (representing 12 consecutive covalent bonds along the chain backbone, see Fig. 2) is mapped onto two effective monomers. Taking the geometrical centers of the carbonate group and of the isopropylidene group as the sites of the effective monomers, it is actually advantageous to allow for a difference between these effective monomers: thus, the coarse-grained chain representing BPA-PC is not a bead-spring model of a homopolymer $A, A, A, A \ldots$, but rather a bead-spring model of an alternating copolymer $ABAB \ldots$. The crucial point of all these models, however, is such that on the coarse-grained length scale the effective potentials $U(\ell)$, $V(\vartheta)$ are much smaller than the atomistic potentials for the lengths of covalent bonds or the angles between them, and torsional potentials on the coarse-grained level can be neglected completely. As a result, on the coarse-grained level all potentials are on the same scale of energy, unlike the

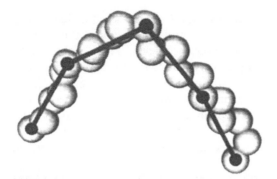

FIG. 3 Illustration of the coarse-graining procedure, starting out from a united-atom model for polyethylene as an atomistic description (the larger spheres show a segment of polyethylene at a temperature $T = 509K$ taken out of a molecular dynamics melt). One coarse-grained bond (straight line) represents the end-to-end distance of $n = 5$ consecutive united-atom bonds. The effective monomers (black dots) actually are represented by the eight sites of an elementary cube of the simple cubic lattice, if a mapping on the bond fluctuation model is used; if a mapping on a bead-spring chain is used, the effective monomers (black dots) are represented by the repulsive part of a suitable Lennard-Jones potential, and the effective bonds are "anharmonic springs" described by the FENE potential. From Tries et al. [9].

atomistic level where the scales for the potentials of covalent bond lengths and angles are much higher than the weak Lennard-Jones type nonbonded forces. Hence, while on the atomistic scale a Molecular Dynamics time step on the femtosecond scale is needed, the molecular dynamics time step for a coarse-grained bead-spring model can be several orders of magnitude larger. As a result, one can reach equilibrium for distinctly lower temperatures using the coarse-grained model than with the atomistic simulation, and somewhat larger length scales are accessible, too.

Nevertheless, one must admit that this step emphasized in Figs. 2 and 3 is only a modest first step towards the goal of bridging the gap between the length and time scales of atomistic simulations and the scales of macroscopic phenomena in materials science. The same caveat applies to other interesting alternatives to this mapping procedure, which we shall not describe in detail here, such as mapping to ellipsoidal rather than spherical effective monomers [14,15], or generating configurations of dense polymer melts by constructing an ensemble of curves of constant curvature, using stream lines of a vector field that never intersect [16,2]. But how can one get from the scales of, say, 10 nm and 100 μsec, that are only accessible to the coarse-grained models, to still much larger scales?

One line of attack to this problem is to try to use a next mapping step from a bead-spring chain to a still much cruder model. For instance,

Murat and Kremer [17] try to map whole chains of soft ellipsoidal particles, which then must be allowed to overlap strongly in the melt. This approach seems particularly useful for a description of heterophase structures in polymeric materials, such as polymer blends that are unmixed on nanoscopic scales. It seems promising to combine this model or related approaches [18] with efficient techniques for describing the time evolution, making use of the fact that very soft potentials between the effective particles allow rather large time steps, such as dissipative particle dynamics [19]. While these approaches clearly are very promising avenues for further research, we feel it is premature to review them here simply because this field is at its beginning and at the same time rapidly developing.

Another line of attack is the idea of bypassing this step-to-step mapping from one scale to the next by directly making connections between the atomistic scale and the macroscopic continuum scale by performing calculations on both scales simultaneously and appropriately connecting them [20,21]. We shall briefly describe the essential ideas of this approach, emphasizing mechanical properties, in Section III of this chapter (a first application to plastic deformation of bisphenol-A-polycarbonate can be found in Chapter 12 of this book). In the following section, we shall explain the mapping alluded to in Figs. 2 and 3 in some more detail, while Section IV summarizes a few conclusions.

II. MAPPING OF ATOMISTIC MODELS TO THE BOND FLUCTUATION MODEL

In this section, we shall focus on polyethylene exclusively because this is a chemically simple polymer and for this case the mapping procedure has been investigated in greater detail than for any other polymer [9]. Using a united-atom description for the CH_2 groups, a potential of the following type results [22,23]

$$\mathcal{H} = \sum_{j=1}^{N_p-1} \mathcal{H}_\ell(l_j) + \sum_{j=1}^{N_p-2} \mathcal{H}_\theta(\theta_j) + \sum_{j=2}^{N_p-2} \mathcal{H}_\phi(\phi_j) + \sum_{j\neq i} \mathcal{H}_{LJ}(\vec{r}_{ij}) \tag{1}$$

where $\mathcal{H}_\ell(\ell_j) \propto \delta(\ell_j - \ell_{cc})$ with a fixed bond length $\ell_j = \ell_{cc} = 1.53\,\text{Å}$, while the potential for the cosine of the bond angles θ_j was assumed harmonic,

$$\mathcal{H}_\theta(\theta_j) = \frac{f_\theta}{2}(\cos\theta_j - \cos\theta_0)^2, \quad \theta_0 = 110°, \quad f_\theta = 120\,\text{kcal/mol} \tag{2}$$

The torsional potential was parameterized as follows,

$$\mathcal{H}_\phi(\phi_j) = \sum_{k=0}^{3} a_k \cos^k \phi_j \tag{3}$$

with constants $a_0 = 1.736$, $a_1 = 4.5$, $a_2 = 0.764$, and $a_3 = -7$ (in units kcal/mol).

Finally, the nonbonded interaction is expressed in terms of the well-known Lennard-Jones potential with parameters $\sigma = 4.5\,\text{Å}$ (position of the minimum), $\varepsilon(CH_2\text{–}CH_2) = 0.09344\,\text{kcal/mol}$, $\varepsilon(CH_3\text{–}CH_3) = 0.22644\,\text{kcal/mol}$, and $\varepsilon(CH_2\text{–}CH_3) = \sqrt{\varepsilon(CH_2\text{–}CH_2)\varepsilon(CH_3\text{–}CH_3)}$. This model is a very good representation of reality, as the direct comparison of molecular dynamics simulations for $C_{100}H_{202}$ at $T = 509\,\text{K}$ with neutron spin echo data shows [24].

The statistical mechanical concept underlying the coarse-graining approach is the same as for the renormalization group treatment of critical phenomena. Suppose we have a microscopic model with degrees of freedom $\vec{x}$ (e.g., the bond lengths, bond angles, etc. of a chemically realistic polymer model). The canonical partition function of this model is given as

$$Z = \sum_{\{\vec{x}\}} \exp\{-\beta\mathcal{H}(\vec{x})\} \tag{4}$$

where $\beta = 1/k_B T$ and the sum is to be interpreted as an integral for continuous degrees of freedom. Let us further assume that we have chosen a set of mesoscopic degrees of freedom (e.g., the lengths and angles of the coarse-grained bonds in Fig. 3), denoted by $\vec{m}$. Then we can write

$$Z = \sum_{\{\vec{m}\}} \sum_{\{\vec{x}\}_{\vec{m}}} \exp\{-\beta\mathcal{H}(\vec{x})\} \tag{5}$$

where the sum over $\{\vec{x}\}$ is restricted to a fixed set of mesoscopic variables $\vec{m}$. Introducing a generalized free energy $F(\vec{m}, T)$ by writing

$$Z = \sum_{\{\vec{m}\}} \exp\{-\beta F(\vec{m}, T)\} \tag{6}$$

we see that this defines a mapping from the probabilities for the microscopic configurations $p(\vec{x}) = (1/Z)\exp\{-\beta\mathcal{H}(\vec{x})\}$ to those for the degrees of freedom $p(\vec{m}) = (1/Z)\exp\{-\beta F(\vec{m}, T)\}$.

On the mesoscopic scale we are dealing with a free energy and not with a Hamiltonian, so that such a mapping has to be done for each temperature separately. The generalized free energy on the mesoscopic scale in general has to be determined numerically from the partial trace $\exp\{-\beta F(\vec{m}, T)\} = \sum_{\{\vec{x}\}_{\vec{m}}} \exp\{-\beta \mathcal{H}(\vec{x})\}$.

In the case of melt polymers we are in the fortunate situation, that the large scale statistical properties of such chains are the same as for isolated chains under Theta conditions, i.e., those of a random walk. We can therefore generate Theta chains for the atomistic model and evaluate the probability distribution $p(L, \theta)$ for adjacent coarse-grained bond lengths and bond angles (see Fig. 4).

Using Eqs. (1)–(3), short single chains are simulated by simple sampling type Monte Carlo methods, as described by Baschnagel et al. [4,25]; see also Chapter 4 by K. Binder et al., "Polymer Models on the Lattice," and Chapter 3 by R. Hentschke, "Single Chain in Solution," for a general explanation of the simple sampling method for polymers. However, a problem is that a cutoff in the Lennard-Jones interaction along the chain must be introduced in order for an isolated chain to show asymptotic Gaussian Theta behavior. One must include the interaction of each unit with its fourth-nearest neighbor along the chain to account for the pentane effect, while further distant neighbor interactions are eliminated. Now n' successive CH_2 groups are mapped onto m effective bonds. To some extent, the choice of the pair of integers (n', m) is arbitrary, but it is clear that neither one of the integers n'/m and m should be too large, in order that the method is practically feasible. The optimal choice of n' and m so that the crude coarse-grained model can represent the atomistic details indirectly as faithfully as possible, is a subtle problem, and clearly will depend on the chemical structure of the polymer that is studied. For polyethylene, the choice $m = 2$, $n' = 10$ (i.e., five CH_2 groups map onto an effective bond, anticipated in Fig. 3) was found to be the best choice, by trial and error [9]. One should note that the freedom in this choice is constrained by the fact that a reasonable volume fraction of occupied sites (such as $\phi = 0.5$) must correspond to melt densities, and the m effective bonds with the bond angle $\vartheta = 0$ must reproduce the length of a piece of n' units of an all-trans chain, for geometric consistency.

With these conditions, the experimental density of amorphous polyethylene at room temperature leads to a correspondence between one lattice spacing and a physical length of $a = 2.03$ Å. Figure 4 shows the probability distributions of the coarse-grained bond length L and the bond angle ϑ between two such bonds, choosing $n' = 10$ C–C bonds [9]. Instead of using these probability distributions directly (as was done in [12,13]) we choose to characterize them by their first two moments, taking into account cross-correlations, and record the moments $\langle L \rangle$, $\langle L^2 \rangle$, $\langle \vartheta \rangle$, $\langle \vartheta^2 \rangle$, $\langle L\vartheta \rangle$ over a wide range of temperatures from $T = 250$K to $T = 800$K. These moments are used

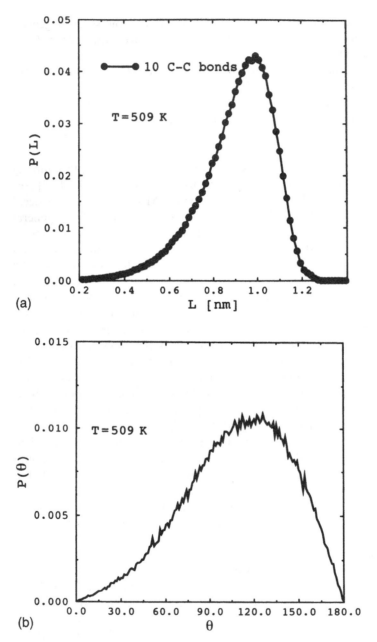

FIG. 4 Probability distribution of (a) the coarse-grained bond length L and (b) the bond angle θ between two such coarse-grained bonds, choosing $n' = 10$, $m = 2$. From Tries et al. [9].

as an input in a nonlinear fitting procedure, by which the parameters of the effective potentials $U(\ell)$, $V(\vartheta)$ for the length ℓ of the effective bonds and the angles ϑ between them are determined. We parameterize these potentials as follows

$$U(\ell) = u_0(\ell - \ell_0)^2 + (1/T - \langle 1/T \rangle)u_1(\ell - \ell_1)^2 \tag{7}$$

$$V(\vartheta) = v_0(\cos \vartheta - c_0)^2 + (1/T - \langle 1/T \rangle)v_1(\cos \vartheta - c_1)^2 \tag{8}$$

where $\langle 1/T \rangle$ denotes an average over the temperatures where input information from the atomistic calculation (such as shown in Fig. 4) is used [8].

From this fit the eight parameters u_0, u_1, ℓ_0, ℓ_1, v_0, v_1, c_0, c_1 are obtained, as well as the time constant τ_0 that is necessary to translate the Monte Carlo time units [= one attempted Monte Carlo step (MCS) per effective monomer] into a physical time. As is well known, the "random hopping" algorithm for the bond fluctuation model, where one selects an effective monomer randomly and tries to displace it in a randomly chosen lattice direction, can represent the Rouse-like or reptation-like dynamics of dense polymer melts in a rough way [2,26]; see also Chapter 4. But the Monte Carlo method never has an intrinsic physical time scale of its own, and just as the density of polyethylene needs to be used to give the length unit of the lattice model a physical meaning, one also needs physical input for the time unit. It is desirable to map the mobility of the lattice model onto the average jump rate of the torsional degrees of freedom of the chain, since these motions dominate the relaxation of its overall configuration. Thus a temperature-dependent time unit $\tau_{MC}(T)$, which one attempted MCS per monomer corresponds to, is introduced via

$$\frac{1}{\tau_{MC}(T)}\left\langle \min\left[1, \exp\left(-\frac{\Delta \mathcal{H}}{k_B T}\right)\right]\right\rangle = \frac{1}{\tau_0}\frac{1}{N_{tor}}\sum_{tor} A_{tor}(T) \tag{9}$$

Here $\Delta \mathcal{H}$ is the energy change in the Monte Carlo move, to be calculated from Eqs. (7) and (8), so $\langle \ldots \rangle$ is just the average acceptance factor, using the Metropolis transition probability for Monte Carlo moves as usual [26]. The attempt frequency of torsional jumps in the atomistic models is denoted as $1/\tau_0$, N_{tor} being the number of torsional degrees of freedom in the coarse-grained unit and A_{tor} their average activated jump probability. We define an average energy barrier $\langle \Delta E \rangle$ from $\sum_{tor} A_{tor}(T)/N_{tor} \equiv \exp(-\langle \Delta E \rangle k_B T)$, and write

$$\tau_{MC}(T) = A_{BFL}(\infty)\tau_0 \exp(\Delta E_{min}/k_B T) \tag{10}$$

where $A_{BFL}(\infty)$ is the acceptance probability of the Monte Carlo moves in the athermal ($T=\infty$) bond fluctuation model, and ΔE_{min} the smallest energy barrier in the torsional potential. Equations (9) and (10) yield a further mapping condition that the potentials $U(\ell)$, $V(\vartheta)$ must fulfill, namely

$$\langle\min[1,\ \exp(-\Delta\mathcal{H}/k_BT)]\rangle = A_{BFL}(\infty)\exp[-(\langle\Delta E\rangle - \Delta E_{min})/k_BT] \quad (11)$$

Including this information on the energy barriers of the torsional potential into the mapping that fixes the parameters of the potential $U(\ell)$, $V(\vartheta)$ implies that not all the many choices of bond lengths and bond angles occurring in the bond fluctuation model can correspond to minima of the torsional potential. On the other hand, for temperatures in the range from $T=250$K to $T=800$K geometrical properties such as $P(L)$ and $P(\theta)$, cf. Fig. 4, would be dominated by the minima of the torsional potential only. Figure 5 shows the time rescaling factor as obtained from Eq. (10), determining the absolute scale by equating the chain self-diffusion coefficient measured in the simulation at 450K to the experimentally known value [27]. While at $T=509$K $\tau_{MC}\approx 10^{-11}$ sec, one sees that at $T=200$K

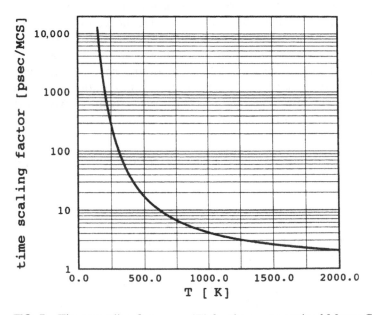

FIG. 5 Time rescaling factor $\tau_{MC}(T)$ for the coarse-grained Monte Carlo model for polyethylene plotted vs. temperature. The constant τ_0 in Eq. (10) was fixed by fitting the self-diffusion constant D_N for $C_{100}H_{202}$ to experiment at $T=450$K. From Tries et al. [9].

$\tau_{MC}(T)$ has increased to about a nanosecond. Since Monte Carlo runs of the order of 10^7 MCS are easily feasible, this means that relaxation processes on the millisecond time range can be explored. Figure 6 shows that in this way a reasonable prediction for the temperature dependence of both static and dynamic properties of polyethylene melts over a wide temperature range is obtained, including the melt viscosity [27], with no further adjusted parameters. Of course, this modeling accounts neither for the "breaking" of covalent bonds at high temperatures, nor for crystallization of polyethylene (that occurs for $T < T_m = 414$K), and a further weak point at the present stage is that attractive intermolecular interactions have been omitted as well. The off-lattice coarse-grained models [12,13] have not included attractive intermolecular forces either, and thus none of these approaches yields a

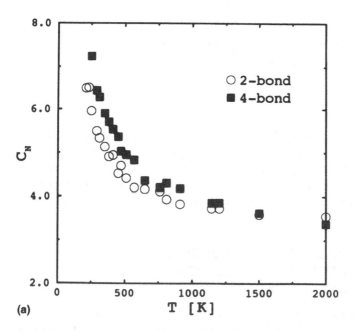

FIG. 6 (a) Characteristic ratio of polyethylene plotted vs. temperature, for $N = 20$ effective monomers (representing $C_{100}H_{202}$). Symbols denote two versions of the mapping (an exact one, needing four subsequent bonds, and an approximate two-bond procedure). (b) Logarithm of the viscosity of $C_{100}H_{202}$ plotted vs. inverse temperature, where full dots are from a constant density simulation of the bond fluctuation model with the potentials of Eqs. (7) and (8), using simulated self-diffusion constants D_N in the Rouse model formula $\eta = (\phi/8a^3)(\langle Rg^2 \rangle/6)k_B T/ND_N$. Open triangles are experimental data obtained by Pearson et al. [27] for $C_{100}H_{202}$ at constant (ambient) pressure. From Tries et al. [9].

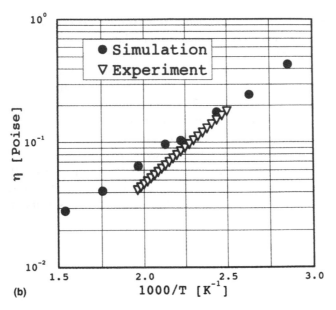

(b)

FIG. 6 Continued.

reasonable equation of state. It appears that using the pressure of the system as an additional input variable into the mapping of bead-spring type off-lattice models to fix the strength of the attractive part of the Lennard-Jones interaction between nonbonded effective monomers is a promising recipe to eliminate this weakness of current coarse-graining procedures [28].

III. ATOMISTIC-CONTINUUM MODELS: A NEW CONCEPT FOR THE SIMULATION OF DEFORMATION OF SOLIDS

While an explicit treatment of the microscopic, atomistic degrees of freedom is necessary when a locally realistic approach is required, on a macroscopic level the continuum framework is perfectly adequate when the relevant characteristics are scalar, vector, or tensor fields (e.g., displacement fields for mechanical properties). A combination of the two levels of description would be useful [29,30]. Here we focus on the deformation of solids: The elastic mechanics of homogeneous materials is well understood on the atomistic scale and the continuum theory correctly describes it. Inelastic deformation and inhomogeneous materials, however, require techniques that bridge length and time scales.

A number of approaches to connect multiple-scale simulation in finite-element techniques have been published [31–34]. They are able to describe macroscopically inhomogeneous strain (e.g., cracks)—even dynamic simulations have been performed [35]—but invariably require artificial constraints on the atomistic scale [36]. Recently, an approach has been introduced that considers a system comprising an inclusion of arbitrary shape embedded in a continuous medium [20]. The inclusion behavior is described in an atomistically detailed manner [37], whereas the continuum is modeled by a displacement-based Finite-Element method [38,39]. The atomistic model provides state- and configuration-dependent material properties, inaccessible to continuum models, and the inclusion in the atomistic-continuum model acts as a magnifying glass into the molecular level of the material.

The basic concept to connect both scales of simulation is illustrated in Fig. 7. The model system is a periodic box described by a continuum, tessellated to obtain "finite elements," and containing an atomistic inclusion. Any overall strain of the atomistic box is accompanied by an identical strain at the boundary of the inclusion. In this way, the atomistic box does not need to be inserted in the continuum, or in any way connected (e.g., with the nodal points describing the mesh at the boundary of the inclusion). This coupling via the strain is the sole mechanism to transmit tension between the continuum and the atomistic system. The shape of the periodic cells is described by a triplet of continuation (column) vectors for each phase (see also [21]), $\mathbf{A}$, $\mathbf{B}$, and $\mathbf{C}$ for the continuous body, with associated scaling matrix $\mathbf{H} = [\mathbf{ABC}]$, and analogously $\mathbf{a}$, $\mathbf{b}$, and $\mathbf{c}$ for the atomistic inclusion, with the scaling matrix $\mathbf{h} = [\mathbf{abc}]$ (see Fig. 8).

The material morphology is specified by a set of nodal points in the continuum description. The inclusion boundary is defined by a mesh of vertices $\mathbf{x}_i^b$ (b for boundary). The exterior of the inclusion contains the vertices $\mathbf{x}_i^c$ (c for continuum). Inside the atomistic system, the (affine) transformations obtained by altering the scaling matrix from $\mathbf{h}_0$ to $\mathbf{h}$ can be expressed by the overall displacement gradient tensor matrix $\mathbf{M}(\mathbf{h}) = \mathbf{hh}_0^{-1}$. The Lagrange strain tensor [40] of the atomistic system is then

$$\varepsilon = \frac{1}{2}(\mathbf{M}^T\mathbf{M} - 1) \tag{12}$$

As is convenient and customary in infinite periodic systems, scaled coordinates are employed for the continuum part as well as for the atomistic part (see Fig. 8). In the continuum part, $\mathbf{s}_i^b = \mathbf{H}^{-1}\mathbf{x}_i^b$ and $\mathbf{s}_i^c = \mathbf{H}^{-1}\mathbf{x}_i^c$ of the nodal points are used as degrees of freedom (the natural state variables of the continuum are, thus, $\mathbf{s}_i^b$, $\mathbf{s}_i^c$, and $\mathbf{H}$). Similarly, the scaled coordinates $\mathbf{s}_i^a$ of

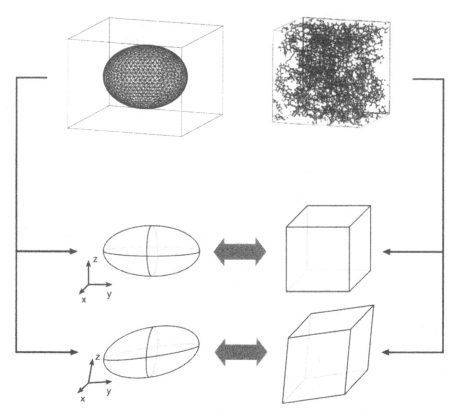

FIG. 7 The key idea of the atomistic-continuum model is to connect the inclusion boundary behavior and the atomistic cell behavior via the strain transformation they undergo. The strain behavior of the atomistic box is identical to the strain behavior of the inclusion boundary.

atom i are related to the Cartesian coordinates through the scaling matrix, $\mathbf{s}_i^a = \mathbf{h}^{-1}\mathbf{x}_i^a$.

Since, as stated above, our model [20] requires that an overall strain of the atomistic box is accompanied by an identical strain at the boundary of the inclusion, we constrain the vertices $\mathbf{x}_i^b$ of the inclusion boundary and the atomistic scaling matrix $\mathbf{h}$ to behave in concert (the vertices $\mathbf{x}_i^c$ in the matrix are not affected) according to $\mathbf{x}_i^b(\mathbf{h}) = \mathbf{M}(\mathbf{h})\mathbf{x}_{0i}^b$, where $\mathbf{x}_{0i}^b$ are the original coordinates of the boundary vertices i. Note that the coordinates $\mathbf{x}_i^b(\mathbf{h})$ are no longer degrees of freedom, because they depend on the scaling matrix $\mathbf{h}$. Changes in the system cell matrix $\mathbf{H}$ act as homogeneous deformations of the entire continuous system [20] and the nodal points of the continuum and the atomic positions in the inclusion should be displaced accordingly. This

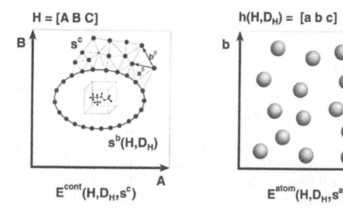

FIG. 8 Sketch of the four types of degrees of freedom used in the atomistic-continuum model: $\mathbf{H}$, $\mathbf{D_H}$, and the scaled coordinates $\mathbf{s}^c$ and $\mathbf{s}^a$ of the nodes in the continuum and the atoms in the atomistic phase. The system unit cell matrix $\mathbf{H}$ and the scaled coordinates $\mathbf{s}^c$ and $\mathbf{s}^b$ entirely determine the finite-element configuration of the continuum. $\mathbf{h}$ and $\mathbf{s}^a$ are necessary to determine the atomistic configuration. The nodal points on the inclusion boundary $\mathbf{s}^b$ and the atomistic scaling matrix $\mathbf{h}$ are calculated from the variables $\mathbf{H}$ and $\mathbf{D_H}$.

can be achieved by introducing a new set of degrees of freedom $\mathbf{D_H}$ (nine scalars), through [20]

$$\mathbf{h} = \mathbf{H}(1 + \mathbf{D_H})\mathbf{H}_0^{-1}\mathbf{h}_0 \tag{13}$$

Both $\mathbf{H}$ and $\mathbf{h}$ undergo concerted changes (a change in $\mathbf{H}$ from the reference state $\mathbf{H}_0$ directly affects the atomistic cell $\mathbf{h}$ and affinely displaces the inclusion boundary vertices according to $x_i^b = \mathbf{HH}_0^{-1}x_{0i}^b$). The reference state, where $\mathbf{H} = \mathbf{H}_0$ and $\mathbf{h} = \mathbf{h}_0$, corresponds to $\mathbf{D_H} = 0$.

The energy of the overall system is the sum of contributions from the atomistic part and the continuum part. The atomistic part contributes a potential energy determined by the interatomic potentials (the "force field") and a kinetic energy from the momenta of the atoms. The continuum system contributes an elastic energy as the sum of the elastic deformations of all finite elements:

$$E^{cont}(\mathbf{H}, \mathbf{s}_i^b, \mathbf{s}_i^c) = \frac{1}{2}\sum_\beta V_\beta(\varepsilon^\beta)^T C^\beta \varepsilon^\beta \tag{14}$$

where β denotes a finite element and V_β is the volume of the undeformed tetrahedron β [41]. The overall system can be described by the multiscale

NtT ensemble distribution function [42].

$$P_{NtT, multi} \propto \exp\left\{ -\frac{\alpha_V(E^{atom} - \langle E^{atom}\rangle_0 + K^{atom}) + E^{cont} + K^{cont} + W^{ext}}{kT} \right\}$$

(15)

where E^{atom} is the potential defined by the atomistic description of the system, $\langle E^{atom}\rangle_0$ its value at the undeformed (reference) state, K^{atom} the atomistic kinetic energy, E^{cont} the continuum elastic strain energy of Eq. (14), K^{cont} the associated kinetic energy (with contributions from each finite element—here the mass of the continuum is distributed onto the nodes of all finite elements and the kinetic energy is estimated as if it were atomistic), W^{ext} the external work done on the system, and the parameter $\alpha_V = V^{inclusion}/V^{atom}$, the ratio of the inclusion and the atomistic cell volume, accounts for the difference between the inclusion volume and the atomistic system periodic cell contained therein.

The elements of the coupling matrix $\mathbf{D_H}$ between the periodic boxes of overall continuum system, $\mathbf{H}$, and atomistic system, $\mathbf{h}$ [see Eq. (13)] are simply minimization variables for the total energy when minimum-energy simulations are carried out, i.e., calculations at $T = 0$ [where the two kinetic-energy terms in Eq. (15) are zero]. At nonzero temperature, when the kinetic-energy terms in Eq. (15) must be taken into account, the coupling is driven by the balance between the internal stress of the atomistic cell and the external stress acting on the inclusion from the resistance of deformation of the surrounding medium [42].

To date, results have been obtained for minimum-energy type simulations of elastic deformations of a nearest-neighbor face-centered cubic (fcc) crystal of argon [20] with different inclusion shapes (cubic, orthorhombic, spherical, and biaxially ellipsoidal). On bisphenol-A-polycarbonate, elastic constant calculations were also performed [20] as finite deformation simulations to plastic unit events (see [21]). The first molecular dynamics results on a nearest-neighbor fcc crystal of argon have also become available [42]. The consistency of the method with thermodynamics and statistical mechanics has been tested to a satisfactory extent [20]; e.g., the calculations with different inclusion shapes all yield identical results; the results are independent of the method employed to calculate the elastic properties of the system and its constituents (constant-strain and constant-stress simulations give practically identical values).

A test-case application to the elastic constants of bisphenol-A-polycarbonate shall give an idea of the manageable complexity of this multiscale approach. Here, a single polymer was simulated as both the atomistic and the continuum material in order to check for possible inconsistencies. The polycarbonate was taken to be at the experimental density of $1.2 \, \text{g/cm}^3$ (300K). The atomistic system considered was composed of three polycarbonate chains of twenty-five repeat units each, i.e., of 2481 atoms in an atomistic cell of edge length 300 Å, and was constructed to conform to the overall torsion-angle statistics of the rotational-isomeric state model of the polymer [43] with the PolyPack algorithm [44]. The overall atomistic-continuum model was chosen to have an atomistic inclusion shape and the inclusion volume fraction was taken to be 18 percent (the atomistic box accounts for about one fifth of the system energy, since the energy is distributed approximately uniformly). The edge length of the overall system cell was taken to be 50 Å so that $\alpha_V = 0.83$ in Eq. (15). The atomistic-continuum system comprised a total of 26,235 scalar degrees of freedom (3×827 atoms and 6259 independent nodal points in the continuum, which give 26,220 coordinates, and the two scaling matrices, $\mathbf{H}$ with 6 and $\mathbf{D_H}$ with 9 degrees of freedom). This is easily within the range of modern molecular simulations.

IV. CONCLUSIONS

The recently proposed atomistic-continuum model is useful for modeling the deformation of three-dimensional solids. In this approach, an inclusion of arbitrary shape is surrounded by a continuum, which is modeled by a variable-metric "finite-element" method. The inclusion behavior is directly driven by that of an atomistic system modeled in detail by molecular-modeling techniques. In this manner, a solution is given to the problem of finding a constitutive description for a material. The atomistic-continuum model comprises four types of degrees of freedom, the appropriate combination of which permits calculation of the configuration of both the atomistic system and the mesh elements of the continuum. The behavior of both levels of modeling is described by their corresponding potential C- and kinetic-energy functions which are added to give the total energy of the system.

The major drawback of the atomistic-continuum approach lies in the limitation to uniform strain on the level of the atomistic periodic box and, hence, the inclusion perimeter. Crack propagation, for instance, cannot be treated with this method.

A partial compensation for this weakness might lie in the flexibility of the atomistic-continuum concept. The approach can be extended, in principle, to

- any number of atomistic inclusions in a continuum phase; also, one can imagine continuous atomistic phases cohabiting with the continuum;

- other overall properties than stress and strain; among the properties that can be simulated are those describable by scalar fields in a continuum [39], such as dielectric constants, thermal conductivity, electrical conductivity, solute permeability, etc.

Also, the approach provides the potential for constant-stress molecular dynamics [40,45] with realistic retracting forces acting on the atomistic periodic box. This would provide an advantage over the current situation, where constant-stress molecular dynamics is weakened by the fact that the device, which maintains constant stress, (a) acts directly and without time delay on each atom, and (b) requires an arbitrary "wall mass." The question of efficiency for atomistic-continuum constant-stress molecular dynamics must be answered first, of course.

ACKNOWLEDGMENTS

We thank Jörg Baschnagel and Andrei A. Gusev for many illuminating discussions and Michael Rozmann and Volker Tries for their help and kind support. S. S. and U. W. S. also gratefully acknowledge the financial support provided by the Swiss National Science Foundation (Schweizerischer Nationalfonds zur Förderung der wissenschaftlichen Forschung). K. B. and W. P. acknowledge support from the Bundesministerium für Bildung und Forschung (BMBF, Germany) under Grants No 03N8008C and 03N6015.

REFERENCES

1. Binder, K. Macromol. Chem.: Macromol. Symp. **1991**, *50*, 1.
2. Baschnagel, J.; Binder, K.; Doruker, P.; Gusev, A.A.; Hahn, O.; Kremer, K.; Mattice, W.L.; Müller-Plathe, F.; Murat, M.; Paul, W.; Santos, S.; Suter, U.W.; Tries, V. Adv. Polymer Sci. **2000**, *152*, 41.
3. Kremer, K. Multiscale problems in polymer simulations. In *Multiscale Simulations in Chemistry and Physics*, Proceedings NATO ARW, Eilat/Israel, April 5–8, 2000, Brandt, A., Bernholc, J., Binder, K., Eds.; IOS Press: Amsterdam, 2001.
4. Baschnagel, J.; Binder, K.; Paul, W.; Laso, L.; Suter, U.W.; Batoulis, J.; Jilge, W.; Bürger, T. J. Chem. Phys. **1991**, *95*, 6014.
5. Paul, W.; Binder, K.; Kremer, K.; Heermann, D.W. Macromolecules **1991**, *24*, 6332.
6. Carmesin, I.; Kremer, K. Macromolecules **1988**, *21*, 2819.
7. Deutsch, H.-P.; Binder, K. J. Chem. Phys. **1991**, *94*, 2294.
8. Paul, W.; Pistoor, N. Macromolecules **1994**, *27*, 1249.
9. Tries, V.; Paul, W.; Baschnagel, J.; Binder, K. J. Chem. Phys.**1997**, *106*, 738.

10. Rapold, R.F.; Mattice, W.L. Macromolecules **1996**, *29*, 2457.
11. Doruker, P.; Mattice, W.L. Macromolecules **1997**, *30*, 5520.
12. Tschöp, W.; Kremer, K.; Batoulis, J.; Bürger, T.; Hahn, O. Acta Polymerica **1998**, *49*, 61.
13. Tschöp, W.; Kremer, K.; Batoulis, J.; Bürger, T.; Hahn, O. Acta Polymerica **1998**, *49*, 75.
14. Zimmer, K.M.; Heermann, D.W. J. Computer-Aided Mater. Design **1995**, *2*.
15. Zimmer, K.M.; Linke, A.; Heermann, D.W.; Batoulis, J.; Bürger, T. Macromol. Theory Simul. **1996**, *5*, 1065.
16. Santos, J. From Atomistic to Continuum Modeling of Polymers-Multiscales Paradigms. Ph.D. thesis, Eidgenössische Technische Hochschule Zürich, Dies No 12941, 1998.
17. Murat, M.; Kremer, K. J. Chem. Phys. **1998**, *108*, 4340.
18. Louis, A.A.; Bolhuis, P.G.; Hansen, J.P.; Meijer, E.J. Phys. Rev. Lett. **2000**, *85*, 2522.
19. Warren, P.B. Current Opinion in Colloid & Interface Science **1998**, *3*, 620.
20. Santos, S.; Yang, J.S.; Gusev, A.A.; Jo, W.H.; Suter, U.W. submitted for publication.
21. Yang, J.S.; Jo, W.H.; Santos, S.; Suter, U.W. Plastic deformation of bisphenol-A-polycarbonate. In *Simulation Methods for Polymers*, Kotelyanskii, M., Theodorou, D., Eds.; Marcel Dekker, 2004, Chapter 12.
22. Smith, G.D.; Yoon, D.Y. J. Chem. Phys. **1994**, *100*, 649.
23. Paul, W.; Yoon, D.Y.; Smith, G.D. J. Chem. Phys. **1995**, *103*, 1702.
24. Paul, W.; Smith, G.D.; Yoon, D.Y.; Farago, B.; Rathgeber, S.; Zirkel, A.; Willner, L.; Richter, D. Phys. Rev. Lett. **1998**, *80*, 2346.
25. Baschnagel, J.; Qin, K.; Paul, W.; Binder, K. Macromolecules **1992**, *25*, 3117.
26. Binder, K.; Paul, W. J. Polym Sci., Part B: Polymer Phys. **1997**, *35*, 1.
27. Pearson, D.S.; Fetters, L.J.; Graessley, W.W.; ver Strate, G.; von Meerwall, E. Macromolecules **1994**, *27*, 711.
28. Müller-Plathe, F.; Biermann, O.; Faller, R.; Meyer, H.; Reith, D.; Schmitz, H. *Autimatic multiscaling*, Paper presented at the ESF-SIMU-Workshop *Multiscale Modeling of Macromolecular Systems*, Mainz, Sept. 4–6, 2000, unpubl.
29. Phillip, R. Journal of The Minerals Metals & Materials Society **1995**, *47*, 37.
30. Vellinga, W.P.; Hosson, J.T.M.D.; Vitek, V. Acta Mater. **1997**, *45*, 1525.
31. Ortiz, M. Computational Mechanics **1996**, *18*, 321.
32. Tadmor, E.B.; Phillips, R.; Ortiz, M. Langmuir **1996**, *12*, 4529.
33. Tadmor, E.B.; Ortiz, M.; Phillips, R. Philosophical Magazine A **1996**, *73*, 1529.
34. Shenoy, V.B.; Miller, R.; Tadmor, E.B.; Phillips, R.; Ortiz, M. Phys. Rev. Letters **1998**, *80*, 742.
35. Rafii-Tabar, H.; Hua, L.; Cross, M. J. Phys.: Condens. Matter **1998**, *10*, 2375.
36. Miller, R.; Phillips, R. Philosophical Magazine A **1996**, *73*, 803.
37. Allen, M.P.; Tildesley, D.J. *Computer Simulation of Liquids*; Clarendon Press: Oxford, 1989.
38. Gusev, A.A. J. Mech. Phys. Solids **1997**, *45*, 1449.

39. Gusev, A.A. Macromolecules **2001**, *34*, 3081.
40. Parrinello, M.; Rahman, A. Journal of Chemical Physics. **1983**, *76*, 2662.
41. Thurston, R.N. In *Physical Acoustics: Principles and Methods*, Mason, W.P., Ed.; Academic Press: New York, 1964.
42. Beers, K.J.; Ozisik, R.; Suter, U.W. in preparation.
43. Hutnik, M.; Argon, A.S.; Suter, U.W. Macromolecules. **1991**, *24*, 5956.
44. Müller, M.; Santos, S.; Nievergelt, J.; Suter, U.W. J. Chemical Physics **2001**, *114*, 9764.
45. Parrinello, M.; Rahman, A. Journal of Applied Physics **1981**, *52*, 7182.

16

CONNFFESSIT: Simulating Polymer Flow

MANUEL LASO Polytechnic University of Madrid, Madrid, Spain

HANS CHRISTIAN ÖTTINGER Eidgenössische Technische Hochschule (ETH), Zurich, Switzerland

I. INTRODUCTION

The present chapter introduces a nontraditional approach to the numerical calculation of complex flows of complex fluids, with especial application to polymeric systems such as melts and solutions.

The applicability of the CONNFFESSIT (Calculation Of Non-Newtonian Flows: Finite Elements and Stochastic Simulation Technique) in its present form is limited to the solution of fluid mechanical problems of incompressible fluids under isothermal conditions. The method is based on a combination of traditional continuum-mechanical schemes for the integration of the mass- and momentum-conservation equations and a simulational approach to the constitutive equation.

The macroscopic equations are solved by a time-marching scheme in which the stress tensor is obtained from a micro simulation and treated as a constant body force. The microscopic simulations reflect the dynamics of the specific fluid at hand and yield the stress tensor for a given velocity field.

The required integration over deformation histories is accomplished by integrating numerically microscopic particle trajectories for large global ensembles simultaneously with the macroscopic equations of mass and momentum conservation. The term "trajectories" in the previous sentence refers to both real space trajectories, i.e., positions $r_i(t)$ and to configurational phase space trajectories, i.e., in the case of a dumbbell model, connector vector Q.

Given the very large ensemble sizes required, the feasibility of such calculations depends crucially on algorithmically optimal data structures

511

and computational schemes. In the present chapter a major effort has been devoted to the detailed description of such techniques and of their implementation on advanced computer architectures. Nontrivial two-dimensional calculations for a simple noninteracting polymer dumb-bell model are presented in order to demonstrate the practical use of the method.

Although CONNFFESSIT opened up the door to so-called micro–macro rheological calculations some ten years ago (Öttinger and Laso 1992), it is but the first member of a group of methods that have appeared along the same line since 1992 in the context of computational non-Newtonian fluid mechanics. Most of the newer methods rely on the concept of variance reduction (Kloeden and Platen 1992, Öttinger 1995, Melchior and Öttinger 1995, 1996) in order to achieve a great improvement on the noise level of micro–macro calculations. Although the field of micro–macro methods (Laso 1995) is still very much in flux, some of the newer and highly original methods (see Hulsen et al. 1997, Halin et al. 1998, Bonvin and Picasso 1998, Bell et al. 1998) can be up to two or three orders of magnitude more efficient for a given level of precision. And improvements are appearing all the time. Nowadays the main applicability of CONNFFESSIT as formula-ted originally is in situations where fluctuations do play an important role. For typical macroscopic problems, the newer techniques are far more efficient and thus the methods of choice.

A. Some Definitions

Throughout the succeeding sections some terms are used frequently, sometimes in a nonstandard way. The following definitions are included in order to avoid confusion:

- *Domain*: a region of d-dimensional space in which we seek the solu-tion to the equations of conservation of mass and momentum coupled to a description of the fluid (constitutive equation).
- *Cell*: in d dimensions, the simplest d-dimensional geometric object in which the domain is subdivided.
- *Element*: synonymous with cell.
- *Finite element*: a *cell* or *element* augmented with nodal points and shape functions.
- *Particle*: the smallest entity of the micro calculation that does not directly interact with other similar entities and which contributes to the stress; it can be as simple as a Hookean dumbbell, or as complex as a whole Brownian Dynamics simulation of a colloidal dispersion under periodic continuation or boundary conditions.

- *Global ensemble*: the set of all *particles* in the *domain*.
- *Local ensemble*: the set of all *particles* in a *cell*.
- *Complex fluid*: a (possibly non-Newtonian) fluid in which microscopic *particles* are explicitly resolved.

Finally, throughout this work, and unless stated otherwise, the term "microscopic" refers either i) to simplified mechanical models for polymer molecules in which no chemical detail is explicitly present or ii) to models for particulate fluids in which the individual particles are explicitly treated and in which the solvent is treated stochastically in an average sense and not resolved in individual molecules. This usage of the term microscopic is adopted here in order to be consistent with the meaning of the term "micro" as used in the literature on methods that combine continuum mechanics with a more detailed description of the flowing material. Strictly speaking, models falling into category i) are more "microscopic" than those under ii). Atomistic and *ab initio* (density functional) polymer modeling reside at increasingly more fundamental levels.

II. OVERVIEW OF RELATED FIELDS

CONNFFESSIT cuts across two different and, up to recently, largely unrelated fields: traditional continuum-mechanical computational rheology and stochastic dynamic methods for polymers. Brief reviews are presented separately. A third subsection is devoted to alternative methods in Newtonian and non-Newtonian CFD which have a direct bearing on the subject of this work.

A. Computational Rheology

Although it is young, the field of continuum-mechanical computational rheology has a lively and eventful history. What was initially expected to be little more than an extension of Newtonian CFD methods to fluids with more complex but in principle harmless constitutive equations turned out to be a source of frustration and of unexpectedly challenging mathematical problems in spite of the vast store of knowledge on numerical techniques for Newtonian fluids that had been accumulating during the past half century. The analysis and solution of what has come to be known as the high Weissenberg-number* problem have engaged the attention of

---

*Rheologists resort to dimensionless groups such as the Weissenburg and Deborah numbers and the stress ratio S_R to characterize the elasticity of the flow. High W_S corresponds to high fluid elasticity.

a large number of both applied mathematicians and engineers and spurred a great amount of analytical and numerical diagnostic work. The book by Crochet et al. (1984) and the review articles by Crochet (1989) and Keunings (1989) give comprehensive if slightly dated overviews of the field; current research in computational rheology is reported in Moldenaers and Keunings (1992), Gallegos et al. (1994), and Ait-Kadi et al. (1996).

For the purposes of the present work the current situation in non-Newtonian CFD can be summarized as follows:

- Converged solutions for non-Newtonian CFD problems can be obtained nowadays for a large variety of geometries and constitutive equations (Debbaut and Crochet 1986, Keunings 1986a, 1986b, Van Schaftingen and Crochet 1984, Mendelson et al. 1982, Brown et al. 1986, Marchal and Crochet 1987, Papanastasiou et al. 1987, Bush et al. 1984, Malkus and Webster 1987), including free-surface flows (Nickell et al. 1974, Keunings and Bousfield 1987).

- Most calculations are performed in two dimensions, under isothermal conditions and at steady state, although transient (Josse and Finlayson 1984, Lee et al. 1984, Van Schaftingen 1986, Phan-Thien et al. 1987, Bousfield et al. 1988) and nonisothermal (Keunings 1986a, 1986b, Sugeng et al. 1987, McClelland and Finlayson 1988, Srinivasan and Finlayson 1988) flow calculations have been reported.

- The high Weissenberg problem has been recognized to actually have several independent causes such as irregular points in the solution family, boundary layers and singularities, loss of evolution, and change of type of the PDEs. Although some of the problems originally encountered still remain, a good deal of these initial difficulties have been overcome by sophisticated numerical schemes (Keunings 1989 and references therein, Franca and Hughes 1993, Baaijens 1998).

Finally, there exists a trend away from physically simple but numerically problematic constitutive relations towards more realistic and algorithmically more demanding ones with better numerical behavior. Some CEs, prominently those most appropriate for melts (Doi–Edwards, Curtiss–Bird, reptating rope, modified reptation models) are and will remain for a long time to come well beyond the capabilities of current continuum-mechanical methods. Further improvements in the polymer dynamics at the microscopic level leads to models intractable via the standard CE approach.

The systematic use of increasingly more complex CEs calls for an alternative way of handling this increased complexity while at the same time allowing the equations of change to be solved in complex geometries.

B. Stochastic Dynamic Methods for Polymers

Stochastic dynamics or stochastic differential equations arise naturally in polymer physics as a consequence of the coarse-graining required when dealing with the many degrees of freedom and the wide spectrum of time scales characteristic of polymers. Brownian Dynamics was first applied in the late 1970s (Fixman 1978, McCammon 1978) and has experienced a colossal growth since then. Although there is abundant literature on stochastic and kinetic-theoretical topics (Van Kampen 1981, Gardiner 1983, Gard 1988, Risken 1989, Honerkamp 1990, Kloeden and Platten 1992, Doi and Edwards 1986, Bird et al. 1987b), a unified approach to the use of stochastic methods in polymer dynamics has been missing until recently. Besides filling this gap, the book by Öttinger (1995) offers a comprehensive overview of the state of current research.

As far as the present work is concerned, the relevance of numerical stochastic methods for polymer dynamics in micro/macro calculations resides in their ability to yield (within error bars) exact numerical solutions to dynamic models which are insoluble in the framework of polymer kinetic theory. In addition, and mainly as a consequence of the correspondence between Fokker–Planck and stochastic differential equations, complex polymer dynamics can be mapped onto extremely efficient computational schemes. Another reason for the efficiency of stochastic dynamic models for polymer melts stems from the reduction of a many-chain problem to a single-chain or two-chain representation, i.e., to linear computational complexity in the number of particles. This circumstance permits the treatment of global ensembles consisting of several tens of millions of particles on current hardware, corresponding to local ensemble sizes of $O(10^3)$ particles per element.

C. Particle Methods

Besides the Finite-Difference and Finite-Element mainstream, there has appeared a handful of alternative computational methods in Newtonian and non-Newtonian CFD that can be collectively characterized as "particulate," in the sense that they involve discrete computational entities in addition to or in lieu of the standard continuum-mechanical discretization devices. In some of these methods (the first four and the last one in the following list), the computational particles have an almost exclusively mathematical character, whereas in the other two it is possible to attach a well-defined physical meaning to the particles.

- As early as 1973 Chorin (1973, 1989, 1994) introduced the two-dimensional random vortex method, a particle method for the solution of the Navier–Stokes equations. These particles can be thought of as carriers of vorticity. Weak solutions to the conservation equations are obtained as superpositions of point vertices, the evolution of which is described by deterministic ODEs. A random walk technique is used to approximate diffusion, and vorticity creation at boundaries to represent the no-slip boundary condition. The extension to three dimensions followed in 1982 (Beale and Majda 1982). An important improvement in stability and smoothness was achieved by Anderson and Greengard (1985) by removing the singularities associated with point vertices. Anderson and Greengard (1988) and Marchioro and Pulvirenti (1984) have written comprehensive reviews of the method.
- Smoothed-Particle Hydrodynamics (SPH) was introduced by Gingold and Monaghan (1977) and Lucy (1977). SPH is a three-dimensional free-Lagrange algorithm that in its most basic form is truly grid-free. SPH treats fluid elements as extended clouds of material; their centers of mass move according to the conservation laws of fluid mechanics. It was first applied to the solution of complex astrophysical problems, such as colliding planets and stars. In its most widely used version, SPH is based on integral interpolants for information at nonordered points. The fluid is represented by spherically symmetric particles, each having a mass that has been smoothed out in space with a density distribution given by a suitable interpolating kernel or distribution. Ordinary differential equations for the smoothed quantities at the location of the particles are obtained by multiplying the original governing equations by the kernel and integrating over the domain. A critical evaluation of the quantitative performance of SPH has been presented in Cloutman (1991). The method seems to be most useful for exploratory studies where modest accuracy is acceptable.
- Another technique, widely used in a different context (kinetic theory, plasma simulation) is the so-called Vortex-in-Cell Method (Hockney and Eastwood 1981, Birdsall and Langdon 1985). In the Vortex-in-Cell Method a two-dimensional computational domain is divided into cells and the vorticity is "counted" in each cell. The Poisson equation relating the vorticity and the Laplacian of the stream function is subsequently solved on the grid by an FTT algorithm, the velocity

is computed on the grid and finally interpolated at the vortex locations. This approach bears some resemblance to SPH.

- Cellular-automata and lattice-gas models are in a sense a diametrically opposed alternative to the pure grid-free NEMD method. They rely on the existence of a regular lattice, each site of which can have a finite number of states usually represented by Boolean variables. Each cellular automaton evolves in time in discrete steps according to deterministic or nondeterministic rules for updating based on the states of the neighboring automata. The average behavior of a collection of cellular automata is found to describe the behavior of physical systems obtained by using models with smooth continuous variables. Lattice-gas models with fluid dynamical features were first introduced by Kadanoff and Swift (1968) and the first fully deterministic lattice-gas model was introduced by Hardy et al. (1976). The main value of cellular-automata methods seems to be in the qualitative study of turbulence (Frisch et al. 1986), although they have been used to investigate other fluid-flow phenomena with varying degrees of success (Rothman and Keller 1988, Balasubramanian et al. 1987, Baudet et al. 1989).

- A grid-free alternative to discretization methods is offered by Non-Equilibrium Molecular Dynamics. The NEMD approach to fluid dynamics was pioneered by Ashurst and Hoover in 1975 for the investigation of momentum transport in homogeneous flows. Some of the first calculations for nonhomogeneous flows in complex geometries were performed by Hannon, Lie, and Clementi (1986). A large number of MD simulations has been performed since then. Most of the very satisfactory features of such calculations arise from the fundamental description of the interactions between fluid particles, which in this case are the atoms or molecules, and the absence of a grid. The drawback of this approach is that it is limited to extremely small length- and time-scales, or conversely, to extremely high rates of deformation, $O(10^{13})\,\mathrm{s}^{-1}$ typically. The same approach has been applied by Hoover et al. (1992) to plastic flow in nonlinear solid mechanics.

III. TWO- AND THREE-DIMENSIONAL TECHNIQUES

A. One-Dimensional vs. Multidimensional Problems

The first application of CONNFFESSIT was in the context of computational non-Newtonian fluid mechanics in $1 + 1$ dimensions (one spatial + one temporal) (Laso and Öttinger 1993). The practical applicability of the method relies on its success in more dimensions $(2+1, 3+1)$. The introduction of one additional spatial dimension entails both a quantitative and a qualitative increase in the complexity of the problem.

The obvious quantitative increase is due to the higher number of particles and of spatial cells or elements required to describe the integration domain, typically between one and two orders of magnitude more per additional spatial dimension.

The qualitative increase stems from the need to perform particle tracking, a task which has no equivalent in one dimension,* since particles never leave the element in which they are initially located as they are carried by the velocity field. The generic term *particle tracking* actually involves a number of nontrivial individual subtasks, which must be tackled efficiently if CONNFFESSIT is to be viable. The key to practicable micro/macro computations is the invention of efficient schemes of reduced algorithmic complexity.

A further increase to three spatial dimensions brings about another increase in the size of the problem. The tracking techniques are, however, not more complex than in the two-dimensional case. Furthermore, some of the algorithms to be presented perform proportionally better than classical continuum-mechanical tools the larger a problem is. For this reason, the two-dimensional case is especially critical: it involves particle tracking in its full complexity, yet typical 2D problem sizes are not yet in the range where the CONNFFESSIT approach can even remotely compete with traditional methods.

Before presenting procedural details, an overview of the general scheme for the micro/macro approach in more than one dimension is in order. Although it is possible to devise micro/macro methods to obtain steady-state solutions by a procedure that draws heavily from traditional tracking schemes for integral constitutive equations (Feigl et al. 1995), the natural implementation of the micro/macro strategy calls for a time-marching integration of the unsteady problem. This approach allows us to obtain, in addition to the full dynamic behavior of the system, the steady-state solution (when one exists) by continuing integration for a sufficiently long time.

The usage of the term micro/macro is justified by the two-level description of the fluid that lies at its core: a microscopic approach permits evaluation of the stress tensor to be used in closing the macroscopic conservation equations. The macro part can be thought of as corresponding to the Eulerian description of the fluid, while the micro part has Lagrangian character.

---

*Except for some multiparticle dynamic models in which particle diffusion across streamlines must be considered (Öttinger 1992).

The overall CONNFFESSIT procedure for the isothermal problem is eminently simple and is illustrated in Figs. 1 and 2 as a block diagram. This approach allows the use of standard continuum-mechanical codes to solve the macroscopic equations of conservation (mass and momentum). Micro/macro interfacing takes place at the points marked A and B in Fig. 1.

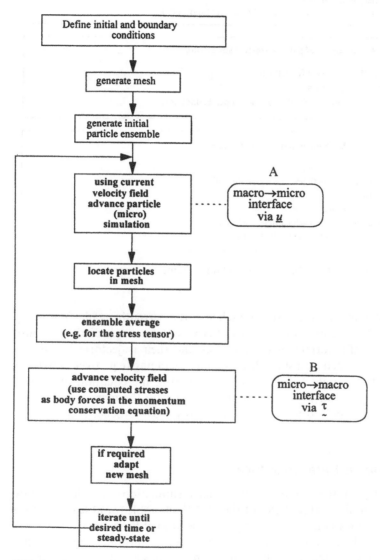

FIG. 1 Basic CONNFFESSIT time-marching scheme.

Generation of initial particle ensemble

- initialize particle initial configurations
    (possibly at equilibrium or from specified pre-history)
- initialize positions
- create cell neighbor table
- obtain auxiliary information for
    particle location (e.g. handedness of cells, cell centroids, etc.)

Advancement of particle (micro) simulation

- advance particle internal configurations
- advance particle trajectories
- introduce/remove particles (for domains with open boundaries)

Location of particles in the mesh

- find particles that left the integration
    domain ("lost" particles) due e.g. to discretization error
- rescue lost particles
- update cell neighbor list
- determine cell number for each particle using point-
    inclusion algorithms and neighbor list

FIG. 2   Basic CONNFFESSIT time-marching scheme (continued).

At a given step in the time-marching scheme, the currently available velocity field is passed to the microscopic part at A where it is used to advance both the internal configurations of the particles and their trajectories. In B, the stress tensor obtained by a suitable ensemble-averaging (e.g., over individual cells) is used as a right hand side (body force) in the equation of conservation of momentum, which is solved together with the mass-conservation equation using standard techniques (Keunings 1989, Crochet et al. 1984).

B.   The Basic Data Structure

Independently of the details of the micro simulations, the information pertaining to individual particles of the global ensemble (or at least pointers to that information) must be kept in a suitable kind of *linear* array. As we will see in Sections IV.A and IV.B.1 the organization of this indexing array into an element-wise sorted array has far-reaching consequences on computational efficiency and, above all, on vectorization and parallelization.

Several micro/macro models, CONNFFESSIT among them, share the common feature that the particles appearing in the micro simulation are entrained by the macroscopic flow and therefore move through a succession of elements. As a consequence, the indices of the particles residing in a given element (the local ensemble) change with time. This circumstance makes physically contiguous particles be noncontiguous in the linear array containing them.

The objective of constructing an element-wise sorted array is to provide indirect indexing capability, which can be used to bypass the noncontiguity problem. Let the array WHERE initially contain the number of the element in which a given particle is located. A small-scale typical arrangement (12 particles in 3 elements) is given in the schematic below where particles are noncontiguous:

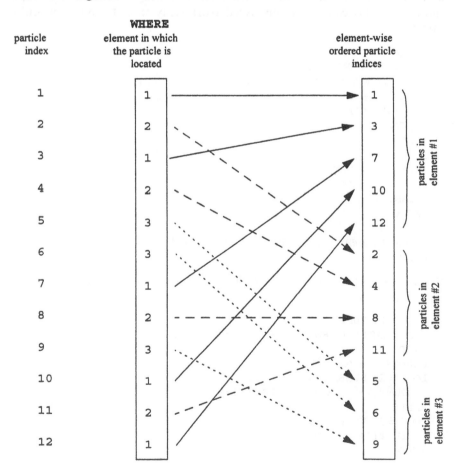

For several reasons to be seen later, it is more convenient to have physically neighboring particles (i.e., particles in the same cell or element) placed in neighboring positions in memory, i.e., particles 1, 3, 7, 10, and 12, etc. While it would be possible to initialize the positions of particles so as to ensure contiguity at the start of a calculation, the macroscopic flow would render this initial order useless at every time step. It is however possible to create an element-wise sorted array ORDWHERE that holds the indices of the particles contained in each element (of which there are NUMEL), plus two pointer arrays FIRST and LAST that mark the start and the end of the sequence of particle indices for each element within the sorted array. Sorting can be accomplished by any efficient algorithm, like heapsort (Knuth 1968) or, having vectorizability in mind, by the algorithm presented in Section IV.B.1. The following pseudo-code excerpt shows how this can be accomplished (sort array WHERE and copy onto ORDWHERE):

```
J = 0
DO I = 1 , NUMPR - 1
  IF (ORDWHERE(I) .NE. ORDWHERE(I+1)) THEN
    J = J + 1
    COUNT(J) = I
  ENDIF
ENDDO
J = J + 1
COUNT(J) = NUMPR
DO I = 1 , NUMEL + 1
  LAST(I) = 0
ENDDO
DO I = 1, J
  LAST(ORDWHERE(COUNT(I))) = COUNT(I)
ENDDO
```

Special care of empty elements (if any) must be taken and the array FIRST created:

```
DO I = 2, NUMEL + 1
  IF (LAST(I) .EQ. 0) LAST(I) = LAST(I-1)
ENDDO
COUNT(1) = LAST(1)
DO I = 2, NUMEL + 1
  COUNT(I) = LAST(I) - LAST(I-1)
ENDDO
DO I = 1, NUMEL + 1
  FIRST(I) = LAST(I) + 1 - COUNT(I)
ENDDO
```

This compact algorithm performs the required task and is fully vectorizable. The indexing array and the indirect addressing capability offered by the information contained in the FIRST and LAST pointer arrays (above) act de facto as an interface that makes the particles appear as if they actually were contiguous:

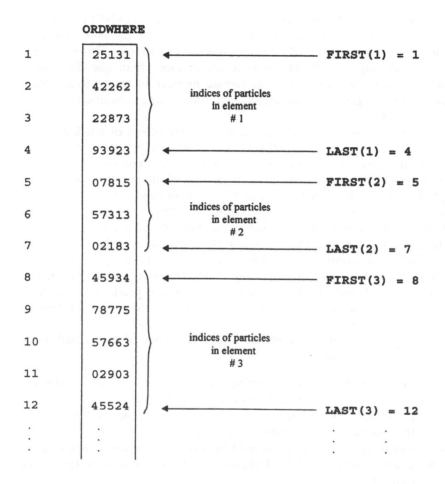

The crucial point of this procedure is that it relies on efficient sorting for restoring contiguity. As we will see in Section IV.B.1, particle index sorting using the binary radix sort has optimal $O(N_{pr} \log N_{pr})$ complexity. Besides, indirect addressing is fully supported (and in many cases hardware-executed) by all vector computers. Thus, the computational overhead associated with reordering the particles at each time step is perfectly acceptable.

C. Point-Inclusion Algorithm

The second key operation lying at the heart of a CONNFFESSIT calculation consists in determining if a particle, the spatial coordinates of which are known, is located within or outside an element. This seemingly minor task acquires immense importance in more than one dimension, since it has to be executed an inordinate number of times: it is almost always the single most-frequently performed basic operation invoked in CONNFFESSIT.

Generalized point-inclusion algorithms are the object of much research in computational geometry (Mulmuley 1994, Preparata and Shamos 1990); the goal of research in this field is the creation of optimal algorithms with respect to the number of sides of the polygon, i.e., how to optimally determine whether or not a point is included in a polygon of a very large number of sides. An extensive search of available literature showed that, somewhat surprisingly, the question of how to determine whether a point is included in a triangle or quadrilateral in the fastest possible way is amply ignored. The only algorithm found in the literature turned out to be a geometrically elegant one frequently used in computer graphics and based on the evaluation of a vector product for each of the sides but which can certainly not compete in speed with the following procedure (Fig. 3).

During the initialization phase, before the start of the actual time-marching computation:

- For each element k, determine the coordinates of some arbitrary interior point, for example, its centroid (X_c^k, Y_c^k).
- For each side of the element, determine and store the coefficients a_i^{kj} of the equations representing the sides $f_j^k(a_i^{kj}, x, y)$, where $j = 1, N_{sides}$, which are in most cases straight lines or polynomials of low degree.
- Evaluate $\mathrm{sign}(f_j^k(a_i^{kj}, X_c^j, X_c^j))$ and store it for each of the sides of each element.

For subsequent use, i.e., in order to determine if the lth particle at (x_l, y_l) is located within the kth element:

- Evaluate $\mathrm{sign}(f_j^k(a_i^{kj}, x_l, y_l))$; if and only if $\mathrm{sign}(f_j^k(a_i^{kj}, x_l, y_l))$ and $\mathrm{sign}(f_j^k(a_i^{kj}, X_c^j, X_c^j))$ are equal for all j, is the particle within the element.

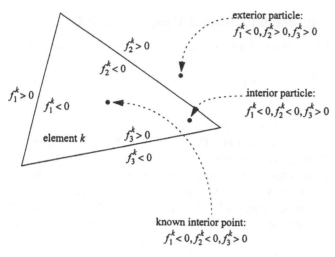

FIG. 3 Particle-inclusion algorithm in two dimensions. The same scheme is applicable to cells with higher number of edges or in higher dimensions.

This sequence of steps corresponds to determining whether the particle in question lies on the same side (left or right) of the sides of the element as a known interior point (like the centroid).

Once the coefficients of the equations defining the sides are known (the computation of which amounts to a very minor overhead during initialization), point-inclusion can be determined with at most N_{sides} function evaluations and the same number of logical comparisons. In particular, an inclusion test for a particle in a triangular element involves at most* three multiplications, three additions (or arithmetic comparisons), and three Boolean comparisons.

On some vector computers like the NEC SX-3 and SX-4 that execute several overlapping vector instructions in parallel, these operations are performed concurrently by multiple functional arithmetic units within a vectorized loop over element number. The algorithm just presented achieves on such hardware for the point-inclusion test on triangular elements the ideal maximum speed of one point-inclusion test per clock cycle.

While this algorithm can conceivably fail in the case of an isoparametric element for which one or more of its sides has extreme curvature, in practice, the probability of such an occurrence is vanishing.

---

*Since the particle fails the inclusion test as soon as a difference in sign is detected.

D. Scalar Velocity-Biased Ordered Neighbor Lists

In the micro/macro method discussed in this work, the macroscopic conservation equations are solved by discretization (typically weighted residuals) methods. The first step in such an approach is the generation of a mesh or grid made up of simple geometric objects like triangles or quadrilaterals in 2D, tetrahedra or parallelepipeds in 3D. The topic of mesh generation for FE is itself an area of intensive research (Wheatherill et al. 1994, Knupp and Steinberg 1993, Castillo 1991). Nowadays there are many efficient mesh generators which are able to subdivide or "enmesh" the integration domain. This mesh is represented by the set of coordinates of the nodal points (loosely speaking the vertices, or other important points of the simple geometrical objects) and information on connectivity and on the occurrence of boundaries. This information is then employed in the WR solution of the macroscopic equation of motion. In micro/macro methods the grid must be additionally used as the stage in which the individual particles evolve and through which they move. As a consequence, in contrast to continuum-mechanical methods, in a micro/macro calculation it is necessary to keep track of the position of each particle, both as coordinates and as the index of the element (or the indirect addressing pointer to the element) in which it is located (Section IV).

Particle neighbor lists are a device widely used in MD and MC simulations of complex fluids to speed up the calculation of interactions (Allen and Tildesley 1987). They contain, for each particle in a simulation, a list of those other particles that fall within a cutoff distance from the particle under consideration (possibly under periodic continuation conditions) and which have to be taken into account in the computation of interactions. Since the number of neighbors depends exclusively on the value of the cutoff and not on the total number of particles, the complexity of the calculation of pairwise interactions is reduced from $O(N_{pr}^2)$ to $O(N_{pr})$.

The meaning of a neighbor list is different in the present context: a cell neighbor list contains, for each element or cell in the mesh, the indices of the elements that share at least one vertex (first neighbors) with the cell under consideration. Second neighbors of a given cell share at least one vertex with its first neighbors. Since the typical particle displacement during a time step is much smaller than the size of a cell, particles that have left a cell after a time step are to be found in most cases in a first neighbor cell or, at most, a second neighbor. It is not necessary to check all elements in the mesh. Just as in the case of the neighbor lists used in MD, this data structure allows a reduction of the complexity of the localization of particles from $O(N_{pr})O(N_{el})$ to $O(N_{el})$.

In micro/macro methods, the grid generator is more or less a black box; normally the generator comes from another application and in principle no information is available about its inner workings, nor is it possible to control in any way the numbering scheme and the resulting connectivity of the elements. Any algorithm to be developed for particle tracking must therefore be valid and efficient for arbitrarily generated and numbered grids. For the sake of discussion we will assume a structured two-dimensional grid made up of convex quadrilaterals; none of these assumptions limits the applicability of the algorithms to be developed in this section and in Section IV. The quadrilaterals are of different sizes and shapes in order to adapt to the geometry and resolve the details of the domain and the velocity and stress fields. The grid is defined as two lists of X and Y coordinates specifying the positions of the vertex points, and cells or elements in the grid are specified by an integer array of the form VERTEX(NUMEL,4), giving the numbers or indices of the four vertices of each cell. Actual cell and vertex numbers are more or less random, in the sense of being unpredictable. The numbering is determined by the internal logic of the grid generator, with cell and vertex numbers assigned as needed, and no guarantee that cell i has cells $i + 1$ and $i - 1$ as two of its nearest neighbors.

Since the characteristic size (length) of a cell or element is very much larger than the displacement of a typical particle in a single time step, the particles that leave an element at a given time step are almost always to be found either in nearest or at most in second-nearest neighbors. The task of locating all particles after every time step is thus greatly simplified if neighbor lists are generated. These lists of first and second neighbors can be generated in a very compact way using the information stored in the array VERTEX(NUMEL, 4): two cells are nearest neighbors if and only if they have one or more vertices in common. The most straightforward algorithm for a static mesh* compares for each cell, each of the four vertices with each of the four vertices for all the other cells. The following code implements this idea in a compact way: it determines the nearest neighbors for all NUMEL cells, the number of neighbors of each cell NOFNEI1 (which in general will vary depending on whether the cell is at a boundary or not, or for internal cells in unstructured meshes), and creates a list of these neighbors NEI1 in a single pass. The following implementation is such that most vector compilers, including the finicky *f77sx* on the NEC-SX series, can handle the innermost four IF's and two DO loops efficiently by unrolling, thus yielding a count for the vectorized DO loop equal to the number of cells in

---

*A *static mesh* remains the same throughout the calculation, both the coordinates of the nodal points and the numbering of the cells.

the mesh, typically an order of magnitude larger than the hardware vector length:

```
DO I = 1, NUMEL
  J = 1
  DO ICORNER = 1, 4
    DO II = 1, NUMEL
      IF (II .NE. I) THEN
        DO IICORNER = 1, 4
          IF (VERTEX(II, IICORNER) .EQ. VERTEX(I, ICORNER)) THEN
            REPEAT = .FALSE.
            DO ICHECK = 1, J
              IF (II .EQ. NEI1(ICHECK, I)) THEN
                REPEAT = .TRUE.
              ENDIF
            ENDDO
            IF (.NOT. REPEAT) THEN
              NOFNEI1(I) = NOFNEI1(I) + 1
              NEI1(NOFNEI1(I), I) = II
            ENDIF
            J = MAX(NOFNEI1(I), 1)
          ENDIF
        ENDDO
      ENDIF
    ENDDO
  ENDDO
ENDDO
```

Although FORTRAN 77 does not contain recursive constructs, it is possible to emulate them via a call-stack (Kießling and Lowes 1985). Recursive application of the above procedure would then allow us to find second neighbors without additional coding. However, it is worth spending some additional effort in developing a slightly more complex procedure to determine second neighbors NEI2 and their number NOFNEI2 in a nonrecursive way:

```
DO I = 1, NUMEL
  J = 1
  DO INEI = 1, NOFNEI1(I)
    ITEST = NEI1(INEI, I)
    DO ICORNER = 1, 4
      DO II = 1, NUMEL
        IF ((II .NE. ITEST) .AND. (II .NE. I)) THEN
          REPEAT2 = .FALSE.
          DO ICHECK2 = 1, NOFNEI1(I)
            IF (II .EQ. NEI1(ICHECK2, I)) THEN
```

```
              REPEAT2 = .TRUE.
            ENDIF
          ENDDO
          IF (.NOT. REPEAT2) THEN
            DO IICORNER = 1, 4
            IF (VERTEX(II,IICORNER).EQ.
              VERTEX(ITEST,ICORNER)) THEN
                REPEAT = .FALSE.
                DO ICHECK = 1, J
                  IF (II .EQ. NEI2(ICHECK, I)) THEN
                    REPEAT = .TRUE.
                  ENDIF
                ENDDO
                IF (.NOT. REPEAT) THEN
                  NOFNEI2(I) = NOFNEI2(I) + 1
                  NEI2(NOFNEI2(I), I) = II
                ENDIF
                J = MAX(NOFNEI2(I), 1)
              ENDIF
            ENDDO
          ENDIF
        ENDIF
      ENDDO
    ENDDO
  ENDDO
ENDDO
```

The slightly greater complication of this procedure stems from the need to avoid double counting the neighbors of the first neighbors. The reward for the more complex code is that again, in spite of the deep nesting, the DO over the number of cells is well vectorizable and by far outperforms the recursive construct.

The lists of first neighbors (NEI1) and second neighbors (NEI2) generated in this way can be used *as is* for the task of relocating the particles that have flowed out of a given cell. The search needs only be performed in the cells contained in the neighbor lists.

However, if the velocity field is known, the search can be further sped up by ordering the neighbors so that those that are downstream of the given cell are searched first. This is the idea behind the velocity-biased neighbor list. The natural way to implement it is to sort the list of neighbors in order of increasing absolute value of the angle formed between the velocity vector at some characteristic point of the central cell (e.g., the centroid) and the vector joining the centroid of the central cell with the centroid of a neighboring cell (angle θ in Fig. 4). Thus, for deterministic particle

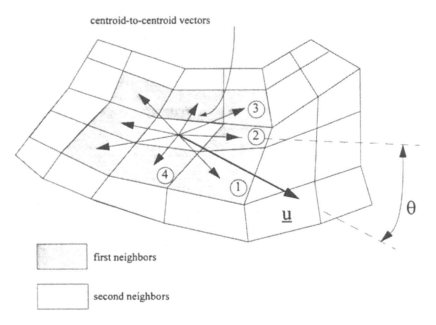

FIG. 4 Velocity-biased neighbor lists. Numbers in circles indicate the searching order in a velocity-biased neighbor list.

trajectories, all particles that have left a cell will necessarily be located in downstream cells; during the search only a small fraction of the neighbor list (seldom more than the first three elements) needs to be scanned. The scheme just presented is equally valid for stochastic particle trajectories, although due to their inherent variability it is no longer guaranteed that particles leaving a cell will be located in cells downstream. However, given the smallness of the typical displacement per step with respect to the cell size, the velocity-biased neighbor list should be just as efficient as in the deterministic case.

As a final remark it should be said that although the procedures for first and second neighbor lists just presented fully exploit vectorizability, they are still $O(N_{el}^2)$ algorithms. Their use is justified for fixed meshes because they are straightforward, well vectorizable, and have to be invoked only once. For moving meshes, however, the neighbor lists may have to be generated at every integration step. In this case, $O(N_{el}^2)$ is unacceptable. As a matter of fact, the experience gathered so far shows that, unless the neighbor list problem is dealt with efficiently, neighbor list generation and the subsequent particle location will become the limiting steps in large-scale calculations (Section IV.B).

IV. MOVING PARTICLES AND REMESHING

While in a one-dimensional geometry the limit to the computationally feasible is set by the number and complexity of the molecular models or stochastic processes to be simulated, the situation is rather different in two or three dimensions. In these cases, the particle or computational model, be it deterministic or stochastic, will move throughout the computational domain following the macroscopic velocity field. Furthermore, particles may enter or leave the domain, or both. The immediate consequence of the macroscopic flow is the need to track the particles' positions at all times concurrently with the integration of the equations of change and of the equations controlling the evolution of the microscopic degrees of freedom. The need to perform particle tracking entails two attendant computational tasks, given in Sections IV.A and IV.B.

A. Integration of Particle Trajectories

The first part of particle tracking consists in the integration of the trajectories particles as they are carried by the macroscopic flow of the fluid. Since the velocity field is known at all integration times, this is a time-consuming but in principle straightforward problem. Although particle trajectories will in most cases be deterministic, for some generalized classes of constitutive equations (Öttinger 1992) intrinsically stochastic trajectories need to be considered.

The integration of deterministic particle trajectories is not problematic since it is possible to draw from the vast body of known algorithms for ODEs. However, whereas exact deterministic trajectories must always remain within the integration domain (except at open boundaries), the discretized versions obtained by any numerical scheme will be subjected to a finite discretization error. This error has two main consequences with far-reaching computational effects in micro/macro methods.

The first effect is a long-time, progressive cumulation of discretization errors (and possibly truncation errors if compressed-variable representation is used). This cumulative effect is quite harmless for trajectories along which the residence time t_R of particles is low, in the sense of $|(t_R/\Delta t)\Delta \mathbf{u}| \ll |\mathbf{u}|$. However, in flow fields where regions of intense recirculation exist, the number of integration steps along a given trajectory $(t_R/\Delta t)$ is large enough for the condition above not to be fulfilled. This situation is reflected for example in the failure of what should be closed-loop trajectories (at steady state) to close exactly, thus leading to inward- or outward-spiraling trajectories. This failure is not specific to micro/macro methods but is common to other tracking techniques, such as those used in viscoelastic flow

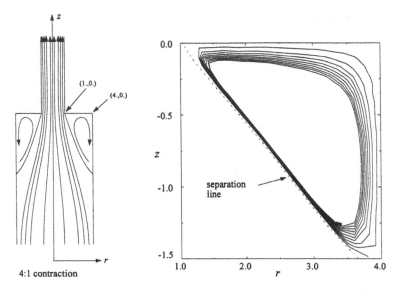

FIG. 5 Cumulation of integration errors during particle tracking resulting in nonclosing recirculation trajectories.

calculations for integral constitutive equations. Figure 5 shows the trajectory of a particle within the recirculation region in the 4:1 contraction geometry obtained by tracking it in a constant steady-state velocity field for an Oldroyd-B model.

In spite of the use of a special type of macroelement (the MDF Finite Element of Nagtigall et al. 1979, Mercier 1979, Bernstein et al. 1985) which allows analytical tracking, the result of the cumulation of round-off errors in an otherwise exact technique is striking. Given the low velocities and shear rates in this particular instance, the consequences of tracking errors are very minor.

However, for flows with strong recirculation, the erratic behavior of the discretized trajectories leads to a rapid depletion or enrichment of particles in certain regions. For sufficiently strong recirculation rates and sufficiently long integration times, the spurious migration of particles with the wrong flow history into the wrong elements and the spurious depletion/enrichment effect render calculations meaningless. The only remedy for this ailment is the use of smaller time steps or higher-order numerical schemes for ODEs, or both.

The use of higher-order schemes is often the preferred solution, in spite of requiring several function evaluations per time step for each of the particle trajectories being integrated. This additional effort may pay back since:

- Trajectory integration does not constitute a large fraction of the overall computational load.
- Relatively inexpensive velocity-field updates can be performed independently of and between two consecutive particle configuration updates with higher frequency than the latter if so required.

The optimal order of the tracking algorithm is thus determined by the ratio of particle tracking to particle simulation.

Figure 6 shows a comparison of a trajectory integration similar to that depicted in Fig. 5, but using first-order Euler (Fig. 6a), second- order, fixed-step trapezoidal rule (Fig. 6b), and 12th-order predictor-corrector Adams–Bashforth with self-adaptive step, (Fig. 6c) schemes for a fixed macroscopic velocity field. While the increase in order does result in increased stability, even the phenomenally accurate 12th-order scheme fails to yield closed-loop trajectories. For simple dynamic models including a few degrees of freedom (Oldroyd-B, Curtiss–Bird) on average-size meshes (a few thousand cells) with simple elements (P^1-C^0) and for the current hardware generation, the best ratio of tracking accuracy to computational time is obtained for second-order, fixed-step explicit schemes. Higher-order schemes are advantageous only for very smooth velocity fields, where a very large integration step can be employed.

The noisy character of micro/macro methods makes it very unlikely that higher-order schemes will significantly increase overall performance. The same applies to richer FEs for which velocity updates are increasingly more expensive.

The second effect of discretization and truncation is not related to the long-time cumulation of errors, but makes itself observable at every time step, as particles close to impenetrable boundaries (which can only be approached asymptotically by trajectories) actually cross them due to a finite time integration step. In general, the existence of a nonzero outward-pointing component of the velocity of a particle leads to its leaving the integration domain if the particle is sufficiently close to the boundary.

This effect is in general of an order of magnitude commensurate with the time step: it acts on a fraction of all particles given by the ratio of the perimeter ∂ (or bounding surface in 3D) of the domain times the average outward-pointing velocity component times the integration step to the total area (or volume in 3D) of the domain Ω. This is a small ratio for any nonpathological domain shape. In most cases, the fraction of particles that leave the integration domain at any given time step due to the discretization error produced by a first-order explicit Euler scheme with

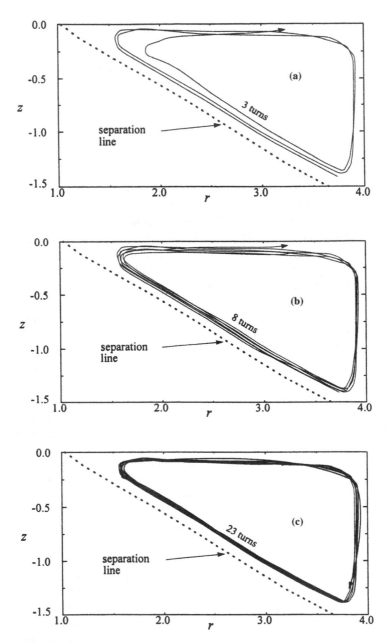

FIG. 6 Effect of order of tracking algorithm on closedness of trajectories: (a) first-order Euler, (b) trapezoidal rule, (c) 12th-order adaptive Adams–Bashforth PC.

reasonable Δt lies between 10^{-4} and 5×10^{-3} and is therefore of no consequence for the numerical accuracy of the solution. Particles that are not in the region close to the boundary do not exit the domain. Their trajectories will drift, however, towards or away from the boundary at a rate dictated both by the local outward- or inward-pointing velocity component and by the discretization error.

Figure 7 shows this effect for three particles close to the reentrant corner in the contraction geometry for P^1-C^0 quadrilaterals. The closer the initial position of the particle to the separation line, the more difficult it is to turn around the corner. Low-order algorithms lead to particle trajectories that intersect impenetrable boundaries.

Unfortunately, even if only a small fraction of all particles leave the domain at each time step, these events cannot be ignored. Even though they have a negligible impact on the numerical values of the stresses computed as element-wise averages, if a fraction of $O(10^{-4}$–$10^{-3})$ of the total number of particles leave the domain at each time step due to discretization and particles drift toward the boundary at a comparable rate, it does not take more than a few thousand steps for a large fraction of particles to disappear from the computational domain.

The most straightforward solution to this question consists in reflecting back into the domain all lost trajectories, i.e., those that have left it at a given time step. This reflection is acceptable and does not result in a reduction of the order of the algorithm. However, the crucial aspect of the problem is that it utterly destroys the vectorizable character we strive for, since such particle-exit events occur in a geometrically localized and structured region (the boundary region), but purely at random when considered from the viewpoint of the particle numbering scheme. Hence the need to develop a scheme to recover and replace such lost particles.

An efficient method to handle this difficulty and retain vectorizability must: i) make use of the numbering schemes, neighbor lists, and point-inclusion techniques explained in Section III, in order to efficiently determine which particles have left the domain and where they are located and ii) obtain this information within and extract it from a loop for subsequent processing without jeopardizing vectorizability. An optimal two-step algorithm works in the following way:

- Determine the indices of all particles that have left the element in which they were located at the previous time step. This is performed in the double-loop construct already explained in Section III.D using indirect addressing to point at all particles (inner loop in the following FORTRAN excerpt) within a given element (outer loop). If a particle has left the element it was in, add its address to the list of particles

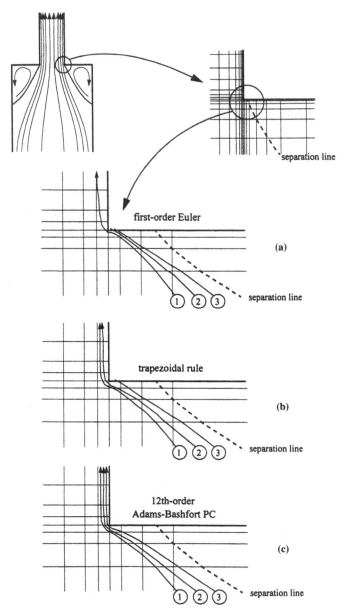

FIG. 7 Effect of order of tracking algorithm on particle trajectories close to impenetrable boundaries: (a) first-order Euler, (b) trapezoidal rule, (c) 12th-order Adams–Bashforth PC.

that left their elements (LEFT1). A fraction of these will be in first or second neighbors; the rest will have left the domain altogether.

```
COUNTLEFT1 = 0
  DO I = 1, NUMEL
  DO II = FIRST(I), LAST(I)
     IHOLD = ORDWHERE(II)
     IF (particle IHOLD is in element I) THEN
       WHERE(IHOLD) = I
     ELSE
       COUNTLEFT1 = COUNTLEFT1 + 1
       LEFT1(COUNTLEFT1) = IHOLD
     ENDIF
  ENDDO
ENDDO
```

The second double loop finds which particles are in first neighbors and spots those particles that have left the domain:

```
LOST = 0
DO I = 1, COUNTLEFT1
  IHOLD = LEFT1(I)
  DO II = 1, NOFNEI1(WHERE(IHOLD))
    III = NEI1(II, WHERE(IHOLD))
    IF (particle IHOLD is in first neighbor element III)
       THEN
      WHERE(IHOLD) = III
      GOTO 10
    ENDIF
  ENDDO
  LOST            = LOST + 1
  NOTFOUND(LOST) = IHOLD
  10 CONTINUE
ENDDO
```

This code is vectorizable, determines where all particles are located, and, with a minimum overhead, spots those particles that have left the domain. The array NOTFOUND contains the list of indices for all particles that have left the domain.

- In the second step a simple pass is performed over all particles that have left the domain in order to reflect their positions back into the domain.

```
DO I = 1, LOST
  IHOLD = NOTFOUND(I)
```

```
ELEMENT = WHERE(IHOLD)
  reflect position of particle back into domain
ENDDO
```

This would seem to complete the procedure, since all particles must have either been located in nearest neighbors or they must have been reflected back into the domain. We are, however, not done yet: any reflection procedure has a loophole which is a consequence of the finite integration time step: if a particle leaves the domain through an element boundary that is close to an external nodal point (see Figs. 8 and 9), there will always exist the possibility that the position after reflection is still outside of the domain or within the domain but not in the element the particle was originally in.

These two possibilities make it necessary to perform a third and last pass (not explicitly presented here because of its trivial vectorizability) in order to check that all reflected particle positions are indeed within the domain and within the intended element.

The root of both of these failures is that, within the discretization error, it is not possible to decide whether the particle moved from $\mathbf{r}_i(t_j)$ to $\mathbf{r}_i(t_{j+1})$ outside the domain directly from element n or through element $n+1$. The assignment of a new position to the lost particle is therefore arbitrary, but since the number of such failures has always a higher error order, it is irrelevant where the position of such particles should be reinitialized, apart from obviously poor choices like $\mathbf{r}_i(t_j) = \mathbf{r}_i(t_{j+1})$.

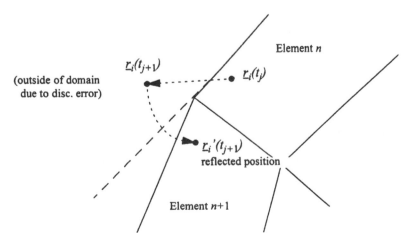

FIG. 8   Failed particle reflection, first case.

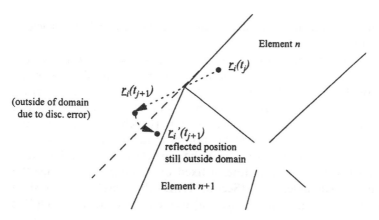

FIG. 9   Failed particle reflection, second case.

We thus see that solving the simple problem of deterministic trajectory integration requires some degree of sophistication if an efficient algorithm is to result. As will often be the case, it is the exceptional (in this instance, particles exiting a closed domain) with little physical significance that usually makes it necessary to invest a disproportionate algorithmic effort and to resort to rather convoluted techniques in order to reach vectorizability. At any rate, the algorithmic complexity of trajectory integration can be kept as $O(N_{pr})$, the size of the mesh being irrelevant for integration of particle trajectories.

The integration of stochastic particle trajectories can be performed using the same techniques for SDEs used to advance the internal configurations of stochastic models (Kloeden and Platten 1992). Given the intrinsic variability of stochastic particle trajectories, the problems associated with particles leaving the domain are more acute in this case, although they can be handled using the same techniques. However, the imposition of reflecting boundary conditions on stochastic trajectories is in general a nontrivial matter (Öttinger 1995).

B.   Particle Localization in the Mesh

On the one hand, the integration of particle trajectories is in principle, and with the use of proper algorithms as well in practice, an $O(N_{pr})$ algorithm, since the information required to integrate a given trajectory is purely local and does not depend on the size of the mesh. Even the treatment of the anomalous cases considered in the last part of the previous section involves no more than the immediate vicinity of the element in which a particle is located at a given time.

On the other hand, the second part of particle tracking consists in the localization of the particle in the FE mesh (i.e., given the coordinates of the particle and of the vertices defining all cells, find in which cell the particle is located). After a time step the particle may still be within the same cell or element where it was at the previous step (which will be the most frequent case) or it may have moved to another element. In both cases it is necessary to determine in which element the particle is located. This task presents no great difficulties for either a fixed or a moving mesh once the neighbor lists are available, since (Section III.D) the search is reduced to the immediate environment of the original element. It is again an $O(N_{pr})$ procedure.

Particle localization in its simplest form calls for checking whether a given particle lies within a given element, using some suitable point-location criterion. This brute-force approach is $O(N_{pr})O(N_{el})$ and therefore unsuitable for intensive use in a time-marching scheme. As soon as particles have been located once (for example when initializing the ensemble), the use of a neighbor list is vastly more efficient. The use of the brute-force approach is justified, at most, during the initialization phase. Even then, an element-by-element fill-up scheme easily bypasses the problem.

Since for fixed meshes, there is no need to update the neighbor list at every time step, the overall effort of advancing trajectories and relocating particles remains $O(N_{pr})$ throughout the time integration. For moving grids, however, the domain has to be reenmeshed after a certain number of time steps [typically $O(10^0-10^2)$], or even after each step (Zegeling 1993, Trease et al. 1991), a task not to be overlooked: gridding a domain is isomorphic to Voronoi-tesselating it, for which the best algorithms are $O(N_{el}^{5/4})$ (Mulmuley 1994).

The reenmeshment makes it necessary to reconstruct the neighbor lists and to relocate all particles in the new mesh. In this case, unlike during a time step in a static mesh, a large number of $[O(N_{pr})$ and possibly all] particles will be located in a cell with a different index, even though the cell itself may be geometrically quite similar and close or even identical to the one before the reenmeshment. Reenmeshment thus has a devastating effect on the efficiency of a micro/macro method: it requires new neighbor lists and particle relocation from scratch. Furthermore, a new neighbor list is *a priori* not particularly useful.

If the neighbor list is reconstructed using the neighbor list algorithms presented in Section III.D and brute-force particle location is used, the combined effort is of order $O(N_{el}^2) + O(N_{pr})O(N_{el})$ (the first term corresponding to neighbor list generation, the second to particle location, which cannot be alleviated by an element-by-element fill-up scheme as during initialization). As a result, the generation of the neighbor lists and relocation of particles may become the bottleneck of the whole micro/macro simulation. Both of these obstacles can be overcome.

1. An Optimal Neighbor List Generator

The algorithms presented in Section III.D for neighbor list generation are $O(N_{el}^2)$. The following scheme reduces the order of complexity to an optimal $O(N_{el} \log N_{el})$. The key to the successful reduction of complexity is the transformation of the double-loop construct in Section III.D into a series of sorting operations. As in Section III.D, we assume that cells or elements in the grid are specified by an integer array of the form VERTEX(NUMEL,4), giving the numbers or indices of the four vertices of each cell.

First the array VERTEX(NUMEL,4) is inverted in order to obtain an array CELL(NVERTS,4) giving the cell numbers associated with each vertex. Further, NBITS is the number of bits needed to hold the maximum vertex number, which is roughly equal to $\log_2$(NVERTS). For any nonpathological two-dimensional mesh, NVERTS will be only slightly larger than NUMEL, the difference between both stemming from vertices belonging to elements at the domain boundary. The array inversion requires an intermediate storage and sorting array VC(NUMEL*4). We next generate all vertex-cell pairings. For this task, the FORTRAN 77 intrinsic routines lshift, ior, ibits come in handy* (they are not essential for the present algorithm, but make its implementation simpler and usually more efficient):

```
INDEX = 0
DO J = 1, 4
  DO I = 1, NUMEL
    INDEX = INDEX + 1
    VC(INDEX) = IOR(LSHIFT(VERTEX(I,J), NBITS), I)
  ENDDO
ENDDO
```

---

*Although not ANSI-FORTRAN, the FORTRAN bitwise Boolean functions are intrinsics from MIL-STD-1753. They are available on most FORTRAN compilers nowadays, including those by IBM, NEC, Silicon Graphics, and Sun.
lshift(M,K) returns the value of the first argument shifted left or right, respectively, the number of times specified by the second (integer) argument. lshift is implemented as a logical shift, i.e., zeros are shifted in.
ior(M,N) performs a bitwise logical OR of M and N.
ibits(M,K,LEN) extracts a subfield of LEN bits from M starting with bit position K and extending left for LEN bits. The resulting field is right justified and the remainig bits are set to zero.

Left shifting the vertex index by NBITS and bitwise ORing it with the cell number, produces an array VC, the elements of which contain:

<- - - - - - - - - - - - - one word - - - - - - - - - - - ->		
0	vertex number	cell number
0	vertex number	cell number
0	vertex number	cell number
...	...	...
0	vertex number	cell number

Sorting VC on the vertex number automatically sorts the corresponding cell numbers. The most efficient general sorting algorithms are $O(N_{el} \log N_{el})$. Having vector computers in mind, a good way of sorting the array VC is to use the following modification of the binary radix sort:

- Starting with the rightmost bit of the word to be sorted, i.e., with each element of VC, determine whether the corresponding bit position in each word is a 0 or a 1. Compress all numbers that have 0 as the rightmost key-bit in a second scratch array of the same length. Following this, compress out all numbers that have 1 as the rightmost key-bit into the remainder of the scratch array. Finally, copy the scratch array back into the original array. Both the scratch array and the original array now contain all the words or numbers in the original array, sorted by their least significant bit in ascending order.
- Proceed in the same fashion with the second least significant bit of each element in the array: all numbers with 0 as the second-rightmost key-bit are gathered into the second array. Following this, compress out all numbers with 1 as the second-rightmost key-bit into the remainder of the scratch array and subsequently copy the scratch array back into the original array.
- The original array now contains all the numbers sorted on their rightmost two key-bits; first will be all numbers with rightmost bits 00, followed by all numbers with rightmost bits 01, followed by all numbers with rightmost bits 10, followed by all numbers with rightmost bits 11. This process continues from right to left until the numbers have been sorted on each of the key-bit positions.

The entire array is now sorted with the keys in ascending order. In the following tables we see the way the binary radix sort operates on a sequence of three-bit numbers (which may represent a three-bit sorting key in a longer word).

011	→	01 1	sort →	00 0	→	000
111		11 1		00 0		000
101		10 1		11 0		110
000		00 0		01 1		011
000		00 0		11 1		111
101		10 1		10 1		101
110		11 0		10 1		101
000	→	0 0 0	sort →	0 0 0	→	000
000		0 0 0		0 0 0		000
110		1 1 0		1 0 1		101
011		0 1 1		1 0 1		101
111		1 1 1		1 1 0		110
101		1 0 1		0 1 1		011
101		1 0 1		1 1 1		111
000	→	0 00	sort →	0 00	→	000
000		0 00		0 00		000
101		1 01		0 11		011
101		1 01		1 01		101
110		1 10		1 01		101
011		0 11		1 10		110
111		1 11		1 11		111

Although in this example the binary radix sort operates on all bits of a word it is, however, perfectly possible to sort words or numbers based on only a few or just one of its bits by applying the bit sort to these bit positions only.

The following simple implementation of the binary radix sort vectorizes well on all vector architectures and extraordinarily well on processors where the compress operations are done by hardware vector instructions (such as the NEC SX-3 and the deceased ETA family). A(N) is the array to be sorted, B(N) is the scratch array, and RIGHT and LEFT are the positions of the bit segment (key) on which sorting should take place (for example, on a 64-bit-word machine, they could take any values that fulfill $1 \leq \text{LEFT} \leq \text{RIGHT} \leq 64$). BTEST is a bit field manipulation intrinsic from MIL-STD-1753* that tests for the state of a given bit.

```
DO I = RIGHT, LEFT
  K = 0
  DO J = 1, N
    IF (BTEST(A(J), I)) THEN
    ELSE
      K = K + 1
      B(K) = A(J)
      ENDIF
    ENDDO
    DO J = 1, N
      IF (BTEST(A(J), I)) THEN
        K = K + 1
        B(K) = A(J)
      ENDIF
    ENDDO
    DO J = 1, N
    A(J) = B(J)
  ENDDO
ENDDO
```

A machine-specific version would avoid the copying from the scratch array to the original array for each bit position and might require only a single BTEST function for each word at the expense of more complicated indexing.*,†

The time to sort a list is therefore proportional to the number of entries (four entries per cell, given the four vertices per cell) and the width of the key in bits. If the vertices are numbered from 1 to NVERTS, where NVERTS is not much larger than the number of cells, the number of bits of the key is proportional to the logarithm of the number of cells. Thus the binary radix sort operates in this case on a permutation of the integers from 1 to NUMEL

---

*btest(N,K) tests the Kth bit of argument N. The value of the function is .TRUE. if the bit is 1 and .FALSE. if the bit is 0.

†Even so, while not the most efficient available, this routine runs at a good fraction of the maximum possible speed for this algorithm on the Cray Y-MP and the NEC SX-3.

with each of them appearing up to four times (for quadrilateral cells) and becomes an $O(N_{el} \log N_{el})$ process.

Most vertices are connected to four cells and have four entries in the array. Those on the edges of the mesh, however, will have fewer. Vertices in the corners of the grid (if any) may have only one. It is now necessary to determine where in the sorted array VC each vertex begins and ends. The following code finds each place where the vertex number changes and enters this information into two index arrays: ILASTI, giving the last index into the array VC for each vertex, and IGATHR, which initially is set to the first index minus one for each vertex. IGATHR is used to extract successive cells connected to each vertex:

```
ENTRY = 0
DO I = 1, INDEX - 1
  IF (IBITS(VC(I),NBITS,NBITS) .NE. IBITS(VC(I+1),NBITS,
    NBITS)) THEN
    ENTRY = ENTRY + 1
    ILASTI(ENTRY) = I
  ENDIF
ENDDO
ENTRY = ENTRY + 1
ILASTI(ENTRY) = INDEX
IGATHR(1) = 0
DO I = 2, ENTRY
  IGATHR(I) = ILASTI(ENTRY-1)
ENDDO
```

The next step consists in building up a four-column table: for each separate vertex enter one of the connected cells in a separate column. If a vertex is shared by fewer than four cells, repeat the last cell number for that vertex as necessary. In a last pass, any such duplicated cell numbers are set to zero:

```
DO I = 1, 4
  DO J = 1, ENTRY
    IF (IGATHR(J) .LT. ILASTI(J)) SIGATHR(J) = IGATHR(J)+1
    CTABLE(J) = IBITS(VC(IGATHR(J), 0, NBITS))
  ENDDO
ENDDO
DO I = 4, 2 -1
  DO J = 1, ENTRY
    IF (CTABLE(J, I) .EQ. CTABLE(J, I-1)) THEN CTABLE(J, I) = 0
  ENDDO
ENDDO
```

For each cell number, the only nearest neighbor cells are those that occur in one or more rows along with it in the array CTABLE. The following code generates a table `CELNBR` that contains all cell-neighbor pairs and only cell-neighbor pairs:

```
PAIRS = 0
DO I = 1, 4
  DO J = 1, 4
    IF (I .NE. J) THEN
      DO K = 1, ENTRY
        IF ((CTABLE(K,I) .NE. 0) .AND. (CTABLE(K,J) .NE. 0)) THEN
          PAIRS = PAIRS + 1
          CELNBR(PAIRS) =
          IOR (LSHIFT(CTABLE(K,I),NBITS),CTABLE(K,J))
        ENDIF
      ENDDO
    ENDIF
  ENDDO
ENDDO
```

The resulting array `CELNBR` contains:

<- - - - - - - - - - - - - one word - - - - - - - - - - - - - ->		
0	cell number	neighbor
0	cell number	neighbor
0	cell number	neighbor
...	...	neighbor
0	cell number	neighbor

There will now be some duplicates since some cell pairs (those where two cells share more than one vertex) will appear in `CELNBR` twice. Additionally, since the cell numbers will be in no particularly useful order, sort the array `CELNBR` using the cell number in each entry as the key (in the middle of the word).

The sorted `CELNBR` contains at most twelve neighbor entries for each cell.* A neighbor that shares only a corner appears once, a neighbor that shares two corners (an edge) appears twice. Now eliminate the duplicates by sorting on the entire cell-neighbor pair rather than on the cell number only. But if we take into account that duplicate entries can be no more than eleven words apart (in the event they are first and last in the neighbor list), we can

---

*In the case of a structured mesh, where the number of neighbors is fixed.

replace this sorting by the more elegant device of comparing each entry in the array, as sorted on cell number only, with the previous eleven entries:

```
DO I = 1, 11
   DO J = I+1, PAIRS
      IF (CELNBR(J) .EQ. CELNBR(J-I)) CELNBR(J) = 0
   ENDDO
ENDDO
UNIQUE = 0
DO I = 1, PAIRS
   IF (CELNBR(I) .NE. 0) THEN
      UNIQUE = UNIQUE + 1
      CELNBR(UNIQUE) = CELNBR(I)
   ENDIF
ENDDO
```

The last step consists in finding the first and last entries for each cell number, which can be done in exactly the same fashion as for the array VC sorted on vertex number and will therefore not be repeated here. After that, reformatting the table of nearest neighbors into whatever format we wish is a trivial matter.

We thus see that while a frontal attack results in the elegant and straightforward but $O(N_{el}^2)$ algorithm of Section III.D, recasting the problem as a quite convoluted sequence of sorting and indirect addressing yields an $O(N_{el} \log N_{el})$ process and therefore optimal.* The break-even point between the two algorithms will obviously depend on the particular mesh being used, in particular on the ratio of boundary cells to internal cells. The figures in Table 1 apply to the two-dimensional eccentric cylinder meshes.

TABLE 1 Peformance Comparison for Increasingly Refined Meshes

Mesh	Number of vertices	Number of cells	CPU-time ratio (Cray Y-MP): algorithm section III.D/ algorithm section IV.B.1	CPU-time ratio for scalar processor
M0	55	50	0.81	0.75
M1	220	200	2.5	2.2
M3	840	800	8.3	8.0
M4	1680	1600	25	23
M5	16800	16000	130	110

---

*The algorithm just described is a constructive proof of the equivalence of building a neighbor list and a search.

We thus see that the break-even point lies at about a few tens of cells, i.e., except for the coarsest grids, which in practice never occur, the algorithm just presented is significantly more efficient. For realistic 2D computations and, above all in three dimensions, where grid sizes of $O(10^4-10^5)$ cells are standard, the algorithm of Section IV.B.1 is at least two orders of magnitude faster and the only viable alternative. Finally, it is worth mentioning that the growing CPU-time ratios apply to absolute CPU-times that in turn grow with problem size.

Interestingly, the effort invested in developing vectorizable algorithms pays off on scalar architectures as well. The CPU ratio between the nested-loop algorithm and the algorithm just presented is slightly more favorable for the latter on vector machines since the nested DO loop constructs of Section III.D contain comparatively more or deeper IF's and low-count DO's that must be unrolled.

2. Particle Relocation

As we saw in IV.B, the problem of determining in which element a given particle is located after the mesh has been created, renumbered, or modified is of $O(N_{pr})O(N_{el})$, i.e., quadratic in problem size, instead of $O(N_{pr})$ as is the case when the mesh is static and neighbor lists can be used. The reason is that knowledge of the neighboring cells is purposeless if the location of the particle at the previous time step is unknown, as is the case when the domain is reenmeshed.* A partial remedy for the $O(N_{pr})O(N_{el})$ complexity could be achieved by reenmeshment procedures that keep renumbering to a minimum. Although the amount of literature on reenmeshing procedures is overwhelming, little attention has been given to the question of minimum-renumbering algorithms.

While the quadratic complexity of particle relocation is unavoidable the very first time a domain is initialized,[†] it is possible in practice to bypass this complexity in subsequent steps. The reconstructed neighbor list can be used in order to avoid brute-force particle location no matter how extensive the renumbering is. The two-step idea is straightforward and does not require pseudo-code to illustrate it:

- Once the updated neighbor list is available, a brute-force search is started for inclusion in all cells of a small number of particles (typically 20 to 100) belonging to the original cell. Since the indices

---

*Even if the reenmeshment would only renumber cells, the new cell number would bear no relationship whatsoever with the number before the reenmeshment.
[†]Even in this case, particle relocation can be avoided by systematic cell fill-up, resulting in $O(N_{pr})$.

of these particles are contiguous in the array ORDWHERE (see Section III.D) this search is very fast; furthermore, the effort spent in this tentative search is the same, no matter what the size of the local ensemble is. Given that most of the particles will not have left the original cell (now renumbered beyond recognition), as soon as the particles are tested for inclusion in the original cell (but with the updated index), a large majority of the particles test positive. This is all that is required to identify the new index of the original cell.

- As soon as the new cell index is known, the search is restarted, this time for all the particles that were in the original cell but using the reconstructed neighbor cell, thus reducing the problem to the same one encountered for static meshes (Section III.D), i.e., to $O(N_{pr})$.

The objective has been accomplished: combined use of the optimal neighbor list generator and the relocation technique just described reduces the $O(N_{el}^2) + O(N_{pr})O(N_{el})$ complexity to $O(N_{el} \log N_{el}) + O(N_{pr})$.

The core of the computationally crucial part of CONNFFESSIT calculations has been described in great detail in Sections III and IV. The remaining elements in a CONNFFESSIT calculation are relatively straightforward. Two recent references describe in detail specific applications to the journal bearing problem (Laso et al. 1997) and to the planar contraction flow (Laso 1998) as well as the implementation on parallel and vector machines.

V. CONCLUSIONS AND PERSPECTIVES

The CONNFFESSIT approach is a feasible novel method for the investigation of the fluid mechanics of complex fluids which cannot be dealt with satisfactorily within the continuum-mechanical framework. In order to keep the computational effort involved in the calculation of two-dimensional time-dependent flows in complex geometries under control, efficient algorithmic tools and data structures are required.

Even so, CONNFFESSIT cannot compete in speed with the most recent micro–macro methods but it is still the tool of choice when fluctuations are important, typically in restricted geometries. Its efficient implementation, while not complex, requires that care be exercised. The techniques presented here are certainly not the only ones, but they are reasonably close to algorithmic optimality. Further improvements are certainly possible through the use of increasingly complex shortcuts, although the gain will probably be only marginal.

The recent micro–macro methods, CONNFFESSIT among them, can, in spite of their youth, already be seriously considered a tool complementary to those of classical computational rheology.

APPENDIX A: SYMBOLS

a_i^{kj}	ith coefficient in Cartesian equations for jth side of kth element
f_j^k	Cartesian equation of the jth side of the kth element
$f_{nl}, \ldots$	inflow factors
l	length along boundary between two cells (m)
n	number density of particles (m^{-3})
$\mathbf{n}$	unitary normal vector
N_{el}	number of mesh cells or elements (NUMEL in FORTRAN excerpts)
N_p	number of computational nodes
N_{pr}	number of particles in the global ensemble (NUMPR in FORTRAN excerpts)
N_{sim}	number of particles per cell or element
N_{ver}	number of vertices of the mesh (NVERTS in FORTRAN excerpts)
N_{sides}	number of sides of an element
$\mathbf{Q}$	connector vector (m)
$\mathbf{r}_i(t)$	position vector of particle i (m)
T	temperature (K)
t	time (s)
Δt	integration time step, or average time step (s)
u	velocity field (m/s)
$\Delta \mathbf{u}$	discretization error in macroscopic velocity field per time step (m/s)
Ws	Weissenberg number
X_i^j	x-coordinate of ith vertex of jth element (m)
Y_i^j	y-coordinate of ith vertex of jth element (m)
X_c^j	x-coordinate of centroid of jth element (m)
Y_c^j	y-coordinate of centroid of jth element (m)
x_i	x-component of position vector of particle i (m)
y_i	y-component of position vector of particle i (m)
z_i	z-component of position vector of particle i (m)
δ	unit tensor
τ	extra-stress or deviatoric part of the stress tensor (Pa)
Ω	integration domain; its area (2D) (m^2) or volume (3D) (m^3)

∂ boundary of the integration domain; its perimeter (2D) (m) or area (3D) (m^2)

APPENDIX B: ABBREVIATIONS

CE	Constitutive Equation
CFD	Computational Fluid Dynamics
FE	Finite Elements
MC	Monte Carlo
MD	Molecular Dynamics
NEMD	Non-Equilibrium Molecular Dynamics
ODE	Ordinary Differential Equation
PC	Predictor-Corrector
PDE	Partial Differential Equation
SDE	Stochastic Differential Equation
SPH	Smoothed-Particle Hydrodynamics
WR	Weighted Residuals

REFERENCES

Ait-Kadi, A.; Dealy, J.M.; James, D.F.; Williams, M.C., Eds., Proceedings of the XIIth International Congress on Rheology, Laval University, Quebec City, 1996.

Allen, M.P., Tildesley, D.J. *Computer Simulation of Liquids*; Clarendon Press: Oxford, 1987.

Anderson, C.; Greengard, C. SIAM J. Math. Anal. **1985**, *22*, 413–440.

Anderson, C.; Greengard, C., Eds., *Vortex Methods*; Springer: Berlin 1988.

Ashurst, W.T.; Hoover, W.G. Phys. Rev. A **1975**, *11*, 658–667.

Baaijens, F.P.T. J. Non-Newtonian Fluid Mech. **1998**, *75*, 119–138.

Balasubramanian, K.; Hayot, F.; Saam, W.F. Phys. Rev. A **1987**, *36*, 2248–2252.

Baudet, C.; Hulin, J.P.; Lallemand, P.; d'Humieres, D. Phys. Fluids A **1989**, *1*, 507–518.

Beale, J.T.; Majda, A. Math. Comp. **1982**, *39*, 1–27.

Bell, T.W.; Nylond, G.H.; de Pablo, J.J.; Graham, M.D. Macromolecules **1997**, *30*, 1806–1812.

Bernstein, B.; Malkus, D.S.; Olsen, E.T. Int. J. Num. Meth. Fluids **1985**, *5*, 43–70.

Bird, R.B.; Curtiss, C.F.; Armstrong, R.C.; Hassager, O. *Dynamics of Polymeric Liquids*; John Wiley: New York, 1987b; Vol. II.

Birdsall, C.K.; Langdon, A.B. *Plasma Physics via Computer Simulation*; McGraw-Hill: London, 1985.

Bonvin, J.C.; Picasso, M. J. Non-Newtonian Fluid Mech. **1999**, *84*, 191–215.

Bousfield, D.W.; Keunings, R; Denn, M.M. J. Non-Newtonian Fluid Mech. **1988**, *27*, 205–221.

Brown, R.A.; Armstrong, R.C.; Beris, A.N.; Yeh, P.W. Comp. Meth. App. Mech. Eng. **1986**, *58* (2), 201–226.

Bush, M.B.; Milthorpe, J.F.; Tanner, R.I. J. Non-Newtonian Fluid Mech. **1984**, *16*, 37–51.

Castillo, J.E., Ed., *Mathematical Aspects of Numerical Grid Generation*; Society for Industrial and Applied Mathematics: Philadelphia, 1991.

Chorin, A.J. J. Fluid Mech. **1973**, *57*, 785–796.

Chorin, A.J. *Computational Fluid Mechanics: Selected Papers*; Academic Press: New York, 1989.

Chorin, A.J. *Vorticity and Turbulence*; Springer: Berlin, 1994.

Cloutman, L.D. An evaluation of smoothed particle hydrodynamics. In *Advances in the Free-Lagrange Method*; Trease, H.E., Fritts, M.J., Crowley, W.P., Eds.; Springer: Berlin, 1991.

Crochet, M.J. Rubber Chem. Technol. **1989**, *62* (3), 426–455.

Crochet, M.J.; Davies, A.R.; Walters, K. *Numerical Simulation of Non- Newtonian Flow*;. Elsevier: Amsterdam, 1984.

Debbaut, B.; Crochet, M.J. J. Non-Newtonian Fluid Mech. **1986**, *20*, 173–185.

Doi, M.; Edwards, S.F. *The Theory of Polymer Dynamics*; Clarendon Press: Oxford, 1986.

Feigl, K.; Laso, M.; Öttinger, H.C. Macromolecules **1995**, *28* (9), 3261–3274.

Franca, L.P.; Hughes, T.J.R. Comp. Meth. Appl. Mech. Eng. **1993**, *105*, 285–298.

Frisch, U.; Hasslacher, B.; Pomeau, Y. Phys. Rev. Lett. **1986**, *56*, 1505–1509.

Gallegos, C.; Guerrero, A.; Muñoz, J.; Berjano, M., Eds., Proceedings IVth European Rheology Conference, Sevilla, 1994.

Gard, T.C. *Introduction to Stochastic Differential Equations*; Marcel Dekker: New York, 1988.

Gardiner, C.W. *Handbook of Stochastic Methods for Physics, Chemistry and the Natural Sciences*; Springer: Berlin, 1983.

Geurts, B.J.; Jongschaap, R.J.J. J. Rheol. **1988**, *32*, 353–372.

Gingold, R.A.; Monaghan, J.J. Month. Not. R. Astr. Soc. **1977**, *181*, 375–387.

Halin, P.; Lielens, G.; Keunings, R.; Legat, V. J. Non-Newtonian Fluid Mech. **1998**, *79*, 387–403.

Hannon, L.; Lie, G.C.; Clementi, E. Phys. Lett. **1986a**, *57*, 695–699.

Hardy, J.; de Pazzis, O.; Pomeau, Y. Phys. Rev. A **1976**, *13*, 1949–1954.

Hockney, R.W.; Eastwood, J.W.; *Computer Simulation using Particles*; McGraw-Hill: London, 1981.

Honerkamp, J. *Stochastische Dynamische Systeme*; VCH: Weinheim, 1990.

Hoover, W.G.; De Groot, A.J.; Hoover, C.G. Computers in Phys. **1992**, *6*, 155–167.

Hulsen, M.A.; van Heel, A.P.G.; van den Brule, B.H.A.A. J. Non-Newtonian Fluid Mech. **1997**, *70*, 79–101.

Josse, S.L.; Finlayson, B.A. J. Non-Newtonian Fluid Mech. **1984**, *16*, 13–36.

Kadanoff, L.P.; Swift, J. Phys. Rev. **1968**, *165*, 310–314.

Keunings, R. J. Computat. Phys. **1986a**, *62*, 199–220.

Keunings, R. J. Non-Newtonian Fluid Mech. **1986b**, *20*, 209–226.

Keunings, R. Simulation of viscoelastic fluid flow. In *Fundamentals of Computer Modeling for Polymer Processing*; Tucker, C.L., Ed.; Carl Hanser Verlag: Munich, 1989; 402–470.

Keunings, R.; Bousfield, D.W. J. Non-Newtonian Fluid Mech. **1987**, *22*, 219–233.

Kießling, I.; Lowes, M. *Programmierung mit FORTRAN 77*; B.G. Teubner: Stuttgart, 1985.

Kloeden, P.E.; Platen, E. *Numerical Solution of Stochastic Differential Equations*; Springer: Berlin, 1992.

Knupp, P.; Steinberg, S. *Fundamentals of Grid Generation*; CRC Press: Boca Raton, FL, 1993.

Knuth, D.E. *The Art of Computer Programming*; Addison-Wesley: Reading, MA, 1968.

Laso, M. AIChEJ **1994**, *40* (8), 1297–1311.

Laso, M. Habilitationsschrift, Eidgenössische Technische Hochschule Zürich, 1995.

Laso, M. In *Dynamics of Complex Fluids*; 72 Adams, M.J., Mashelkar, R.A., Pearson, J.R.A., Rennie, A.R., Eds.; Imperial College Press, The Royal Society: London, 1998.

Laso, M.; Öttinger, H.C. J. Non-Newtonian Fl. Mech. **1993**, *47*, 1–20.

Laso, M.; Picasso, M.; Öttinger, H.C. AIChEJ **1997**, *43*, 877–892.

Lucy, L.B. Astrophys. J. **1977**, *82*, 1013–1022.

Malkus, D.S.; Webster, M.F. J. Non-Newtonian Fluid Mech. **1987**, *25*, 221–243.

Marchal, J.M.; Crochet, M.J. J. Non-Newtonian Fluid Mech. **1987**, *26*, 77–114.

Marchioro, C.; Pulvirenti, M. *Vortex Methods in Two-Dimensional Fluid Mechanics*; Springer: Berlin, 1984.

McCammon, J.A.; Northrup, S.H.; Karplus, M.; Levy, R. Biopolymers **1980**, *19*, 2033, 2045.

McClelland, M.A.; Finlayson, B.A. J. Non-Newtonian Fluid Mech. **1988**, *27*, 363–374.

Melchior, M.; Öttinger, H.C. J. Chem. Phys. **1995**, *103*, 9506–9509.

Melchior, M.; Öttinger, H.C. J. Chem. Phys. **1996**, *105*, 3316–3331.

Mendelson, M.A.; Yeh, P.W.; Brown, R.A.; Armstrong, R.C. J. Non-Newtonian Fluid Mech. **1982**, *10*, 31–54.

Mercier, B. Int. J. Num. Meth. Eng. **1979**, *14*, 942–945.

Moldenaers, P., Keunings, R., Eds., Proceedings of the XIth International Congress on Rheology, Brussels, Elsevier: Amsterdam, 1992.

Mulmuley, K. *Computational Geometry: An Introduction Through Randomized Algorithms*; Prentice Hall: Englewood Cliffs, NJ, 1994.

Öttinger, H.C. Rheol. Acta **1992**, *31*, 14–21.

Öttinger, H.C. *Stochastic Processes in Polymeric Fluids*; Springer-Verlag: Berlin, 1995.

Öttinger, H.C.; Laso, M. In *Theoretical and Applied Rheology*, Moldenaers, P., Keunings, R., Eds., Proceedings XIth Int. Congress on Rheology, Brussels, Elsevier: Amsterdam, 1992; 286–288.

Papanastasiou, A.C.; Scriven, L.E.; Macosko, C.W. J. Non-Newtonian Fluid Mech. **1987**, *22*, 271–288.

Preparata, F.P.; Shamos, M.I. *Computational Geometry: An Introduction*; Springer: New York, 1990.

Risken, H. *The Fokker-Planck Equation*; Springer Series in Synergetics: Berlin, 1989.

Rothman, D.H.; Keller, J.M. J. Stat. Phys. **1988**, *52*, 1119–1128.

Srinivasan, R.; Finlayson, B.A. J. Non-Newtonian Fluid Mech. **1988**, *27*, 1–15.

Sugeng, F.; Phan-Thien, N.; Tanner, R.I. J. Rheol. **1987**, *31*, 37–58.

Trease, H.E.; Fritts, M.F.; Crowley, W.P., Eds.; *Advances in the Free-Lagrange Method*; Springer: Berlin, 1991.

Van Kampen, *Stochastic Processes in Physics and Chemistry*; North-Holland: Amsterdam, 1981.

Van Schaftingen, J.J., The Linear-Stability of Numerical-Solutions of Some Elementary Viscoelastic Flows, 1986.

Van Schaftingen, J.J.; Crochet, M.J. Int. J. Num. Meth. Fluids **1984**, *4*, 1065–1081.

Weatherill, N.P. et al. Ed., *Numerical Grid Generation in Computational Fluid Dynamics and Related Fields*; Pineridge Press: Swansea, UK, 1994.

Zegeling, P.A. Moving-Grid Methods for Time-Dependent Partial Differential Equations, CWI Tract, Amsterdam, 1993. Also available at http://www.math. uu.nl/people/zegeling/publist.html.

BIBLIOGRAPHY

Ackerson, B.J.; Pusey, P.N. Phys. Rev. Lett. **1988**, *61*, 1033–1036.

Aggarwal, R.; Keunings, R.; Roux, F.-X. J. Rheology **1994**, *38* (2), 405–419.

Akl, S.G. *Parallel Sorting Algorithms*; Academic Press: New York, 1985.

Allen, M.P. Simulations and phase behaviour of liquid crystals. In *Observation, Prediction and Simulation of Phase Transitions in Complex Fluids*, Baus, M., Rull, L.F., Ryckaert, J.-P., Eds.; NATO ASI Series, 1995; 557–591.

Ballal, B.Y.; Rivlin, R.S. Trans. Soc. Rheol. **1976**, *20*, 65–101.

Bartlett, P.; van Megen, W. *Physics of Hard-Sphere Colloidal Suspensions*, Mehta, A. Ed.; Springer: Berlin, 1994.

Beris, A.N.; Armstrong, R.C.; Brown, R.A. J. Non-Newtonian Fluid Mech. **1983**, *13*, 109–148.

Beris, A.N.; Armstrong, R.C.; Brown, R.A. J. Non-Newtonian Fluid Mech. **1984**, *16*, 141–171.

Beris, A.N.; Armstrong, R.C.; Brown, R.A. J. Non-Newtonian Fluid Mech. **1986**, *19*, 323–347.

Beris, A.N.; Armstrong, R.C.; Brown, R.A. J. Non-Newtonian Fluid Mech. **1987**, *22*, 129–167.

Bird, R.B.; Armstrong, R.C.; Hassager, O. *Dynamics of Polymeric Liquids*; John Wiley: New York, 1987a; Vol I.

Bird, R.B.; Dotson, P.J.; Johnson, N.L. J. Non-Newtonian Fluid Mech. **1980**, *7*, 213–235.

Brady, J.F.; Bossis, G. J. Fluid Mech. **1985**, *155*, 105–129.

Burdette, S.R. J. Non-Newtonian Fluid Mech. **1989**, *32*, 269–294.

Chen, L.B.; Ackerson, B.J.; Zukoski, C.F. J. Rheol. **1994**, *38*, 193–202.

Clemençon, C.; Decker, K.M.; Endo, A.; Fritscher, J.; Masuda, N.; Müller, A.; Rühl, R.; Sawyer, W.; de Sturler, E.; Wylie, B.J.N.; Zimmermann, F. SPEEDUP Journal, 1994.

Cook, R.; Wilemski, G. J. Phys. Chem. **1992**, *96*, 4023–4027.

Cooke, M.H.; Bridgwater, J. Powder Bulk Solids Technol. **1979**, *3*, 11–19.

Curtiss, C.F.; Bird, R.B. Physica. **1983**, *118A*, 191–204.

Curtiss, C.F.; Bird, R.B.; Hassager, O. Adv. Chem. Phys. **1976**, *35*, 31–117.

Davies, M.J.; Walters, K. In *Rheology of Lubricants*, Davenport, T.C., Ed.; Halsted Press, 1973.

Doi, M.; Edwards, S.F. J. Chem. Soc., Faraday Trans. II **1978**, *74*, 1789–1832.

Doi, M.; Edwards, S.F. J. Chem. Soc., Faraday Trans. II **1979**, *75*, 38–54.

Doi, S.; Washio. T.; Muramatsu, K.; Nkata, T. Preprints of Parallel CFD'94, Kyoto Institute of Technology, Kyoto, Japan, May 1994, pp. 31-36.

Ermak, D.L.; McCammon, J.A.; J. Chem. Phys. **1978**, *69* (4), 1353–1360.

Erpenbeck, J.J. Phys. Rev. Lett. **1984**, *52* (15), 1333–1335.

Evans, D.J.; Morriss, G.P. Phys. Rep. **1984**, *1*, 297–344.

Evans, D.J.; Morriss, G.P. Phys. Rev. Lett. **1986**, *56* (20), 2172–2175.

Fix, G.J.; Paslay, P.R. J. Appl. Mech. Trans. ASME **1967**, *34*, 579–589.

Fixman, M. J. Chem. Phys. **1978**, *69*, 1527–1538.

Fixman, M., Macromolecules **1981**, *14*, 1710–1717.

Fletcher, C.A.J. *Computational Techniques for Fluid Dynamics*; Springer: Berlin, 1991.

Fuentes, Y.O.; Kim, S. AIChEJ **1992**, *38*, 1059–1078.

Goodwin, J.W. Ed., *Colloidal Dispersions*; Royal Society of Chemistry: London, 1982.

Hanley, H.J.M.; Morriss, G.P.; Welberry, T.R.; Evans, D.J. Physica **1988**, *149A*, 406–414.

Hoffman, R.L. Trans. Soc. Rheol. **1972**, *16*, 155–163.

Hopkins, A.J.; Woodcock, L.V. J. Chem. Soc. Faraday Trans. II **1990**, *86*, 2121–2132.

Hwa, T.; Kardar, M. Phys. Rev. Lett. **1989**, *62*, 1813–1820.

Irving, J.H.; Kirkwood, J.G. J. Chem. Phys. **1960**, *18*, 817–829.

Joseph, D.D.; Saut, J.C. J. Non-Newtonian Fluid Mech. **1986**, *20*, 117–141.

Kalus, J.; Lindner, P.; Hoffmann, H.; Ibel, K.; Münch, C.; Sander, J.; Schmelzer, U.; Selbach, J. Physica B **1991**, *174*, 164–169.

Kamal, M.M. J. Basic Engng. ASME **1966**, *88*, 717–724.

Keunings, R. *Computers and Chemical Engng.*; in press, 1995.

Keunings, R.; Halin, P. personal communication, 1995.

Kim, S.; Karniadakis, G.; Vernon, M.K. *Parallel Computing in Multiphase Flow Systems Simulations*; Eds.; The Fluids Engineering Division, ASME, 1994.

Kirkwood, J.G. *Macromolecules*; Gordon & Breach, 1967.

Krieger, I.M. Adv. Coll. Interface Sci. **1982**, *3*, 111–127.

Lakshmivarahan, S.; Dhall, S.K.; Miller, L.L. Parallel sorting algorithms. *Advances in Computers*, **1984**, *23*, 295–354.

Laun, H.M.; Bung, R.; Hess, S.; Loose, W.; Hess, O.; Hahn, K.; Hädicke, E.; Hingmann, R.; Schmidt, F.; Lindner, P.; J. Rheol. **1992**, *36* (4), 743–787.

Lawler, J.V.; Muller, S.J.; Brown, R.A.; Armstrong, R.C. J. Non-Newtonian Fluid Mech. **1986**, *20*, 51–92.

Lee, S.J.; Denn, M.M.; Crochet, M.J.; Metzner, A.B.; Rigins, G.J. J. Non-Newtonian Fluid Mech. **1984**, *14*, 301–325.

Lees, A.W.; Edwards, S.F. J. Phys. C: Solid State Phys. **1972**, *5*, 1921–1929.

Lesieur, M. *Turbulence in Fluids: Stochastic and Numerical Modelling*; Kluwer: Dordrecht, 1990.

Lighthill, J. *An Informal Introduction to Theoretical Fluid Mechanics*; Oxford Science Publications: Oxford, 1986.

Mochimaru, Y., J. Non-Newtonian Fluid Mech. **1983**, *12*, 135–152.

MPIF (Message Passing Interface Forum), Technical Report CS-94-230, University of Tennessee, Knoxville, TN, USA, April 1994; also J. of Supercomputer Applications **1994**, *8*, also available from: http://www.mcs.anl.gov/mpi/mpi-report/mpi-report.html

Nagtigaal, J.; Parks, D.M.; Rice, J.R. Comp. Meths. Appl. Mech. Eng. **1974**, *4*, 153–178.

Nickell, R.E.; Tanner, R.I.; Caswell, B. J. Fluid. Mech. **1981**, *65*, 219–248.

Olivari, D. *Introduction to the Modeling of Turbulence*, Von Karman Institute for Fluid Dynamics, 1993.

Öttinger, H.C. J. Chem. Phys. **1990**, *92*, 4540–4549.

Öttinger, H.C.; Laso, M. Bridging the gap between molecular models and viscoelastic flow calculations. In *Lectures on Thermodynamics and Statistical Mechanics*; World Scientific, Singapore, 1994; Chapter VIII, 139–153.

Parrinello, M.; Rahman, A. Phys. Rev. Lett. **1980**, *45*, 1196–1199.

Parrinello, M.; Rahman, A. J. Appl. Phys. **1981**, *52*, 7182–7190.

Parrinello, M.; Rahman, A. J. Chem. Phys. **1982**, *76*, 2662–2666.

Peterlin, A. Makr. Chem. **1961**, *44–46*, 338–346.

Peterlin, A. Kolloid-Zeitschrift **1962**, *182*, 110–115.

Pironneau, O. *Finite Element Methods for Fluids*; John Wiley: New York, 1989.

Phan-Thien, N.; Tanner, R.I. J. Non-Newtonian Fluid Mech. **1981**, *9*, 107–117.

Phan-Thien, N.; Sugeng, F.; Tanner, R.I. J. Non-Newtonian Fluid Mech. **1984**, *24*, 301–325.

Powell, R.L. J. Stat. Phys. **1991**, *62*, 1073–1094.

Pusey, P.N. Colloidal suspensions. In *Liquids, Freezing and Glass Transition*; Hansen, J.P., Levesque, D., Zinn-Justin, J., Eds.; North-Holland, Amsterdam, 1991; 765–942.

Rajagopalan, D.; Byars, J.A.; Armstrong, R.C.; Brown, R.S. J. Rheol. **1992**, *36* (7), 1349–1375.

Reiner, M.; Hanin, M.; Harnoy, A. Israel J. Tech. **1969**, *7*, 273–279.

Rigos, A.A.; Wilemski, G. J. Phys. Chem. **1992**, *96*, 3981–3986.

Rivlin, R.S. J. Non-Newtonian Fluid Mech. **1979**, *79*, 79–101.

Roberts, G.W.; Walters, K. Rheol. Acta **1992**, *31*, 55–62.

Russel, W.B. *The Dynamics of Colloidal Systems*; University of Wisconsin Press, 1987.

Schofield, A.N.; Wroth, C.P. *Critical State Soil Mechanics*; McGraw-Hill: London, 1968.

Sommerfeld, A. Z. Math. Physik **1904**, *50*, 97–155.

Tanner, R.I. Aust. J. Appl. Sci. **1963**, *14*, 129–136.

Temam, R. *Navier-Stokes Equations*; North-Holland: Amsterdam, 1977.

ter Horst, W.J. MSC internal report MEAH-172. Laboratory for Aero and Hydrodynamics. Delft University of Technology, 1998.

Townsend, P. J. Non-Newtonian Fluid Mech. **1980**, *6*, 219–243.

Ungar, L.H.; Brown, R.A. Philos. Trans. R. Soc. London **1982**, *A306*, 347–370.

Van Megen, W.; Snook, I. J. Chem. Phys. **1988**, *88*, 1185–1191.

Warner, H.R. Ind. Eng. Chem. Fundamentals **1972**, *11*, 379–387.

Wilcox, D.C. *Turbulence Modeling for CFD*, DCW Industries, 1994.

Wilemski, G. J. Stat. Phys. **1991**, *62* (5/6), 1239–1253.

Woodcock, L.V. Phys. Rev. Lett. **1985**, *54* (14), 1513–1516.

Xue, W.; Grest, G.S. Phys. Rev. A **1989**, *40* (3), 1709–1712.

Xue, W.; Grest, G.S. Phys. Rev. Lett. **1990**, *64* (4), 419–422.

Yamada, T.; Nosé, S. Phys. Rev. A **1990**, *42*, 6282–6288.

Yeh, P.-W.; Kim, E.M.E.; Armstrong, R.C.; Brown, R.A. J. Non-Newtonian Fluid Mech. **1984**, *16*, 173–194.

Zone, O.; Vanderstraeten, O.; Keunings, R. A parallel direct solver for implicit finite element problems based on automatic domain decomposition. In *Massively Parallel Processing: Applications and Development*, Dekker, L., Smit, W., Zuidervaart, J.C., Eds.; Elsevier, Amsterdam, 1994; 809–816.

17

Simulation of Polymers by Dissipative Particle Dynamics

W. K. DEN OTTER   University of Twente, Enschede, The Netherlands

J. H. R. CLARKE   University of Manchester Institute of Science and Technology, Manchester, United Kingdom

I.   INTRODUCTION

Dissipative particle dynamics (DPD) was introduced a decade ago by Hoogerbrugge and Koelman [1,2] as a particle-based off-lattice simulation method for the flow of complex fluids. Developed for mesoscopic length and time scales, i.e., much larger than atomic but still much smaller than macroscopic, DPD faithfully reproduces the relevant thermodynamics, hydrodynamics, and the omnipresent thermal noise. It was quickly recognized that DPD offered new and exciting opportunities for modeling polymers, materials that are rich in physical phenomena characterized by their mesoscopic scale. Already it has made significant impact on our understanding of microphase separation in block copolymers and phase behavior of polymer solutions [3].

   The DPD particles loosely represent a number of co-moving fluid molecules, much like the familiar fluid elements of classical hydrodynamics. Because of this coarse-grained interpretation of the particle, in both the spatial and the temporal sense, the conservative force acting between any two neighboring particles represents the time average of the sum of all atomic interactions between the atoms constituting the two particles, resulting in a smooth potential. In DPD the conservative force is taken to be of a purely repulsive nature, causing the particles to spread out evenly over the periodic simulation box. The effects of fluctuations around the averaged interaction, due to the subparticle dynamics, are represented by dissipative and random forces, in analogy with the Langevin model for the Brownian particle and the more general Mori–Zwanzig formalism for separating slow and fast variables [4,5]. The strength of the DPD method lies in the softness of the forces, and in the specific construction by which all forces act between particle pairs following Newton's third law. Because of this latter condition, momentum is conserved locally. Hydrodynamic interactions, therefore, arise automatically during the simulation, so the system will—when viewed on a sufficiently large time and length scale—behave like any regular fluid [4]. As it turns out, surprisingly few particles are needed in practice to recover hydrodynamic phenomena.

Hoogerbrugge and Koelman tested their method on a quiescent fluid, a fluid undergoing Couette flow, and a fluid flow around an array of cylinders [1]. The objects in the latter simulation were made by locally immobilizing the liquid. But their *pièce de résistance* was calculating the viscosity of a colloidal suspension (with rigid groups of particles acting as moving and rotating colloids) as a function of the colloid volume fraction, in good agreement with the celebrated Einstein expression at low fractions [2]. Research in this area has since continued in the group of Coveney and Boek [6]. Using DPD, liquid mixtures have been simulated by the simple expedient of assigning a "color" to every particle [7–9]. A rapid quench to temperatures below the spinodal was then mimicked by instantaneously increasing the repulsion between unlike particles to a value larger than that between like particles. The scaling law for the growth of phase-separated domains with time agreed well with the theoretical predictions for systems with hydrodynamic interactions [7–9].

Polymers, the topic of this book, have been simulated by introducing springs between the DPD particles, as in a coarse-grained bead-spring model [10,11]. The results of these simulations are discussed in detail below. In a related area, the rich phase behavior of aqueous surfactants has been characterized; these molecules were modeled as a rigid unit of one hydrophobic and one hydrophilic particle [12]. Even cell membranes [13] and vesicles [14] have been simulated, using surfactants with flexible tails. Thermotropic liquid crystalline phases have been characterized for a system of rigid rods [15]. A list of all articles relating to DPD can be found in [16].

In Section II we look more closely at the computational aspects of DPD, before focusing attention on the specific application to polymer systems. Section III describes the matching of simulation parameters to the properties of real polymer systems, with an emphasis on the relation between the conservative force field and the common Flory–Huggins χ parameter for mixtures. The dynamics of individual polymer chains in a solvent and in a melt are discussed in Section IV, and the ordering dynamics of quenched block copolymer systems is described in Section V. A summary and conclusions are given in Section VI.

II. DISSIPATIVE PARTICLE DYNAMICS

The interactions between DPD particles consist of three parts, each of which locally conserves momentum by conforming to Newton's third law. The conservative force between any pair of neighboring particles reads

$$\mathbf{f}_{ij}^c = a\omega_c(r_{ij})\hat{\mathbf{r}}_{ij} \tag{1}$$

where $r_{ij} = |\mathbf{r}_{ij}|$ and $\hat{\mathbf{r}}_{ij} = \mathbf{r}_{ij}/r_{ij}$ are, respectively, the length and the direction of the difference vector $\mathbf{r}_{ij} = \mathbf{r}_i - \mathbf{r}_j$. Here ω_c describes the distance dependence of the force, which is usually taken to be $\omega_c(r) = (1 - r_{ij}/r_c)$, with r_c the cutoff distance of the potential. The parameter a controls the magnitude of the force. In the case of polymers an additional conservative force $\mathbf{f}_{ij}^s$ is added to model the bond connecting the two particles. The force constant and the equilibrium length of these harmonic or FENE springs are chosen such that the average bond length coincides with the first peak in the radial distribution function.

Between every pair of particles, whether they are bound or not, there also acts a friction force

$$\mathbf{f}_{ij}^d = -\gamma\omega_d(r_{ij})(\mathbf{v}_{ij} \cdot \hat{\mathbf{r}}_{ij})\hat{\mathbf{r}}_{ij} \tag{2}$$

with $\mathbf{v}_{ij} = \dot{\mathbf{r}}_{ij}$, and a random force

$$\mathbf{f}_{ij}^r = \sigma\omega_r(r_{ij})\theta_{ij}(t)\hat{\mathbf{r}}_{ij} \tag{3}$$

The strength and range of these two forces are related by a conventional fluctuation-dissipation theorem, which in the current context reads as [17]

$$2\gamma k_B T \omega_d(r_{ij}) = \sigma^2\omega_r^2(r_{ij}) \tag{4}$$

with Boltzmann's constant k_B and temperature T. The common choice for the distance dependence is $\omega_r(r) = \omega_c(r)$. In Eq. (3) $\theta_{ij}(t)$ denotes a random number of zero mean, unit variance, uncorrelated between particle pairs, and devoid of memory, i.e., $\langle\theta_{ij}(t)\rangle = 0$ and

$$\langle\theta_{ij}(t)\theta_{kl}(t')\rangle = \left(\delta_{ik}\delta_{jl} + \delta_{il}\delta_{jk}\right)\delta(t - t') \tag{5}$$

DPD codes are usually simple modifications of molecular dynamics (MD) programs, and since this latter correlation is inconvenient for algorithms with a finite time step h, the actual random number to be used in the nth step becomes $\zeta_{ij}(n) = \int_{(n-1)h}^{nh} \theta_{ij}(t)\, dt$. Clearly, $\langle\zeta_{ij}(n)\rangle = 0$ and $\langle\zeta_{ij}^2(n)\rangle = h$, while

there still are no correlations between particle pairs, nor memory effects
between consecutive steps.

Note how the friction and random forces differ from their counterparts
in Brownian Dynamics (BD), where these forces act relative to a fixed
background. In BD, therefore, a bulk flow is not sustained, and the absence
of momentum conservation leads to a screening of the hydrodynamic
interactions. By making all forces explicitly Galilean invariant, as in Eqs. (1)
through (3), these deficiencies are avoided. The friction and random forces,
in addition to their physical interpretation, also serve as a thermostat to the
simulation; see Eq. (4). More elaborate models in which the particles are
equipped with an internal energy have been proposed [18,19], allowing for
a temperature gradient. This may prove a particularly valuable addition
to DPD, as it augments mass and momentum conservation with energy
conservation to obtain the complete set of five hydrodynamical equations
describing any real liquid [4,20]. Particles with an angular momentum
[21,22] and internal elasticity [23] have also been suggested.

The equations of motion, obtained from the above force expressions in
combination with Newton's second law,

$$\frac{d\mathbf{x}_i}{dt} = \mathbf{v}_i \tag{6}$$

$$m\frac{d\mathbf{v}_i}{dt} = \sum_j (\mathbf{f}_{ij}^c + \mathbf{f}_{ij}^d + \mathbf{f}_{ij}^r + \mathbf{f}_{ij}^s) \tag{7}$$

are commonly integrated with a Verlet leapfrog algorithm [1,24]. Because of
the softness of the particle potential, the algorithm is stable up to fairly large
time steps. Unfortunately, the assumption that the friction and random
forces can be treated as constants during the time step leads to a thermostat-
induced spurious increase of the temperature [25]. A number of algorithms
have been proposed to counter this effect, with mixed results [24–28]. We
think that an Ermak-type approach, the de facto standard in BD, offers the
best performance [28]. But if one is willing to accept a temperature rise of,
say, two percent, it is perhaps most convenient to stick with an ordinary
leapfrog algorithm: the conversion of an existing MD code into a DPD
code is then straightforward. An interesting alternative with an Andersen
thermostat has been described by Lowe [29].

III. PARAMETERIZATION AND RELATION TO
FLORY–HUGGINS THEORY

The most common system of units in DPD, as introduced by Groot and
Warren [24], is to define the mass of the particle as the unit of mass, the

cutoff radius of the potential as the unit of length, and the thermal energy $k_B T = \frac{1}{3} m \langle v_i^2 \rangle$ as the unit of energy. The unit of time is then implicitly defined; a unit of temperature is not needed, since the temperature only appears as the product $k_B T$. These units, together with a field parameter a and a friction γ of the order ten, suffice for qualitative results, like the role played by hydrodynamics in the microphase separation of block copolymers. A further tuning is required for more quantitative results. We will summarize here the method proposed by Groot and Warren [24], including modifications of recent date [13].

One first chooses n_{map}, the number of atoms to be represented by a single DPD particle, thereby fixing the mass of the particle. Next one chooses a density of ρ particles per r_c^3, which combines with the weight density of the real fluid to give a value for r_c. In simulations typical values for ρ are 3, 4, or higher, to ensure a sufficiently high number of nearest neighbor interactions for a fluid-like structure.

The strength of the conservative force field can be obtained from the requirement that the density fluctuations of the real fluid should be reproduced. As the static structure factor for a coarse-grained system is smaller than that of the atomic system by a factor n_{map} for low wave numbers, it follows that $(\rho k_B T \kappa_T)_{DPD} = n_{map}^{-1} (\rho k_B T \kappa_T)_{atom}$. The isothermal compressibility coefficient κ_T of the DPD fluid is readily obtained from its surprisingly simple equation of state [24], $p \approx \rho k_B T + 0.1 a \rho^2$, and this set of equations is then solved for a.

Finally a friction coefficient is chosen, and the unit of time is set by comparing the self-diffusion coefficient D of a particle with that of n_{map} atoms. Typical values obtained this way for water with $n_{map} = 3$, $\rho = 3[l^{-3}]$, and $\sigma = 3[mlt^{-3/2}]$ are $a = 80[ml^2 t^{-2}]$ and the units $1[l] = 3 \text{ Å}$, $1[t] = 90 \text{ ps}$, and $1[m] = 9 \times 10^{-26} \text{ kg}$ [13]. As the implications of n_{map} for the conservative field have been realized only recently, most simulations have actually been run using softer a values in the range of 10 to 25.

This procedure for choosing the simulation parameters is by no means unique. One could, for instance, in the above scheme opt to define the unit of time by matching the thermal velocity of a particle with that of n_{map} atoms. The friction coefficient would then have to be tuned by matching the diffusion coefficient, using the readily understood observation $D^{-1}(\gamma) = D^{-1}(0) + g(6 k_B T / \gamma)^{-1}$, where both $D(0)$ and g are determined by the conservative field. Note that there will be no physically meaningful solution if the conservative force field itself already gives rise to a diffusion coefficient below the experimental value. An altogether different approach, using theoretical predictions [22,30–32] for the ideal fluid ($a = 0$) to match dynamical properties such as the viscosity and the sound velocity, has also been proposed [33].

Although most simulations to date assume a linear dependence of ω_c and ω_r on r, as introduced in the original DPD for numerical reasons, there is little evidence to suggest that this is the required form of the potential for a coarse-grained fluid. Kong et al. [34] tried an alternative simple potential, but found it to have little effect on the overall results of the simulation. In this context we mention recent attempts to place DPD on a firmer footing through a strict definition of a particle as the matter contained in a smoothed Voronoi cell. Dynamical equations for these cells have been obtained in "bottom-up" (starting from the atomic motion [35,36]) and "top-down" (starting from hydrodynamics [37]) approaches, in each case including a fluctuation-dissipation theorem and the aforementioned energy conservation.

An important parameter in the theory of diluted polymers, polymer mixtures, and block copolymers is the Flory–Huggins χ parameter, as it determines the propensity of the system to mix or de-mix. In DPD, the χ parameter is taken into account by systematically varying the conservative interaction parameter between unlike particles, whilst the repulsion between like particles remains unchanged to maintain a homogeneous compressibility. Hence, for a mixture of A and B particles one has $a_{AA} = a_{BB}$ and $a_{AB} = a_{AA} + \Delta a$. Groot and Warren [24] argued on the basis of the DPD equation of state that $\chi \propto \Delta a$. This was also found by simulating phase-separated monomeric mixtures at various values of Δa, using the fraction A in the B-rich phase and vice versa to calculate the corresponding value of χ. Recent simulations using Gibbs ensemble Monte Carlo, to avoid interfacial perturbations, have confirmed these findings [38]. For polymer–solvent mixtures an "identity change" Monte Carlo move was introduced, exchanging a chain of polymer particles by an equal number of solvent particles. This time the agreement with Flory–Huggins was less satisfactory, particularly near the critical point and in the low polymer concentration phase. These discrepancies were attributed to the deficiencies of the mean field model, and an eight parameter master equation was proposed to describe all simulation data, as well as experimental data [38].

In DPD, as in the Flory–Huggins model, the coexistence curve ends at an upper critical point. Above this critical temperature the system just will not phase separate. There exist experimental systems, however, that only phase separate above a critical temperature. In order to simulate a lower critical point, van Vliet et al. [39] turned the conservative force parameters into functions of the global temperature and pressure of the system. Pagonabarraga and Frenkel [40] extended this idea, making the conservative energy a function of the local density, $E_i = \psi^{ex}[\sum_j \omega(r_{ij})]$. The weight function ω represents the spatial spread-out of the atom cloud of particle j neighboring particle i. Any phase diagram can now be reproduced by dialing in its excess free energy ψ^{ex} as a function of density and temperature. This also opened up the possibility of studying liquid–vapor interfaces with DPD [41]. Note the similarities between this potential and the one used in smoothed particle hydrodynamics (SPH), an alternative off-lattice particle-based method for simulating fluid flow [21,42].

IV. ROUSE AND ZIMM DYNAMICS

The first application of DPD to polymers was presented by Schlijper, Hoogerbrugge, and Manke in 1995 [43]. They simulated a single polymer, a chain of N particles connected by $N-1$ harmonic springs, in a solvent of identical particles. The average bond length was made to coincide with the location of the first peak in the radial distribution function of the monomer fluid, $r_p \approx 0.85\, r_c$, by tuning the strength k and the equilibrium length l_{eq} of the springs. There was no explicit bending potential, so the beads are best viewed as representing fragments of a polymer at least as big as a Kuhn segment. The radius of gyration was found to follow the scaling law $\langle R_g \rangle \propto (N-1)^\nu$ for both bond models considered, with N ranging from 2 to 30. At $\nu \approx 0.52$, the exponent is slightly above the 0.5 characteristic of a Gaussian chain, but well below the 0.588 of a chain in a good (e.g., athermal) solvent [10,11]. Given the repulsive interactions between polymer beads, the potential energy of two coinciding beads was about five times the thermal energy in this model, a more pronounced indication of excluded volume effects was to be expected. The scaling law for the first relaxation time, $\tau_1 \propto N^\alpha$, yielded $\alpha \approx 1.95$ for the weak bonds ($l_{eq} = 0$, small k) and 1.83 for the strong bonds ($l_{eq} = r_p$, large k). These results lie below the value of 2 obtained in Rouse theory (in which hydrodynamics is completely ignored), but above the $\alpha = 3\nu$ predicted by Zimm theory (which includes hydrodynamics in a preaveraging approximation) [10,11], indicating that hydrodynamic interactions do play a role in the DPD polymer model. On the basis of this result, the strong spring model appeared to be better than the weak spring model, hence it was used in successive simulations by these authors. Later simulations showed a much better agreement with theory, even for the weak springs, as discussed below.

The strong spring polymer model was further analyzed by looking at the dynamic structure factors,

$$S(\mathbf{k}, t) = \frac{1}{N}\left\langle \left(\sum_{j=1}^{N} e^{i\mathbf{k}\cdot\mathbf{r}_j(t)}\right)\left(\sum_{j=1}^{N} e^{i\mathbf{k}\cdot\mathbf{r}_j(0)}\right)^* \right\rangle \tag{8}$$

For a 20 bead chain, the plot of $\ln S(k,0)$ vs. $\ln k$ showed a region of linear decay with a slope corresponding to $\nu \approx 0.61$ [43]. This is a clear indication of swelling due to excluded volume interactions; the overshoot of the theoretical value is typical for short chains. Scaling theory predicts that for $kR_g \gg 1$ the rescaled dynamic structure factors $S(k, k^x t)/S(k, 0)$ coalesce for $x = 3$ (Zimm) or $x \approx 4$ (Rouse) [10,11]. Although the numerical value of $x \approx 2.7$ was lower than expected, in the good company of previous polymer simulations, it evidences the impact of hydrodynamic interactions on the motion of the polymer. A further indication hereof was found in the distribution of relaxation times in the decay of the conformational time correlation function,

$$C(t) = \frac{1}{N} \sum_{i=1}^{N} \langle [\mathbf{r}_i(t) - \mathbf{R}_{com}(t)] \cdot [\mathbf{r}_i(0) - \mathbf{R}_{com}(0)] \rangle \tag{9}$$

where $\mathbf{R}_{com}$ denotes the center of mass position. In a nutshell, their approach was to fit this function with

$$C(t) \propto \sum_i i^{-2+h} \exp\left(-\frac{t}{\tau_1 i^{-2+h}}\right) \tag{10}$$

by varying h and τ_1. The values $h = 0$ and $h = 0.5$ correspond to the Rouse and to the Zimm model, respectively. In their appendix, the authors warned that for short chains, $N \leq 10$, a noticeable deviation from the long N predictions of the Zimm and Rouse models is to be expected.

The effect of solvent quality on the conformational distribution of a solvated polymer was investigated by Kong, Manke, Madden, and Schlijper [34]. While all interactions between like particles were kept constant, $a_{ss} = a_{pp}$, the interaction between polymer and solvent beads was varied, $a_{sp} = a_{ss} + \Delta a$. For negative Δa the polymer–polymer interaction was more repulsive than the polymer–solvent interaction, and the polymers were observed to swell to $\nu \approx 0.61$. The opposite case of positive Δa saw the polymers collapse to $\nu \approx 0.31$ (theory: 1/3). A smooth transition between the two extremes was observed by varying Δa. Defining the expansion coefficient $\lambda = R_g/R_g^\theta$, with R_g^θ the radius of gyration in the theta solvent, a qualitative agreement was found with Flory's expression,

$$\frac{\lambda^5 - \lambda^3}{\sqrt{N-1}} = \begin{cases} A(\frac{1}{2} - \chi) & : \quad \text{theory, } A \approx 4/3 \\ 3.0(0.05 - \Delta a/a_{ss}) & : \quad \text{simulations} \end{cases} \tag{11}$$

This suggests a simple linear relationship between χ and Δa, as was also observed by Groot and Warren [24].

The effect of solvent quality on the dynamics of the polymer was studied by calculating the conformational time correlation function as a function of N and Δa [34]. While the scaling exponent α decreased from nearly 2 in the good solvent to 0.9 in the bad solvent, the ratio ν/α remained relatively constant between 3 and 3.3 (theory: 3). A sheared simulation box with Lees–Edwards boundary conditions was used to calculate the impact of a single dissolved polymer on the viscosity of the box [44–46]. The largest zero-shear viscosities were observed for the good solvent, the lowest for the bad solvent. Upon increasing the shear rate $\dot{\gamma}$, the solution was found to shear-thin in the region surrounding $\dot{\gamma}\tau_1 \approx 1$, the effect being most pronounced in the good solvent and nearly absent in the bad solvent. All of these static and dynamic results, save for the $\nu \approx 0.61$ from the static structure factor $S(k,0)$, confirmed the rather unsatisfactory result that the athermal solvent acted as a "near-theta" solvent.

The case of a single polymer in a quiescent athermal solvent was recently revisited by Spenley [47], using chains of 2 to 100 beads in a weak spring model. For the end-to-end vector he found the exponents $\nu = 0.58 \pm 0.04$ and $\alpha = 1.80 \pm 0.04$, both in excellent agreement with the predictions for a polymer with excluded volume and hydrodynamic interactions. It is not clear what caused this marked difference with the above discussed results. The verification of the scaling law for the self-diffusion coefficient, $D \propto N^\mu$, was more involved, as D is rather sensitive to the size of the simulation box. After correcting for this complication, a satisfying agreement with theory was observed.

In the melt [47] the polymers adopted a Gaussian distribution, $\nu = 0.498 \pm 0.005$, as predicted by Flory. The dynamical coefficients, $\alpha = 1.98 \pm 0.03$ and $\mu = -1.02 \pm 0.02$, and the viscosity $\eta \propto N$, all followed the Rouse model, showing that the interpolymer interactions were sufficiently strong to screen the hydrodynamic interactions, despite the softness of the potential. In fact, the potential was so soft that polymers were seen to cross one another [47,48]; a transition for large N from Rouse dynamics to de Gennes reptation is therefore not to be expected. In cases where the entanglement of polymers is an issue, DPD is readily combined with recently developed algorithms which explicitly reintroduce the uncrossability lost in coarse-grained soft potentials [49,50].

We have performed simulations at polymer concentrations ranging from a dilute system to a melt, for polymers of 20 to 100 beads, to study the impact of interpolymer interactions on the dynamics [51]. The polymer–solvent repulsion was tuned to a theta point condition, rendering the exponent ν and the radius of gyration at fixed N virtually independent of the concentration. For the lowest concentrations we found a reasonable agreement with the Zimm theory of the theta state. When the polymer bead concentration c became of the order of the overlap concentration, $c^* = N/\frac{4}{3}\pi\langle R_g\rangle^3$, the interpolymer contacts became numerous enough to change the scaling laws into $\tau_1 \propto N^\alpha(c/c^*)^l$ and $D \propto N^\mu(c/c^*)^m$. This behavior is reminiscent of semidilute polymer systems, though the chains are arguably too short to have reached this limit of combining low concentration with many interpolymer contacts. At $\alpha \approx 1.7$, $l \approx 0.2$, $\mu \approx -0.65$, and $m \approx -0.4$, the exponents are much closer to the swollen state (α and μ as for the ideal system, $l = \frac{1}{4}$ and $m = -\frac{1}{2}$) than the theta state ($l = 1 = -m$), though there was no other direct indication of swelling. Our theoretical understanding of these phenomena is still incomplete.

A number of articles have discussed polymers with limited freedom of motion. For a polymer between two parallel walls, the component of the radius of gyration perpendicular to the wall started to collapse as the gap is reduced to about five times the unperturbed radius of gyration, while the relaxation time of this component simultaneously strongly increased [45,52]. A recent study described the forced flow of solvated polymers through a square tube of comparable dimensions, showing that the polymers had a preference for the center of the flow [53]. Polymers in a grafted polymer brush were found to predominantly orient perpendicular to the substrate; their ordering increased with chain length and surface coverage [54]. The effects of (oscillatory) shear on a brush have been studied recently [55,56]. Colloidal particles present in the solvent above the brush were absorbed by the brush when the colloid–polymer interaction was less repulsive than the colloid–solvent interaction [48,57]. The absorption was effected by the polymer density and the quality of the solvent with respect to the polymer. Attaching short polymers to the surface of spherical particles strongly increased the viscosity and shear thinning of a solution of these particles, the more so in a good solvent [46]. In a bad solvent the particles even aggregated [48].

V. BLOCK COPOLYMERS

In an important series of articles the Unilever group have described the use of DPD to characterize microphase separation phenomena in linear diblock copolymers of the general formula A_nB_m [24,58,59], making detailed comparisons with both experiments and the predictions of self-consistent field theory [60]. Block copolymers are both interesting and useful because of their property of forming complicated aggregated structures, as the natural tendency for segregation of the blocks composed of A and B monomers is frustrated by the connectivity of the copolymer. The system can therefore only reduce its free energy by coalescing the A- and B-rich domains, forming e.g., sheets, rods, or spheres. The phase diagram is fully determined by just two parameters, $N\chi$ and the fraction f of A beads.

The basic parameterization used was as discussed above, with $\chi \propto \Delta a$. Groot et al. [24,58] showed that in DPD the interfacial tension between the component homopolymers followed the same scaling law, as a function of $N\chi$ and ρ, as experimental data for PS-PMMA. This is an important result, since it is the interfacial tension which drives microphase separation. It also provides validation for the comparison of experimental data on long chains with DPD data on short chains (but correspondingly much larger χ).

With $N\chi = 41$, the simulations were well outside the weak segregation regime, and when the melts were quenched the expected order of equilibrium structures was found as the copolymers become successively more asymmetric—lamellae ($f \geq 0.37$), perforated lamellae ($0.31 \leq f \leq 0.37$), hexagonally arranged rods ($0.26 \leq f \leq 0.31$), fluid peanut-shaped micelles (at $f = 0.2$), and a disordered phase (at $f = 0.1$). These stable structures emerged via nontrivial pathways, forming a series of metastable states before equilibrium is reached. Discrepancies between the exact positions of phase transition points and those predicted by theory were explained in terms of the short length of the DPD copolymer chains, $N = 10$, which effectively reduced the Flory–Huggins parameter to $N\chi \approx 20$ [58]. The finding of a hexagonally perforated lamellar phase is interesting since this structure has been identified in recent experiments, while current mean field theory predicts a gyroid phase in this region. Finite sample size effects cannot, however, be discounted as a reason for the nonappearance of the gyroid phase in the simulations.

In a later article [59] the importance of hydrodynamic interactions in the formation of certain microphases was demonstrated by close comparison of simulations using DPD and Brownian dynamics (BD). Whilst both simulation methods describe the same conservative force field, and hence should return the same equilibrium structure, they differ in the evolution algorithms. As mentioned above, DPD correctly models hydrodynamic interactions, whereas these are effectively screened in BD. Distinctly different "equilibrium" structures were obtained using the two techniques in long simulations of a quenched 2400 A_3B_7 polymer system, as shown in Fig. 1. Whilst DPD ordered efficiently into the expected state of hexagonal tubes, BD remained trapped in a structure of interconnected tubes. By way of contrast, both DPD and BD reproduced the expected lamellar equilibrium structure of A_5B_5 on similar time scales, see Fig. 2.

There is experimental evidence [61] that the formation of microphases occurs on at least three different time and length scales. In the first stage there is phase separation on the level of the interacting beads. This is

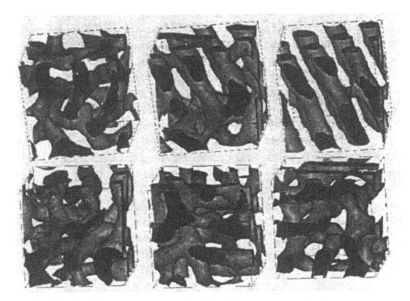

FIG. 1 Comparison of the time evolution of a quenched A_3B_7 block copolymer as obtained by DPD (top three snapshots) and BD (bottom). Shown are the surfaces surrounding areas of more than 50% A, after 1500, 7500, and 15,000 units of time. (From Ref. [59].)

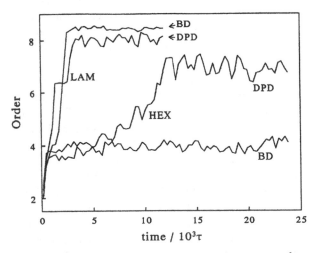

FIG. 2 The global order parameter $\int S(\mathbf{k}) \ln S(\mathbf{k}) \, d^3\mathbf{k}$ as a function of time for quenched block copolymers, as obtained in DPD and BD simulations. For A_3B_7 (HEX) DPD shows an activated transition from a disordered network of tubes to a hexagonal phase of rods, whereas BD remains stuck in the former. The A_5B_5 (LAM) immediately form a lamellar structure in both DPD and BD. (From Ref. [59].)

followed rapidly by a second stage in which the initially formed clusters grow out into micelles or highly connected rod-like or planar networks, depending on the composition. Finally there is the much slower macroscopic ordering of these percolating structures into a symmetric superstructure, which in the case of A_3B_7 may occur via a nematic-like transition.

Groot et al. [59] rationalized the different behavior of the A_3B_7 and A_5B_5 copolymer in terms of different evolution mechanisms for the final slow ordering process. They argued that the formation of hexagonal rods appeared to proceed by a nucleation-and-growth mechanism, starting from the gyroid-like and therefore presumably metastable disordered network of tubes formed in the second stage. The hydrodynamical interactions are essential in the growth of the hexagonally ordered structure. In BD, by contrast, hexagonal domains were formed locally, but instead of growing to encompass the entire simulation box, they disappeared again. The lamellar phase is formed by spinodal decomposition (i.e., this process does not involve surmounting a free energy barrier), and thus does not rely as much on hydrodynamics for its growth. Interestingly, DPD simulations of large boxes with short chains (A_1B_3 and A_2B_2) grew a patchwork of ordered structures of limited dimensions. This was also observed experimentally; macroscopic ordering is achieved only in sheared systems.

Recently the ordering effects of steady state shear flow on microphase separated copolymer melts has been investigated using DPD and the Lees–Edwards boundary conditions for Couette flow [62]. A 150 A_3B_{17} melt transformed under high shear, $\dot{\gamma}\tau_1 = 5.0$, from a bcc micellar structure into cylinders aligned along the flow direction. Cylinders of A_6B_{14} aligned with the flow at moderate shear, $\dot{\gamma}\tau_1 = 0.5$, and were torn into lamellae at higher shear rates. These lamellae, as well as those formed by the symmetric $A_{10}B_{10}$, were oriented perpendicular to the neutral direction. All these systems showed shear-thinning of the viscosity.

VI. CONCLUSIONS

Dissipative particle dynamics has proved to be a very versatile simulation technique for complex flows on the mesoscopic scale, where thermodynamics, hydrodynamics, and thermal noise play important roles. The polymer simulations reviewed here cover but a part of the phenomena that have been, and will be, studied using DPD.

The conservative force field in DPD is extremely soft compared to those used in atomic simulations, yet after parameterization using the compressibility or the Flory–Huggins parameter it is able to capture many of the essential properties of polymers at the mesoscopic level. Thermodynamic phenomena such as the swelling of polymers in a good solvent and the phase diagram of block copolymers are thus correctly reproduced. But DPD's most remarkable feature is that it faithfully combines thermodynamics with fluctuating hydrodynamics. This is achieved by expressing all forces, including the Brownian forces that thermostat and randomize the system, as pair interactions between the particles. Numerous simulations have confirmed the presence and importance of hydrodynamic interactions. In the context of polymers, we mention the observed Zimm dynamics of solvated polymers and the insights DPD has given in regard to the pathways followed by microphase separating block copolymer systems, including the role played therein by hydrodynamics.

REFERENCES

1.   Hoogerbrugge, P.J.; Koelman, J.M.V.A. Europhys. Lett. **1992**, *19*, 155.
2.   Koelman, J.M.V.A.; Hoogerbrugge, P.J. Europhys. Lett. **1993**, *21*, 363.
3.   Warren, P.B. Curr. Opin. Colloid Interface Sci. **1998**, *3*, 620.
4.   Español, P. Phys. Rev. E **1995**, *52*, 1734.
5.   Español, P. Phys. Rev. E **1996**, *53*, 1572.
6.   Boek, E.S.; Coveney, P.V.; Lekkerkerker, H.N.W.; vander Schoot, P. Phys. Rev. E **1997**, *55*, 3124.
7.   Coveney, P.V.; Novik, K.E. Phys. Rev. E **1996**, *54*, 5134.
8.   Coveney, P.V.; Novik, K.E. Phys. Rev. E **1997**, *55*, 4831.
9.   Jury, S.I.; Bladon, P.; Krishna, S.; Cates, M.E. In *High Performance Computing*; Allen, R.J. et al. Ed.; Kluwer Academic: New York, NY, 1999; 185–192 pp.
10. Doi, M.; Edwards, S.F. *The Theory of Polymer Dynamics*; Oxford University Press: Oxford, UK, 1998.
11. Grosberg, A.Y.; Khokhlov, A.R. *Statistical Physics of Macromolecules*; AIP Press: New York, NY, 1994.
12. Jury, S.; Bladon, P.; Cates, M.; Krishna, S.; Hagen, M.; Ruddock, N.; Warren, P. Phys. Chem. Chem. Phys. **1999**, *1*, 2051.
13. Groot, R.D.; Rabone, K.L. Biophys. J. **2001**, *81*, 725.
14. Yamamoto, S.; Maruyama, Y.; Hyodo, S. J. Chem. Phys. **2002**, *116*, 5842.
15. Sunaidi, A.; den Otter, W.K.; Clarke, J.H.R. *In preparation*.
16. URL http://www.fisfun.uned.es/~mripoll/Invest/dpd.html.
17. Español, P.; Warren, P. Europhys. Lett. **1995**, *30*, 191.
18. Español, P. Europhys. Lett. **1997**, *40*, 631.
19. Bonet Avalos, J.; Mackie, A.D. Europhys. Lett. **1997**, *40*, 141.
20. Ripoll, M.H.E.M.; Español, P. J. Chem. Phys. **2001**, *115*, 7271.
21. Espanol, P. Europhys. Lett. **1997**, *39*, 605.
22. Español, P. Phys. Rev. E **1998**, *57*, 2930.
23. ten Bosch, B.I.M. J. Non-Newtonian Fluid Mech. **1999**, *83*, 231.

24. Groot, R.D.; Warren, P.B. J. Chem. Phys. **1997**, *107*, 4423.
25. den Otter, W.K.; Clarke, J.H.R. Int. J. Mod. Phys. C **2000**, *11*, 1179.
26. Pagonabarraga, I.; Hagen, M.H.J.; Frenkel, D. Europhys. Lett. **1998**, *42*, 377.
27. Besold, G.; Vattulainen, I.; Karttunen, M.; Polson, J.M. Phys. Rev. E **2000**, *62*, R7611.
28. den Otter, W.K.; Clarke, J.H.R. Europhys. Lett. **2000**, *53*, 426.
29. Lowe, C.P. Europhys. Lett. **1999**, *47*, 145.
30. Marsh, C.A.; Backx, G.; Ernst, M.H. Phys. Rev. E **1997**, *56*, 1676.
31. Masters, A.J.; Warren, P.B. Europhys. Lett. **1999**, *48*, 1.
32. Español, P.; Serrano, M. Phys. Rev. E **1999**, *59*, 6340.
33. Dzwinel, W.; Yuen, D.A. Int. J. Mod. Phys. C **2000**, *11*, 1.
34. Kong, Y.; Manke, C.W.; Madden, W.G.; Schlijper, A.G. J. Chem. Phys. **1997**, *107*, 592.
35. Flekkøy, E.G.; Coveney, P.V.; Fabritiis, G.D. Phys. Rev. E **2000**, *62*, 2140.
36. Español, P.; Serrano, M.; Zúñiga, I. Int. J. Mod. Phys. C **1997**, *8*, 899.
37. Serrano, M.; Español, P. Phys. Rev. E **2001**, *64*, 046115.
38. Wijmans, C.M.; Smit, B.; Groot, R.D. J. Chem. Phys. **2001**, *114*, 7644.
39. van Vliet, R.E.; Hoefsloot, H.C.J.; Hamersma, P.J.; Iedema, P.D. Macromol. Theory Simul. **2000**, *9*, 698.
40. Pagonabarraga, I.; Frenkel, D. J. Chem. Phys. **2001**, *115*, 5015.
41. Pagonabarraga, I.; Frenkel, D. Mol. Sim. **2000**, *25*, 167.
42. Monaghan, J.J. Annu. Rev. Astron. Astrophys. **1992**, *30*, 543.
43. Schlijper, A.G.; Hoogerbrugge, P.J.; Manke, C.W. J. Rheol. **1995**, *39*, 567.
44. Schlijper, A.G.; Manke, C.W.; Madden, W.G.; Kong, Y. Int. J. Mod. Phys. C **1997**, *8*, 919.
45. Kong, Y.; Manke, C.W.; Madden, W.G.; Schlijper, A.G. Tribol. Lett. **1997**, *3*, 133.
46. Zhang, K.; Manke, C.W. Mol. Sim. **2000**, *25*, 157.
47. Spenley, N. Europhys. Lett. **2000**, *49*, 534.
48. Gibson, J.B.; Zhang, K.; Chen, K.; Chynoweth, S.; Manke, C.W. Mol. Sim. **1999**, *23*, 1.
49. Terzis, A.F.; Theodorou, D.N.; Stroeks, A. Macromol. **2000**, *33*, 1397.
50. Padding, J.T.; Briels, W.J. J. Chem. Phys. **2001**, *115*, 2846.
51. den Otter, W.K.; Ali, I.; Clarke, J.H.R. *In preparation.*
52. Kong, Y.; Manke, C.W.; Madden, W.G.; Schlijper, A.G. Int. J. Thermophys. **1994**, *15*, 1093.
53. Willemsen, W.M.; Hoefsloot, H.C.J.; Iedema, P.D. J. Stat. Phys. **2002**, *107*, 53.
54. Malfreyt, P.; Tildesley, D.J. Langmuir **2000**, *16*, 4732.
55. Irfachsyad, D.T.D.; Malfreyt, P. Phys. Chem. Chem. Phys. **2002**, *4*, 3008.
56. Wijmans, C.M.; Smit, B. Macromolecules **2002**, *35*, 7138.
57. Gibson, J.B.; Chen, K.; Chynoweth, S. J. Colloid Interface Sci. **1998**, *206*, 464.
58. Groot, R.D.; Madden, T.J. J. Chem. Phys. **1998**, *108*, 8713.

59. Groot, R.D.; Madden, T.J.; Tildesley, D.J. J. Chem. Phys. **1999**, *110*, 9739.
60. Matsen, M.W.; Schick, M. Curr. Opin. Colloid Interface Sci. **1996**, *1*, 329.
61. Balsara, N.P.; Garetz, B.A.; Newstein, M.C.; Bauer, B.J.; Prosa, T.J. Macromolecules **1998**, *31*, 7668.
62. Zhang, K.; Manke, C.W. Comput. Phys. Comm. **2000**, *129*, 275.

18

Dynamic Mean-Field DFT Approach for Morphology Development

**A. V. ZVELINDOVSKY, G. J. A. SEVINK, and
J. G. E. M. FRAAIJE** Leiden University, Leiden, The Netherlands

I. INTRODUCTION

For a long time, chemical engineers have analyzed macroscale properties using a variety of continuum mechanics models. In the last decade molecular modeling has grown to an essential part of research and development in the chemical and pharmaceutical industries. Despite the considerable success of both molecular and macroscale modeling, in the past few years it has become more and more apparent that in many materials mesoscale structures determine material properties to a very large extent. Mesoscale structures are typically of the size of 10 to 1000 nm. The industrial relevance of mesoscale modeling is obvious, nevertheless the necessary general purpose computational engineering tools are absent.

We are developing a general purpose method for mesoscale soft-condensed matter computer simulations, based on a functional Langevin approach for mesoscopic phase separation dynamics of complex polymer liquids. This project aims to consider topics of utmost importance in chemical engineering, such as chemical reactions, convection and flow effects, surfaces and boundaries, etc.

The morphology formation in complex liquids has been studied by many authors using time dependent Landau–Ginzburg models [1–5]. These models are based on traditional free energy expansion methods (Cahn–Hilliard [6], Oono–Puri [7], Flory–Huggins–de Gennes [8]) which contain only the basic physics of phase separation [9] and are not well suited for specific application to the different complex industrial and biological liquids. In contrast to these phenomenological theories we use dynamic density functional theory [9–14] where we do not truncate the free energy at a certain level, but rather retain the full polymer path integral by a numerical procedure (see Appendix A). Recently Kawakatsu and Doi started to use a similar approach [15,16].

Although calculation of polymer path integrals is computationally very intensive, it allows us to describe mesoscopic dynamics of *specific* complex polymer liquids [17].

Our approach uses essentially the same free energy functional as in many self-consistent field (SCF) calculations [18–21] (see also [22] for an excellent overview). In the past decade the SCF methods have been widely used for static calculations in various polymeric systems. The dynamic DFT approach complements the static SCF calculations by providing a detailed dynamical picture of system evolution, which is crucial, for example, for the investigation of metastable intermediate mesophases and for the study of the influence of shear and other processing conditions.

The prediction of the mesoscale morphology of complex polymer systems is very important for the final product properties. The application area of the proposed method includes computer simulation of such processes as emulsion copolymerization, copolymer melts and softened polymer melts, polymer blends, polymer surfactant stabilized emulsions, and adsorption phenomena in aqueous surfactant systems.

In this chapter we demonstrate only several types of modulation of self-assembly in complex polymer systems. They are: shearing of concentrated aqueous solution of amphiphilic polymer surfactant, shearing of symmetric diblock copolymer blend, reactions in polymer mixture, surface directed phase separation in copolymer melt.

II. DYNAMIC DENSITY FUNCTIONAL THEORY

We give a short outline of the theory of the mesoscopic dynamics algorithms. For more details see [11]. We consider a system of n Gaussian chains of N beads of several different species (for example, $A_{N_A}B_{N_B}$, $N = N_A + N_B$ for a diblock copolymer, $E_{N_E}P_{N_P}E_{N_E}, N = N_P + 2N_E$ $N = N_P + 2N_E$ for a symmetric triblock copolymer, etc.). The solvent can easily be taken into account [17]. The volume of the system is V. There are concentration fields $\rho_I(\mathbf{r})$, external potentials $U_I(\mathbf{r})$, and intrinsic chemical potentials $\mu_I(\mathbf{r})$.

Imagine that, on a course-grained time scale, there is a certain collective concentration field $\rho_I(\mathbf{r})$ of the beads of type I (say, A or B). Given this concentration field a free energy functional $F[\rho]$ can be defined as follows:

$$\beta F[\rho] = -n \ln \Phi + \ln n! - \beta \sum_I \int U_I(\mathbf{r})\rho_I(\mathbf{r})d\mathbf{r} + \beta F^{nid}[\rho] \tag{1}$$

Here Φ is the partition functional for the ideal Gaussian chains in the external field U_I, and $F^{nid}[\rho]$ is the contribution from the nonideal interactions. The free energy functional (1) is derived from an optimization criterion (see [11]) which introduces the external potential as a Lagrange multiplier field. The external potentials and the concentration fields are related via a density functional for ideal Gaussian chains:

$$\rho_I[U](\mathbf{r}) = n \sum_{s'=1}^{N} \delta_{Is'}^K \mathrm{Tr}_c \psi \delta(\mathbf{r} - \mathbf{R}_{s'}) \tag{2}$$

Here $\delta_{Is'}^K$ is a Kronecker delta function with value 1 if bead s' is of type I and 0 otherwise. The trace Tr_c is limited to the integration over the coordinates of one chain

$$\mathrm{Tr}_c(\cdot) = \mathcal{N} \int_{V^N} (\cdot) \prod_{s=1}^{N} d\mathbf{R}_s$$

$\mathcal{N}$ is a normalization constant. ψ is the single chain configuration distribution function

$$\psi = \frac{1}{\Phi} e^{-\beta\left[H^G + \sum_{s=1}^{N} U_s(\mathbf{R}_s)\right]} \tag{3}$$

where H^G is the Gaussian chain Hamiltonian

$$\beta H^G = \frac{3}{2a^2} \sum_{s=2}^{N} (\mathbf{R}_s - \mathbf{R}_{s-1})^2 \tag{4}$$

with a the Gaussian bond length parameter. The density functional is bijective; for every set of fields $\{U_I\}$ there is exactly one set of fields $\{\rho_I\}$. Thus there exists a *unique* inverse density functional $U_I[\rho]$. There is no known closed analytical expression for the inverse density functional, but for our purpose it is sufficient that the inverse functional can be calculated efficiently by numerical procedures (see Appendices A and B).

We split the nonideal free energy functional formally into two parts

$$F^{nid}[\rho] = F^c[\rho] + F^e[\rho]$$

where F^e contains the excluded volume interactions, and F^c the cohesive interactions. The intrinsic chemical potentials μ_I are defined by the functional derivatives of the free energy:

$$\mu_I(\mathbf{r}) \equiv \frac{\delta F}{\delta \rho_I(\mathbf{r})} = -U_I(\mathbf{r}) + \frac{\delta F^c}{\delta \rho_I(\mathbf{r})} + \frac{\delta F^e}{\delta \rho_I(\mathbf{r})} \tag{5}$$

$$= -U_I(\mathbf{r}) + \mu_I^c(\mathbf{r}) + \mu_I^e(\mathbf{r}) \tag{6}$$

Here we have introduced the cohesive potential $\mu_I^c(\mathbf{r})$ and the excluded volume potential $\mu_I^e(\mathbf{r})$. For the cohesive interactions we employ a two-body mean field potential:

$$F^c[\rho] = \frac{1}{2} \sum_{IJ} \int \int \epsilon_{IJ}(|\mathbf{r} - \mathbf{r}'|)\rho_I(\mathbf{r})\rho_J(\mathbf{r}') \, d\mathbf{r} \, d\mathbf{r}' \tag{7}$$

$$\mu_I^c(\mathbf{r}) \equiv \frac{\delta F^c}{\delta \rho_I} = \sum_J \int_V \epsilon_{IJ}(|\mathbf{r} - \mathbf{r}'|)\rho_J(\mathbf{r}') \, d\mathbf{r}' \tag{8}$$

where $\epsilon_{IJ}(|\mathbf{r} - \mathbf{r}'|) = \epsilon_{JI}(|\mathbf{r} - \mathbf{r}'|)$ is a cohesive interaction between beads of type I at $\mathbf{r}$ and J at $\mathbf{r}'$, defined by the Gaussian kernel

$$\epsilon_{IJ}(|\mathbf{r} - \mathbf{r}'|) \equiv \epsilon_{IJ}^0 \left(\frac{3}{2\pi a^2}\right)^{3/2} \exp\left[-(3/2a^2)(\mathbf{r} - \mathbf{r}')^2\right] \tag{9}$$

The excluded volume interactions can be included via the correction factor or insertion probability for each bead, c [12]:

$$\beta F^e[\rho] = -\sum_I \int_V \rho_I(\mathbf{r})\ln c(\mathbf{r}) \, d\mathbf{r} \tag{10}$$

The insertion probability is interpreted as the effective fraction of free space. The lower the fraction of free space, the lower the insertion probability and the higher the excess free energy. We have studied several models (van der Waals, Flory–Orwoll–Vrij, Carnahan–Starling) for the excess free energy function. We found that the phenomenological Helfand penalty function [last term in Eq. (13)] provides a numerically and mathematically simple way to account for compressibility effects in the system [12]. In equilibrium $\mu_I(\mathbf{r})$ is constant; this yields the familiar self-consistent field equations for Gaussian chains, given a proper choice for F^{nid}. When the system is not in equilibrium the gradient of the intrinsic chemical potential $-\nabla\mu_I$ acts as a thermodynamic force which drives collective relaxation processes. When the Onsager coefficients are constant the stochastic diffusion equations are of the following form

$$\frac{\partial \rho_I}{\partial t} = -\nabla \cdot \mathbf{J}_I \tag{11}$$

$$\mathbf{J}_I = -M\nabla\mu_I + \tilde{\mathbf{J}}_I \tag{12}$$

where M is a mobility coefficient and $\tilde{\mathbf{J}}_I$ is a noise field, distributed according to the fluctuation-dissipation theorem.

In Appendix B we provide some details of the numerical solution of these equations.

III. APPLICATION

A. Pluronics in Water Mixtures

The aqueous solution of triblock copolymer Pluronic L64 $EO_{13}PO_{30}EO_{13}$ [23] is modeled as a compressible system consisting of ideal Gaussian chain molecules in a mean field environment. The free energy is a functional of a set of particle concentrations $\{\rho\}$ [11,12]:

$$\begin{aligned}
F[\{\rho\}] = &-kT \ln \frac{\Phi_p^{n_p} \Phi_s^{n_s}}{n_p! n_s!} - \sum_I \int_V U_I(\mathbf{r})\rho_I(\mathbf{r})\, d\mathbf{r} \\
&+ \frac{1}{2}\sum_{I,J} \int_{V^2} \epsilon_{IJ}(\mathbf{r} - \mathbf{r}')\rho_I(\mathbf{r})\rho_J(\mathbf{r}')\, d\mathbf{r}\, d\mathbf{r}' \\
&+ \frac{\kappa_H}{2}\int_V \left(\sum_I \nu_I(\rho_I(\mathbf{r}) - \rho_I^0)\right)^2 d\mathbf{r}
\end{aligned} \tag{13}$$

where n_p (n_s) is the number of polymer (solvent) molecules, Φ is the intra-molecular partition function, I is a component index (EO, PO, or solvent), and V is the system volume. The external potential U_I is conjugate to the particle concentration ρ_I via the Gaussian chain density functional (2). The average concentration is ρ_I^0 and ν_I is the particle volume. The cohesive interactions have kernels ϵ_{IJ} (7) [11]. The Helfand compressibility parameter is κ_H [12].

The dynamics of the system is governed by the diffusion-convection equation (11) with the periodic boundary conditions as described in [11,17]. The dynamic equations are closed by the expression for the free energy (13) and the Gaussian chain density functional (2). Seven dimensionless parameters enter the numerics: three exchange parameters $\chi_{IJ} \equiv (\beta/2\nu)[2\epsilon^0_{IJ} - \epsilon^0_{JJ} - \epsilon^0_{II}]$, the dimensionless time $\tau \equiv kTMh^{-2}t$ (h is the mesh size), a noise scaling parameter Ω ($\Omega = 100$ in the simulation) [13], the grid scaling $d \equiv ah^{-1} = 1.1543$ [a is the Gaussian chain bond length (4)] [14], and the compressibility parameter $\kappa' \equiv \beta\kappa_H\nu = 10$. For simplicity we use identical mobility coefficients and particle volumes for all components.

From comparison of Random Phase Approximation (RPA) [24] and molecular force-field single chain structure factors (generated by Monte Carlo), we found that the structure factor is well represented by an $E_3P_9E_3$ Gaussian chain [17], which corresponds to 3 to 4 monomers per bead. The solvent molecule is represented as a single bead.

The solvent–polymer interaction parameters were calculated from vapor pressure data of aqueous homopolymer solutions [25], using the Flory–Huggins expression [26] $\chi_{IJ} = \theta^{-2}\{\ln p/p^0 - \ln(1-\theta) - (1-1/N)\theta\}$, where p is the vapor pressure and θ is the polymer volume fraction. The chain length N was determined using 13/3 (EO) or 30/9 (PO) monomers per bead. This gives for the interaction parameters $\chi_{ES} = 1.4$, $\chi_{PS} = 1.7$ (here S denotes solvent). For the EO–PO interaction parameter from group contribution methods [27] we estimated $\chi_{EP} = 3.0$.

We have simulated the time evolution of the Pluronic–water mixture in a cubic box $64 \times 64 \times 64$. The simulation was started from the homogeneous solution. The resulting morphologies are represented by Figs. 1 and 2. Four phases are clearly observed in good agreement with the experiments [23]: lamellae, bicontinuous (gyroid), hexagonal cylindrical, and micellar. The gyroid phase appeared to be metastable, which will be shown in Section III.C.

B. Multicolor Block Copolymers

Multicolor block copolymers are a very fascinating example which have recently become a hot topic in polymer simulation [21]. We demonstrate just one example obtained by our method. Figure 3 shows mesophases in a three-color quadrublock copolymer $A_3B_3C_3A_3$. The side chain block A forms lamellae and middle blocks C and D form two grids of alternating cylinders. All blocks have equal interaction strength with each other, $\chi_{AB} = \chi_{AC} = \chi_{BC} = 5.4$. Our 3D results correspond well to the 2D simulation of [21] but give a more realistic picture, being 3D results.

An even more sophisticated case is demonstrated in Fig. 4, a four-color four arm star block copolymer. Again, all interactions are taken of the same strength. The system develops into four interpenetrating micellar grids with a deformed dodecahedral space arrangement.

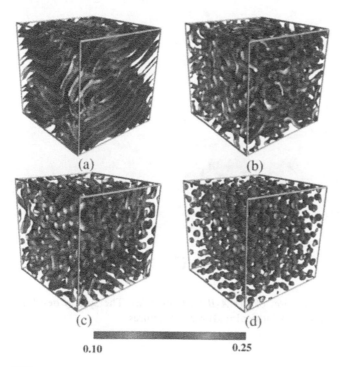

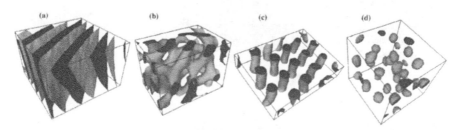

FIG. 1 Isosurface representation of PL64 in water mixture for 70% (a), 60% (b), 55% (c), and 50% (d) polymer content. The isolevel is $\theta_{EO} = v\rho_{EO} = 0.3$.

FIG. 2 Detail of the simulation snapshots from Fig. 1. (a) LAM, (b) GYR, (c) HEX, and (d) micellar phases are clearly visible.

C. Modulation by Shear

It is known that flow fields affect mesoscale structures in complex liquids and polymer systems, giving rise to global orientation [28,29]. Because of its

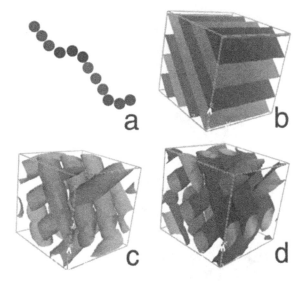

FIG. 3 Isosurface representation of the $A_3B_3C_3A_3$ melt (a). The system develops into one lamellar (b) and two cylindrical (c,d) mesostructures.

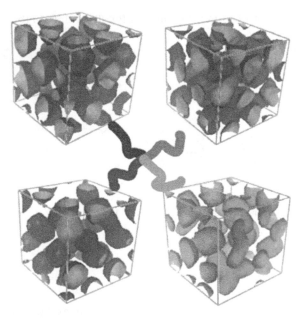

FIG. 4 Isosurface representation of the star block copolymer $(A)_8(B)_8(C)_8(D)_8$ melt.

industrial importance, the behavior of complex polymer liquids under shear is an intensively studied topic, both experimentally and theoretically [29–32]. Most of the work is dedicated to the stability analysis of patterns and to phase transitions in complex polymer liquids under shear.

The time evolution of morphologies in complex liquids under shear was also studied in computer simulations using Landau type free energies in 2D geometries such as a square cell [3–5,28], a rectangle [1,2], and in a two-dimensional Couette flow [1]. Recently the shear effect in a 2D polymer system was studied using the path integral formalism for the kinetic coefficient [15]. However, this was carried out for a model with simple periodic boundary conditions and a conventional phenomenological free energy.

The time evolution of the density field $\rho_I(\mathbf{r})$ under simple shear flow, $v_x = \dot{\gamma}y$, $v_y = v_z = 0$, can be described by dynamic Eq. (11) with a convective term

$$\dot{\rho}_I = M_I \nabla \cdot \rho_I \nabla \mu_I - \dot{\gamma}y\nabla_x\rho_I + \eta_I$$

where $\dot{\gamma}$ is the shear rate (the time derivative of the strain γ). For an $L \times L \times L$ cubic grid we use a sheared periodic boundary condition [33,34]:

$$\rho(x, y, z, t) = \rho(x + iL + \gamma jL, y + jL, z + kL, t)$$

Figure 5 illustrates the application of our method for a 3D melt of block copolymers A_8B_8 under simple steady shear flow. Applying shear speeds up the lamellar formation in a diblock copolymer melt enormously. The alignment qualitatively differs from the 2D case—so-called "perpendicular" lamellae are formed in 3D. From experiments and stability analysis this orientation is well known to be the most stable one (see, e.g., [31]). The structure remains stable after switching off the shear.

Figure 6 demonstrates the formation of a perfectly aligned hexagonal cylindrical phase in aqueous solution of 55% triblock copolymer Pluronic L64 (cf. Figs. 1c and 2c).

An even more spectacular effect of shearing can be seen in Fig. 7. The gyroid structure of 60% PL64 from Figs. 1b and 2b was taken as a starting structure before applying shear. Shearing breaks the system into coexistence of two phases, lamellae and cylinders, Fig. 7 (top). Cessation of shear does not bring the system back, but it stays as a coexistence of two phases. This point of the phase diagram is now in perfect agreement with the experiment [23]. The gyroid phase for this particular system was a metastable one [32].

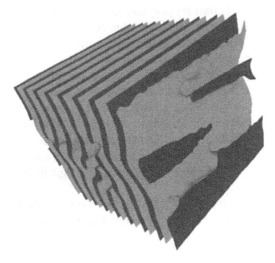

FIG. 5 Isosurface representation of the A_8B_8 melt at $\tau = 75{,}000$ and shear flow $v_x = \dot\gamma y$, $v_y = v_z = 0$, using the structure at $\tau = 500$ as a starting structure. The isolevel is $\theta_A = v\rho_A = 0.3$. One can clearly observe the global lamellar orientation. The x- and y-axes are indicated.

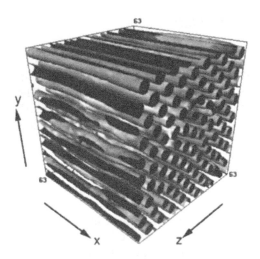

FIG. 6 Mesophases of 55% Pluronic L64 in water under shear at time $\tau = 12{,}500$, using the structure at $\tau = 2500$ from Figs.1c and 2c as a starting structure before shearing. The isosurfaces are at $\theta_P \equiv v_P\rho_P = 0.33$. The x-axis is the velocity direction, the y-axis is the velocity gradient direction, and the z-axis is a neutral direction.

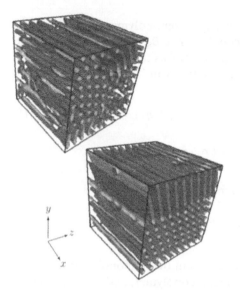

FIG. 7 Mesophases of 60% Pluronic L64 in water at the end of shearing (top) and after a long relaxation after cessation of shear (bottom).

D. Modulation by Reactions

In dynamic mean field density functional theory, the dynamics of the polymer melt under simple steady shear is described by the time evolution of density fields ρ_I. The dynamics is governed by a diffusion-convection equation with sheared periodic boundary conditions and can readily be extended with (second order) reaction terms in the following manner:

$$\frac{\partial \rho_I}{\partial t} = M\nabla \cdot \rho_I \nabla \frac{\delta F}{\delta \rho_I} - \dot{\gamma}y\nabla_x\rho_I + \sum_{J=1,K=1}^{N} k_{JK}\rho_J\rho_K + \eta_I \tag{14}$$

Here M is a mobility parameter, $\dot{\gamma}$ is the shear rate, which is zero if no shear is applied, η_I is a stochastic term which is distributed according to a fluctuation-dissipation theorem [13], and k_{JK} is the reaction rate, which can be either negative for reactants or positive for products. Notice that the reactive noise can be neglected here. Different order reactions or multiple reaction terms can be added without any difficulties, but as a proof of principle we focus here on the above type of reactions. In this subsection we study the effect of combined microphase separation, shear, *and* reaction to gain insight in the mechanisms that are important in pathway-controlled morphology formation and in particular in reactive blending.

We therefore study a reaction in which a homopolymer A_8 couples to a homopolymer B_8 to form a diblock copolymer A_8B_8. The reaction is limited to the end groups and we assume that if two blocks that can couple are close enough, the reaction takes place. This reaction can be modeled as follows:

$$\frac{\partial \rho_{hA}}{\partial t} = M\nabla \cdot \rho_{hA}\nabla \frac{\delta F}{\delta \rho_{hA}} - \dot{\gamma}y\nabla_x\rho_{hA} - k\rho_{hA}\rho_{hB} + \eta_{hA} \tag{15}$$

$$\frac{\partial \rho_{hB}}{\partial t} = M\nabla \cdot \rho_{hB}\nabla \frac{\delta F}{\delta \rho_{hB}} - \dot{\gamma}y\nabla_x\rho_{hB} - k\rho_{hA}\rho_{hB} + \eta_{hB}$$

$$\frac{\partial \rho_{dA}}{\partial t} = M\nabla \cdot \rho_{dA}\nabla \frac{\delta F}{\delta \rho_{dA}} - \dot{\gamma}y\nabla_x\rho_{dA} + k\rho_{hA}\rho_{hB} + \eta_{dA}$$

$$\frac{\partial \rho_{dB}}{\partial t} = M\nabla \cdot \rho_{dB}\nabla \frac{\delta F}{\delta \rho_{dB}} - \dot{\gamma}y\nabla_x\rho_{dB} + k\rho_{hA}\rho_{hB} + \eta_{dB}$$

Here, ρ_{hA} (ρ_{hB}) is the density of A (B) beads in the homopolymer, ρ_{dA} (ρ_{dB}) is the density of A (B) beads in the diblock copolymer, and k is the reaction rate of the coupling reaction.

Figure 8 gives an example of formation of double layer droplets in a sheared reactive polymer system. Initially the A/B blend was subject to shear which resulted in the formation of elongated droplets. Then reaction on the surface of the droplets took place after switching off the shear. That leads to relaxation of the elongated shape of the droplets towards spherical. The excess of polymer formed at the interface goes inside the droplets which forms a double layer structure.

E. Modulation by Geometry Constraints

The polymer melt is modeled as a compressible system, consisting of Gaussian chain molecules in a mean field environment. The free energy functional for copolymer melts has a form that is similar to the free energy that was used before:

$$F[\{\rho\}] = -kT\ln\frac{\Phi^n}{n!} - \sum_I \int_V U_I(\mathbf{r})\rho_I(\mathbf{r})d\mathbf{r} + \frac{1}{2}\sum_{I,J} \int_{V^2} \epsilon_{IJ}(\mathbf{r}-\mathbf{r}')\rho_I(\mathbf{r})\rho_J(\mathbf{r}')d\mathbf{r}d\mathbf{r}'$$

$$+ \frac{1}{2}\sum_I \int_{V^2} \epsilon_{IM}(\mathbf{r}-\mathbf{r}')\rho_I(\mathbf{r})\rho_M(\mathbf{r}')d\mathbf{r}d\mathbf{r}'$$

$$+ \frac{\kappa_H}{2}\int_V \left(\sum_I v_I(\rho_I(\mathbf{r}) - \rho_I^0)\right)^2 d\mathbf{r} \tag{16}$$

except for an extra fourth term that contributes only in the direct vicinity of the filler particles. This accounts for the interaction of a polymer melt with

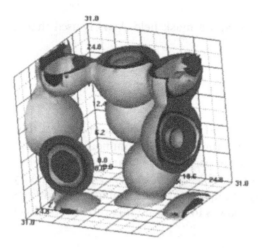

FIG. 8 Three-dimensional simulation of a homopolymer blend 90%/10% A_8/B_8. Before the reactions were switched on, 28,000 steps of shear were performed ($\Delta\tau\dot\gamma = 0.001$) on the blend. The shear was stopped at $\tau = 28,000$ and 2500 reaction steps were performed ($\Delta\tau k = 0.1$). In the figure the isodensity surfaces at different levels are depicted at $\tau = 30,500$ of ρ_{hB}, ρ_{dA}, and ρ_{dB}. The total volume fractions at this time level are 80.5% A_8, 0.5% B_8, and 19% $A_8 B_8$.

surfaces. In this equation, n is the number of polymer molecules, Φ is the intramolecular partition function for ideal Gaussian chains in an external field U, I is a component index, ρ_I are the density fields of the different bead types I, and V is the system volume. Inside the filler particles, the densities ρ_I of the different bead types are equal to zero. Since the density ρ is present in all integrals in the definition of the free energy (16), integrals over the entire volume V are equal to the integrals restricted to V/V^0, standing for the total volume V with exception of the volume taken by the filler particles, denoted as V^0. The filler particles considered here are constrained to the condition of stationary position in time. The constant density field ρ_M (M represents beads of the filler particle type) that appears in Eq. (16) is defined as $\rho_M(\mathbf{r}) = 1$ for $\mathbf{r} \in V^0$ and $\rho_M(\mathbf{r}) = 0$ for $\mathbf{r} \in V/V^0$. The average concentration is ρ_I^0 and v_I is the particle volume. The surface interactions have kernels ϵ_{IM}. The Helfand compressibility parameter is κ_H [12].

The ensemble average particle density $\rho_s(\mathbf{r})$ of a certain bead s at position $\mathbf{r}$ in space is

$$\rho_s[U](\mathbf{r}) = C\mathcal{M}(\mathbf{r}) \int_{V^N} \psi(\mathbf{R}_1, \dots, \mathbf{R}_N)\delta(\mathbf{r} - \mathbf{R}_s)\, d\mathbf{R}_1, \dots, d\mathbf{R}_N \qquad (17)$$

where C is a normalization constant and a mask field $\mathcal{M}(\mathbf{r})$ is used that is defined as

$$\mathcal{M}(\mathbf{r}) = \begin{cases} 0 & \mathbf{r} \in V^0 \\ 1 & \mathbf{r} \in V/V^0 \end{cases}$$

The time evolution of the density field $\rho_I(\mathbf{r})$ can be described by a time dependent Landau–Ginzburg type equation (11). The boundary conditions that are used on the simulation box are periodic boundary conditions. For the diffusion flux in the vicinity of the filler particles, rigid-wall boundary conditions are used. A simple way to implement these boundary conditions in accordance with the conservation law is to allow no flux through the filler particle surfaces, i.e.,

$$\nabla \mu_I \cdot \mathbf{n} = 0 \tag{18}$$

where $\mathbf{n}$ is the normal pointing towards the filler particle. The same boundary conditions apply to the noise η_I. Figure 9 demonstrates formation of lamellar structure in an A_8B_8 melt in the presence of interacting walls.

In Fig. 10 one can see the same system but confined between neutral walls. This confinement leads the system to form "perpendicular" to the walls' lamellae.

Thickness of the slit/film can have a drastic effect on the polymer morphology. In Fig. 11 the same 55% PL64 in water system was confined in between two different slits. The bulk morphology of this system is cylindrical, Figs. 1c, 2c, and 6. Adopting conformational freedom the system develops into cylinders, which are either parallel or perpendicular to the wall, depending on the slit width.

IV. DISCUSSION AND CONCLUSION

In this chapter we described the theoretical basis and some applications of a new model for computer simulation of time evolution of mesoscale structures of complex polymer liquids. Here we describe just a few possible technologically oriented applications of the proposed method.

One of the main questions in industrial emulsion polymerizations is to produce a latex or polymer with any desired morphology, composition, sequence distribution, molecular weight distribution, etc. A specific example is given by the core-shell techniques in which one type of (co)polymer is grown around a core of another type. In paints the core polymer may provide gloss and mechanical stability, whereas the shell might contain a

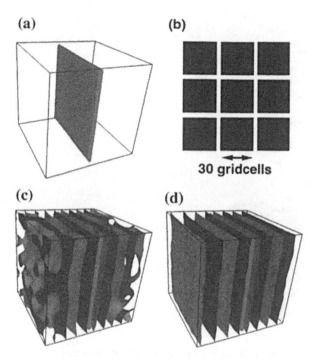

FIG. 9 Lamellar formation of an A_8B_8 copolymer melt in the presence of square plates of one grid-cell thickness. The interaction of polymer blocks with the surface is $\beta\epsilon_{AM}v^{-1}=-1.0$ and $\beta\epsilon_{BM}v^{-1}=1.0$. (a) View of filler particle in simulation box, (b) space filled with filler particles (the slots between filler particles are drawn as white lines), (c) morphology of A beads (isolevel $v\rho_A=0.5$) in one simulation box at $\tau=500$, (d) the same for $\tau=2000$.

rubbery polymer to provide a uniform surface coating. The final morphology that is obtained in the production process determines the quality of the product to a very large extent and hence prediction of the morphology based on production process parameters is desirable. The morphology is controlled by both thermodynamic (the microphase separation process) and kinetic (the reaction process) principles. Several practical applications are within direct reach. An example of a one-stage core-shell technique that has been described in the literature concerns the mixing of silicone oils (containing Si-H and vinyl groups) and vinyl monomers emulsified in water. After a (cross-linking) reaction of Si-H and Si-CH=CH$_2$ (kinetics) the hydrophilic monomers are excluded to the surface layer (thermodynamics) and a core-shell morphology results.

Another potential application is the investigation of the stability of polymer surfactant protected emulsions. In particular, many industrial

(e)   (f)

(g)   (h)

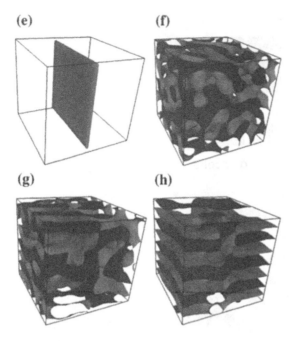

FIG. 10   Lamellar formation of an A_8B_8 copolymer melt in the presence of the same filler particle as Fig. 9. Moreover, there is no interaction between the polymer beads and boundaries of the filler particles. (e) View of filler particle in simulation box, (f) morphology of A beads (isolevel $\nu\rho_A = 0.5$) in one simulation box at $\tau = 500$, (g) the same for $\tau = 4000$, (h) final morphology at $\tau = 10,000$.

FIG. 11   Cylinder formation of 55% PL64 in water mixture in two different slits, $12h$ (left) and $15h$ (right), forming parallel (left) or perpendicular (right) cylinders.

systems contain a specified mixture of different surfactants, so as to provide a certain stability of the interphase. Thus the application of the emulsions is controlled by the nature of the surfactant mixture. Within the described method it is relatively easy to study the synergetic effects of surfactant composition and architecture, and the way processing conditions influence the emulsion stability and morphology.

APPENDIX A: NUMERICAL IMPLEMENTATION OF THE PATH INTEGRAL

The method presented here is off-lattice in nature. It means that the calculations are carried out on a (cubic) lattice, but the chain conformations are not restricted to lattice positions. The Wageningen school [35–38] has shown that for lattice models, where the chains can only occupy lattice positions, the number of conformations is finite and summation amounts to an exact enumeration. In this case numerical integration is rapid and relatively simple. The off-lattice nature of our method introduces an extra difficulty in deriving quadrature (or numerical integration) rules. Common rules, such as used in lattice methods, are found to give problems. First, the conformation restriction inherent in the lattice models can lead to unphysical singularities in the inverse structure factor for copolymer melts. As a consequence, the mapping between particle density field and external potential may have a null space and is no longer bijective (for more details see [9]). Second, we found that the cubic lattice chain structure factor for copolymer melts has strong intrinsic anisotropies, especially on bond length scales. These anisotropies are an important source for a well known numerical plague of lattice models, so-called "lattice-artifacts."

In this appendix we describe a *stencil* algorithm which avoids many of the drawbacks of quadrature rules used in classical lattice models, while the extra computational cost is modest. The derivation consists of finding a unique and optimal set of stencil coefficients for a convolution with a Gaussian kernel, adapted to the special case of off-lattice density functional calculations. Stencil coefficients are the multipliers of the function values at corresponding grid points.

The familiar Feynman decomposition of the path integral (2) results in the well known "modified diffusion" algorithm [39,40]. In our case, where sequence space is discrete, the density functional can be expressed in terms of Green propagators

$$\rho_s(\mathbf{r}) \propto G_s(\mathbf{r})\sigma[G_{s+1}^{inv}](\mathbf{r}) \tag{19}$$

The set of once integrated Green's functions $G_s(r)$ and $G_{s+1}^{inv}(r)$ are related by the recurrence relations

$$G_s(\mathbf{r}) = \mathcal{M}(\mathbf{r})e^{-U_s(\mathbf{r})}\sigma[G_{s-1}](\mathbf{r})$$
$$G_s^{inv}(\mathbf{r}) = \mathcal{M}(\mathbf{r})e^{-U_s(\mathbf{r})}\sigma[G_{s+1}^{inv}](\mathbf{r}) \tag{20}$$

with $G_0(\mathbf{r}) = G_{N+1}^{inv}(\mathbf{r}) = 1$. The linkage operator $\sigma = \sigma[f](\mathbf{r})$ is defined as a convolution with a Gaussian kernel

$$\sigma[f](\mathbf{r}) = \left(\frac{3}{2\pi a^2}\right)^{2/3} \int_V \exp[-(3/2a^2)(\mathbf{r} - \mathbf{r}')^2] f(\mathbf{r}') \, d\mathbf{r}' \tag{21}$$

Note that the Green propagator method to calculate the density functional (2) yields an *exact* answer.

Now, the crucial point of the algorithm is a numerical representation of the linkage operator σ, given a cubic uniform grid with mesh width h, which is both efficient and accurate. Since we work on large cubic grids and use domain decomposition for the parallelization, FFTs and Gaussian Quadrature rules are not considered. Efficiency is crucial, since the linkage operation has to be repeated $2(N-1)$ times for a single density functional calculation. Furthermore, if the numerical representation of the linkage operator is not accurate, its repeated application will result in error accumulation, especially in the small $\mathbf{q}$ range, i.e., on chain length scales. Large errors in the high $\mathbf{q}$ range are also undesirable, since they may disturb the bijectivity of the mapping between density field and external potential.

A 27-point stencil is the most compact stencil that meets all constraints. We reduce the number of parameters by invoking symmetry rules, similar to the polyhedra rules used in multidimensional integration of symmetric kernels [41]. The linkage operator on a uniform grid is expressed as:

$$\sigma[f](\mathbf{r}) = c_0 f(\mathbf{r}) + \sum_i \frac{c_i}{2}[f(\mathbf{r} + \mathbf{r}_i) + f(\mathbf{r} - \mathbf{r}_i)] \tag{22}$$

where f is an arbitrary field, $\mathbf{r}_i$ is a lattice vector, and c_i is a stencil coefficient. It is easy to see that, due to the symmetry of the linkage operator, there is considerable redundancy in the stencil coefficients. For a 27-point stencil only four reduced coefficients remain: a central coefficient c_0 and three coefficients for each symmetry-reduced stencil direction; c_1 in direction $(1, 0, 0)$, c_2 in direction $(1, 1, 0)$, and c_3 in direction $(1, 1, 1)$.

In Fourier space the linkage operator is:

$$\int_V \sigma(f) e^{-i\mathbf{q}\cdot\mathbf{r}} \, d\mathbf{r} = \sigma_{\mathbf{q}} f_{\mathbf{q}}$$

where the Fourier multipliers $\sigma_{\mathbf{q}}$ are given by

$$\sigma_{\mathbf{q}} = \begin{cases} e^{-(a^2|\mathbf{q}|^2)/6} & \text{continuum} \\ \sum_{i=0}^{3} c_i \sum_{j=1}^{d_i} \cos[\mathbf{q} \cdot \mathbf{r}_{ij}] & \text{discrete} \end{cases}$$

Here, d_i is the number of independent lattice vectors in symmetry-reduced stencil direction i ($d_0 = 1, d_1 = 3$, $d_2 = 6$, and $d_3 = 4$), $\mathbf{r}_{ij}$ denotes the jth lattice vector in stencil direction i [e.g., $\mathbf{r}_{01} = \mathbf{0}$, $\mathbf{r}_{12} = (0, h, 0)$, $\mathbf{r}_{23} = (0, h, h)$, and $\mathbf{r}_{34} = (h, -h, -h)$].

First, we guarantee maximum isotropy on the grid by imposing that the continuum and discrete (grid-restricted) linkage operators are identical for

$$\mathbf{q} \in \{(0, 0, 0), (\pi/h, 0, 0), (\pi/h, \pi/h, 0), (\pi/h, \pi/h, \pi/h)\}$$

In this way, we transfer the isotropy that should have resulted from an isotropically chosen grid (for instance, honeycomb) to the stencil. Second, we guarantee that the long length scalings of the continuum and discrete Fourier multipliers are identical by also equating the curvatures in the point $\mathbf{q}=(0, 0, 0)$. This assures the bijectivity of the density-external potential mapping. Because of symmetry, it is sufficient to equate the two curvatures in direction $(1, 0, 0)$. With the help of these five conditions, we determine the four stencil coefficients c_i and a/h. The unique solution of the nonlinear system is: $c_0 = 0.171752$, $c_1 = 0.137231$, $c_2 = 0.0548243$, $c_3 = 0.0219025$, and $a/h = 1.15430$. As a result, the stencil depends on the ratio a/h. For other ratio, we consider different stencil coefficients [9].

APPENDIX B: NUMERICAL SCHEME FOR SOLVING THE DYNAMICAL EQUATIONS

In Appendix A we described the numerical implementation of the Gaussian chain density functions that retains the bijectivity between the density fields and the external potential fields. The intrinsic chemical potentials $\mu_I = \delta F/\delta \rho$ that act as thermodynamic driving forces in the Ginzburg–Landau equations describing the dynamics, are functionals of the external potentials and the density fields. Together, the Gaussian chain density functional and the partial differential equations, describing the dynamics of the system, form a closed set.

Since the external potentials are highly nonlinear functionals of the density fields (they cannot even be inverted analytically), the partial differential equations we have to solve numerically are in themselves highly nonlinear. We should be very careful in choosing a method, since there are apparent risks of introducing numerical errors (for instance by taking the time steps too large). The Crank–Nicolson (CN) method, that aims to solve differential equations by mixing implicit and explicit parts, is known to be rather stable for this kind of problem. We use periodic boundary conditions.

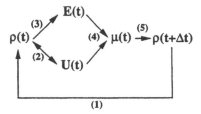

FIG. 12 Schematic representation of the iterative scheme. We have two nested iterative loops: the time loop (1) for updating the density fields ρ, and within each time iteration an iterative loop (2) for updating the external potential fields U. We start with an initial guess for the external potential fields. We use Eq. (17) to generate a set of unique density fields. The cohesive chemical potential E [relation (3)] can be calculated from the density fields by Eq. (8). The total chemical potential $\mu(4)$ can now be found from Eq. 6. We update the density fields (5) [by using the old and updated fields in Eq. (23)] and accept the density fields if the condition (26) is satisfied. If this is not the case, the external potential fields are updated by a steepest descent method.

The way the time integration is carried out is as follows: we start with a homogeneous distribution of Gaussian chains $\rho_I = \bar{\rho}_I$ (where $\bar{\rho}$ is the mean density of the fields) and $U_I = 0$. We solve the CN equations (to be defined below) by an iterative process in which the external potential fields are updated according to a steepest descent method. The updated external potential fields are related, by the chain density functional, to unique density fields. We calculate the intrinsic chemical potential from the set of external potential fields and density fields and update the density fields. The density fields are only accepted as an update for the next time step if the L_2 norm of the residual is below some predefined upper boundary. An overview of this scheme is shown in Fig. 12.

The general expression of the CN equations for this problem is

$$\frac{\rho_I^k - \rho_I^{k-1}}{\Delta t} = \omega_1 M h^{-2} \mathcal{D}(\rho_I^k \mathcal{D}(\mu_I^k)) + (1 - \omega_1) M h^{-2} \mathcal{D}(\rho_I^{k-1} \mathcal{D}(\mu_I^{k-1})) + \eta_I^k$$

(23)

where k is the time index, h the mesh size, ω_i the CN parameters determining the degree of explicitness versus implicitness of the method, and $h^{-2}\mathcal{D}(f\mathcal{D}g)$ the discrete operation on the grid representing $\nabla \cdot f \nabla g$. The computational procedure for the div-gradient operation is again important because of the risk of introducing grid artifacts. We have developed a procedure similar to the procedure for the numerical integration. As computational cell, again a

27-point stencil is used. Isotropy and scaling behavior should be considered as conditions for the stencil weights. A more detailed overview of this scheme can be found in [13].

In dimensionless parameters $\theta_I = \nu\rho_I$ (ν is the bead volume), $\phi_I = \beta\mu_I$ ($\beta = kT$), and $\tau = \beta^{-1}Mh^{-2}t$ we can rewrite the CN equations as

$$res_I^k = \theta_I^k - \omega_1\Delta\tau\mathcal{D}(\theta_I^k\mathcal{D}(\phi_I^k)) - \theta_I^{k-1} - (1-\omega_1)\Delta\tau\mathcal{D}(\theta_I^{k-1}\mathcal{D}(\phi_I^{k-1})) - \eta_I^k = 0 \tag{24}$$

Since we do not have an explicit expression for the external potential fields in terms of the densities, we solve this equation for the external potential fields. The external potential fields U_I are found by an iterative scheme

$$\begin{aligned} U_{I,0}^k &= U_{I,\text{final}}^{k-1} \\ U_{I,p}^k &= U_{I,p-1}^k + \alpha res_{I,p-1}^k \quad p \geq 1 \end{aligned} \tag{25}$$

where p is the iteration index and α is a parameter in a bisection method that is employed in order to find update fields U_p along the direction of search with a smaller residuals norm than the residual of the previous fields U_{p-1}. The problem is solved for time step k when the norm of the residual satisfies

$$\left\| \sum_I res_I^k \right\| < \epsilon \tag{26}$$

In general, the CN parameter is chosen equal to $1/2$. In this case the starting value of the residual norm for every new time iteration is $\| \sum_I \eta_I^k \|$, and we consider as a rule of thumb an ϵ such that $\epsilon \approx 0.01\| \sum_I \eta_I^k \|$. When this condition is satisfied, the updates of the density and external potential fields are accepted as solutions at time step k and used as starting values for the new iterative update scheme at time step $k+1$.

REFERENCES

1. Qiwei He, D.; Nauman, E.B. Chem. Engng. Sci. **1997**, *52*, 481–496.
2. Gonnella, G.; Orlandini, E.; Yeomans, J.M. Phys. Rev. Lett. **1997**, *78*, 1695–1698.
3. Pätzold, G.; Dawson, K. J. Chem. Phys. **1996**, *104*, 5932–5941.
4. Kodama, H.; Komura, S. J. Phys. II France **1997**, *7* (1), 7–14.
5. Ohta, T.; Nozaki, H.; Doi, M. Phys. Lett. A **1990**, *145* (6,7), 304–308.

6. Cahn, J.W.; Hilliard, J.E. J. Chem. Phys. **1958**, *28*, 258–267.
7. Oono, Y.; Puri, S. Phys. Rev. Lett. **1987**, *58*, 836–839.
8. de Gennes, P.G. J. Chem. Phys. **1980**, *72*, 4756–4763.
9. Maurits, N.M.; Fraaije, J.G.E.M. J. Chem. Phys. **1997**, *106*, 6730–6743.
10. Fraaije, J.G.E.M. J. Chem. Phys. **1993**, *99*, 9202–9212.
11. Fraaije, J.G.E.M.; van Vlimmeren, B.A.C.; Maurits, N.M.; Postma, M.; Evers, O.A.; Hoffmann, C.; Altevogt, P.; Goldbeck-Wood, G. J. Chem. Phys. **1997**, *106* (10), 4260–4269.
12. Maurits, N.M.; van Vlimmeren, B.A.C.; Fraaije, J.G.E.M. Phys. Rev. E **1997**, *56*, 816–825.
13. van Vlimmeren, B.A.C.; Fraaije, J.G.E.M. Comp. Phys. Comm. **1996**, *99*, 21–28.
14. Maurits, N.M.; Altevogt, P.; Evers, O.A.; Fraaije, J.G.E.M. Comp. & Theor. Pol. Sci. **1996**, *6*, 1–8.
15. Kawakatsu, T. Phys. Rev. E **1997**, *56*, 3240–3250.
16. Hasegawa, R.; Doi, M. Macromolecules **1997**, *30*, 3086–3089.
17. van Vlimmeren, B.A.C.; Maurits, N.M.; Zvelindovsky, A.V.; Sevink, G.J.A.; Fraaije, J.G.E.M. Macromolecules **1999**, *32*, 646–656.
18. Fleer, G.J., et al. *Polymers at Interfaces*; Chapman & Hall: London, 1993.
19. Doi, M. *Introduction to Polymer Physics*; Oxford University Press: Oxford, 1996.
20. Matsen, M.W.; Schick, M. Phys. Rev. Lett. **1994**, *72*, 2660–2663.
21. Drolet, F.; Fredrickson, G.H. Phys. Rev. Lett. **1999**, *83*, 4317–4320.
22. Hamley, I.W. *The Physics of Block Copolymers*; Oxford University Press: Oxford, 1998.
23. Alexandridis, P.; Olsson, U.; Lindmann, B. Macromolecules **1995**, *28*, 7700–7710.
24. Leibler, L. Macromolecules **1980**, *13* (6), 1602–1617.
25. Malcolm, G.N.; Rowlinson, J.S. Trans. Faraday Soc. **1957**, *53*, 921–931.
26. Hill, T.L. *An Introduction to Statistical Thermodynamics*; Addison-Wesley: Reading, MA, 1962.
27. Krevelen, D.W. *Properties of Polymers, Their Correlation with Chemical Structure; Their Numerical Estimation and Prediction from Additive Group Contribution*; Elsevier: Amsterdam, third edition, 1990.
28. Kodama, H.; Doi, M. Macromolecules **1996**, *29* (7), 2652–2658.
29. Fredrickson, G.H. J. Rheol. **1994**, *38*, 1045–1067.
30. Zvelindovsky, A.V.; Sevink, G.J.A.; van Vlimmeren, B.A.C.; Maurits, N.M.; Fraaije, J.G.E.M. Phys. Rev. E **1998**, *57*, R4879–R4882.
31. Doi, M.; Chen, D. J. Chem. Phys. **1989**, *90*, 5271–5279.
32. Ohta, T.; Enomoto, Y.; Harden, J.L.; Doi, M. Macromolecules **1993**, *26*, 4928–4934.
33. Chen, Z.R.; Kornfield, J.A.; Smith, S.D.; Grothaus, J.T.; Satkowski, M.M. Science **1997**, *277*, 1248–1253.
34. Zvelindovsky, A.V.; Sevink, G.J.A.; Fraaije, J.G.E.M. Phys. Rev. E **2000**, *62*, R3063–R3066.
35. Scheutjens, J.M.H.M.; Fleer, G.J. J. Phys. Chem. **1979**, *83*, 1619–1635.

36. Scheutjens, J.M.H.M.; Fleer, G.J. J. Phys. Chem. **1980**, *84*, 178–190.
37. Leermakers, F.A.M.; Scheutjens, J.M.H.M. J. Chem. Phys. **1988**, *89*, 3264.
38. Evers, O.A.; Scheutjens, J.M.H.M.; Fleer, G.J. Macromolecules **1990**, *23*, 5221–5233.
39. de Gennes, P.-G. *Scaling Concepts in Polymer Physics*; Cornell University Press: Ithaca, NY, 1979.
40. Doi, M.; Edwards, S.F. *The Theory of Polymer Dynamics*; Clarendon: Oxford, 1986.
41. Zwillinger, D. *Handbook of Integration*; Jones and Barlett Publishers: Boston, 1993.

Index